COMPOSITE STRUCTURES

Proceedings of the 1st International Conference on Composite Structures, held at Paisley College of Technology, Scotland, from 16 to 18 September 1981, organised in association with the Institution of Mechanical Engineers and the National Engineering Laboratory.

COMPOSITE STRUCTURES

Edited by

I. H. MARSHALL

Department of Mechanical and Production Engineering,
Paisley College of Technology, Scotland

APPLIED SCIENCE PUBLISHERS
LONDON and NEW JERSEY

APPLIED SCIENCE PUBLISHERS LTD
Ripple Road, Barking, Essex, England
APPLIED SCIENCE PUBLISHERS INC.
Englewood, New Jersey 07631, .USA

British Library Cataloguing in Publication Data

International Conference on Composite Structures
(*1st: 1981: Paisley College of Technology*)
Composite structures.
1. Composite materials—Congresses
2. Composite constructions—Congresses
I. Title   II. Marshall, I. H.
III. Institution of Mechanical Engineers
IV. National Engineering Laboratory
624.1′8   TA664

ISBN 0-85334-988-6

WITH 74 TABLES AND 362 ILLUSTRATIONS

© APPLIED SCIENCE PUBLISHERS LTD 1981

Printed in Great Britain by Galliard (Printers) Ltd, Great Yarmouth

Preface

The papers contained herein were presented at the First International Conference on Composite Structures held at Paisley College of Technology, Paisley, Scotland, in September 1981. This conference was organised and sponsored by Paisley College of Technology in association with The Institution of Mechanical Engineers and The National Engineering Laboratory (UK).

There can be little doubt that, within engineering circles, the use of composite materials has revolutionised traditional design concepts. The ability to tailor-make a material to suit prevailing environmental conditions whilst maintaining adequate reinforcement to withstand applied loading is unquestionably an attractive proposition. Significant weight savings can also be achieved by virtue of the high strength-to-weight and stiffness-to-weight characteristics of, for example, fibrous forms of composite materials. Such savings are clearly of paramount importance in transportation engineering and in particular aircraft and aerospace applications.

Along with this considerable structural potential the engineer must accept an increased complexity of analysis. All too often in the past this has dissuaded the designer from considering composite materials as a viable, or indeed better, alternative to traditional engineering materials. Inherent prejudices within the engineering profession have also contributed, in no small way, to a certain wariness in appreciating the merits of composites.

However, the potential benefits of composite materials are inescapable. The last two decades have seen a phenomenal increase in the use of composites in virtually every area of engineering, from the high technology

v

aerospace application to the less demanding structural cladding situation. Research and development in this field have been unparalleled in history with the rate of advance of knowledge seeming to increase daily. Such advances can only be fully utilised if shared and discussed with others who have similar interests.

With this in mind, and with a conscious need for dissemination of knowledge between users, manufacturers, designers and researchers involved in structures manufactured using composite materials, the present international conference was organised.

Authors from thirteen countries combine with delegates from virtually every major industrial nation in the world to make this a truly international gathering of specialists in an ever-expanding technology. Topics under discussion range from the possible uses of natural fibre composites to precision fabrication techniques employed in the space shuttle programme, each contribution relating to a particular aspect of composite structural engineering.

An international conference can only succeed in making a contribution to knowledge through the considerable efforts of a number of enthusiastic and willing individuals. In particular, thanks are due to the following:

The Conference Steering Committee

Professor J. Anderson (Chairman)	Paisley College of Technology
Dr W. S. Carswell	National Engineering Laboratory
Dr J. Rhodes	University of Strathclyde
C. I. Phillips	Scott Bader Co. Ltd
J. A. Wylie	Centre for Liaison with Industry and Commerce, Paisley College of Technology
Dr E. J. Smith	Pilkington Brothers Ltd

The International Advisory Panel

L. N. Phillips, OBE	Royal Aircraft Establishment (UK)
Professor R. M. Jones	Institute of Technology, Southern Methodist University, Dallas (USA)
Dr A. R. Bunsell	Ecole des Mines de Paris (France)
Professor S. W. Tsai	Air Force Materials Laboratory, Ohio (USA)
Dr W. M. Banks	University of Strathclyde (UK)
Dr C. Patterson	Institute of Physics, Stress Analysis Group (UK)
Dr E. Anderson	Battelle Laboratories, Geneva (Switzerland)

The Local Organising Committee
   G. Macaulay
   J. S. Paul
   J. Kirk
   F. A. Allen
   The Conference Secretary, Mrs C. MacDonald

We are also grateful to many other individuals who contributed to the success of this event. A final thanks to Nan and Simon for their support during the conference.

I. H. MARSHALL

Contents

Session V: Research and Development: Marine Applications
(*Chairman*: L. N. PHILLIPS, |OBE|, *Royal Aircraft Establishment, Farnborough, England*)

Session VI: Research and Development: Modelling Techniques
(*Chairman*: A. W. LEISSA, *The Ohio State University, USA*)

Session VII: Physical and Mechanical Characteristics (1)
(*Chairman*: D. HULL, *University of Liverpool, England*)

Session VIII: Structural Analysis: Platework Systems
(*Chairman*: D. W. WILSON, *Center for Composite Materials, University of
Delaware, USA*)

Session IX: Structural Analysis: Structural Systems
(*Chairman*: R. M. JONES, *Southern Methodist University, Dallas, USA*)

Session X: Physical and Mechanical Characteristics (2)
(*Chairman*: L. S. NORWOOD, *Scott Bader Co. Ltd, Wellingborough,
England*)

Session XI: Structural Evaluation Techniques
(*Chairman*: E. ANDERSON, *Battelle Laboratories, Geneva, Switzerland*)

1

Composite Materials Education in the United States

ROBERT M. JONES

School of Engineering Applied Science, Southern Methodist University, Dallas, Texas 75275, USA

ABSTRACT

Composite materials is a rapidly maturing technology, with emerging applications in a broad range of industries far beyond the aerospace domain, where composites first became popular. However, educational institutions have not uniformly recognized this new status by appropriately changing the engineering curricula. Most educational programs in composites are still at the embryo stage. Some institutions have taken the lead in reshaping what is presented to students. This revision of engineering education is reviewed at three levels: undergraduate, graduate, and after graduation. Formal courses at all levels are addressed, as well as special design projects which involve students to a very high degree in activities that enable them to experience a broad spectrum of topics in composites. Most importantly, the various philosophies for teaching composites are discussed including courses, design projects, research projects, seminars, and other activities that constitute an appropriate composite materials educational experience.

INTRODUCTION

The modern revolution in composite materials began in the fifties with the development of filament winding techniques for glass fibers to make rocket motor cases. The usefulness of composites expanded drastically in the sixties with the advent of the so-called advanced fibers boron and then graphite. These fibers had sufficiently better mechanical properties to allow

1

significant weight reduction and, in time, cost reduction in comparison to structures made of conventional metals. These accomplishments were achieved by a relatively small number of engineers and scientists. Now with a technology that is increasingly attractive to an ever-broadening group of industries, the problem is to educate a larger number of engineers. They must have an appropriate background in analysis, design, and fabrication of structures and mechanical devices made of composite materials that are available now. Moreover, the new engineers must be prepared to treat materials that are sure to be developed in the future.

Engineering educators, like engineers in general, are typically conservative and do not respond rapidly to changes in their environment. Specifically, the engineering curricula have not been revised by *all* institutions to respond to the obvious industrial demand for engineers with part of their education devoted to composite materials. However, some institutions have been quite innovative in developing strong programs in composite materials at both the undergraduate and the graduate levels. Moreover, after graduation education activities in the form of short courses and seminars have played an important role in the development of composites technology. The various current forms of education at the three levels, undergraduate, graduate, and after graduation, are reviewed in the following three sections. There, philosophies of education in composites are described along with examples of active educational programs.

UNDERGRADUATE EDUCATION

Introduction

Undergraduate education is the level of education which in the future will have the most impact on widespread acceptance and use of composite materials. By sheer numbers alone, because undergraduates far outnumber graduate students, the composites community will benefit most by enlarged educational activities at the undergraduate level. More students should be made aware of and become familiar with at least the advantages of composites, if not the analysis tools to use them. Those graduates will then think of composites as a viable alternative to conventional metals when they design as practising engineers. However, there is no uniform agreement on how undergraduates should be educated in composite materials technologies.

First, the elements or phases of composites education are examined along with the prerequisites that must (or should) be met prior to treatment of composites. Then, the facilities needed for the various phases of composites education are developed. Finally, examples of educational philosophies are discussed.

Elements of Composites Education

The basic elements or phases of composites education are listed in Fig. 1. There, the logical first step is observation of composite material behavior. Naturally, we must first know what we are dealing with. The fact that composites have different stress–strain response in different directions is but the beginning. We must examine all stress–strain response as well as how the response is changed by different proportions of the composite constituents.

- Observation of material behavior
- Modeling material behavior
- Analysis of structural elements
- Analysis of structures
- Design of composite structures
- Fabrication of composite structures

FIG. 1.   Elements of composite materials education.

The next element in Fig. 1 is modeling the observed material behavior. The most common and simplest model, or idealization of behavior, that of a linear elastic orthotropic material, is a topic that can be readily studied at the undergraduate level (even though many students do not seem to recall the relation between Young's modulus and the shear modulus for isotropic materials). Development of linear elastic idealizations of composite material behavior is probably the realistic limit of complexity for an undergraduate student (except for truly bright students who always manage to accomplish more than we expect anyway). More complex models such as various kinds of nonlinearities or viscoelastic behavior might best be left to the graduate level.

Analysis of structural elements is the next step in Fig. 1 in the logical development of composites education. Laminated beams can easily be addressed at the undergraduate level with some possible progression to laminated plates and shells. The difficulty here is the lack of preparation of the student to address plate problems. Such preparation would typically occur in an advanced mechanics of materials course which is usually taught as an optional course in the senior (fourth) year. Accordingly, composites as a course would logically follow an optional course, i.e. as an elective after an elective. Such a tenuous chain of circumstances does not lead to the probability that great numbers of undergraduates will be educated in composite materials. The alternative to enforcing a prerequisite is to introduce plates in a composites course when laminated plates are addressed. An additional difficulty is that students will not necessarily have had a structural or mechanical design course prior to composites. Thus, students do not get the full impact of contrasting composites design with metals design because metals design has not yet 'become a part of them', i.e. they are not yet really familiar with metals design, so they cannot fully appreciate the advantages of composites over metals.

Perhaps the most effective manner in which to include composites in an undergraduate curriculum is to *treat composites topics simultaneously with metals topics in the same courses.* This approach would require extensive revision of many current 'standard' courses as well as the corresponding textbooks. Currently, only brief consideration is given to composite materials topics in a very few undergraduate mechanics of materials textbooks. If such an extensive rethinking and revision of undergraduate mechanics courses were to take place, then we would be able to guarantee that all undergraduates have a significant exposure to composites. That is, we would have brought composite materials topics into the mainstream of the mechanics courses that form an essential part of civil, mechanical, and aerospace engineering curricula. Such a full integration of composites will probably take at least a few more years, if not longer.

The topics of analysis of structures and then design of composite structures in Fig. 1 are subject to the same, if not stronger, preparation difficulties as analysis of structural elements. Namely, all those topics have barely been digested or understood by undergraduates, so the substantial complication of composite materials is a difficult step for them to take. Real progress in these topics will probably await graduate study or independent study after graduation.

The final topic in Fig. 1 of fabrication of composite structures is perhaps the easiest for students to understand of all the topics in Fig. 1. That is, for some fabrication processes, no specific knowledge of advanced analysis topics is necessary if you merely want to observe the process. However, the meaningful control of fabrication processes to achieve specific structural performance goals does require substantial knowledge of the analysis process as well as the design process. Thus, some fabrication concepts can be introduced at the undergraduate level, but many must wait until later. Perhaps the most important factor controlling what fabrication is taught is the availability of fabrication facilities, as addressed in the next section (you certainly cannot do much more than introduce filament winding if you do not have a filament winding machine!).

Facilities for Instruction
The facilities necessary for instruction in various topics in composite materials have a wide range of possibilities. As a lower limit at which only mechanics of composite materials is addressed, no facilities are required. However, such a purely theoretical or textbook approach has serious limitations and does not constitute a balanced, well-developed educational experience. Measurements, experiments, and fabrication can be described with the aid of sketches and pictures, but there is no true understanding without the actual experience of seeing an experiment being performed or, better yet, performing it yourself! The desirable instructional facilities are now quite briefly summarized. To start with, the usual hydraulic testing machines found in almost every laboratory can be used for some of the simple characterization tests such as normal stress versus normal strain. However, more specialized testing devices are necessary to perform general mechanical characterization of composites, e.g. shear testing with a rail shear rig. For fabrication of epoxy matrix composites, vacuum bagging or pressure bagging is an absolute minimum capability with a more reasonable minimum being a heated press or, better yet, an autoclave. Advanced fabrication facilities include filament winding and compression molding with little likelihood of tape laying machines, pultruders, or complex weaving machines being available at the university level. Fabrication of metal matrix composites is seemingly too complex and expensive to be considered at the undergraduate level. Finally, various levels of computers are essential for analysis as well as for control of experiments and reduction of experimental data.

Philosophies of Undergraduate Education
Currently, three different philosophies of undergraduate education are operational in the United States. First, composites are addressed solely in formal coursework as a supplement to the normal curricula. Second, design project experience is the first contact a student has with composites. Third, a combination of coursework and design project experience is available to the student. Examples of each of the philosophies are discussed in the following paragraphs. In all the discussed implementations of these philosophies, fabrication of composites is included, even in the formal course.

Formal coursework

At the Georgia Institute of Technology, a course 'Fundamentals of Fiber Reinforced Composites' is cosponsored by the Schools of Aerospace Engineering, Civil Engineering, Engineering Science and Mechanics, Mechanical Engineering, and Textile Engineering. Obviously, the course has widespread support among the schools and, moreover, is externally supported by a grant from Owens/Corning Fiberglas. Fundamental analysis techniques are studied along with manufacturing processes and fabrication as well as design. The course is quite popular (it attracted nearly fifty students at its first offering). The fundamental motive for the course is to provide a forum for addressing the special characteristics of composite materials outside the usual mechanics courses.

Design project experience

At the Massachusetts Institute of Technology, undergraduate students are able to participate in design projects as well as in research projects as junior members of a team headed by a graduate student. Through the mechanism of seminars and weekly discussions, all members of each team are made familiar with composites, with the work of other members of their team, and with the work of other teams. Thus, all participants in the program learn from each other. Analysis, fabrication, and testing are possible activities for all teams.

Combination of coursework and design project experience

Two schools follow the philosophy of combining formal coursework with design project experience. These schools are the Rensselaer Polytechnic Institute and the University of Delaware.

At RPI, the program in composite materials is a strong interdisciplinary effort involving the Civil Engineering and Materials Engineering Departments as well as the Mechanical Engineering, Aerospace Engineering, and Mechanics Department. Many composites topics are addressed in some of the customary courses, i.e. *the usual courses have been modified to address composites*. The first all-composites contact a student might have is in a senior design project called CAPGLIDE (Composite Aircraft Program Glider) sponsored by the National Aeronautics and Space Administration (NASA) and the Air Force Office of Scientific Research (AFOSR). This design project is an alternative to the senior laboratory course. The project is popular because, by the senior year, students are anxious to actually build something (in contrast to the usually disjointed purely theoretical or purely experimental activities prior to the senior year). Students enter the project without specific coursework background in composites. However, their experience on the project during the fall semester usually entices them to take a formal composites course in the spring. Still, the biggest problem for students is obtaining the requisite technical background early enough to have an impact on the design project. As in the MIT program, weekly discussions and seminars are an important part of the information transfer in the project. These discussions are the formal mechanism for teaching the students about composites. During the fall, students might design, analyse, fabricate, and test a fitting for the glider. Or they might make parts in preparation for the spring phase. Or they might develop a fabrication technique. During the spring, the glider design is refined, and the glider is fabricated. Simultaneously, students might take a formal composites course. Typically, in accordance with the experience across the country for complicated design projects, the project activity might extend into the summer (and beyond). The ability to continue the project stems from the involvement of juniors and even sophomores in the design teams. Thus, year-to-year continuity is maintained.

The University of Delaware also has a combination coursework and design project program based in the Mechanical and Aerospace Engineering Department with strong participation from the Chemical Engineering and Civil Engineering Departments. The courses 'Experimental Mechanics for Composite Materials' and 'Composite Materials' follow a plates and shells course offered during the junior year. During the senior year, fully half of the required design projects involve composites. For example, a high speed bicycle was developed, and the prototype placed third in forty-three nationwide entries. Also, a Mini-Baja off-the-road cart-like vehicle was designed and constructed of composites and entered in a regional competition.

GRADUATE EDUCATION

Introduction

Graduate education in composite materials must include the fundamental elements of composites education displayed in Fig. 1, but at a higher level of sophistication, completeness, and accomplishment. Generally, by the graduate level, most difficulties with prerequisites to courses have disappeared. For example, an advanced mechanics of materials course should have been completed at the beginning of a typical graduate program (if not before). Moreover, the uniformly higher maturity of the graduate student as compared to the undergraduate student allows a more detailed examination of topics in composites. In particular, graduate students have a more advanced mathematics background than undergraduates. Moreover, some form of design coursework and/or project is part of the background of the typical graduate student. Thus, graduate students are generally prepared to address the complexities of composites.

In this section, the various philosophies of graduate education currently being pursued in American institutions are briefly discussed. Then, the specific activities in what are termed major programs are described. Finally, other programs of a smaller magnitude are briefly mentioned.

Philosophies of Graduate Education

The philosophies of graduate education in composites depend both on the school and on the degree pursued. Three major elements are prominent in the current philosophies: coursework, the design project, and the research project.

Virtually all schools with any activity in composites have some formal composites coursework. Moreover, many such schools have modified other courses which are not predominantly about composites to include some composites topics that are complementary to the main theme of the course. For example, laminated plates can be treated in the latter stages of a plates course (with similar developments in a shells course). Similarly, anisotropic elasticity can be addressed as part of an elasticity course.

A second important, but not uniformly accepted, element in a philosophy for graduate education in composites is the design project. The importance of a design project seems to lie most clearly at the Master's degree level as opposed to the Ph.D. level. That is, a Master's degree can be a program developed to prepare a designer or to give an analyst significant design experience both to motivate analysis and to put analysis efforts in proper perspective. In contrast, a Ph.D. program would probably be focused on a more restricted topic (than a design) which might be attacked from either an analysis or an experimental point of view or both.

The third element in a philosophy for graduate education in composites is the research project. Traditionally, research is an intimate part of every Ph.D. program. And, the opportunities for research in composites are now even more exciting than in more mature areas simply because the area of composite materials constitutes an emerging technology. Research activities are specifically excluded from this discussion, but would be an appropriate subject for another paper.

The three elements, coursework, the design project, and the research project, are blended into strong programs at only a few American universities. Those programs are called 'major' programs in this paper. No criticism of other programs is intended; the point is that several programs have achieved special distinction and deserve attention as possible models for other programs. Both major programs and other programs are described in the following paragraphs.

Major Composite Education Programs

Anyone who attempts to categorize activities involving people runs the risk of being criticized for his/her definitions used to distinguish between the various categories. With this risk clearly in mind, I propose the following qualifications for a major program in composites education:

(1) More than one formal course in composites is offered regularly.

(2) More than just a few of the faculty participate in the program, and several of the faculty devote a large percentage of their time to composites (as opposed to only a partial interest in composites).

(3) Design experience is a prominent part of the program in either or both of the undergraduate or beginning graduate levels.

(4) Extensive research activity exists.

The United States institutions which I feel have major programs are: Virginia Polytechnic Institute and State University (VPI), University of Delaware, Rensselaer Polytechnic Institute (RPI), University of Oklahoma (one of the oldest continuous programs), University of Florida, Texas A & M University, and University of Wyoming. Unless I am incompletely informed, these universities are the only ones whose programs satisfy the foregoing obviously imperfect criteria to be called a major program. The first three mentioned programs are the largest and are described in the following paragraphs.

Virginia Polytechnic Institute and State University (VPI)

With the strong participation of National Aeronautics and Space Administration (NASA) personnel over the past twenty years, VPI has built up an impressive program in structural mechanics. During the past ten years, substantial growth has occurred in the number of faculty members engaged in composites research and education to the point where about a dozen faculty members now work solely on composites. Some of this growth has resulted from a cooperative program between NASA and VPI at the graduate level. Most of the growth stems from extensive sponsored research activity.

The NASA–Virginia Tech. Composites Program is part of the VPI program and has two major objectives. First, graduate students are educated in preparation for careers in research, development, design, or teaching. Second, research is conducted on problems of current interest. Both objectives are in the national interest, hence the NASA involvement. Two phases exist for both the M.S. and Ph.D. degrees: on-campus study and a research residency at NASA Langley Research Center. For the M.S. program, on-campus study in formal classes lasts for twelve months followed by at least a six-month research residency. For the Ph.D. degree, the on-campus phase consists of 18 months of courses plus initiation of the research activity. Then, the research residency is at least 12 months at NASA Langley Research Center. The research can be either of a fundamental nature or of an innovative structural application.

VPI offers five quarter-long courses in composites:

ESM 4040	'Mechanics of Composite Materials'
ESM 5150	'Stress Analysis of Composites'
EMS 5070	'Mechanics of Composite Structures'
MATE 5100	'Modern Composite Materials'
ESM 6100	'Failure in Composite Materials'

where ESM stands for Engineering Science and Mechanics and MATE stands for Materials Engineering. The first four courses are required in the M.S. program along with the usual structural mechanics and materials courses. The last course is required in the Ph.D. program (as are the first four for persons entering at the Ph.D. level) in addition to the usual courses.

University of Delaware

The University of Delaware has a strong program in composites featuring a multidisciplinary organization, the Center for Composite Materials. The Center has the usual objectives of education and research, plus the objectives of documentation of the state of the art and dissemination of information. One of the activities in the documentation objective is the creation of a composites design guide. The research activities are sponsored by various governmental agencies and companies. Overall sponsorship of the industrial research program of the Center comes from over a dozen companies, some of which are listed in Fig. 2. Each company pays the same fee each year as a continuing sponsor. In addition, some other companies and government agencies sponsor specific research projects of a short term nature.

- Owens/Corning Fiberglas
- International Harvester
- Pittsburgh Plate Glass
- General Motors
- General Electric
- Graftek/Exxon

- Celanese
- Rockwell
- Hercules
- Dupont
- Xerox
- Ford

FIG. 2.   University of Delaware Center for Composite Materials industrial research program sponsors.

The formal courses in composites at the University of Delaware include:

MAE 410   'Experimental Mechanics for Composite Materials'
MAE 617   'Composite Materials'
MAE 817   'Composite Materials'

where MAE stands for Mechanical and Aerospace Engineering. Some of the courses are designed and numbered for both undergraduate and graduate students with the latter being required to prepare a term paper and/or do some experimental research.

The experimental and fabrication facilities and capabilities include:

Ultrasonic Nondestructive Evaluation
Tension–Torsion Fatigue System
Injection Molding Machine
High Strain Rate System
Electron Microscope
Autoclave
Extruder

plus some of the more usual facilities of mechanical, aerospace, and chemical engineering departments.

Rensselaer Polytechnic Institute (RPI)

The strong RPI interdisciplinary program at the undergraduate level is continued into the graduate level also with the sponsorship of NASA and the Air Force Office of Scientific Research (AFOSR). Formal coursework includes 'Composite Materials' and 'Designing with Composite Materials' as well as the addition of composites topics to the usual structural mechanics courses.

The NASA–AFOSR Program includes three activities: CAPCOMP, COMPAD, and INSURE. CAPCOMP stands for Composite Aircraft Program Component. Currently, the Boeing 727 elevator actuator attachment is being designed in graphite–epoxy composites to replace an aluminum alloy attachment. COMPAD stands for Computer Aided Design. This phase of the program is in its initial stages with activities in enhancement of the SPAR finite element computer program, investigation of structural optimization programs, and development of preprocessors and postprocessors for finite element analysis programs. INSURE stands for Innovations and Supporting Research and is a collection of small research projects supporting the overall composites program. Examples of such research projects are resin characterization and optimization, moisture effects, optimization of laminated plates under shear loading, fatigue, ultrasonic nondestructive testing, and metal–matrix composites. Of course, RPI has other research activities sponsored by organizations besides NASA and AFOSR. One of the attractive features of the RPI program is that professors work across the usual department boundaries without the aid (or constraint) of a formal composites organization. Weekly meetings of the more than a dozen interested faculty members from the various participating departments keep the program moving.

Other Programs

The remaining composites education programs in the United States are classified as nonmajor because they have the following characteristics:

(1)   Usually only a single formal course in composites exists, and that course is typically mechanics of composite materials. Such a course might or might not be offered on a regular basis.
(2)   Usually only a few faculty are involved in composites, and, of those few, perhaps only one is dedicated solely to composites.
(3)   Little if any design experience results from the program.
(4)   The research activity underlying the advanced educational levels is small to moderate (because of the small number of faculty members dedicated to composites).

● Iowa State	● Wisconsin
● Univ. of Illinois (Chicago & Urbana)	● Ohio state
● Massachusetts Institute of Technology	● Hartford
● Georgia Institute of Technology	● Cincinnati
● Illinois Institute of Technology	● Purdue
● Air Force Institute of Technology	● Drexel
● Univ. of California at Berkeley	● Kansas
● Univ. of California at Los Angeles	● Dayton
● Wichita State	● Tulsa
● Southern Methodist	● Texas

FIG. 3.   'Other' programs in composite materials education.

Without any further discussion, the so-called 'other' programs in composites education are listed in Fig. 3. Other programs of various sizes surely do exist, and I apologize for overlooking them. I would be pleased to have more information about all programs mentioned and especially about programs that I do not list in Fig. 3.

AFTER GRADUATION EDUCATION

Education apart from formal full-term courses offered by universities has played a very important role in the development and application of advanced composite materials. Historically, the first formal courses on composites might be the short courses organized by Dr Stephen W. Tsai under the sponsorship of the University of California at Berkeley in the late sixties. Of course, Professor Albert G. H. Dietz of the Massachusetts Institute of Technology and others lectured on composites in the late forties and early fifties. However, those efforts preceded the advanced fiber development that is the foundation for the current rapid expansion of composites applications.

The present after graduation educational activities are of two types: (1) open short courses or seminars and (2) in-plant short courses. A list of the major open short courses is given in Fig. 4. Three major universities sponsor

Open short courses or seminars
- Univ. of California at Berkeley (Tsai)
- Univ. of California at Los Angeles (Beaumont)
- George Washington Univ. (Noor)
- Technomic Publishing Co. (Salkind)
- Jones Mechanics Research (Jones)

In-plant short courses
- Bert
- Vinson, Chou and Pipes
- Jones

FIG. 4.   After graduation composite materials courses.

open short courses organized by the gentlemen whose names are in parentheses. Also, two private ventures have sponsored composites short courses in recent years. Other composites short courses have been offered from time to time, but only those listed have been offered regularly. In-plant short courses have been presented by Dr Charles Bert, Dr Jack Vinson, Dr Tsu-Wei Chou and Dr Byron Pipes, as well as by the author. My in-plant courses are listed in Fig. 5 to show the interest by industry in the education of engineers in composite materials. Course sizes ranged from twenty to

- Naval Air Development Center (Aug. 71)
- Naval Weapons Center (June 75)
- Bell Helicopter (fall 75 and spring 79)
- Rockwell—Tulsa Div. (fall 77)
- Thiokol (June 79)
- Naval Ship R & D Center (July 79)
- Lockheed Missiles & Space (Dec. 79)
- Boeing Aerospace (April 80)
- Naval Air Rework Facility, Cherry Point, NC (July 80)
- Vought Corporation (fall 80)
- Celanese Research (Jan. 81)

FIG. 5.   In-plant composite materials short courses by Robert M. Jones.

over fifty people. Some companies, such as General Dynamics in Fort Worth, Texas, organize and present their own composites courses with emphasis on topics of special importance in company applications.

CONCLUDING REMARKS

Composite materials education in the United States is reviewed at three different levels: undergraduate, graduate, and after graduation. Programs now exist at all levels in contrast to the situation only a few years ago. Rapid progress is being made toward integration of composite materials topics into the mainstream of traditional engineering education.

At the undergraduate level, courses in composites have been introduced, and some existing courses have been modified to incorporate composites topics in parallel to similar metals topics. These changes are merely recognition that composite materials are here to stay and, in fact, are expanding rapidly in application. However, as is typical with conservative engineers, the applications have occurred before the formal treatment of the new topic in engineering curricula. And, there is a compelling need to treat composite materials at the undergraduate level because the B.S. degree is tending to become a terminal degree. Thus, educators might have only one early opportunity to have an impact on the lifetime educational background of a typical engineer.

Graduate education continues to be the primary source of highly educated contributors in research, analysis, and design of structural and mechanical systems made of composite materials. The various research activities are a fundamental part of graduate education and are now quite lively.

Many engineers are redirecting their careers toward composite materials because of recent developments in engineering technology and because of design philosophy changes forced by decreased energy availability. The drive to increase the energy efficiency of all vehicles from airplanes to automobiles has put pressure on engineers to use composite materials. Engineers with traditional educational backgrounds have several approaches available to help them redirect their careers accordingly. Short courses are an ever more popular way to start a new speciality on a formal basis (as opposed to self-study) in a short period of time (in contrast to the always available alternative of going back to school for a Master's degree). Some companies are either bringing in consultants to teach such short courses or developing company courses.

The various levels of composites education are assisted by the availability of the textbooks listed in the bibliography. The list is restricted to American textbooks, in consonance with the topic of this paper. More advanced books are expected to appear as composites technology matures and the need for more advanced courses becomes apparent and widespread.

Some composite materials education topics are ignored in this paper, the basic technical focus of which is on the structural mechanics of composite materials. Perhaps the most obvious missing technical topic is polymer chemistry and its influence on mechanical properties. Also, processing technology is only obliquely alluded to. Both topics, and others not even mentioned, are important parts of developing composites technology. Perhaps others more able than I in those areas will see fit to address how those topics are treated in the American educational system.

BIBLIOGRAPHY OF AMERICAN COMPOSITE MATERIALS TEXTBOOKS

CALCOTE, L. R., *The analysis of laminated composite structures*, New York, Van Nostrand Reinhold, 1969.
ASHTON, J. E., HALPIN, J. C. and PETIT, P. M., *Primer on composite materials: Analysis*, Westport, Connecticut, Technomic, 1969.
ASHTON, J. E. and WHITNEY, J. M., *Theory of laminated plates*, Westport, Connecticut, Technomic, 1970.
VINSON, J. R. and CHOU, T.-W., *Composite materials and their use in structures*, New York, Halsted Press (Wiley), 1975.
JONES, R. M., *Mechanics of composite materials*, New York, McGraw-Hill, 1975.
CHRISTENSEN, R. M., *Mechanics of composite materials*, New York, Wiley, 1979.
TSAI, S. W. and HAHN, H. T., *Introduction to composite materials*, Westport, Connecticut, Technomic, 1980.
AGARWAL, B. D. and BROUTMAN, L. J., *Analysis and performance of fiber composites*, New York, Wiley, 1980.

2

Engineering Plastics—Some Factors Affecting Technology Transfer

JOHN HUMPHREYS

Pilkington Brothers Limited, Research and Development Laboratories, Lathom, Ormskirk, Lancashire L40 5UF, England

ABSTRACT

This paper holds as its basic tenet that market growth for any new engineering material results from being able to provide the design engineer not only with parts that work but also with sufficient design technology for his continuing use of that material in like applications. The higher the design stresses and risk the greater is the quantity and quality of technology demanded. Since with many of today's composite materials much of this technology is still in the hands of materials suppliers, toolmakers and moulders it is suggested that actions by these to ensure as rapid as possible a transfer of technology to the designer is an essential prerequisite to growth. Naturally the desire of designers to acquire such information varies by country and by company, involving many factors such as legislation, regulation, consumer demands, investment needs, etc., as well as the state of technology itself. It is therefore further suggested that the preferred route for generating and transferring technology is by attacking actual, specific, design needs through collaborative developments. Examples from the automotive industry are cited.

INTRODUCTION

'Engineering Plastics and Composites' covers an extremely wide spectrum of candidate materials, thermosets versus thermoplastics, filled versus unfilled, reinforced versus unreinforced additive modified, etc. These are not only in competition with each other where reduced weight and energy

16

consumption are required, but with the newer metal products of HSLA steel/plastic or aluminium/plastic laminates, and thin wall zinc die casting.

Pilkington Brothers Ltd currently promotes glass reinforced systems in automotive application, but the specific constraints here are, it is believed, no more than typical of the problems to be faced generally. Automotive applications are the biggest single market area identified for engineered plastics. Conferences and papers (references 1–6) on the various merits and problems are legion. Substitution by plastics of virtually every part on a car has been shown to be technically possible and yet they still account for only 5% or less of total vehicle weight. Why should this be so and what has technology yet to achieve?

TECHNOLOGY TRANSFER—AN OVERVIEW

The normal design sequence is to prepare schemes on paper, detail to fulfill a specific duty, prepare a model, select materials and process, produce prototypes, refine and finally manufacture; i.e. the component designer is in full control making value judgements based on past experience. With a new material this is not possible, much of the technology resides with the developer of that material and with its subsequent processors. In the case of composite plastics the situation is more complex (Fig. 1) in that there are not only many choices, each with its own mix of properties, but the basic technologies of polymer chemistry and fibre physics are completely alien to the currently practising engineer. Training of engineers in these technologies is obviously vital (as perhaps is the training of chemists and physicists in engineering) and—as Bob Jones, in our other plenary paper, has shown—is receiving attention. But what of today's engineers, how do they learn on the run and gain sufficient confidence in these new materials as to use them in *ab initio* design? While it is always possible to pass information along the supply chain the submission is that the only way which has so far proved to achieve any real success is through collaborative developments. Technologists, from the material suppliers, compounders, moulders, toolmakers and customers, must be prepared to act in unison to promote new materials into new applications. It is not sufficient merely to make parts that work, some of the technology must be transferred as well. Although perhaps an altruistic belief, supplier companies must clearly separate activities geared to making a larger marketplace from those intended to establish or increase market share. This is not to say that, in an industrial environment, one must lose sight of potential sales, but rather to demonstrate firstly that a material works and secondly that it works

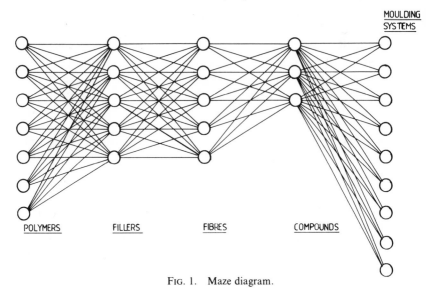

MOULDING
SYSTEMS

POLYMERS   FILLERS   FIBRES   COMPOUNDS

FIG. 1.   Maze diagram.

because of one's particular ingredient. Put it the other way round and immediately the bogey of sole sourcing arises. It is equally important to break down these materials and markets and identify those applications which are easy from those which are not, and to define what developments are necessary for the latter to happen. There will be no overnight or across-the-board changes; gaining of technical expertise, knowledge and hence confidence is a gradual thing and only when the designer has that confidence will new materials be adopted. This is of course galling for the eager supplier, but is nevertheless confirmed by history.

TECHNOLOGY TRANSFER—THE AUTOMOTIVE INDUSTRY

While there is quite clearly an automotive industry, the all-embracing concept of auto-design is extremely elusive. It does however segment readily into:

BODY DESIGN—STRUCTURAL (chassis, unitary); APPEARANCE (doors, wings, bonnet).
STRUCTURAL PARTS—sub-frames, suspension, wheels.
MECHANICAL PARTS—engine, transmission.
ELECTRICAL PARTS—ignition, lighting, controls.
INTERIOR COMPONENTS—dashboard, seats, trim.

which crudely, is in descending order of mechanical duty, and hence risk. There is of course a wide spread of duty for all components within each category too. The history of developments in each group so far shows that initial trials were with low risk parts fitted to low volume models, spreading progressively to larger volumes and higher risks.

The penetration of plastics into each sector is very different and the prime motivators are, it is suggested, also different. Considering these in reverse order we see:

Interior components. A fully mature sector with plastics as the preferred materials both in trim and in semi-structural items such as dashboards, using mainly PVC and urethane foam systems. These are now highly cost-effective materials and it is not often remembered that many are composite systems. At least part of the initial motivation was through Government regulation for occupant safety in an accident.

Electrical components. Again a fully mature plastics sector with virtually every polymer system represented. In this sector the initial motivator was almost entirely technical, electrical resistivity, and for this reason plastics (including rubber) has been the preferred material from inception.

In both these sectors the basic design requirements of the materials are well understood by the auto-engineer who has full confidence in specifying them.

Body appearance panels, lightly stressed mechanical parts. Here the market is fragmented both as between the US and Europe and as between high and low volume production.

The low volume market has long recognised the design flexibility and low part cost of glass reinforced systems, initially with hand lay-up and more recently with matched die moulding. General Motors, Lotus and now de Lorean are the obvious proponents of all plastic, high technology parts. The truck industry is equally a large user of matched-die moulded plastics in body parts; both sectors specify plastics systems 'as standard'.

High volume production is today's major talking point. Under Government regulatory pressure to improve low speed impact, and to reduce fuel consumption, and with assumed adverse consumer reaction to smaller cars, the US auto industry and its suppliers have invested heavily in the technology of plastic body parts, with the result that GM and Ford are today direct major producers of parts in SMC, DMC and latterly RRIM urethanes,[7-10] with large in-house research and development resources. Government standards were such that the industry was initially prepared to pay for weight savings (figures up to $1 per lb were quoted). Meanwhile in

Europe, long subject to energy conscious scales of vehicle and petrol taxation, vehicles were already more energy efficient and the main plastics developments were in impact enhanced bumpers using mainly glass reinforced thermosets (Renault 5, Fiat Strada, Talbot Alpine) or steel reinforced thermoplastics (Volkswagen, Renault) although plastic rear doors are used in some lower volume models (Volkswagen). Quite clearly as a result of higher activity in manufacturing and development there has been a major transfer of technology in the US, but much less so in Europe. With the recent down-sizing of US cars developments over the next few years will be most interesting. Both markets are now being supplied with cars of similar shape, size and weight.

Neither market will pay for weight reduction, plastic parts have to be lighter and cheaper to justify further capital write-off of metal-working machinery; their technical objectives are similar, to reduce unit costs through:

(i) improved compounding control to reduce material variability;
(ii) improved moulding control to reduce component variability;
(iii) improved mechanical handling to reduce cycle times;
(iv) controlled anisotropy;
(v) improved painting capability;
(vi) improved test methods for materials and components.

In addition, action to reduce the tooling costs for early prototypes will increase the number of components evaluated.

This sector is expected to be the growth sector for the 1980s with SMC and RRIM as currently preferred systems. Apart from the recently announced Talbot Murena and GM P cars, most progress is nevertheless expected to be through part by part, model by model substitution. The major difference will probably be that in the US this growth will be led and fed by the auto-industries' own R. & D. effort, whereas in Europe much of the auto-maker's needs will continue to be fed for several years either by its suppliers, or by direct import of US technology.

Highly stressed body and mechanical parts. Perhaps the most imaginative development here is the 'plastic engine' developed by Polimotor Research Inc. comprising 27 kg of resin, 7 kg of carbon fibre and 11 kg glass fibre in a 2·3 litre engine of 77 kg total weight; this is half the weight of a traditional US 'cast iron' engine. Power outputs from limited running, are comparable with conventional engines of similar capacity (> 100 bhp). While far beyond any cost effective criteria, at $2 on cost per kg of weight saved, it does demonstrate technical capability and provide a base for practical value

engineering. At the opposite end of the technical spectrum, and empha-
sising the part by part approach, Mitsubishi have a glass reinforced
phenolic carburettor body fitted to the production Galant engine, while
GM have a glass reinforced epoxy leaf spring on their 1981 Corvette.
Glass reinforced nylon is now beoming the norm for radiator header or side
tanks, and more recently for valve or rocker box covers.

It is clear, therefore, that over a wide range of duties, the automotive
industry has already acquired an appreciation of the technology of
composites (Fig. 2), today's applications are no more than extending this
into new sectors with new designers. The part by part approach, which has
worked well in the past, is likely to be successful in the future, and where
significant new materials technology still resides with the supplier(s) it
behoves that supplier to undertake collaborative development with the
auto-industry to ensure that not only is a successful part produced but also
that some of the technology is transferred, until the designer has sufficient
knowledge and confidence to specify these new materials. The efforts of the
Polymer Engineering Directorate (SRC) and the Polyester Compounds
Group to promote these ends are worthy of our support.

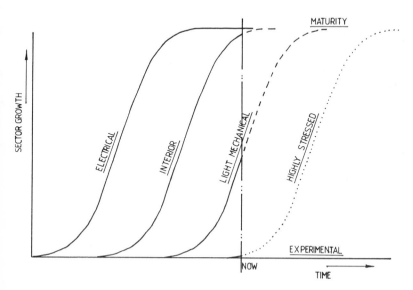

Fig. 2.   Auto-market sector penetration diagram.

TRANSFER OF TECHNOLOGY—SOME EXAMPLES

In support of these beliefs, the Pilkington Group has assigned significant R. & D. effort to promoting the use of glass fibre reinforced engineering composites, primarily in automotive applications. A total system resource from material selection, through to moulding of prototype components in thermosets, with support from laboratory test facilities, has been established. It works closely with resin suppliers, compounders, moulders, toolmakers and auto-engineers. Obviously, individual prototype components are unique and confidential to collaborating partners as is any detailed material or processing specification, but much of the general information gained is being incorporated into published data[12-15] Direct facilities available allow for production of prototypes up to about $1\frac{1}{2}$ kg in weight, and not unexpectedly substitution of cast aluminium or zinc die castings falls readily within this ambit.

TABLE 1
GM materials characterisation chart—required FRP material property matrix

Property	Condition				
	23 °C	90 °C	150 °C	−40 °C	90 °C 100 % RH
Tensile strength (MPa)					
Tensile modulus (GPa)					
Poisson's ratio			—	—	—
Strain to failure (%)					
'Knee' stress (MPa)					
'Knee' strain (%)					
Compression strength (MPa)					—
Compression modulus (GPa)					—
Strain to failure (%)					—
Shear strength (MPa)					—
Shear modulus (GPa)					—
Strain to failure (%)					—
Flex modulus (GPa)		—	—		—
Flex strength (MPa)		—	—		—
Impact energy (kJ/m^2)			—		—
Matrix crazing strain (%)		—	—	—	—
Stress level for 10^6 (T–T^a)			—		—
Fatigue life (MPa) (Flex)		—	—	—	—
Creep				—	
CTE (μm/m/ °C)			—	—	—

TABLE 1—*contd.*

Property	23 °C	90 °C	Condition 150 °C	−40 °C	90 °C 100 % RH
Specific gravity		—	—	—	—
Fibre percent (%)		—	—	—	—
Resin percent (%)		—	—	—	—
Filler percent (%)		—	—	—	—
Interlaminar shear strength (MPa)					
Moisture expansion (%)		—	—	—	—
Residual strengths:	—	—	—	—	—
Moisture (MPa)		—	—	—	—
Gasoline (MPa)		—	—	—	—
Brake fluid (MPa)		—	—	—	—
Transmission fluid (MPa)		—	—	—	—
Motor oil (MPa)		—	—	—	—
Antifreeze (MPa)		—	—	—	—
5 % Salt solution (MPa)		—	—	—	—
1·5 mm Notch strength (MPa)		—	—	—	—
3·0 mm Notch strength (MPa)		—	—	—	—
6·0 mm Notch strength (MPa)		—	—	—	—
Thermal cycle (MPa)		—	—	—	—
Bolt strength (MPa)			—	—	—
Bond strength (MPa)			—	—	—

—: Test conditions which may be omitted.
[a] T–T = Tension–tension fatigue.

Many of these parts have common factors; for example, they:

(i)   are exposed to petrol, oil, water, antifreeze and hydraulic fluid;
(ii) are subject to compressive bolt loadings;
(iii) experience thermal cycling;
(iv) have hang-on parts providing steady loads;
(v)   experience fatigue induced through engine and road motions;
(vi) have circular upstands for pipe connections; and
(vii) are mounted on resilient gaskets.

Material characterisation involves extensive testing, as typified by the GM Specification for Random Fire Reinforced SMC shown in Table 1, extracted from reference 7.

However one cannot simply assume that all SMC or DMC systems have similar properties. Our own work, Figs 3 and 4, shows dependence upon both the resin system used and the environment so that what is 'best' in water is not necessarily 'best' in petrol. To date we have found water or water/antifreeze to be, generally, the largest detractor from start property although we expect ethanol or methanol based fuels to be much more aggressive than current petrols.

Additionally, the properties obtaining in a moulded part are usually significantly lower than ideal, either in total or locally, due to such factors as (i) fibre attrition during processing, (ii) fibre orientation during flow through the mould and (iii) flow fronts and fibre starvation.

Injection moulded DMC parts are notorious in this respect, as compared to similar SMC parts (Fig. 5). In order to discriminate between the performance of like-materials in use, we have found it essential to operate test methods which can rapidly assess fibre length, dispersion and orientation, and relate these to observed mechanical properties.

Our recent work on test method development has two main objectives:

(i)   to obtain a better appreciation of the fibre contribution to the dynamics of failure, i.e. not just the gross material performance;

(ii)   to generate test data of use both to the material supplier and the component designer or producer.

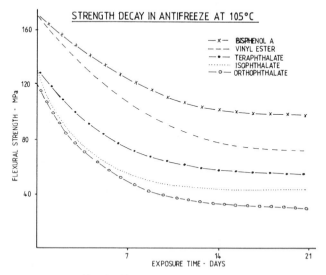

FIG. 3.   Exposure tests—antifreeze.

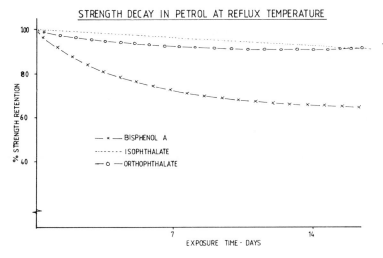

FIG. 4.   Exposure tests—petrol.

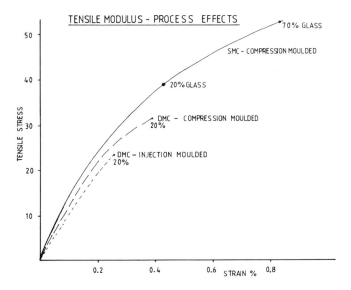

FIG. 5.   SMC versus DMC modulus.

John Humphreys

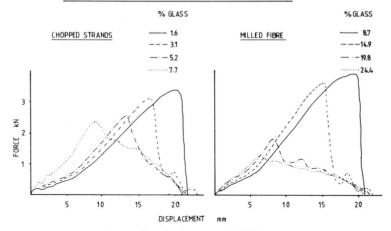

FIG. 6. Impact on RRIM.

We have found instrumented impact measurement and acoustic emission techniques valuable tools. In Fig. 6 we see the effect on impact of increasing concentration of chopped and milled glass fibres on RRIM urethane. The respective concentrations required to give impact equivalence are quite clearly seen to be in the ratio 1:3. With AE we have been able not only to reproduce this equivalence ratio but also to determine anisotropy due to fibre orientation (Fig. 7). Extending this to measurements on components

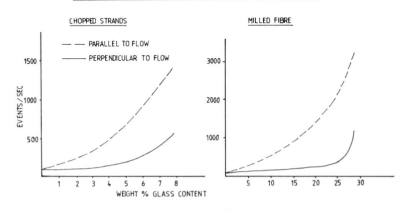

FIG. 7. Orientation in RRIM using AE.

we can also demonstrate variations in the ability of lugs, on a compression moulded SMC part, to withstand bolt forces. These compare well with measured variations in the failure loads although we are as yet unclear as to the true significance of the AE data in such an application. This work does, we believe, parallel the existing use of AE to predict or monitor the suitability of parts for their specified duty.

Many automotive parts are intended to be removed and replaced to facilitate routine maintenance and are therefore usually bolted on. Allowance must here be made not only for normal use but for abuse by the unskilled mechanic or owner.

Bolt (or screw) sizes and applied torques are often specified to make the bolt the weakest link. Direct substitution with a GRP part, particularly where a thick resilient gasket is employed, frequently results in matrix cracking due to local overstraining and, in an under-bonnet environment, with oil and grease, such cracks are readily seen. Our laboratory simulation (Fig. 8) attempts to provide data on suitable bolt/washer/section thickness combinations, and shows that for a given thread size mushroom-head screws are preferable to hexagon-head bolts and large washers preferable to small (Fig. 9). Alternatively section thicknesses around bolt holes can be increased (Fig. 10).

LABORATORY DETERMINATION OF BOLT DOWN CAPABILITY

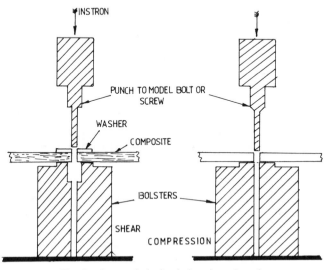

FIG. 8.   Instron bolt simulation rig and results.

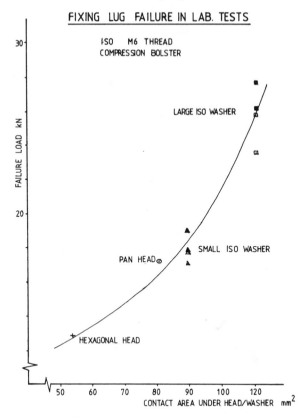

FIG. 9.   Bolt/screw/washer effects.

Finally, by combining points from several projects into two generalised examples it can be demonstrated that through consideration of the total system of materials/moulding/component design, not normally possible without collaboration, it is possible to obtain acceptable results.

Failure Under Applied Load

A component of cross-section as shown in Fig. 11 cracked at point A under a low applied load, P. Finite element analysis predicted a much higher loading should be possible, based on normalised material failure stress. Measurements of fibre attrition showed that the moulding process was breaking down the fibres. Experiments with changed processing conditions reduced the attrition and increased the failure load, but not

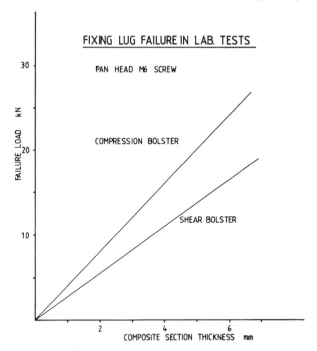

FIG. 10.   Thickness effect on bolt down.

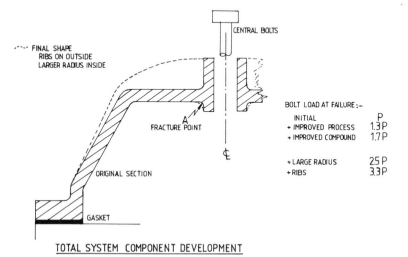

FIG. 11.   Generalised cross-section of component.

sufficiently. Further experiments with changes to the compound formulation and incorporated fibres gave further improvements, but still not adequate for the specified duty. However, fibre attrition was now minimal and the actual failure stress was comparable with that predicted; any further improvements would have to be from changed component design. Again using finite element analysis several options were considered, the increased load capability being balanced against added weight, until an acceptable result was achieved. The tooling was modified and the new component confirmed predictions. Changing formulation, processing conditions and component design collectively would have been extremely

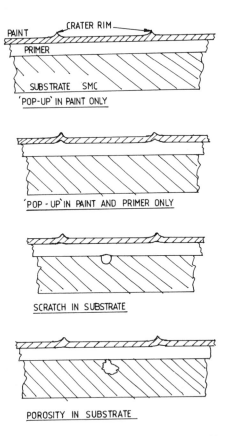

SECTIONS THROUGH VISIBLE PAINT FAULTS

Fig. 12.   Painting.

difficult without the full cooperation and technical appreciation of resin and fibre suppliers, toolmakers, moulders, and components designers.

Painting

There have been many reports on the problems of painting compression-moulded SMC, usually expressed as 'pop-up' or 'mooncraters' in the painted surface, and blame is often laid at the door of the compounder or moulder. Inspection, under a scanning electron microscope (SEM), of local sections taken through the faults in painted plaques showed four quite distinct causes producing the same visual effect (Fig. 12):

(i)   top coat faults not reproduced in the primer;
(ii) primer faults not reproduced in the substrate;
(iii) substrate faults: (a) surface scratches from mishandling and (b) surface porosity from compound/moulding.

Clearly not only had the causes to be separated but potential solutions were in quite different steps in the production process.

The real point to be made is that this is an area not strictly relevant to the expertise of a fibre supplier and solutions were not expressly under our control, but we did own an SEM. Because of our participation in other developments we were invited and happy to assist. I hope our small contribution helped those involved to identify what was and was not under whose control and hence actively promoted attacks on all the real causes.

CONCLUDING REMARKS

In addition to all the quantifiable factors which must be satisfied before composite materials are used as regular engineering materials is the intangible, but daunting factor, of what can be called the 'technology gap' which if not bridged engenders a total lack of designer confidence. With composite materials there is a 'materials maze' in which much of the technology presently resides with the material suppliers and/or moulders and is often of a form or a technical language not readily assimilated by the busy engineer. It is believed that the only successful route so far proven is for all parties to participate in specific applications developments. In this way technology is transferred progressively through the period of that development, leading to requests for further joint work. Historical evidence from within the auto-industry indicates that once the designer has acquired sufficient confidence in a particular material he is more than capable of generating the customer pull which reflects in growth and increased penetration.

It may be thought that all the foregoing is essentially correct and so patently obvious that it need not have been said or written. Conversations with those at the other end of the 'materials maze' indicates, unfortunately, otherwise.

ACKNOWLEDGEMENTS

This paper is published with the permission of the Directors of Pilkington Brothers and Mr A. S. Robinson, Director of Group Research and Development. Thanks are also due to those in the fibre, polymer, toolmaking, moulding and automotive industries who have contributed in any way to composite materials processing through collaborative developments.

REFERENCES

1. PALERMO, J. C., Glass fibre reinforced plastics in the European and North American transportation market, *Proceedings B.P.F. Conference, Brighton,* 1980.
2. CHARLESWORTH, D., Potential uses of plastics in automobiles, *PRI International Conference Plastics on the Road, July 1980.*
3. SCOTT, P., International developments in fibre reinforced plastics for land transport, RAPRA Members Report No. 47.
4. BEST, J. R., Reinforced plastics to minimise energy consumption over life cycle of an automobile, *36th Annual Conference, RP/C, SPI, Washington, 1981.*
5. SHELTON, J. A., GRP in automobiles, *Fulmer Research Institute, Meeting on Fibre Reinforced Materials in the Motor Industry.*
6. HABLITZEL, H. and JOHNKE, K. D., The use of plastics in European cars—an update, *SAE Passenger Car Meeting, Dearborn, June 1980.*
7. SANDERS, B. A. and RIEGNER, D. A., Fiber reinforced plastics test specification. General Motors Report No. MD-006, G.M. Tech. Center, Warren, Michigan, March 1979.
8. SANDERS, B. A. and RIEGNER, D. A., A characterisation study of automotive continuous and random glass fiber composites, G.M. Report No. MD 79-023, presented to *SPI 1979 National Technical Conference, Detroit, November 1979.*
9. MIKULEC, M. J., RRIM—A new process for the automotive industry, *34th Annual Technical Conference, RP/C Institute, SPI, 1979.*
10. MIKULEC, M. J., Refining of the RRIM process, materials and equipment, *International Conference, Strasbourg, France, June 1980.*

11. HARTLEY, J. R., More than just a lightweight, *Automotive Industries*, **63**, 1980, Sept.
12. JOHNSON, A. E. and JACKSON, J. R., The effect of milled and chopped glass fibres on the anisotropy of RRIM composites, *RRIM—What's in it for me? PRI Conference, Solihull, Feb. 1981.*
13. JACKSON, J. R., Acoustic emission from short fibre GRP composites, *Interfaces in Composite Materials, PRI Conference, Liverpool, April 1981.*
14. ROWBOTHAM, M., Achieving the impossible—plastic intake manifold, *BPF Conference, Brighton, 1980.*
15. SMITH, E. J., Ford inlet manifold project, Materials development and selection, Pilkington Group Press Release, Feb. 1980.

3

Analysis of the Shearout Failure Mode in Composite Bolted Joints*

DALE W. WILSON AND R. BYRON PIPES

Center for Composite Materials, College of Engineering, University of Delaware, Newark, Delaware 19711, USA

ABSTRACT

A semi-empirically based strength analysis was developed for the shearout failure mode in composite bolted joints. The failure model utilizes a polynomial 'stress function' in conjunction with a point stress failure criterion to predict strength as a function of fastener size, edge distance and half spacing. The development of the stress function is based on a two dimensional plane stress finite element analysis using quadrilateral elements with orthotropic material properties. The two constants, m and C, for the point stress criterion, were determined using empirical data from bolted joint tests performed for $[45/0/-45/0_2/-45/0/45/0_2/90]_s$ Hercules AS/3501-6 graphite epoxy laminates with fastener sizes ranging from 3·18 mm (0·125 in) to 9·52 mm (0·375 in). Verification of the strength predictions determined from the failure model was accomplished by comparison with experimentally determined shearout strength data.

INTRODUCTION

With advanced composites being considered for many structural applications requiring mechanically fastened joints the importance of effective design and analysis procedures is evident. Material anisotropy, brittleness and heterogeneity intensify the stress concentration effects and provide mechanisms for competing failure modes. The mechanisms and

* Sponsored by NASA/Langley Research Center, Hampton, Virginia 23665, USA.

strengths associated with each failure mode have been studied extensively[1-18] but to date no single analysis procedure has been developed which accurately models bolted joint strength for all failure modes.

The analysis of bolted joint strength requires the analytic determination of the state of stress in the joint and the application of an appropriate failure criterion. However, the closed-form solution for determination of the stress field near a hole loaded by a rigid inclusion in a finite-width, semi-infinite, anisotropic plate is intractable. Hence numerical methods must be used to determine the state of stress. Both finite-element[1-4] and approximate elasticity solutions[4-6] have been developed for plane stress analysis of the stresses in an anisotropic plate loaded by a frictionless pin. The choice of an appropriate failure criterion has been the deficiency of most analyses. Distortional energy,[1,2] Tsai Wu,[3] maximum stress, fracture toughness[5] and the Average Stress Failure criteria[6] have all been employed with various degrees of success. An acceptable degree of accuracy and reliability has not been obtained with any one of the above criteria for all failure modes primarily because there are different failure mechanisms associated with each mode and no single criterion can model them all. This paper concentrates on identification of the appropriate failure criterion for the shearout failure mode.

Since the shearout failure mechanism resembled that exhibited by finite-width notched composites, the point stress failure criterion used in the prediction of notched strength of laminates was the focus of this study. The point stress criterion formulated originally by Whitney[19] assumes that failure occurs when stress at a characteristic distance (d_0) adjacent to the notch reaches the unnotched strength of the laminate. Pipes[20,21] has since generalized this model by showing that the characteristic length, d_0, is not a material property, but varies exponentially with notch size. The characteristic distance must be determined empirically for each laminate configuration and material system.

Application of this criterion to bolted joint analysis can reduce substantially the empirical data base required for a general analysis. Characterization tests need only determine effective laminate properties and the notch sensitivity for bolt loading, thus eliminating the other geometric variables.

PROCEDURE

Analysis of bolted joint shearout strength using the point stress failure criterion requires determination of the state of stress in the joint and

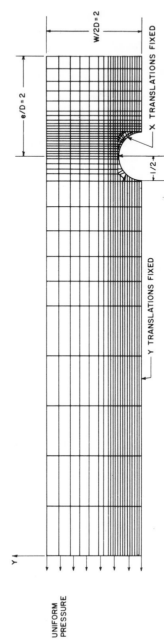

Fig. 1. Finite-element model used to analyze the bolted joint.

experimental evaluation of the characteristic length parameter, d_0. Empirical data characterizing the effects of fastener edge distance and half spacing was developed previously[15,16] and is used in the verification of the proposed strength model.

A finite-element analysis was used to determine the state of stress along the shearout failure plane. A functional representation of the stress profile along this plane including the parametric effects of geometry was developed based upon the finite element results. Strain measurements at three points along the failure plane of a bolted joint test coupon were used to verify the finite-element results.

The Structural Analysis Program SAPV was used to develop the two-dimensional finite-element model of the joint. The model employed the plane stress quadrilateral element with orthotropic material properties to form the mesh subjected to the boundary conditions depicted in Fig. 1. The bolted joint was treated as a loaded hole containing a frictionless rigid inclusion in a finite-width, semi-infinite strip. The X translations were fixed in the region $0 \leq \alpha < \pi/2$ (the point at $\pi/2$ was not fixed) and the load introduced by a pressure loading at the $Z = 0$ edge. Sixteen different geometric variations of the model were developed to analyze the effects of geometric parameters e/D and W/D on the state of stress. The experimentally determined material properties used in the finite-element model and failure criterion are listed in Table 1.

The test specimens for both the material characterization and the fastener size effect studies were fabricated from a single lot of Hercules AS/3501-6 graphite/epoxy prepreg. The laminate configuration was $[45/0/-45/0_2/-45/0/45/0_2/90]_s$ which resulted in a cured panel thickness

TABLE 1
Material and strength properties

Property	Value	
E_x	$10 \cdot 9 \ 10^6$ psi	$75 \cdot 2$ GPa
E_y	$4 \cdot 39 \ 10^6$ psi	$30 \cdot 3$ GPa
v_{xy}	$0 \cdot 397$	
v_{yx}	$0 \cdot 160$	
G_{xy}	$2 \cdot 00 \ 10^6$ psi	$13 \cdot 8$ GPa
$\sigma_{x,tens}^{ult}$	$131 \cdot 0$ ksi	$903 \cdot 2$ MPa
$\sigma_{y,tens}^{ult}$	$45 \cdot 2$ ksi	$311 \cdot 6$ MPa
$\sigma_{x,comp}^{ult}$	$140 \cdot 0$ ksi	$965 \cdot 0$ MPa
τ_{xy}^{ult}	$21 \cdot 7$ ksi	$149 \cdot 6$ MPa

of 3·0 mm (0·118 in) using manufacturer-recommended autoclave processing procedures.

Standardized test procedures were utilized in determination of the basic material properties. Details of the test procedure, fixturing and specimen geometry can be found in the reference section.[22–24] The bolted joint specimen was of the same single fastener design described by Wilson and Pipes[20] in an earlier report and is shown in Fig. 2. The geometric parameters e/D and W/D were selected to result in shearout failures and were held constant for three fastener hole diameters: 3·2 mm, 6·4 mm and 9·5 mm.

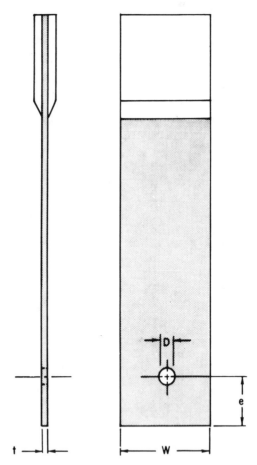

FIG. 2.   Bolted joint test specimen geometry.

Both bolted and pin loading tests were performed. A specially designed clevis fixture was used to simulate bolted load reaction through the hole in the coupon while standard wedge action friction grips introduced load at the tabbed end of the specimen. For pinned tests the load was reacted by a high strength steel pin and no constraining contact was allowed on the laminate surface surrounding the pin. To simulate the out-of-plane constraint caused by a bolt, annular washers with an inside diameter equal to the pin diameter and an outside diameter equal to twice the pin diameter were employed. Care was taken to insure excellent fit of the washer to the pin and the pin to the fastener hole for each test.

RESULTS AND DISCUSSION

The finite-element analysis was performed for sixteen different geometries with values of the e/D and W/D parameters ranging from two to six. Defining the shearout plane as the x-axis shown in Fig. 3, shearout stress distributions were plotted for each geometry studied. Variations in these stress distributions were then determined as functions of e/D and W/D. The point stress criterion was then applied using the parameters defined in Fig. 3.

The stress analysis was verified by normal strain measurements taken at 45° to the loading axis at three locations along the x-axis. A comparison of the measured strains with strains calculated from finite-element analysis results at the same locations is shown in Fig. 4. The comparison showed

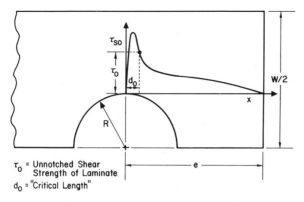

FIG. 3. Description of the shearout point stress failure criterion.

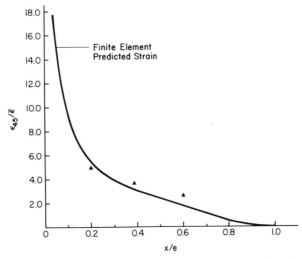

FIG. 4.   Plot of the strain profile along the shearout plane for strains oriented at 45° to the
loading direction.

close agreement of the measured and predicted response which formed the basis for a reasonable degree of confidence in the finite-element analysis.

A simple collocation method was used to fit the shearout stress distribution, τ_{so}, by a polynomial expansion of the space variable in non-dimensional form. The shearout stress distribution fit by this procedure possessed the geometry $W/D = 6.0$ and $e/D = 2.0$. The polynomial was assumed to be of the form:

$$\tau_{so}/\bar{\sigma} = A(x_0/x) + B(x_0/x)^2 + C(x_0/x)^3$$

where A, B and C are constants determined by fitting the polynomial through these points on the shearout stress distribution curve, Fig. 5. The dotted line shows the correlation of the stress polynomial fit with the original finite-element profile. Since the critical distance parameter (d_0) is small, the constants were chosen to provide the best fit for $x/e < 0.3$. Note that the stresses determined from the polynomial are conservative for $x/e > 0.3$, and do not pass through zero at $x/e = 1$. The constants A, B and C are given on Fig. 5.

Parametric studies were conducted using finite-element models to investigate the influence of e/D and W/D on the shear stress distributions. Results from this study are shown in Figs 6 and 7 for W/D and e/D, respectively. Note that the form of the curves in the region of interest (small

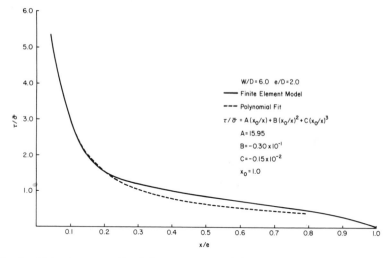

FIG. 5.   Finite-element predicted shearout stress profiles and comparison to approximate polynomial fit.

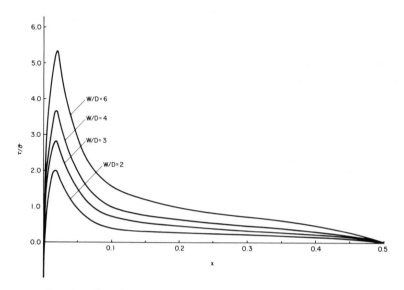

FIG. 6.   Effect of varying W/D on shearout stress profiles for $e/D = 2$.

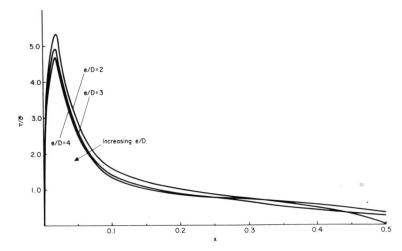

Fig. 7.   Effect of varying e/D on shearout stress profiles for $W/D = 6$.

values of x) doesn't change significantly, only the magnitude of the stress changes. Plotting the maximum shear stress as a function of W/D and e/D results in the curves shown in Figs 8 and 9, respectively. From Fig. 8, τ_{max} is obviously a linear function of W/D. The maximum shear stress is a non-linear function of e/D and asymptotically approaches a constant for $e/D \geq 6$. It was found that this non-linear relationship is exponential in character.

Using these results it was possible to incorporate the W/D and e/D dependence of the shear stress distribution into the polynomial stress function. Since the shape of the curves and the location of the peak shear stress remains constant for all W/D and e/D geometries of interest, the magnitude of the stresses can be adjusted by introducing non-dimensional shift parameters $\xi(W/D)$ and $\eta(e/D)$. These parameters are functions of W/D and e/D respectively and were developed from the data in Figs 8 and 9.

The following forms were found for the parameters:

$$\xi(W/D) = \frac{m \cdot (W/D)}{\tau_{max}|_{W/D=6}} \tag{1}$$

where m = slope of the τ_{max} versus W/D curve in Fig. 8 (m = 0·83) and

$$\eta(e/D) = \alpha e^{(\xi - e/D)} + \beta \tag{2}$$

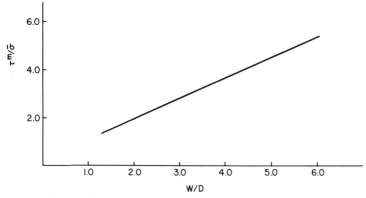

FIG. 8. Variation in shearout stress concentration with W/D.

where α, β and ξ are constants found by fitting the curve in Fig. 9. It was determined that $\alpha = 0.75$, $\beta = 0.84$ and $\xi = 0.50$.

By combining the non-dimensional geometric parameters with the original functional representation of the stress profile, the following 'stress function' incorporating geometric effects was determined.

$$\tau_x/\bar{\sigma} = \xi(W/D)\eta(e/D)\left[A(x_0/x) + B(x_0/x)^2 + C\left(\frac{x_0}{x}\right)^3 \right] \tag{3}$$

This approximate functional relationship for the stress is suitable for incorporation into the failure criterion.

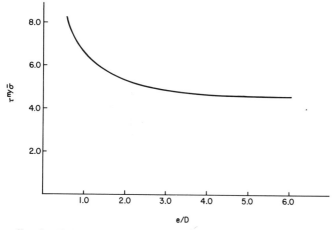

FIG. 9. Variation of maximum shear stress concentration with e/D.

The point stress failure criterion is expressed as

$$\tau_{xy}\big|_{x=d_0} = \tau_0$$

where τ_{xy} is the inplane shear stress along the shearout plane (eqn. (3)), τ_0 is the unnotched strength of the laminate and d_0 is the critical distance.

The critical distance parameter is a function of hole size and has the following form[20]

$$d_0 = \frac{1}{C}\left(\frac{R}{R_0}\right)^m$$

Constants C and m are found from empirical data by measuring τ_0 and determining τ_{xy}^{ult} for two fastener sizes. For the $[45/0/-45/0_2/-45/0/45/0_2/90]_s$ Gr/E laminate tested in this program m and C for shearout failure were found to be 0·114 and 7·12, respectively. A plot of shearout strength as a function of fastener size in Fig. 10 shows the correlation of the predicted failure strengths and experimental data for three fastener sizes. The excellent agreement is not surprising since two of the three data points were empirically fitted to the failure criterion for determination of d_0. As a further test of the model's validity, results from experimental studies characterizing shearout strength as a function of W/D were compared to the strengths predicted by the semi-empirical shearout model. Excellent

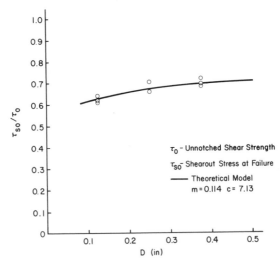

FIG. 10.   Correlation of model predicted shearout strengths with experimental data as a function of hole diameter.

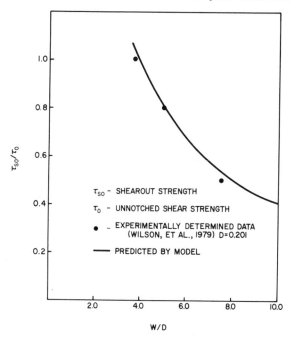

FIG. 11. Correlation of shearout strength as a function of W/D with experimental data.

agreement was found between the measured strengths and predicted strengths as seen in Fig. 11. These results are for a fastener size of 4·77 mm which is not one of the fastener sizes used to determine the constants for the model.

The bolted joint's shearout strength behavior predicted by the model was investigated for several combinations of edge distance and half spacing geometry of practical interest. It is seen that shearout strength decreases sharply with increasing half spacing for W/D geometries between 2·0 and 8·0 (Fig. 12). For $W/D < 4·0$ the shearout failure mode is not predicted ($\tau_{so}/\tau_0 > 1·0$) for any of the e/D cases studied. Failure is predicted for all e/D cases when W/D exceeds 5·25. From a design standpoint this implies that if the half spacing is made large, the tendency for shearout failure is increased. This is clearly seen in Fig. 13 which examines shearout strength behavior as a function of W/D. For $W/D = 6$, shearout failure is indicated for all values of e/D shown. When W/D is less than 4·0 shearout failures are not predicted for any of the e/D geometries studied. These predicted behaviors agree well with experimental findings.

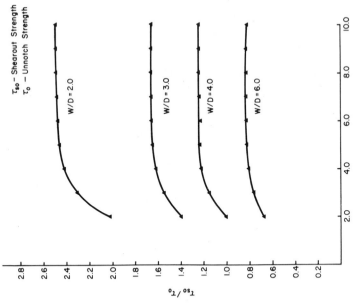

FIG. 13. Shearout model strength predictions as a function of e/D.

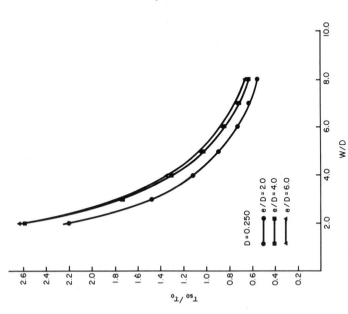

FIG. 12. Shearout model strength predictions as a function of W/D.

CONCLUSIONS

A semi-empirical shearout strength model for analysis of composite bolted joints has been formulated which accounts for effects of joint geometry on strength behavior. The model employs the point stress failure criterion for notched strength in composites to analyze a shear mode failure.

The critical distance parameter d_0 is determined empirically once the shear stress distribution along the shearout failure plane is known. The shear stress distribution was determined using plane-stress, finite-element analysis and was examined in the analytic form by curve fitting to a polynomial expansion.

Comparison of experimentally determined shearout strength data with model predicted failures has substantiated the accuracy of the model. It can thus be concluded that: (1) the point stress criterion may be applied for the shearout failure mode occurring in notched composites loaded by a mechanical fastener when an accurate determination of the shear stress distribution along the failure plane exists; (2) the simple plane stress finite-element analysis predicts the shear stresses along the shearout failure plane with sufficient accuracy for the present analysis and (3) the failure model can incorporate the effects of geometry (e/D and W/D) on strength behavior and predict shearout failure without dependence on empirical data for various e/D and W/D geometries.

The advantage of using the proposed shearout failure model is a reduction in empirical data needed for bolted joint analysis. The basic material properties and a single set of data quantifying shearout strength as a function of fastener size are sufficient for analysis using the proposed model. This eliminates the need for a large series of tests to determine shearout strength for various W/D and e/D geometries.

REFERENCES

1. WASZCZAK, J. P. and CRUSE, T. A. Failure mode and strength predictions of anisotropic bolt bearing specimens, *J. of Composite Materials*, **5**, 1971, July, p. 421.
2. WASZCZAK, J. P. and CRUSE, T. A. A synthesis procedure for mechanically fastened joints in advanced composite materials, *AIAA/ASME/SAE 14th Structures, Structural Dynamics, and Materials Conference, March 1973.*
3. SONI, R. SOM, Failure analysis of composite laminates with a fastener hole, Materials Laboratory, Air Force Wright Aeronautical Laboratories, Technical Report No. AFWAL-TR-80-4010, March 1980.

4. HARRIS, H. G. and OJALVO, I. U. Simplified three-dimensional analysis of mechanically fastened joints, *Proceedings of the Army Symposium on Solid Mechanics, AMMRC MS74-8, September 1974.*

5. JONG, THEO DE, Stresses around pin-loaded holes in elastically orthotropic or isotropic plates, *J. Composite Materials,* **11**, 1977, July, p. 313.

6. OPLINGER, D. W. and GANDHI, R. R. Stresses in mechanically fastened orthotropic laminates, *Proceedings of the 2nd Conference on Fibrous Composites in Flight Vehicle Design,* Dayton, Ohio, May 21–4, 1974.

7. EISENMANN, J. R. Bolted joint static strength model for composite materials, NASA-TM-X-3377, 1976.

8. RAMKUMAR, R. L. Bolted joint design, *Proceedings of ASTM Symposium on Test Methods and Design Allowables for Fibrous Composites, Dearborn, Michigan, October 3–4, 1979.*

9. HART-SMITH, L. J. Bolted joints in graphite–epoxy composites, Douglas Aircraft Company, NASA Contract Report No. NASA CR-144899, June 1976.

10. Advanced composites design guide, Vol. I, *Design,* 3rd Edition, Contract F33615-74-C-5075, North American Rockwell/Los Angeles Division, January 1973.

11. VAN SICLEN, R. C. Evaluation of bolted joints in graphite–epoxy, *Proceedings of the Army Symposium on Solid Mechanics, AMMRC MS74-8, September 1974.*

12. COLLINS, T. A. The strength of bolted joints in multi-directional CFRP laminates, *Composites,* **8**, 1977, January.

13. HYER, M. W. and LIGHTFOOT, M. C. Ultimate strength of high-load-capacity composite bolted joints, *Composite Materials: Testing and Design (5th Conference),* S. W. Tsai (Ed.), ASTM STP 674, American Society for Testing and Materials, 1979, 118–36.

14. QUINN, W. J. and MATTHEWS, F. L. The effect of stacking sequence on the pin-bearing strength in glass fibre reinforced plastic, *J. Composite Materials,* **11**, 1977, April, p. 139.

15. WILSON, D. W. and PIPES, R. B. Behavior of composite bolted joints at elevated temperature, University of Delaware Center for Composite Materials, NASA Contract Report No. 159137, September 1979.

16. WILSON, D. W., PIPES, R. B., WEBSTER, J. W. and RIEGNER, D. L. Mechanical characterization of PMR-15 graphite/polyimide bolted joints, *Proceedings of ASTM Symposium on Test Methods and Design Allowables for Fibrous Composites, Dearborn, Michigan, October 3–4, 1979.*

17. KIM, R. Y. and WHITNEY, J. M. Effect of temperature and moisture on pin bearing strength of composite laminates, *J. Composite Materials,* **10**, 1976, April, p. 149.

18. WILKINS, D. J. Environmental sensitivity tests of graphite-epoxy bolt bearing properties, *Composite Materials: Testing and Design (4th Conference),* ASTM STP 617, American Society for Testing and Materials, 1977, 497–513.

19. NUISMER, R. J. and WHITNEY, J. M. Uniaxial failure of composite laminates containing stress concentrations, *Fracture Mechanics of Composites,* ASTM STP 593, 1975.

20. PIPES, R. B., WETHERHOLD, R. C. and GILLESPIE, J. W. *J. Comp. Mater.,* **12**, 1979, p. 148.

21. PIPES, R. B., GILLESPIE, J. W. and WETHERHOLD, R. C. *Polymer Eng. and Sci.*, **15**, 1979, No. 16.
22. Standard test method for tensile properties of oriented fiber composites, *ASTM Standards, Part 35*, ASTM D-3039-76.
23. HOFER, KENNETH E., RAO, P. N. and HUMPHREYS, V. E. Development of engineering data on the mechanical and physical properties of advanced composite materials, AFML TR-72-205, Part I, IIT Research Institute, 1972.
24. GARCIA, R., WEISSHAAR, T. A. and MCWITHEY, R. R. An experimental and analytical investigation of the rail shear-test method as applied to composite materials, *Experimental Mechanics*, **20**, 1980, No. 8, August.

4

Stress and Strength Analysis of Bolted Joints in Composite Laminates

Som R. Soni

Universal Energy Systems, Inc., 3195 Plainfield Road, Dayton, Ohio 45432, USA

ABSTRACT

Composite laminates with a through-the-thickness fastener hole, free or loaded, have been treated for stress and strength analysis under uniaxial tensile loading conditions. This investigation has been carried out within the framework of laminated plate theory. A finite element method has been utilized to conduct the stress analysis of the laminate with a loaded fastener hole. For this case the hole is assumed to be filled with a rigid core, simulating a bolt, and displacement boundary conditions are applied at the semicircular contact surface. The strength analysis is based on the tensor polynomial failure criterion applied to each ply. The results for a free hole boundary condition are obtained by using closed form solutions. An approximation procedure is suggested to calculate stress levels for multidirectional laminates using the stress fields in individual ply laminate systems. This procedure works for both the loading conditions. The strength predictions for the loaded hole case are close to the experimental results.

INTRODUCTION

With the increasing demand for composite materials and their wide use in highly stressed lightweight constructions, it has become important to develop a deep insight into the failure of composite bolted joints. There have been several investigations on the strength analysis of bolted joints.[1] For the optimum design of bolted joints in composite laminates, a

50

knowledge of stress distribution around the fastener hole due to the applied load is very important. With the variation in ply orientations and volume fractions in the laminate, the stiffness properties change and consequently the stress levels pertaining to the same boundary conditions differ. For optimum strength requirements, one needs to compute stress levels in the laminate for given boundary conditions with different volume fractions and ply orientations. Because, in many practical situations, the closed form elasticity solutions are not available, a finite element method has to be implemented. Conducting this analysis by changing the effective material properties for each laminate in the fine element analysis will be very expensive. In the present investigation, the finite element method has been used to conduct the stress analysis of the laminate for a number of ply orientations and volume fractions. For the free hole boundary condition a closed form solution[2] has been used. A simple averaging procedure has been suggested to approximate the stress levels in the composite laminates, with any combination of ply volume fractions, from the stress distributions in constituent ply laminates with the same set of boundary conditions.

The present study consists of the computation of stress distribution in composite laminates with a free or a loaded fastener hole. The results are calculated for various multidirectional laminates including the constituent angle ply laminates. The stress levels at various points in the constituent angle ply laminates are used to approximate the states of stress for multidirectional composite laminates with different ply volume fractions. There exists a very good agreement between the exact results and the results obtained by the approximation method for multidirectional laminates. The exact results are the values obtained by using the effective material properties of the multidirectional composite laminate.

Strength analysis is conducted by using the tensor polynomial failure criterion.[3] In the bolted joint case the strength of the joint is considered as an applied stress at which the strongest ply at the weakest point fails. This consideration is supported by a recent study by Knight,[4] on strength analysis of composite laminates. His conclusion is that the uniaxial strength of a laminate containing a $0°$-ply is its last ply failure strength and that of one without $0°$-ply is its first ply strength. The strength ratios for two laminates are computed for various values of diameter to width ratio. The mode of failure is also predicted.

In the case of the free hole boundary condition the strength of composite laminates at each point around the hole boundary is computed. Two laminates were considered. It has been shown that the laminate may not fail at the point of stress concentration.

PROBLEM DESCRIPTION

Figure 1 shows a laminate with a fastener hole, co-ordinate axis and possible simple uniaxial loads. The following three combinations of loading conditions can be investigated. (i) Loaded hole ($N' = 0$). (ii) Free hole ($P = 0$). (iii) Partially loaded hole ($N' = 0$, $P = 0$).

The loaded hole condition resembles the bolted joint situation and is of great practical importance in engineering applications. The first two cases of boundary conditions are investigated.

Loaded Hole

There exists no reliable closed form solution for the study of loaded hole laminates. Consequently, a finite element technique has been utilized to conduct the stress analysis of such laminates. During the development of the mathematical model for this analysis, the following assumptions were made.

(1) The laminate obeys the laws of classical laminated theory.
(2) The contact surface between the laminate and the bolt is semicircular.
(3) The hole is filled with a rigid core.
(4) No transverse load, due to the bolt, is acting at the laminate.

The loaded hole boundary conditions are imposed by introducing radial displacement constraints at the semicircular contact surface and a prescribed load on the opposite plane edge. A general purpose finite element computer code, NASTRAN, has been used to conduct the stress analysis of the laminate. Due to the symmetry of the laminate and applied loads about the x-axis, half of the laminate has been modeled for finite element analysis. This part has been divided into 372 quadrilateral and

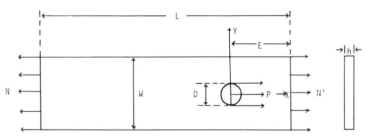

FIG. 1.   Laminate with dimensions and co-ordinate axis. $L = 13\cdot3$ cm, $h = 0\cdot2032$ cm, $W = 2\cdot54$ cm, $E = 0\cdot894$ cm, $D = 0\cdot3175$–$0\cdot9$ cm.

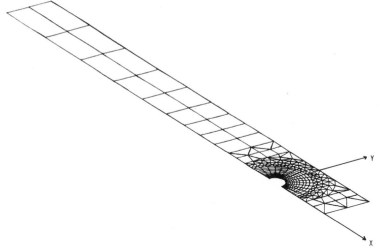

FIG. 2.   Finite element grid of the half of the laminate.

triangular constant strain elements. The effective laminate material properties, based on the laminated plate theory,[3] have been used. A finite element grid plot, as obtained during the NASTRAN computations, is given in Fig. 2. Various numerical exercises with different finite element grids show that the present model is good enough to give acceptable results for all practical purposes.

Free Hole

A closed form solution given by Lekhnitskii[2] for laminates with a free circular hole has been used. The main objective of the present study is to verify the approximation, suggested in the following section, to determine the stress levels for multidirectional laminates from stress distribution in constituent angle ply laminates with the same boundary conditions. For that reason, no finite width correction factor has been included in the computation of results through this formulation.

APPROXIMATION

The stress levels for individual angle ply laminates $(0, 90, (\pm \theta)_s)$ are computed either by the finite element method or closed form solutions. Let these stress components, at a point, be denoted by $(\sigma_x^i, \sigma_y^i, \tau_{xy}^i)$,

$i = 0°, 90°, \ldots (\pm \theta)_s$ for laminates $0°, 90°, \ldots (\pm \theta)_s$. The stress levels at the corresponding points for a multidirectional laminate $(0_m/90_n/(\pm \theta)_p)_s$ are:

$$\sigma_x = \frac{1}{m+n+2p} \{ m\sigma_x^0 + n\sigma_x^{90} + 2p\sigma_x^{(\pm \theta)_s} \}$$

$$\sigma_y = \frac{1}{m+n+2p} \{ m\sigma_y^0 + n\sigma_y^{90} + 2p\sigma_y^{(\pm \theta)_s} \}$$

$$\tau_{xy} = \frac{1}{(m+n+2p)} \{ m\tau_{xy}^0 + n\tau_{xy}^{90} + 2p\tau_{xy}^{(\pm \theta)_s} \}$$

where m, n and p are the number of plys for $0°$, $90°$, $\pm\theta°$ orientations, respectively, and σ_x, σ_y and τ_{xy} are the stress components in the multidirectional laminate.

LAMINATE MATERIAL PROPERTIES

The analysis of composite laminates which have midplane symmetry, i.e. the ply orientations in the lower half thickness of the laminate, are the reflections of those of the upper half plys. Such laminates are assumed to behave like homogeneous anisotropic plates. The procedure of computing the effective material properties of multidirectional laminates is described in references 5 and 6. The effective modulus of the composite laminate is simply the arithmetic average of the modulus of the constituent plys.

RESULTS AND DISCUSSION

Stress distribution is computed for each element around the hole boundary of the laminate. Graphite epoxy T300/5208 is taken as a representative composite material for all the numerical calculations. The dimensions of the laminate considered are shown in Fig. 1. The material properties are as follows.

Longitudinal Young's modulus $E_x = 181$ GPa
Transverse Young's modulus $E_y = 10.3$ GPa
Longitudinal shear modulus $E_s = 7.17$ GPa
Longitudinal Poisson's ratio $v_x = 0.28$
Longitudinal tensile strength $X = 1500$ MPa
Longitudinal compressive strength $X^- = 1500$ MPa
Transverse tensile strength $Y = 40$ MPa
Transverse compressive strength $Y^- = 246$ MPa
Longitudinal shear strength $S = 68$ MPa

The finite element results for the stress field are presented in graphical form. For checking the accuracy of numerical values obtained by the present model, stress component σ_x was calculated for a free hole laminate along the line normal to the x-axis and bisecting the hole for a quasi-istropic laminate. Values of σ_x computed by using infinite plate theory and finite width correction factor were also given. The agreement was very good.[7]

Figure 3 shows the comparison between the exact results and approximate results for a loaded hole laminate $(0/90/\pm 30)_s$. Figures 4 and 5

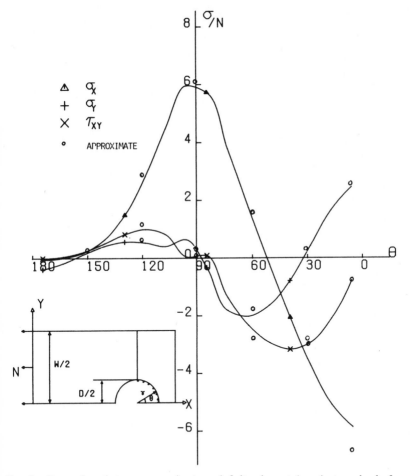

FIG. 3. Comparison between approximate and finite element (exact) stress levels for $(0/90/\pm 30)_s$-laminate, loaded fastener hole, $D = 0.5$ cm.

Som R. Soni

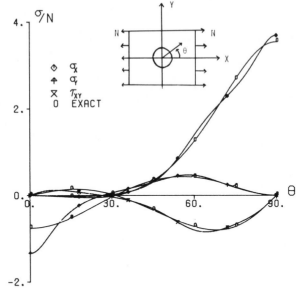

FIG. 4. Comparison between approximate and exact stress levels for $(0/90/\pm 30)_s$-laminate, free fastener hole.

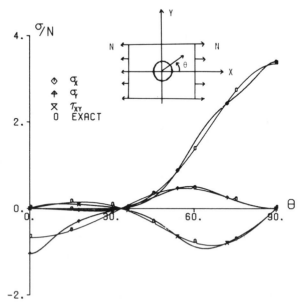

FIG. 5. Comparison between approximate and exact stress levels for $(0/90/(\pm 30)_2)_s$-laminate, free fastener hole.

show the comparison of results for $(0/90/\pm 30)_s$ and $(0/90/(\pm 30)_2)_s$-laminates, free hole boundary condition. Thus it has been shown that, given the stress distribution in angle ply laminates with a loaded or a free fastener hole, the stress levels in a multidirectional composite laminate with any volume fraction of these angle plys can be approximated. It has been demonstrated that the suggested approximation gives results very close to the finite element results obtained by using the effective material properties of the laminate.

Figures 6 and 7 demonstrate the variation in strength of a laminate having a loaded fastener hole. The parameter P ($=$ load/diameter $\times$ thickness) denotes the strength of the notched laminate and N_0 denotes the strength of the unnotched laminate. As indicated earlier, the last ply strength is taken as the strength of the laminate. The laminate ply orientations considered are $(0/90/\pm 45)_s$ and $(0/90/(\pm 45)_2)_s$. Experimental results are also shown in the diagrams and are in close proximity to

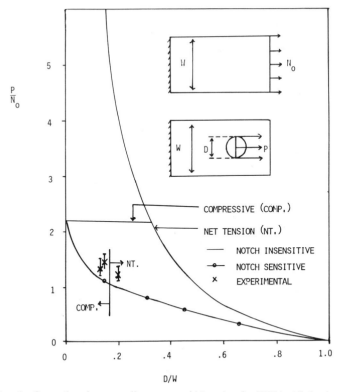

FIG. 6.   Strength ratio versus diameter to width ratio of a $(0/90/\pm 45)_s$-laminate.

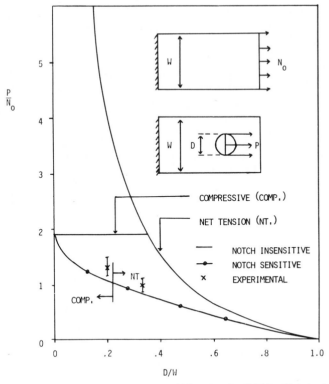

FIG. 7.   Strength ratio versus diameter to width ratio of a $(0/90/(\pm45)_2)_s$-laminate.

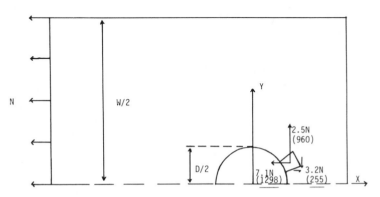

FIG. 8.   Projection of failure mode for $(0/90/\pm45)_s$ laminate, $D/W = 0\cdot125$. Compressive failure.

predicted strengths. In both cases, a demarcation line is drawn that gives the impending mode of failure with the variation in D/W for a given value of E (shown in Fig. 1). The criterion used to decide the mode of failure is shown in Fig. 8. The point of failure is acertained by using the tensor polynomial failure criterion. The inspection of stress components at that point and the laminate strength in the corresponding direction will project the type of failure. This aspect has been explained in detail in reference 1.

Figure 9 gives the first and last ply failure strengths of $(\pm 15)_s$-laminate at each point of the free hole boundary. This diagram is given to show that the point of failure is not at the point of stress concentration. Another case considered in the free hole boundary condition is a laminate with continuous variation in ply angle. The trend of variation in ply orientation

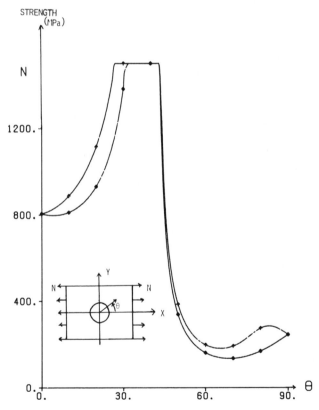

FIG. 9. Variation in strength of the $(\pm 15)_s$-laminate w.r.t. θ, for a free hole boundary. ◇, first ply; +, last ply.

Som R. Soni

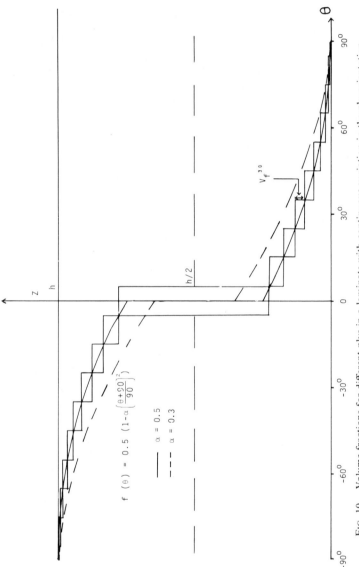

FIG. 10. Volume fractions for different plys in a laminate with continuous variation in the ply orientation.

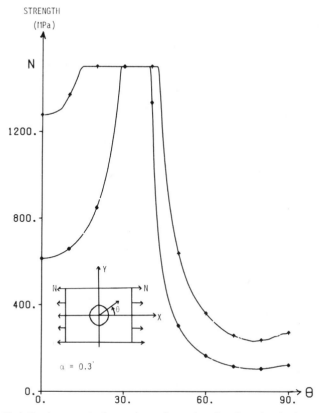

FIG. 11. Variation in strength of a continuously varying ply orientation laminate, w.r.t. θ, for a free hole boundary, $\alpha = 0.3$. $\diamondsuit$, first ply; $+$, last ply.

is given in Fig. 10. The volume fraction of each ply orientation is given by the corresponding blocks. The angle lying at the center of the block represents ply orientation and the height of the block is the corresponding volume fraction. Figure 11 gives the first and the last ply strengths of this laminate for $\alpha = 0.3$ (a parameter given in Fig. 10). There, too, we find that the failure does not occur at the point of stress concentration ($\theta = 90°$).

REFERENCES

1. SONI, S. R., Failure analysis of composite laminates with a fastener hole, *ASTM Special Technical Publication. Joining of Composite Materials*, **STP749** (1981).
2. LEKHNITSKII, S. G., *Anisotropic plates*. Translated from second Russian edition

by Tsai, S. W. and Cherion, T. New York, Gordon and Breach Science Publishers, 1968, pp. 171–5.

3. TSAI, S. W. and WU, E. M., A general theory of strength for anisotropic materials, *J. Comp. Mat.*, **5** (1971), 58–80.

4. KNIGHT, M., Experiments on strength of composite laminates. (Private communication.)

5. TSAI, S. W. and HAHN, H. T., *Introduction to composite materials*, Technomic Publication Co., Westport, Ct., USA, 1980.

6. TSAI, S. W. and HAHN, H. T., *TI-59 magnetic card calculator solution to composite materials*, Air Force Materials Laboratory Report, AFML-TR-79-4040, January, 1981.

7. SONI, S. R., Inplane stress analysis of multidirectional composite laminates with a fastener hole—Using stress distribution in the constituent angle ply laminates. *Intl. Symp. on the Mechanical Behaviour of Structured Media, Carleton University, Ottawa, May, 1981.*

5

Some Environmental and Geometric Effects on the Static Strength of Graphite Cloth Epoxy Bolted Joints*

J. A. BAILIE, L. M. FISHER, S. A. HOWARD and K. G. PERRY

*Lockheed Missiles & Space Company Inc.,
Sunnyvale, California 94088, USA*

ABSTRACT

Results of an extensive test program to determine static strength are correlated with the design method for mechanically fastened composite joints proposed by Hart-Smith. Among the variables covered were the amount of absorbed moisture in the laminates, their layup, bolt spacing, multiple bolt rows, single and double lap configurations. Tension shear testing was conducted at room temperature, 394 K (250°F), and 427 K (310°F) to assess the impact of temperature on strength. Correlation with the design method over this wide range of variables is considered to be highly satisfactory. Of particular interest was the ability of the method to correctly predict interaction effects among closely spaced bolts.

NOTATION

C	Stress concentration relief parameter.
d	Bolt diameter.
F^{bru}	Ultimate bearing strength.
F^{tu}	Ultimate tension strength.
k_{tc}	Stress concentration factor for a composite.
k_{te}	Stress concentration factor for a linear elastic material.

* This work was supported by LMSC Missile Systems Division Independent Development Funds.

P Joint failure load.
p Bolt spacing.
T_g Glass transition temperature.
w Joint width.
σ_b Bearing stress.

INTRODUCTION

The relative brittleness of graphite epoxy laminates, compared with many light alloy structures, means that special attention must be given to the design of bolted joints in these composites. Plastic deformation and the consequent stress redistribution in ductile metals results in lower stress concentrations. This leads to rather uniform load sharing among the bolts at high load levels. Such forgiving behavior occurs to a lesser degree in fiber-dominated composites. In addition, there are several complications, unique to high performance composites, arising in bolted joints in these laminates. Among them are the effects of moisture, layup, interaction among adjacent bolts and many geometric variables. To cover this wide range of designs, various approaches are being developed.[1] These can be broadly considered to range from the empirical methods intended for preliminary design use to very detailed non-linear finite element analyses of specific configurations. This study is intended to produce data to correlate with the empirical approaches suggested by Hart-Smith.[2,3]

In those works, Hart-Smith presents the results of test programs on bolted joints in graphite tape epoxy and utilises them in the derivation of empirical design methods. His techniques were used to explain some of the results in a test series on graphite cloth epoxy joints, including high temperature effects.[4] The present work is an extension of that study. One purpose is to assess the importance of absorbed moisture. The effect of temperature and moisture on the bearing strength of single bolt joints has been studied.[5,6] Those studies demonstrate typical strength degradations of bolt bearing capability at elevated temperatures.

Another item of concern is the incomplete knowledge of bolt interactions as functions of bolt spacing, moisture and temperature. A great deal of effort is currently being devoted to multi-bolt joints at room temperature.[7,8] For some applications, a need arises to provide data at higher temperatures. Because tension and bearing response are affected differently by temperature and moisture, their interaction requires careful study for joints that are designed to be near optimum.

Anomalous results between single and two row tests on single lap joints soon suggested that the difference between single and double lap joints deserved further attention. Agarwal[8] has shown a small difference between the two for joints in tape laminates, tested at room temperature. Most other test results reported are for symmetric double lap conditions where bolt rotation is inherently prevented. In single lap configurations, particularly of the thinner laminates, the bolt rotation can be quite severe, leading to non-uniform loading across the joint thickness. A subset of tests was devoted to evaluating the parameters influencing the importance of this loading eccentricity.

To provide empirical data of interest to design and development engineers, a test program was structured to cover the variables of interest in laminates typical of launch vehicle primary, and aircraft secondary, structures. These results were to be interpreted using Hart-Smith's method to promote further understanding of graphite cloth epoxy joint behavior.

TEST SPECIMENS

The material used is Fiberite HMF330C/34 graphite cloth epoxy. The cloth consists of an 8 HS weave of T300 fibers. The matrix is the 934 resin. All the single lap joints were 8-ply layups giving thicknesses of approximately 0·274 cm (0·108 in) thickness. To assess the impact of layup, a series of tests were run with both $[0]_{4S}$ and $[45/0]_{2S}$ layups. These laminates will provide data on the influence of 45-degree ply percentage. To conserve on the amount of composite material, many tests were conducted using 0·254 cm (0·10 in) thick titanium laps in conjunction with composite members. Failure always occurred in the composite. In the fully composite double lap joints, the inner members were $[45/0]_{2S}$ layups while the outer members were $[45/0]_S$ to provide balanced symmetric configurations. Edge distance was $4d$ to ensure freedom from shear out failures. All bolts were 0·476 cm (3/16 in) diameter titanium and the bolts torque to 3·39 kNm (30 in lb). In multi-row joints, the rows were 2·54 cm (1·0 in) apart.

Evaluation of the importance of moisture conditioning was carried out by testing after three different conditioning cycles. Allowing laminates to remain in ambient atmosphere for between 3 and 4 months prior to testing was shown by weight monitoring to gain roughly 0·4 % water. These are referred to as 'ambient'. Others were conditioned at 60 % relative humidity (RH) until equilibrium was reached. This corresponded to approximately

0·8 % water absorbed. A few specimens, referred to as 'dry', were dried out in an oven until no further weight change was noted.

Tests on the laminates showed the glass transition temperatures (T_g) averages of ten tests to be 443 K (337 °F) when dry, 409 K (277 °F) when ambient and 394 K (250 °F) when conditioned to 60 % RH. Thus the highest structural test temperature was above T_g, except when the material had been dried out. It is hardly likely that a structural material will be subjected to severe loading when above its T_g. These 427 K (310 °F) tests were intended to determine whether a catastrophic strength degradation occurred slightly above T_g.

TEST PROCEDURES

Conventional tension shear loading was utilized with a 'hot box' surrounding the test specimen, as illustrated in Fig. 1. Three replicates of

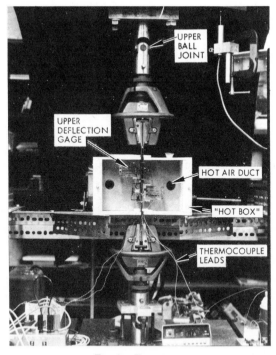

FIG. 1.   Test set up.

every test were performed at the three temperatures 297, 394 and 427 K (75, 250 and 310 °F). Ball joints were employed top and bottom to exclude moment loading. For high temperature testing, two hot-air guns, set to produce air at the test temperature, were utilized. In structural tests, only surface thermocouples were installed. To ensure equilibrium thermal conditions, a few joints not used for strength testing but containing thermocouples on their surface and buried in the laminate midplanes, were employed in preliminary tests. Their objective was to derive the time between the surface and midplane thermocouples reaching test temperature. One minute after test temperature was reached at midplane 1·27 cm (0·5 in) behind the bolts, load application commenced at the rate of 0·05 cm/min (0·02 in/min).

To measure load–deflection data, two methods were employed. One was typical loading head motion—a crude estimate as it includes all the motion within the loading fixture. A much more precise indication was the relative deflection between a bolt and neighboring laminate, obtained using the deflectometer shown in Fig. 2. Operating on the quadraflex principle with a strain gaged arm, it provided the desired information for the double lap joints. A few exploratory tests on single lap unrestrained joints confirmed that rotation under load, inherent in these configurations, precluded the extraction of useful data from these configurations.

FIG. 2. Deflection gage to measure bolt-composite relative motion.

TEST DATA AND ITS INTERPRETATION

The general philosophy behind the test matrix was to derive the data needed for the use of Hart-Smith's methodology.[2,3] The essential ingredients of this type of test program are:

(a)   Bearing strength determination of wide, single-bolt joints.
(b)   Tension strength determination using narrow, single-bolt joints.

The specimens used to determine bearing strength had a d/w ratio of 0·2, or width of 2·38 cm (0·9375 in); those to extract tension strength had a d/w of 0·375 or width of 1·27 cm (0·5 in). In addition, it is necessary to know F^{tu},

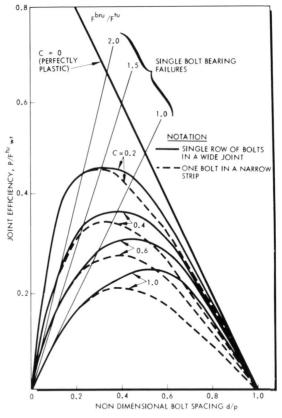

FIG. 3.   Difference in strength between one bolt in a narrow strip and a single row of bolts in a wide joint.

the tension strength of the unnotched laminates. Item (b) provides coefficient C, which defines the degree of stress concentration relief, relative to linear elastic (brittle) materials.[2,3] This latter reference also shows that there is a difference between a single bolt and wide joint containing a single bolt row whose bolt spacing is the same as the single-bolt joint width. This applies to both theoretical stress concentration in brittle materials and its relief in non-brittle materials. This difference is quantified in Fig. 3, where it is seen to be small for lower values of C and when bolt diameter is less than approximately one-third the bolt spacing. It is only significant for very close bolt spacing and highly brittle materials. To check this, specimens with seven bolts in a single row were tested, and the theory was confirmed. Hence, the adequacy, in most design applications, of testing single-bolt joints to obtain C and applying the result to wide joints was substantiated.

SINGLE BOLT TESTS

Bearing behavior is summarized in Fig. 4. The importance of temperature is significant and, as expected, moisture degrades high temperature strength. This Figure also demonstrates a clear difference between single- and double-lap joints. Unfortunately, since the basic references[2,3] are not yet generally available, the bare essentials of the derivation of parameter C are repeated here. For a joint failing in tension, it is postulated that a linear relationship exists between the elastic isotropic stress concentration factor, k_{te}, and that for the composite at failure, k_{te} of this relationship is:

$$k_{tc} - 1 = C(k_{te} - 1) \tag{1}$$

in which k_{tc} is given in terms of the failure load, P, by:

$$k_{tc} = F^{tu}wt(1 - d/w)/P \tag{2}$$

Clearly, for a completely brittle material, C is unity while for a perfectly plastic material it is zero and there is no stress concentration. Knowing k_{te},[2,3] simple tests produce k_{tc} from eqn. (2) and C is extracted from eqn. (1).

Bolted joint tension tests to determine parameter C are summarized in Table 1. Joint strength is well known to be sensitive to the percentage of 45-degree plies, so 0 and 50 % were utilized in this test series, to quantify the effect. Table 1 highlights the fact that decreasing the percentage of 45-degree plies and/or raising temperature over the test range produce significant decreases in stress concentration relief (i.e. raise the value of C).

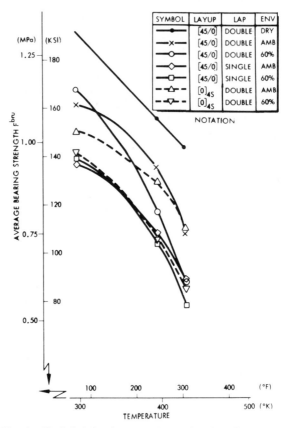

FIG. 4. Single bolt bearing strength as a function of temperature.

Qualitative explanations are as follows: the 45-degree plies diffuse load around the bolt hole and thereby reduce stress concentration at the hole edge. At higher temperatures, resin strength and stiffness are reduced, thereby lessening its ability to support the fibers and transfer load adjacent to the hole.

Table 1 also indicates that differences in tension failures, as reflected in parameter C, obtained from single- and double-lap tests, do not differ very significantly from quasi-isotropic layups. Those with no 45-degree plies appear to be more sensitive, as would be expected in a more brittle material. Rotation under load of multirow configurations is less than for single rows. Hence, it appears reasonable to expect multirow joints, critical in tension, to be less sensitive to whether single- or double-laps were used.

TABLE 1

Stress concentration relief parameter C as a function of layup, temperature and environment HMF330C/34 graphite cloth epoxy

Layup	Lap	Temp.	Parameter C		
			Dry	Ambient	60% RH
$[45/0]_{2S}$	Single	X	0.13	0.18	0.18
$[45/0]_{2S}$	Single	Y	0.14	0.19	0.25
$[45/0]_{2S}$	Single	Z	0.22	0.24	0.32
$[45/0]_{2S}$	Double	X	0.11	0.14	0.14
$[45/0]_{2S}$	Double	Y	0.14	0.12	0.16
$[45/0]_{2S}$	Double	Z	0.22	0.23	0.21
$[0]_{4S}$	Single	X	0.54	0.41	0.45
$[0]_{4S}$	Single	Y	0.58	0.52	0.64
$[0]_{4S}$	Single	Z	0.60	0.70	1.13
$[0]_{4S}$	Double	X	0.42	–	0.39
$[0]_{4S}$	Double	Y	0.44	–	0.53
$[0]_{4S}$	Double	Z	0.58	–	0.65

Notation: X = 297 K (75 °F); Y = 394 K (250 °F); Z = 427 K (310 °F).

MULTIPLE BOLT TESTS

Using Hart-Smith's method, design charts can be produced, one for each value of C. Those for $C = 0.2$ and $C = 0.4$ are presented here as Fig. 5. The joint efficiency is simply the failure load divided by the gross area of the laminate; once it is known, failure load prediction is trivial. The parameter σ_b/F^{tu} is the bearing stress ratio. As the number of bolt rows increases, this value decreases since the load per bolt decreases for a given total load on the joint. For prescribed values of the design parameters σ_b/F^{tu} and d/p, it is clear from these two Figures that the lower the value of C, the greater is the joint efficiency.

To assess the degree of correlation between Hart-Smith's theory and the HMF330C/34 graphite cloth epoxy, a fairly extensive series of multibolt tests were run and predictions made for each one. The parameter range included both layups, all three test temperatures and environmental

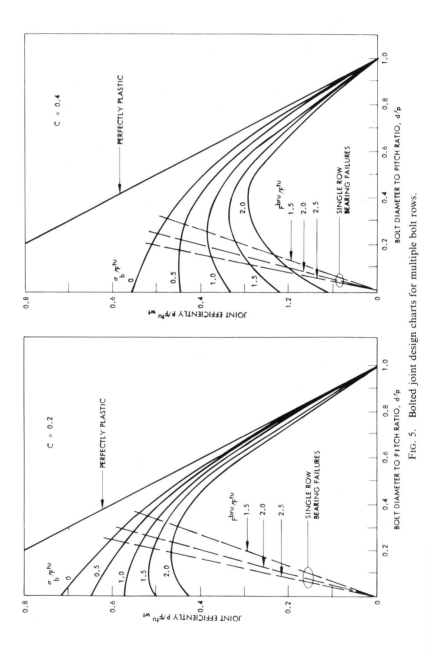

Fig. 5. Bolted joint design charts for multiple bolt rows.

conditions, as well as narrow strips with two and three bolts, two values of d/p in two-row, eight-bolt 'wide' joints. Some were double laps, others single. Results are summarized in Table 2. Counting each test temperature as a separate test, a total of 63 tests were conducted.

In general, agreement between test and prediction is considered to be most encouraging. The test to prediction ratio given in the last column of Table 2 was within 10 % of unity for 46 of the 63 tests. Only in two cases did it exceed 20 %. This is a strong endorsement for the method as a preliminary design tool.

Among the general observations that can be made for the parameter ranges considered are the following:

The differences between the ambient and conditioned specimens are not particularly significant. To see this, compare tests 1 with 2, 3 with 4, 11 with 12, 14 with 15 and 16 with 17. This arises because of the failure mode being dominated by tension strength which has been shown[11] to be only slightly influenced by moisture and is entirely consistent with Agarwal's tests.[8]

Differences in strength per unit width between narrow strips having the same width as bolt spacing in wide joints are small. This is evident by comparing tests 1 with 3, 2 with 4, 5 with 6, 7 with 8 and 9 and 10 with 11 and 12. This is important as it confirms the adequacy of testing narrow joints as a good simulation of larger, more expensive, joints.

As expected from comments made in discussing single bolt data, it is clear that there is relatively little difference between single- and double-lap multirow joints made from quasi-isotropic layups. Comparing tests 3 with 6, 9 with 14 and 10 with 15 illustrates this. However, comparison of tests 8 with 13 and 9 with 19 shows the laminates without any 45-degree plies to be more sensitive to lap type. Hence, different values of C were used in the predictions for tests 8 and 13. Using single lap, single-bolt data to predict failures for the 13 series at room temperature resulted in markedly inferior correlation. The quasi-isotropic joint data suggest that, for realistic joints containing a reasonable percentage of 45-degree plies, only one of the two lap configurations needs to be tested to provide the design data for most joints, regardless of lap details. This, too, is in agreement with Agarwal's work[8] on tape laminates containing bolted joints. It should be noted that these single lap joints were unrestrained against rotation. In practical applications it is highly likely that the joint will be attached to a member offering some degree of rotational restraint about an axis parallel to the bolt lines. Then the agreement between the two is likely to be even closer.

Greater degradation as a function of temperature is evident in the $[0]_{4S}$ joints relative to the quasi-isotropic laminates. This is consistent with Table 1 data and previously derived[11] unloaded hole data which demonstrate higher stress concentration factors in the former material.

Except for tests 16 and 17, which have three bolt rows, all those summarized in Table 2 have two rows. In these two, the failure loads at the lower temperatures are a somewhat smaller fraction of the predicted load than in most other tests. Further testing is required to resolve why the method is overpredicting the stress concentration relief in these situations.

TABLE 2

Multiple bolt test summary

Test No.	Layup	Bolts	Lap	Env	d/p or d/w	Width cm	Width in	Temp	Test Load kN	Test Load k lb	Prediction kN	Prediction k lb	Test Prediction
1	$[45/0]_{2S}$	2	S	A	0.2	2.38	0.94	X	16.95	3.81	17.04	3.83	0.99
								Y	17.26	3.88	16.59	3.73	1.04
								Z	9.74	2.19	10.90	2.45	0.89
2				60%				X	18.15	4.08	17.04	3.83	1.06
								Y	15.66	3.52	16.59	3.73	0.94
								Z	10.81	2.43	10.90	2.45	0.99
3		8		A	9.53	3.75		X	75.66	17.01	68.10	15.31	1.11
								Y	71.75	15.13	66.41	14.93	1.08
								Z	55.73	12.53	58.09	13.06	0.96
4				60%				X	71.48	16.07	68.10	15.31	1.05
								Y	69.88	15.71	66.41	14.93	1.05
								Z	54.62	12.28	58.09	13.06	0.94
5		2	D	A	2.38	0.94		X	16.01	3.60	18.90	4.25	0.85
								Y	15.48	3.48	18.40	4.14	0.84
								Z	12.54	2.82	13.43	3.02	9.94
6		8			9.53	3.75		X	79.91	17.74	75.66	17.01	1.05
								Y	72.86	16.38	73.61	16.55	0.99
								Z	57.78	12.99	53.60	12.05	1.08
7	$[0]_{4S}$	2	S	A	0.286	1.67	0.66	X	13.39	3.01	12.32	2.77	1.09
								Y	11.83	2.66	10.50	2.36	1.13
								Z	7.16	1.61	8.50	1.91	0.84
8		8			6.73	2.63		X	53.69	12.07	49.33	11.09	1.08
								Y	49.51	11.13	44.21	9.94	1.18
								Z	37.14	8.35	33.98	7.64	1.09
9	$[45/0]_{2t}$	2		A	1.67	0.66		X	11.61	2.61	11.48	2.58	1.01
								Y	10.85	2.44	11.21	2.52	0.97
								Z	8.01	1.80	8.01	1.80	1.00

TABLE 2—*contd*

Test No.	Layup	Bolts	Lap	Env	d/p or d/w	Width cm	Width in	Temp	Test Load kN	Test Load k lb	Prediction kN	Prediction k lb	Test Prediction
10	[45/0]$_{2S}$	2	S	60%	0.286	1.67	0.66	X	11.25	2.53	11.48	2.58	0.98
								Y	10.05	2.26	11.21	2.52	0.90
								Z	8.50	1.91	8.01	1.80	1.06
11		8		A		6.73	2.63	X	47.55	10.69	45.90	10.32	1.04
								Y	46.26	10.40	44.75	10.06	1.03
								Z	36.52	8.21	32.03	7.20	1.14
12				60%				X	48.48	10.90	45.90	10.32	1.06
								Y	43.32	9.74	44.75	10.06	0.97
								Z	37.50	8.43	32.03	7.20	1.17
13	[0]$_{4S}$		D					X	61.47	13.82	66.90	15.04	0.92
								Y	52.22	11.74	41.99	9.44	1.24
								Z	37.94	8.53	33.99	7.64	1.12
14	[45/0]$_{2S}$	2	A		1.67	0.66		X	10.68	2.40	11.48	2.58	0.93
								Y	10.05	2.26	11.21	2.52	0.90
								Z	9.03	2.03	8.01	1.80	1.13
15				60%				X	10.63	2.39	11.48	2.58	0.93
								Y	9.74	2.19	11.21	2.52	0.87
								Z	8.81	1.98	8.01	1.80	1.10
16		3	A					X	9.96	2.24	11.88	2.67	0.84
								Y	9.87	2.22	11.61	2.61	0.85
								Z	8.72	1.96	8.58	1.93	1.01
17				60%				X	9.83	2.21	11.88	2.67	0.83
								Y	9.83	2.21	11.61	2.61	0.85
								Z	9.12	2.05	8.58	1.93	1.06
18		2	A	0.375	1.27	0.50		X	9.12	2.05	8.81	1.98	1.03
								Y	8.01	1.80	8.14	1.83	0.99
								Z	7.03	1.58	7.29	1.64	0.96
19	[0]$_{4S}$							X	12.05	2.71	12.14	2.73	0.99
								Y	8.27	1.86	9.16	2.06	0.90
								Z	6.94	1.56	7.61	1.71	0.92
20	[45/0]$_{2S}$		Dry	0.286	1.67	0.66		X	12.85	2.89	13.03	2.93	0.99
								Y	11.65	2.62	12.68	2.85	0.92
								Z	11.25	2.53	11.74	2.64	0.97
21	[0]$_{4S}$							X	15.17	3.41	15.12	3.40	1.00
								Y	15.26	3.43	14.81	3.33	1.04
								Z	12.72	2.86	12.01	2.70	1.06

Notation
D = Double lap joint
S = Single lap joint
A = Ambient conditioning
60% = 60% Relative humidity conditioning
Dry = No moisture
X = 297 K (75 °F)
Y = 394 K (250 °F)
Z = 427 K (310 °F)

While not evident from the Table, it is important to restate that not all double lap joints were of the same construction. Only tests 5 and 6 had all composite members. Tests 13 to 21 had titanium alloy outer members, and 8-ply composite inner members. In all tests 5 and 6 failure was of the 4-ply outer laps with $[45/0]_s$ layups. This was surprising because many failure theories for composites suggest a 'volume' effect in which a flaw size per unit volume is postulated. Hence, thicker laminates have a greater probability of containing a critical flaw, resulting in lower strength. These failures suggest a different mechanism at work. To check into it a series of 2·54 cm (1·0 in) wide 4- and 8-ply quasi-isotropic test specimens, each containing a single hole, were tested in tension. Three hole sizes— 0·476, 0·635 and 1·27 cm (3/16, 1/4 and 1/2 in)—were included. Failure stress ratio of 4- to 8-plies ranged from 0·83 to 0·93, while the unnotched strength ratio was 0·93. This confirms the reduced tension strength of thinner laminates in all cases. Fiber volume measurements and photomicrographs failed to show any discernible differences between the two thicknesses, so no explanation is currently evident.

Thus far, only failure loads have been discussed, without any attention to either failure modes or the load–deflection data. The purpose of obtaining the load–deflection curves was to assess joint linearity and provide data on the failure process. These findings can be briefly summarized as follows. Non-linear response at load levels below approximately 10 % to 20 % of the failure load were indicative of the bolts seating and the joint settling down. When tension failures occurred, response was linear to within less than 5 % of the failure followed by a sudden typical tension failure. In bearing failures, stiffness reduced rapidly, beginning at approximately 90 % to 95 % of the peak load, and became negative as a progressive failure occurred in conjunction with decreasing load. In joints where single-bolt tension failures occurred, there was very little, if any, decrease in stiffness as a function of temperature or moisture conditioning. When bearing failures resulted, there was a clear decrease as a function of temperature, in joint stiffness over the linear load–deflection regime. This amounted to as much as 30 % between RT and 427 K (310 °F) tests.

CONCLUSIONS

This large series of experiments covering numerous geometric and material variations correlates very well with predictions by Hart-Smith's method.[2,3]

These references have previously been used with confidence in a number of unpublished studies as a design method for joints in tape laminates. Similar confidence is now justified for its application to cloth laminates. The ability of the method to account for interaction effects between closely spaced bolts is considered to be particularly important.

It has been demonstrated that the quasi-isotropic layups containing loaded holes exhibit markedly superior stress concentration relief compared with laminates without any 45-degree plies.

As efficiently designed multirow bolts frequently fail in a predominantly tensile mode, the influence of absorbed moisture on their strength was confirmed to be relatively unimportant.

Wide joints having many bolts per row were compared with narrow strips containing a single bolt per row and having a width equal to the bolt spacing of the wide joints. Their strengths per unit width were very similar. This substantiates the simulation of operational wide joints by much cheaper, easier to test narrow strips, when the loading is normal to the joint axis.

A special deflection gage was developed to measure relative motion between a bolt and the neighboring laminate. It provides for more useful data than the conventional crosshead motion measurement which includes the motions of the loading fixture.

The small sample size of three precludes any definitive statements being made about design strengths. More replicates are required to provide an adequate data base for this purpose.

ACKNOWLEDGEMENT

The relative deflection gage used in these tests was developed by A. M. Holmes.

REFERENCES

1. GARBO, S. P. and OGONOWSKI, J. M., *Effect of variance and manufacturing tolerances on the design strength and life of mechanically fastened composite joints*, AFFDL-TR-78-179, December, 1978.
2. HART-SMITH, L. J., *Bolted joints in graphite epoxy composites*, NASA CR144899. June, 1976.
3. HART-SMITH, L. J., Mechanically fastened joints for advanced composites— Phenomelogical considerations and sample analyses, *Proc. of the Fourth Conference on Fibrous Composites in Structural Design, San Diego, CA. November, 1978.* (To be published.)

 4. BAILIE, J. A., DUGGAN, M. F., BRADSHAW, N. C. and MCKENZIE, T. G., Design of graphite cloth epoxy bolted joints at temperatures up to 450 K. *Proc. of ASTM Symposium 'Joining of Composite Structures'. Minneapolis, Minnesota, April, 1980.* (To be published.)
 5. KIM, R. Y. and WHITNEY, J. M., Effect of temperature and moisture on pin bearing strength of composite laminates, *J. Comp. Matls.*, **10** (April, 1976), 149–55.
 6. WILKINS, D. J., Environmental sensitivity tests of graphite epoxy bolt bearing properties. *Composite Materials Testing and Design (Fourth Conference). ASTM STP617*, 1977, pp. 497–513.
 7. RAMKUMAR, R. L., Bolted joint design. *Proc. of Symposium on Test Allowables for Fibrous Composites. ASTM Conference held at Dearborn, MI on October 3rd and 4th, 1979.* (To be published.)
 8. AGARWAL, B. L., Behavior of multifastener bolted joints in composite materials (Paper No. 80-0307). *Proc. AIAA 18th Aerospace Sciences Meeting. Pasadena, CA. January 14–16, 1980.*
 9. MCKENZIE, T. G. and HOWARD, S. A., Unpublished Data. Lockheed Missiles & Space Company, 1979.
10. CROSSLEY, F. A., VOLLERSEN, C. A., GOETZ, A. C. and DAVIS, G. E., Structural test program for a major graphite epoxy composite component for the Trident I Missile. *Proc. 4th Aerospace Testing Seminar. Institute of Environmental Sciences*, 1978.
11. BAILIE, J. A., DUGGAN, M. F., FISHER, L. M. and YEE, R. C., Effect of holes on graphite cloth epoxy laminate's tension strength at temperatures up to 450 K (350 °F). Presented at AIAA/ASME/ASCE/AHS 21st Structures Structural Dynamics and Materials Conference. Seattle, WA. May 12–14, 1980. (To be published.)

6

The Stress-Rupture Behaviour of GRP Laminates in Aqueous Environments

R. C. WYATT

CEGB Scientific Services Department, South Western Region, Bedminster Down, Bristol, England

L. S. NORWOOD

Scott Bader Co. Ltd, Wollaston, Wellingborough, Northamptonshire NN9 7RL, England

AND

M. G. PHILLIPS

School of Materials Science, The University, Claverton Down, Bath BA2 7AY, England

ABSTRACT

To remedy the lack of soundly based design data which has hindered the use of GRP for power station cooling water systems, an extensive programme of creep-rupture testing is in progress at Bath University, funded jointly by CEGB and Scott Bader Co. Ltd.

The paper describes the background and aims of the work and gives details of the test programme and equipment, including a cell for single-sided exposure of a tensile testpiece. The behaviour of a polyester/csm/w.r. laminate in seawater is shown to be little affected by temperature in the range 20–60°C. Seawater and distilled water are indistinguishable in their effects on fully immersed laminates at 40°C. The influence of testpiece width and reinforcement pattern are discussed.

INTRODUCTION

The South Western Region Scientific Services Department of the CEGB has recently embarked on a programme of GRP component design and installation, the objective being to stimulate the future use of the material on the basis of sound design and successful installations at selected sites.[1,2]

Design of hand-lay-up components has been based on BS 4994: 1973, using a derived load design factor (K of BS 4994) of about 15, but certain misgivings are felt about the approach defined in the standard. One point of particular concern is the degree to which account is taken of time-dependency in the presence of water, bearing in mind a CEGB requirement for a thirty-year life on new plant. The design factor specified in BS 4994 to cover long-term behaviour, K_2, has a specified maximum value of 2·0. By contrast, early tests on the GRP material, subsequently used in most of the recent CEGB applications,[1] predicted a thirty-year creep-rupture strength in water at 40 °C, only 10 % of initial strength. This suggests that a design factor of 10 is needed to cover time dependence, leaving a factor of 1·5 to cover all other features.

To establish a more satisfactory basis for design in the long-term, it was decided that further creep-rupture tests were necessary, especially as certain features of the previous tests[1] had been questioned. Scott Bader Company Ltd, suppliers of resin for CEGB's recently fabricated c.w. components, being similarly involved in development in this field, agreed to contribute to the cost of a comprehensive test contract. A third contributor is the Generation Construction and Development Division of CEGB (Barnwood, Gloucester) whose role is to monitor and approve the design of new power stations. A programme was agreed between the three parties and a contract placed with South Western Industrial Research Ltd (SWIRL), the industrial research organisation of Bath University.

TEST PHILOSOPHY AND PROGRAMME DETAILS

Main Programme

Details of the main programme (Programme A) are given in Table 1. The material was prepared in individual panels of size 760×760 mm by W. and J. Tod Ltd, using conventional hand-lay-up techniques. This company has fabricated a number of large components recently installed in CEGB c.w. systems, using similar materials and construction. The company were requested to aim for glass/resin ratios (by weight) of 2:1 for csm and 1:1 for woven-roving—ratios typical of commercial practice.

The main aim of Programme A is to establish a statistically sound creep-rupture strength prediction. To ensure statistical security, the testing specification demands a *minimum* of 30 tests, broadly separated into

TABLE 1
Details of main test programme A

Materials and construction	Test conditions
Resin: Scott Bader Crystic 625 TV	*Immersion medium:* Artificial seawater
Reinforcement: Multemat chopped strand mat (csm) by Fibreglas Ltd Type ECK 25 bi-directional woven-roving (w.r.) by TBA Ltd	*Mode of immersion:* Total
	Specimen width: 25 mm
	Edge conditions: Unsealed
Construction: 1 × 300 g/m² csm	*Test temperatures:*
1 × 830 g/m² w.r. (warp)	20 °C: Code A20
1 × 600 g/m² csm	40 °C: Code A40
1 × 830 g/m² w.r. (weft)	60 °C: Code A60
1 × 300 g/m² csm	
Cure: 5 h at 60 °C	

distinct bands of failure time, such that plotting on a logarithmic time scale will provide a reasonably uniform spread of results over the interval 10^3–$10^{7.8}$ s (17 min–2 years). Five replicate tests are called for at each load level for all but the longest-term tests.

Tests to the above schedule are being carried out at three temperatures: 20, 40 and 60 °C (Programmes A20, A40 and A60 respectively). Since the material is nominally identical to that used in earlier work the results now obtained at 40 °C offer a direct comparison. The tests at 20 °C will provide data at a temperature more relevant to sustained service conditions, whilst those at 60 °C will complete a time/temperature frame-work in which superposition techniques can be examined.

A major criticism of the previous test programme (see the introduction) concerned the width of the test specimens (6 mm). Scepticism regarding the validity of data obtained with such narrow specimens stems from two sources. Firstly, a very rapid take-up of water *throughout* the specimens would be expected because of the exaggerated influence of cut specimen edges and exposed glass fibre ends. Secondly, unpredictable variations in strength might arise from using a width no greater than that of a single bundle in the woven-roving reinforcement. A notable difference introduced in the present programme is therefore the use of a specimen of 25 mm width.

A further criticism of the previous work is that it used distilled water as an immersion medium, whereas seawater would be involved in CEGB applications. It has been argued that osmotic influences, known to apply in polyester resins, would diminish the take-up of water in the latter case and might thus reduce the rate of decay of strength in a creep-rupture test. Hence, to counter such arguments, the present programme uses an artificial seawater of standard composition.

Supplementary Programmes

To exploit fully the comprehensive data obtained in Programme A, supplementary programmes were included to examine certain important materials and test variations. These are summarised below to show the scope of the investigation. Table 2 provides details of the variations upon which results are reported here.

Programme T1 provides a link with the earlier work, reference 1, by using testpieces of the same width, 6 mm.

Programme T2 investigates the effect of sealing cut edges in the testpiece.

As a second link with the earlier programme,[1] T3 uses distilled water as the immersion medium, so that the importance of osmotic effects can be assessed.

The last of this group of supplementary programmes, T4, deals with exposure to water on only one of the two main faces of the laminate. The pattern of water take-up under such conditions will clearly be very different from that in conditions of total immersion, with or without edge sealing. To summarise therefore Programmes T1–T4, together with A40, study the factors governing water ingress, and their influence upon degradation of the laminate.

In Programmes M1–M7, the response of different laminate systems to seawater at 40 °C are studied.

The polyester resin used in M1 is nominally identical to that used in Programme A except that it contains no thixotropic agent. That in M2 is of

TABLE 2
Supplementary programmes: Materials and testing procedures

Programme code	Chosen variation from A40	Experimental details (variations from A40)
T1	Narrow testpiece	*Specimen width:* 6 mm
T3	Distilled water	*Immersion medium:* Distilled water
M3	Epoxy resin	*Resin:* Ciba-Geigy XD 927
		Reinforcement: Type 9504 w.r. from Scandinavian Glassfiber (nom. wt. 820 g/m^2)
M7	All-woven-roving construction	*Construction:* 4×830 g/m^2 w.r. (warp/weft/warp/weft)

a different polyester type expected to provide a greater degree of water resistance.

Programmes M3 and M4 examine the use of resin of different chemical types, epoxy and vinyl ester.

The powder-bound csm of Programme M5 replaces the p.v.a. emulsion-bound csm of the main programme.

The incorporation of Programmes M6 and M7 is an attempt to isolate the performance, in creep-rupture, of the two basic reinforcement types combined in Programme A, i.e. chopped strand mat and woven-rovings.

TESTING EQUIPMENT AND PROCEDURES

Design and Construction of Test Rigs
Load points

A simple first-order lever was employed, with an arm ratio of 20:1. The loads available ranged from 180 to 2000 kgf.

The basic unit comprised two load points symmetrically disposed about the centre of a 54 in length of $3 \times 1\frac{1}{2}$ in channel section. Groups of four units were mounted on base frames as shown in the photograph (Fig. 1). Rectangular tanks in stainless steel were constructed in two sizes, or accepted respectively four to eight load points.

To keep down cost of manufacture and to facilitate small variations in

FIG. 1.   General view of test facility at Bath University.

applied load, hanging weights were made in the form of large boxes, to be filled with gravel or (for the highest loads) lead shot.

To reduce vibrations and bearing damage when a sample breaks, a baulk of timber is located beneath the lever arm, as may be seen in the left foreground of Fig. 1. This is made the basis of a simple system for sample break detection by standing it upon a pressure switch so that contact is made when the sample fails.

Cell for single-sided exposure

Programme T4 requires laminates to be exposed to seawater at 40 °C on one face only. This technique does not appear to have been used before with simple tensile loading, but only with specimens in the form of tubes.

So that use could be made of the facility constructed for the main programme it was decided to construct an 'air-cell', for the protection of one face and the edges of a standard testpiece. Besides excluding water, it was required that the cell exert the least possible influence on the mechanical behaviour of the testpiece.

The design adopted is shown in the photograph (Fig. 2).

A 'perspex' strip, profiled to match the specimen gauge length, carries at one end a 25 mm spacer which forms the base of the air-cell. At the upper end, the cell wall is located against the sample clamping nut. There is thus no restraint on the extension of the testpiece.

After the testpiece has been correctly clamped up, a two-part cold cure adhesive, Bostik 'Boscoprene Cement' No. 2402 is used to attach the lower end of the cell to the GRP, and then to attach flexible side membranes over the profiled edges of testpiece and cell wall. 'Neoprene' rubber sheet 1 mm thick is used. Tensile tests carried out with a water-filled cell in place show no detectable change in UTL and no leakage of water prior to fracture of the GRP.

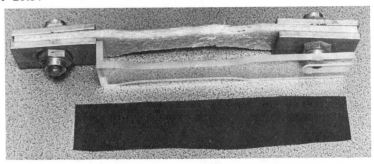

FIG. 2.   Components of cell for single-sided immersion of tensile testpiece.

Experimental Procedure

Sample preparation

For the majority of tests, samples were machined to give a parallel-sided central portion 65 mm long by 25 ± 0.5 mm wide. End tabs were 38 mm wide by 90 mm, and shoulder radius 75 mm.

Samples for Programme T1 were prepared with a central parallel portion 50 mm long by 6 mm wide, but were in respect of end tabs and fitting identical to those described above.

Tensile testing

Ten samples were tested from each category of material, using an Instron 1195 screw-driven machine with crosshead speed 5 mm/min, causing failure to occur within the time range 30–90 s.

Stress-rupture testing

The lever arm was first supported by a small hydraulic jack, in such a position that it would move down toward the horizontal as the sample extended. The sample train was anchored first at the lower end. The size and design of the anchor points was such that this was quite readily achieved, although elbow-length rubber gloves had to be worn when operating in tanks at 60 °C.

After the top anchor point had been attached and slack taken up in the linkage, load was gradually applied by operating the release valve of the hydraulic jack.

RESULTS AND DISCUSSION

Tensile Testing

Table 3 gives the tensile test results, for all categories of material concerned, expressed as Ultimate Tensile Unit Load (UTUL) in units of newtons per millimetre width as recommended in BS 4994, 1973.

As stated, each result is the mean of ten determinations on testpieces selected at random from the laminates. The standard deviation of each sample is reported, and in the worst case the 95 % confidence limits are set at mean value ± 20 %.

The mean and standard deviation for the whole population are reported also. Using these as a basis for comparison and employing Student's t test, it is found that for five of the material variables (i.e. A, M1, M2, M4, M5) involving changes to resin but none to the reinforcement, no significant effect can be detected on the UTUL against the background variability.

TABLE 3
Tensile test results

Code	Description	Glass content (kg m^{-2})	Mean UTUL^a (N/mm)	Standard deviation (N/mm)
A	Crystic 625	2·86	239	19
M1	No thixotropic agent	2·86	269	10
M2	Crystic 397	2·86	262	15
M3	Epoxy resin	2·84	276	18
M4	Vinyl ester resin	2·86	261	10
M5	Powder-bound	2·86	251	17
M6	All csm	3·00	223	10
M7	All-w.r.	3·32	309	29
T1	Narrow sample	2·86	206	17
T2				
T3	As 'A'			
T4				
	Total population		253	31

^{a} Ultimate Tensile Unit Load, BS 4994, 1973.

The increased unit strength associated with epoxy resin (M3) is significant at the 5 % level, i.e. it could arise by chance once in twenty tests.

As would be expected, changes in the reinforcement pattern to all csm (M6) and to all woven-roving (M7) produce more significant changes in unit strength: the reduction in the first case being significant at the 1 % level and the increase in the latter significant at the 0·1 % level.

Finally it may be remarked that for the standard lay-up there is a highly significant reduction in unit strength on changing to a narrow testpiece (T1). This is not unexpected, but it emphasises the need for using testpieces of adequate size to give a realistic assessment of these materials.

Creep Rupture Tests

Creep rupture testing commenced in November 1979. The results reported in this paper are those available at the time of writing, for (a) the main programme (A20, A40, A60) and (b) the four supplementary programmes (M3, M7, T1 and T3).

Main programme

Statistical treatment of results. Figures 3 and 4, show creep rupture plots for Programmes A40, and A20 plus A60 respectively. All experimental

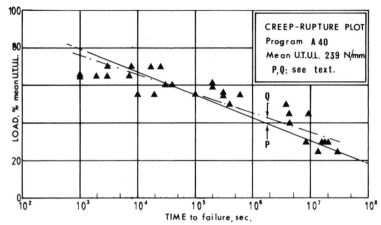

FIG. 3. Experimental results for creep-rupture of polyester/glass laminate in seawater at 40°C.

points are included, and unit loadings are expressed as a percentage of mean (short-term) ultimate tensile unit load. In each case a best-fit straight line is shown, calculated by regression analysis of log time on normalised unit load; i.e. assuming an equation of the form:

$$\log t = A - B(u/u_0) \tag{1}$$

where A and B are constants, t is the time to failure associated with unit load u, and u_0 is Ultimate Tensile Unit Load.

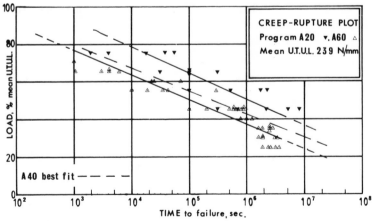

FIG. 4. Experimental results for creep-rupture of polyester/glass laminate in seawater at 20°C and at 60°C, compared with 'best-fit' line for 40°C from Fig. 3.

This equation is appropriate if life predictions are to be made, but it must be used with caution. To illustrate this point, Fig. 3 shows the two regression lines determined from A40. Curve P conforms to the above equation. Curve Q shows the regression of stress upon log time and has an equation

$$\frac{u}{u_0} = M - N \log t \tag{2}$$

where M and N are constants not related in any simple way to A and B of eqn. (1). In this case the two lines are quite close, indicating a high degree of correlation in the data ($r = 0.91$), but nonetheless curve Q (eqn. (2)) gives an unjustifiably optimistic prediction for the thirty-year creep-rupture strength. In making predictions of this type it is important to choose the correct regression line, a fact which some previous workers have not appreciated.

In Figs 3 and 4 the longer-term points suggest a down-turn in the creep-rupture curves at both 40 and 60 °C, and it is proposed to collect more data in these regions.

Temperature effects. Figure 4 compares the best-fit lines for the three test temperatures. The trend is as expected, and is similar to the findings of D. J. Steel[3] who performed flexure tests on similar materials. However, the differences with temperature in the present work are quite small, and may not be significant against the background scatter. Furthermore, unit loads in each case have been normalised to the measured Ultimate Tensile Unit Load at 20 °C. It is probable that the observed effects of temperature in stress-rupture would appear even smaller when normalised against Ultimate Tensile Unit Load measured at the temperature of test.

Supplementary programmes

Exposure conditions. In Fig. 5 are plotted the results for 6 mm wide testpiece (T1) and for distilled water immersion (T3) together with the best-fit line for A40 as a reference.

It appears that the effect on test results of a large reduction in specimen width is negligible. This is a rather surprising conclusion, for which two opposing explanations could be offered. Firstly, it might be argued that, in both cases, water take-up is sufficiently rapid (perhaps totally or partially through cut specimen edges) for saturation to be reached very quickly. Beyond very short times, the creep-rupture performance would then be similar for both specimen types. Alternatively, if water take-up is extremely

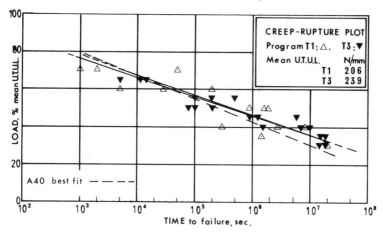

FIG. 5. Effects of sample width and of immersion medium upon creep-rupture behaviour of polyester/glass laminate at 40 °C.

slow, then penetration of water might be insufficient within the time scale of the tests to affect creep-rupture performance. Again the behaviour of the two types of specimen would appear similar. At the time of writing no water uptake measurements are available to resolve this question. The lower initial strength of the 6 mm specimen should be recalled. A consequence is that a comparative plot in terms of absolute (non-normalised) unit load would indicate a noticeably poorer performance in the case of the narrower specimens for the same material.

Changing the immersion medium from seawater to distilled water is shown by Programme T3 to produce no significant change in the rate of degradation. Such a conclusion is perhaps not surprising in view of the low concentrations of soluble salts in the former (e.g. 6 g/litre NaCl, equivalent to a 0·1N solution. Thus distilled water may be used with confidence for future test work, which will greatly reduce experimental difficulties.

Alternative laminates. Data from programme M3 are plotted in Fig. 6. Clearly the epoxy laminate deteriorates more rapidly than does the polyester material. In the preparation of this laminate the woven-roving reinforcement was changed to one known to have an epoxy-compatible finish. No commercially available csm reinforcement with a finish specifically designed as epoxy-compatible could be traced so the material used in Programme A was retained. This was recognised as a potential shortcoming but it was unavoidable. In the present situation a would-be commercial user of epoxy resin for a similar purpose would face the same

R. C. Wyatt et al.

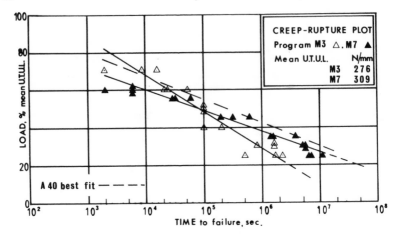

FIG. 6.   Effects of resin type and reinforcement pattern on creep-rupture behaviour of laminates in seawater at 40 °C.

problem. It cannot be deduced from present evidence how significant this incompatibility is.

Figure 6 shows also a plot of data from Programme M7. Comparison of the best-fit straight line with that from Programme A40 reveals what is probably a negligible difference in creep-rupture performance between laminate of all-woven-roving construction and that of the basic csm/woven-roving construction. It should be noted that the absolute unit loads applied are higher in the case of the woven-roving material, because of its higher strength.

CONCLUSIONS

(i)   The stress-rupture behaviour of a polyester/csm/w.r. laminate in seawater is relatively little affected by temperature and that change may be due to the alteration in static strength.

(ii)   At 40 °C, distilled water and seawater are indistinguishable in their effects, upon stress-rupture of the standard laminate.

(iii)   In a comparison between 6 mm and 25 mm wide testpieces, the UTUL of the narrow specimens is lower, but the rate of degradation in stress-rupture strength is indistinguishable.

(iv)   An all-w.r. laminate and a csm/w.r. laminate show little difference in stress-rupture behaviour, in relation to their respective strengths.

ACKNOWLEDGEMENTS

The authors are indebted to Mr N. Heppell who carried out much of the experimental work, W. and J. Tod Ltd who prepared the laminates, the Generation Development and Construction Division of the CEGB for financial support, Mr M. D. Holdstock of that department for his contributions to the preparation of programmes and Mr D. A. Pask, South West Region Director-General of the CEGB, for permission to publish this paper.

REFERENCES

1. BRYAN-BROWN, M. H. and WALKER, D. M. Power station condenser water boxes in GRP, *British Plastics Federation 11th Reinforced Plastics Congress, 1978*, Paper 14.
2. BRYAN-BROWN, M. H., WALKER, D. M. and WYATT, R. C. Advances in the use of GRP for the power industry, *British Plastics Federation 12th Reinforced Plastics Congress, 1980*, Paper 22.
3. STEEL, D. J. The creep and stress-rupture of reinforced plastics, *J. Plastics Inst.*, 1965, October, 161–7.

7

Water Absorption by Glass Fibre Reinforced Epoxy Resin

P. BONNIAU AND A. R. BUNSELL

*Ecole des Mines de Paris, Centre des Matériaux,
B.P. 87, 91003 EVRY Cédex, France*

ABSTRACT

Water absorption by glass fibre reinforced epoxy resin and its influence on tensile strength have been studied and the role of the resin hardener shown to be very important. All tests were conducted under conditions of relative humidity ranging from 0% to 100% RH and in the temperature range 25°–90°C. It was found possible to apply diffusion theories to those materials for which diamine and dicyandiamide hardeners had been used but not when an anhydride hardener had been employed. Damage, as revealed by a fall in tensile strength, was found to occur at relative humidities greater than 60–70% RH. It was seen, both under humid conditions and in immersion tests, that this damage was not related to the quantity of water absorbed by the composite but to the temperature and time of exposure after the water concentration limit was passed.

NOTATION

D	Composite diffusivity ($m^2 s^{-1}$)
t	Time (s)
h	Thickness of plate (m)
$p = \dfrac{\sqrt{Dt}}{h}$	Diffusion parameter (Dimensionless)
W	Weight of plate (kg)
Wd	Weight of dry plate (kg)
Wm	Weight of saturated plate (kg)

$M = \dfrac{W - Wd}{Wd}$ Moisture content percentage (%)

$Mm = \dfrac{Wm - Wd}{Wd}$ Maximum moisture content percentage (%)

Vf Fibre volume fraction (%)

T Temperature (°C)

RH   Relative humidity (%)

α Probability of a molecule of water passing from the combined phase to the free phase (s^{-1})

β Probability of a molecule of water passing from the free phase to the combined phase (s^{-1})

INTRODUCTION

The use of glass fibre reinforced composites is often limited by poor knowledge of their long-term behaviour and one of the most important unknown influencing factors is the effect of absorbed water. The amount of water generally absorbed by these materials is quite small, typically less than 1%; however, earlier studies have shown that this water can cause dramatic loss of mechanical and physical properties.[1,2] Several studies on carbon fibre reinforced epoxy resin indicate that this loss of properties is due primarily to the absorption of water by the body of the resin.[3,4]

Water absorption by epoxy resins seems to occur by instantaneous adsorption at the surface followed by diffusion into the body of the resin. The water is in the form of free molecules or groups of molecules linked loosely by hydrogen bonds to the polymer. This being the case, it has been found that the rate of water absorption can often be correctly described for both the resin[5] and the composites[6] by employing the diffusion laws. In this instance there is simply the diffusion mechanism operating and no degradation of the resin occurs. Above certain thresholds of humidity and temperature other mechanisms due to the presence of the water in the resin, such as swelling[7] and plastification,[3] lead to degradation of the material. These effects have been reported elsewhere but our knowledge of the causes is largely qualitative and they seem to vary with the composite components and the conditions of curing. The primary aim of this study has been to examine the limitation of the diffusion laws for three types of glass reinforced epoxy (grp) under humid and immersion conditions whilst assuming that simple diffusion cannot be considered to be, in itself, a damaging process which would lead to a fall in material strength.

DIFFUSION MODELS AND CURVE FITTING

Two diffusion models have been considered; the first is the classical case of absorption of a single free phase and the second is of the Langmuir type which involves the diffusion of two phases, one free to diffuse and the other linked to the material and unable to diffuse.

Both these models are based on the theoretical ideas which are expressed as Fick's Law and which involve the water concentration gradient as the driving force of the diffusion.

The validity of the models can be determined by observing the water uptake of a plate of thickness, h, initially dry and then exposed to conditions of fixed humidity and temperature. The weight gain, $M\%$, as a function of time, t, may be expressed for the single free phase model as a function of two parameters—the diffusivity, D, and the weight gain at saturation, $Mm\%$.[6-8]

$$\frac{Dt}{h^2} < 0.05 \qquad M\% = Mm\% \cdot \frac{4}{\sqrt{\pi}} \cdot \sqrt{\frac{Dt}{h^2}} \qquad (1)$$

$$\frac{Dt}{h^2} > 0.05 \qquad M\% = Mm\%\left[1 - \frac{8}{\pi^2}\exp\left[\left[-\frac{Dt}{h^2}\cdot\pi^2\right]\right]\right] \qquad (2)$$

In the two-phase model $M\%$ as a function of time is expressed by four parameters—diffusivity, D, $Mm\%$, the probability of a molecule passing from the attached to the free phase, α, and the probability of a molecule passing from the free phase to the bound phase, β.[8-10]

When:

$$\alpha \ll \frac{D\pi^2}{h^2} \quad \text{and} \quad \beta \ll \frac{D\pi^2}{h^2}$$

absorption may be written as:

$$\frac{Dt}{h^2} < 0.05 \qquad M\% = Mm\% \cdot \frac{\alpha}{\alpha+\beta} \cdot \frac{4}{\sqrt{\pi}} \cdot \sqrt{\frac{Dt}{h^2}} \qquad (3)$$

$$\frac{Dt}{h^2} > 0.05 \qquad M\% = Mm\%\left[1 - \frac{\beta}{\alpha+\beta}\exp(-\alpha.t)\right.$$

$$\left. - \frac{\alpha}{\alpha+\beta} \cdot \frac{8}{\pi^2} \cdot \exp\left[-\frac{Dt}{h^2}\cdot\pi^2\right]\right] \qquad (4)$$

It will be seen that the second model reduces to the first when $\alpha = 1$ and $\beta = 0$.

The parameters are calculated using either eqn. (2) or eqn. (4) according to the model employed and for points for which $(Dt/h^2) > 0.05$. The method of minimum variances, as used for curves of an exponential shape, was employed.[11] The calculations were made using a TI 59 calculator and a precision of 0.001 was arrived at in 10 min for the first model and in 1 h for the second.

EXPERIMENTAL DETAILS

The three materials used were in the form of plates made by UDD-FIM and are generally used as insulators in electric motors. All three were similar, being made from glass type E fibre cloth in a Bisphenol A epoxy resin. The only difference between the three series of materials was the hardener used to cure the resin. The hardeners used were diamine, dicyandiamide and anhydride. These are shown in Table 1.

Ten environmental chambers were used in these tests, allowing conditions of 0% to 100% RH and 25–90 °C to be obtained plus immersion in distilled water over the same temperature range (see Fig. 1). Relative humidity was controlled in the chambers by means of saturated salt solutions. Relative humidity was measured with an hygrometer probe of lithium chloride type DMS 100 produced by Richard-Pekly. Temperature

TABLE 1
Characteristics of the three studied materials

		Material 1	Material 2	Material 3
Type of hardener used with Bisphenol A resin		Diamine	Dicyandiamide	Anhydride
Glass fibre cloth	Filament type	E.C.9.68	E.C.9.68	E.C.9.68
	Number of threads in warp per centimetre	17	17	17
	Number of threads in weft per centimetre	13	13	13
Composite	Number of cloth layers per millimetre	7	6	6
	Fibre volume fraction	56%	48%	48%
	Density (g/cm^3)	1.90	1.80	1.80
Usual upper service temperature		155 °C	140 °C	140 °C

P. Bonniau and A. R. Bunsell

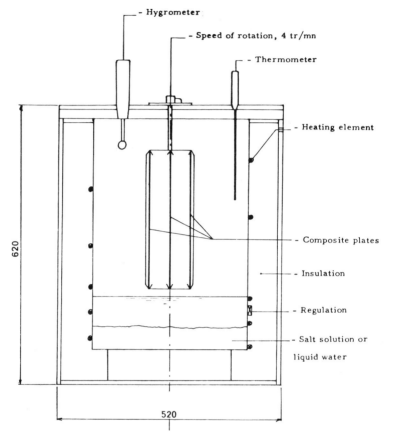

FIG. 1.   Schematic view of an environmental chamber.

measurements were made with a numerical thermometer, type PN 2S, made by A.O.I.P., as well as an electronic transducer, type LX 5600, which also controlled the regulation and which was made by National Semiconductor.

The composites were in the form of thin plates so as to have negligible edge effects and ensure a condition of one-dimensional diffusion normal to the fibre directions. The dimensions of the plates were $n = 210$ mm ± 0.4, $l = 300$ mm ± 0.4 and $h = 1$ mm ± 0.16. A range of plate thicknesses were used with the first material, $h = 0.2$, 0.5, 1 and 2 mm.

The plates were weighed on an electronic balance made by Sartorius, type 1205 MP. The balance was enclosed in a Faraday cage in order to eliminate electrostatic effects on the plates and to act as a draught shield.

The tests were conducted in a total of 29 different environments for material 1 and 13 different environments for the two other materials. In every test either two or three plates were simultaneously exposed to the same conditions.

The tensile tests were conducted at $5 \times 10^{-2}\,\mathrm{mN^{-1}}$ on an Instron tensile machine. Tensile tests were conducted after nine different exposures under humid conditions and for immersions. The shape of the tensile test specimens is shown in Fig. 2. All specimens were initially dried over a period of two weeks at 100 °C.

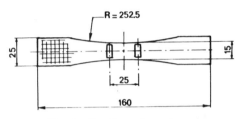

FIG. 2. Tensile specimen shape.

RESULTS AND DISCUSSION

The first material, having a diamine hardened resin, showed excellent agreement with the first model, as shown in Fig. 3. Initially the absorption curve is linear as a function of $\sqrt{t}$ and this is followed by a curving until saturation. Studies with plates of differing thickness showed that the water was uniformly distributed in the saturated material and the behaviour appeared to be perfectly reversible. Figure 4 shows a different behaviour for the dicyandiamide hardened material which shows a double plateau which is not correctly described by the first model. It is necessary to apply the second two-phase model to this material in order to describe the experimental results and it has been possible to calculate the different diffusion parameters.

It was found that the behaviour described by Shen and Springer[6] for carbon fibre reinforced epoxy is also seen in these first two materials in that the diffusivity is a unique function of the temperature and follows a Arrhenius type law (see Fig. 5) and the saturation level is a linear function of relative humidity (see Fig. 6). For the second material it has been possible to show that $\alpha/(\alpha + \beta)$, which corresponds to the proportion of the water concentration in the free phase in the total, is constant and around 0·7.

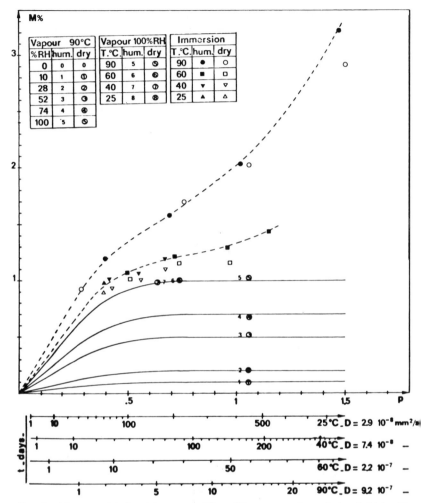

FIG. 3. First material absorption curves. The solid curves are theoretical and obtained by consideration of single free phase diffusion. The experimental points on the curves correspond to the conditions used for the tensile tests.

Knowing the diffusion parameters it was then possible to use the parameter $p = \sqrt{Dt/h}$ in order to represent the curves for the two materials.

It proved impossible to arrive at a stable state of saturation for the third material due to leaching and loss of material which became appreciable above 40 °C, as can be seen in Fig. 7. The principal loss mechanism was due to erosion of the resin at the surface. However, it is to be noted that the

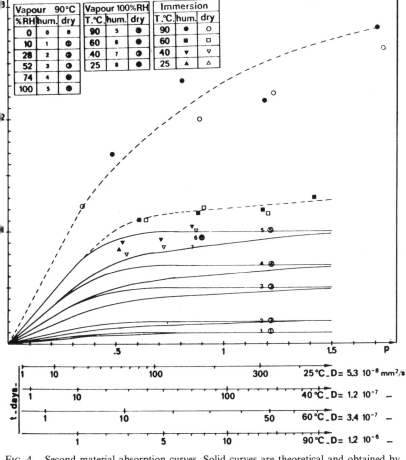

FIG. 4. Second material absorption curves. Solid curves are theoretical and obtained by consideration of two-phase model. The experimental points on the curves correspond to the conditions used for the tensile tests.

curve obtained at 25 °C was similar to that found with the second material and so corresponds to two-phase diffusion; this enables us to again employ the reduced parameter $p = \sqrt{Dt/h}$ in comparing curves.

It is to be noted that the saturation levels and the diffusivities obtained in the study agree well with the results given by Morgan *et al.*[5] for pure resin.

We can see from the above results that in the case of the first material water absorption is accounted for by simple diffusion into the body of the

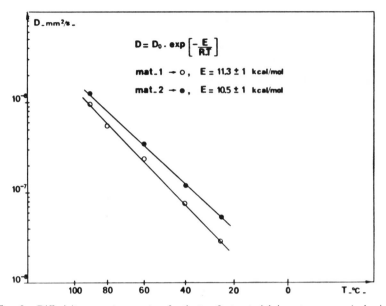

FIG. 5.   Diffusivity versus temperature for the two first materials in water vapour. Arrhenius relationship.

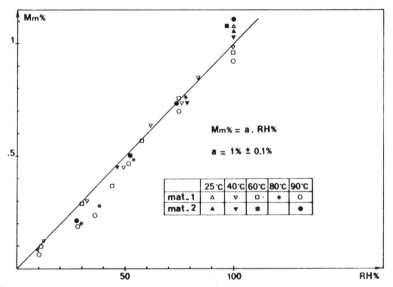

FIG. 6.   Saturation limit versus relative humidity for the two first materials in water vapour.

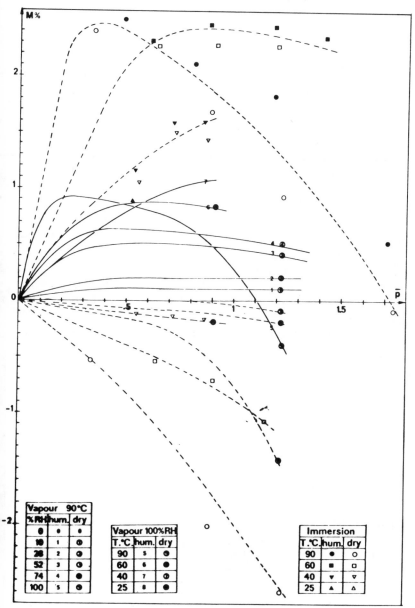

FIG. 7.   Third material absorption curves. The experimental points on the curves correspond to the conditions used for the tensile tests. The negative points show the weight loss due to leaching and were revealed by redrying.

resin where it is loosely attached by hydrogen bonding. This mechanism is also the principal—but not the sole—one in the other two materials considered. In the case of all three materials, however, the strength was seen to fall at values of relative humidity above 60–70% corresponding to a weight gain of 0·6 to 0·7% for the first two materials (see Figs 8, 9 and 10). It seems that at this stage of absorption a concentration limit is reached at which the water concentration in the material is no longer in equilibrium with the surrounding environment.

The result is a swelling and plastification of the resin leading to increased internal stresses. It should be noted that there was not a direct relationship between weight gain and the damage produced. For identical saturation levels ($\simeq 1\%$) it seemed that the strength reduction produced depended on the temperature and also the time of applied load. In addition, the third material shows an increase in breaking strength going from the dry state to the concentration limit (Fig. 10), suggesting a mechanism of stress relaxation due to leaching.

This hypothesis of a limiting concentration agrees well with the results obtained by immersion in that, with all three materials, the observed behaviour did not correspond to the situations covered by the diffusion laws. Above 40 °C the weight gain was continuous and passed the saturation levels found under humid conditions for the first two materials (Figs 3 and

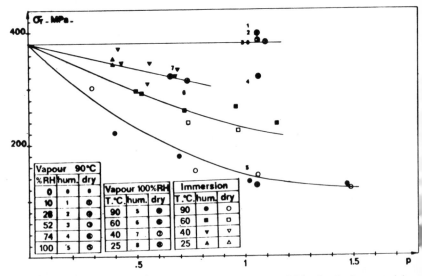

FIG. 8.　Variations of tensile strength under different conditions for the first material.

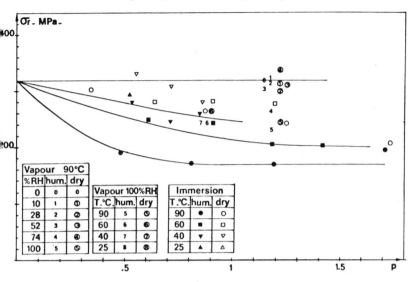

FIG. 9.   Variations of tensile strength under different conditions for the second material.

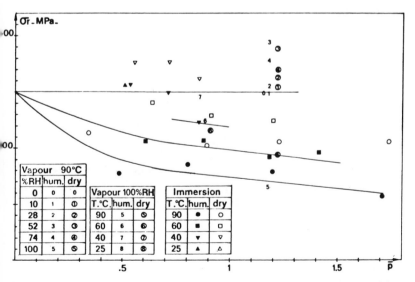

FIG. 10.   Variations of tensile strength under different conditions for the third material.

4). In the case of the third material leaching remained most important (Fig. 7). When large weight gains were detected—for example, of 3 % at 90 °C—this was due to capillary movement of water along the cracks which were produced. In all cases the strength of the material continued to drop, as was found on exposure to humid conditions, as a function of temperature and time and was again seen not to be a simple function of weight gain.

Strength loss produced in the first type of material proved to be irreversible but that produced in the other two materials was partially reversible after drying (see Figs 8, 9 and 10). The complexity of the internal situation of the materials prevents a clear interpretation of the phenomenon; however, it seems that the mechanisms which determine the two-phase absorption in materials 2 and 3 lead to bonding which results in partial recovery when dried.

In all cases the first damage produced seems to be connected with a transition process in the resin. According to Carter and Kibler[12] the observed behaviour due to the presence of water is produced by a lowering of the glass transition temperature and hence greater mobility of the molecular chains. On the other hand, Peyser and Bascom[13] consider that it is more likely that hydrogen bonds are being broken due to the greater mobility of the water molecules. In addition, the presence of glass fibres in the resin complicates the interpretation as they can be attacked by water[14] and failure at the interface allows water to penetrate the material by capillary action.

CONCLUSIONS

It has been shown that a good relationship between the observed water uptake of a glass epoxy composite and the diffusion laws is a necessary but not an intrinsically sufficient condition for no damage to be produced. In the cases of resins hardened with diamine or dicyandiamide a steady saturation level was reached under humid conditions but at greater than 60–70 % RH damage progressed as a function of temperature and time. This behaviour was also seen for the composite hardened with anhydride although surface erosion prevented a quantitative analysis.

The hypothesis that a concentration limit exists agrees with the results obtained by immersion during which damage occurred under all conditions and as a function of temperature and time.

The observed behaviour suggests that a transition process is involved in the resin and that cracks are produced which allow the penetration of liquid water by capillary action. It is only at this latter stage that the possible deterioration of the glass fibres need be considered.

REFERENCES

1. SHEN, C. H. and SPRINGER, G. S. Effects of moisture and temperature on the tensile strength of composite materials, *J. Composite Materials*, I. **11** (January 1977) 2.
2. ISHAI, O. and ARNON, U. Instantaneous effect of internal moisture conditions on strength of glass-fiber-reinforced plastics, *Advanced composite materials—Environmental effects*, ASTM-STP 658 (J. R. Vinson), 1978, pp. 267–76.
3. DEIASI, R. and WHITESIDE, J. B. Effect of moisture on epoxy resins and composites, *Advanced composite materials—Environment effects*, ASTM-STP 658 (J. R. Vinson), 1978, pp. 2–20.
4. SHIRREL, C. D. Diffusion of water vapor in graphite/epoxy composites. *Advanced composite materials—Environment effects*, ASTM-STP 658 (J. R. Vinson), 1978, pp. 21–42.
5. MORGAN, R. J., O'NEAL, J. E. and FANTER, D. L. The effect of moisture on the physical and mechanical integrity of epoxies. *J. Materials Science*, **15** (1970) 751–64.
6. SHEN, C. H. and SPRINGER, G. S. Moisture absorption and desorption of composite materials, *J. Composite Materials*, **10** (January, 1976) 2.
7. HAHN, H. T. and KIM, R. Y. Swelling of composites laminates. *Advanced composite materials—Environmental effects*, ASTM-STP 658 (J. R. Vinson), 1978, pp. 98–120.
8. CRANK, J. *The mathematics of diffusion*. Oxford, Clarendon Press.
9. GURTIN, M. E. and YATOMI, C. On a model for two phase diffusion in composite materials, *J. Composite Materials*, **13** (April, 1979) 126.
10. CARTER, H. G. and KIBLER, K. G. Langmuir-type model for anomalous moisture diffusion in composite resins, *J. Composite Materials*, **12** (April, 1978), 118.
11. LEWIS III, DAVID, Curve fitting techniques and ceramics, *Ceramic Bulletin*, **57** (1978).
12. CARTER, H. G. and KIBLER, K. G. Entropy model for glass transition in wet resins and composites, *J. Composite Materials*, **11** (July, 1977) 265.
13. PEYSER, P. and BASCOM, W. D. The anomalous lowering of the glass transition of an epoxy resin by plasticization with water, *J. Materials Science*, **16** (1981) 75–83.
14. MARTIN, D. M., AKINC, MUFIT and MOO OH, SHIN, Effect of forming and aging atmospheres on E glass strength, *J. American Ceramic Society*, **61** (1978) 308–11.

8

Failure of GRP in Corrosive Environments

P. J. Hogg, D. Hull and M. J. Legg

Department of Metallurgy and Materials Science,
University of Liverpool, PO Box 147, Liverpool L69 3BX, England

ABSTRACT

Stress and strain corrosion of glass fibre–polyester resin materials is characterised by flat fracture surfaces with only small amounts of fibre pull-out. The times to failure and fracture surfaces of hoop wound and $\pm 55°$ helically wound pipe sections tested in diametrical compression are compared with the results of stress corrosion on $\pm 55°$ wound tubes tested under biaxial loading conditions. All the tests were carried out in water or hydrochloric acid. The failure times are strongly dependent on the stresses parallel to the fibres. Transverse cracking, parallel to the fibres, is also important since it affects the flow of acid to the fibres and produces local regions of stress concentration. The results are compared with fracture surface studies on GRP tanks which have failed in service.

INTRODUCTION

GRP is extensively used in the chemical process industry for applications such as pipework, reaction vessels and storage tanks. The conditions are often severe, frequently combining pressure, high temperatures and chemically reactive reagents. In most environments GRP is reasonably inert, especially when the component is not subjected to service loads. Recently, however, it has been shown that even in an aqueous environment in the presence of a sustained load stress corrosion cracking can occur.[1]

106

When the environment is acidic, this cracking can occur very rapidly and may result in catastrophic failure.

The increasing industrial awareness of acidic stress or strain corrosion has led to the use of barrier layers which delay or eliminate cracking of the structural GRP. These layers may be pure resin gel-coats, thermoplastic liners or resin-rich layers reinforced with glass or organic fibres.

The use of barrier layers is expensive and usually serves only to delay the cracking effect. To ensure an adequate service life, generous safety factors have to be used in barrier layer design. Detailed design information is required, therefore, concerning the predicted life of GRP under acidic stress corrosion conditions. Laboratory tests are required to characterise crack initiation and growth, and provide information relevant to the many and varied loading conditions encountered in service.

Most of the work reported in the literature has involved simple test configurations such as three-point bend tests on flat laminates and diametrical compression tests on sections of pipe. All these tests show that the times to failure are strongly dependent on the level of applied tensile loading.[2−5] Hogg and Hull[2] reported that, in unidirectional laminae, cracks nucleate normal to the fibre direction and produce distinctive flat fracture surfaces. This observation emphasises the importance of tensile stresses parallel to the fibre direction. Bailey and Jones[6] have also reported a stress corrosion phenomenon associated with crack propagation parallel to the fibres due to stresses acting transverse to the fibres.

The present work is concerned primarily with the effects of biaxial loading on stress corrosion and is based on internal pressure tests on filament wound pipes. This is particularly relevant to the more complex loading conditions experienced in service conditions and the results are compared with those obtained from diametrical compression tests on pipe sections in the same environmental conditions. The fracture surfaces obtained in both these tests are compared with those found on a vessel which has failed under stress corrosion conditions in an attempt to correlate laboratory test results with service failures in large vessels.

REVIEW OF RING COMPRESSION TEST RESULTS

In this section, which is based on previously published work,[2−4] a brief review is given of the failure processes which occur in ring compression of glass fibre–polyester resin composite materials. In ring compression tests, pipe sections are dammed to contain the corrosive environment and compressed between parallel plates under either constant load (stress corrosion) or constant deflection (strain corrosion) conditions. The process of environmental degradation is monitored continuously by measuring the deflection versus time and load relaxation versus time for stress and strain corrosion respectively. Failure results from crack formation along the bottom of the pipe section in the areas of maximum tensile strain. These cracks eventually result in leakage of the corrosive liquids out of the section.

Failure can be divided into four main stages. After the initial visco-elastic response of the pipe, there is a region of slow degradation during which crack nuclei form and coalesce. This leads to rapid crack propagation and pipe fracture. After leakage of the liquid the fracture processes slow down. The micromechanisms involved are described in reference 2.

To demonstrate the effect of different fibre geometries, loads and corrosive environments, it is necessary to specify some critical stages in the failure process. In this work the times to the onset of rapid crack growth, t_1, and to leakage of the liquid through the pipe wall, t_3, have been used.

The main results obtained from the ring compression tests can be summarised as follows:

(a) Stress and strain corrosion are similar phenomena. Stress corrosion occurs in shorter times than strain corrosion because stress relaxation occurs in the latter test configuration.

(b) In hoop wound pipe sections, the predominant stresses in the material in contact with the corrosive environment are tensile stresses parallel to the fibres $\sigma_\parallel$. The times to failure decrease with increasing $\sigma_\parallel$ as shown in Fig. 1.

(c) In $\pm 55°$ helically wound pipe sections, which experience both longitudinal ($\sigma_\parallel$) and transverse stresses ($\sigma_\perp$), the times to failure show the same form of stress dependence as hoop wound sections but the times are shorter for a given value of $\sigma_\parallel$ (Fig. 1). This is attributed to the formation of transverse cracks (parallel to the fibres) which allow the corrosive environment to penetrate into the material and raise the local value of $\sigma_\parallel$ acting on the underlying fibres.

(d) The times to failure decrease with increasing acid concentration in the range 0·1M to 2·5M and are shorter in mineral acids than in organic acids (Fig. 1).

(e) Both stress and strain corrosion are characterised by flat fracture surfaces with little evidence of fibre pull-out. In hoop wound sections the flat fractures grow normal to the fibres and the

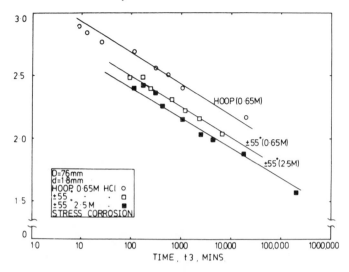

FIG. 1. Effect of longitudinal tensile stress, $\sigma_\parallel$, on the failure times (t_3) of hoop wound and $\pm 55°$ sections of glass fibre–polyester resin composite materials, $V_f = 0.45$.

direction of maximum tensile stress. In $\pm 55°$ laminates fracture is a combination of transverse cracks and 'stress corrosion' cracks. In a given lamina the stress corrosion cracks form at right angles to the fibres and develop a stepped morphology which is associated with the distribution and position of transverse cracks in adjacent laminae.

STRESS CORROSION SERVICE FAILURES

Several GRP plant failures, thought to have been caused by stress corrosion cracking, have been investigated. These structures were typically large tanks for storage of aqueous acids.

In most cases fracture originated at regions of local stress concentration such as the junctions between the tank and connecting pipework. Final failure was catastrophic and consistent with the sudden propagation of a crack. Otherwise, no general weakening of the structure was observed. Some of the failed vessels were found to contain cracks on the inside walls which extended up to 80 % of the wall thickness. These cracks were situated at positions of high local stress.

The fracture surfaces viewed on a macro-scale were reasonably flat throughout. Fairly smooth surfaces were observed at the fracture origin, and the surface became more fibrous in nature away from this area. Scanning electron microscopy of the fracture origin revealed all the features which are characteristic of stress corrosion in the ring compression tests. These features include co-planar fibre–matrix fracture, featureless fibre fracture surfaces and delaminations between poorly bonded layers of fibres. An example is shown in Fig. 2. The fracture surfaces were rougher than in unidirectional laminae presumably due to the many possible fracture paths which are present in the CSM laminates used in tank construction.

FIG. 2. Scanning electron micrograph of the fracture surface at the origin of a crack in a GRP tank used for storing acid after it had failed in service. Tank fabricated from chopped strand mat and woven cloth reinforced polyester.

It is apparent that the planar cracks nucleate at the inside wall of the tank and grow slowly in a plane approximately normal to the pipe wall. Away from the nucleation region the crack surfaces become more fibrous indicating fibre pull-out and more conventional fracture processes. This suggests that as the cracks get longer and grow more quickly the influence of the corrosive environment decreases. Flat fractures appear to be associated with stable crack growth. The transition to fibrous fracture, which is associated with unstable crack growth, is illustrated in Fig. 3.

Some of the service failures occurred in vessels protected by barrier layers, notably gel coats reinforced with organic fibres. Although direct evidence is not available, it is probable that these layers delayed fracture but it is equally clear that they have not prevented it. Examination of the fracture surface markings reveals that in some cases fracture initiated below the barrier layer suggesting that barrier layer damage is not a pre-requisite for crack initiation.

In general the fracture surfaces are consistent with the model proposed for stress corrosion (reference 2). No comparisons of the times to failure of commercial vessels and laboratory test samples have been made because there is insufficient data on service failure history or on laboratory pipe tests using CSM.

FIG. 3. Scanning electron micrograph showing the transition from flat fracture to fibrous fracture. Sample cut from a GRP tank after it had failed in service.

BIAXIAL STRESS CORROSION

Test Procedures

In biaxial stress, where $\sigma_{Hoop} = 2\sigma_{Axial}$ (Mode 2), optimum structural integrity of a filament wound pipe, as defined by the maximum weepage and burst strengths, occurs at helix angles close to $\pm 55°$ for tests in air and water.[7,8] This work has been extended to tests in acidic environments and, to date, has been restricted to pipes wound at the 'so-called' optimum angle (54·7°) using 1M hydrochloric acid. A full description of pipe preparation has been given elsewhere;[7,9] Fibreglass FGRE 20/70 'E' glass roving and Impol T500 polyester resin (Atlac 490-05C) have been used throughout.

The Mode 2 test, described in reference 9, has been modified to provide static fatigue conditions in which a constant pressure is applied for the duration of the test. Certain other refinements were necessary to contain the acid medium over long periods. Three successful types of end closure have been used and are illustrated in Fig. 4. Types (a) and (b) are essentially adaptations of the conventional method,[10] in which the 'O' ring seals have been extended into the pipe away from any possible damage due to clamping. The type (c) end plug is relatively inexpensive and ideal for long-term tests. It relies on the adhesion between the previously roughened pipe bore and a cast epoxy plug to resist the end thrust generated by internal pressure, and has proved successful up to a pressure of 50 bar. The stainless

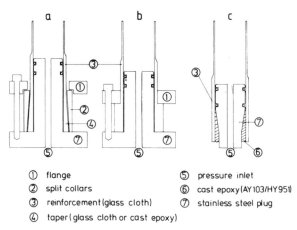

① flange   ⑤ pressure inlet
② split collars   ⑥ cast epoxy (AY 103/HY 951)
③ reinforcement (glass cloth)   ⑦ stainless steel plug
④ taper (glass cloth or cast epoxy)

FIG. 4.   End closures for GRP pipes internally pressurised with acids.

steel plugs are treated with release agent prior to casting; thus it is a simple task to liberate the plugs after testing by cutting off the pipe end and pushing the plugs out. The pipe ends were provided with a resin gel coat and glass surfacing tissue before winding. For long exposures, ethylene propylene rubber rings are used.

A schematic representation of two creep stations capable of subjecting four 50 mm diameter pipes to two different test pressures is shown in Fig. 5. Each station may be isolated enabling other stations on the rig to be used simultaneously without detriment to the other creep tests. Separation of acid and hydraulic fluid is achieved in the single-specimen station by means of the conventional rubber bag (see reference 9) and in the multi-specimen station by PTFE pistons running in honed stainless steel cylinders. The

Station 1 (single specimen)   Station 2 (three specimens)

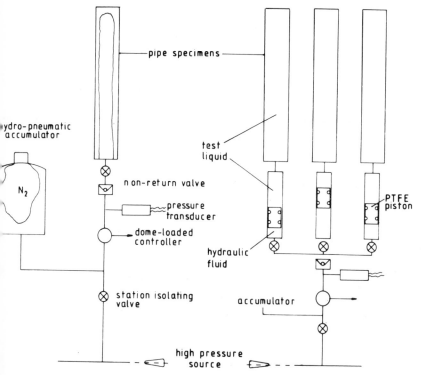

FIG. 5.   Schematic layout of two internal pressure creep stations for GRP pipes.

pistons also serve as shut-down valves so that loss of fluid from one specimen is arrested before it can affect adjacent tests.

Initial weepage in the pipe is detected by a 'rain alarm'. A closely spaced double helix of fuse wire is first wrapped around the pipe; loss of liquid through the pipe wall produces a contact between the wires which is utilised to trigger an audible alarm and stop a timer. Strain measurements have been made using resistance strain gauges bonded to the external surface of the pipe with an epoxy resin.

Results

In short-term tests at 20 °C using a water environment, weepage occurred at $\sigma_H \simeq 105 \, \mathrm{MN \, m^{-2}}$ in the form of fine droplets. Weepage was preceded by creaking and whitening which started at $\sigma_H \simeq 40 \, MN \, m^{-2}$ and was associated with transverse cracking. The amount of transverse cracking increased rapidly with increasing pressure. For $\sigma_H < 105 \, MN \, m^{-2}$ under constant load (i.e. creep) conditions using a water environment delayed weepage occurred. The times to failure in both water and 1M hydrochloric acid are shown in Fig. 6 along with data obtained by Mieras[8] from similar tests on $\pm 50°$ glass fibre–polyester resin pipes. The data in Fig. 6 are plotted in terms of the applied hoop stress σ_H and the initial elastic strains at the start of the test.

At short times the weepage stress in acid was similar to water and weepage occurred as fine droplets. However, for creep tests at lower pressures, the times to weepage were significantly less than in water, as shown in Fig. 6. There was a corresponding change in the weepage process. For hoop stresses in the range 50 to 90 MN m^{-2}, the first stage of weepage was droplet formation but this changed to fine jets which developed into strong jets associated with well defined cracks in the pipe wall. Eventually the rubber bag expanded through the crack in the wall and burst. Below $\sigma_H = 50 \, MN \, m^{-2}$ jet weepage was not observed before exhaustion of the acid through droplet weepage.

The fracture surface appearance of tubes failed after loading in the range $\sigma_H = 50$ to 90 MN m^{-2} is completely different from failure in the absence of a corrosive environment. A general view of the fracture zone is shown in Fig. 7. The line crack AB has formed parallel to one set of fibres in the pipe wall. At each end of the line crack there are regions showing extensive microcracking and delamination. These zones are typical of normal pipe deformation in air and formed when the rubber gag expanded through the pipe wall. They are not directly associated with stress corrosion cracking.

The section of the pipe in the region AB was removed to examine the fracture surface directly. The surface was flat (Fig. 8) and similar to other stress corrosion failures reported elsewhere.[2,4] The four-layer construction of the pipe is clearly visible. A detailed examination of the fracture surface showed that failure started at the inside surface of the pipe due to transverse cracking of the inner layer. The cracks propagated in a plane parallel to the fibres at or near to the fibre–resin interface. Some fibre fracture occurred in this layer owing to the presence of out of plane fibres (see position X, Fig 9(a)). Fracture of the second layer occurred in the same plane as the first by stress corrosion cracking normal to the fibre direction (Fig. 9). Fracture surface markings show that stress corrosion cracks in the second layer nucleate at the tip of the transverse crack in the first layer (Fig. 9(b)). The fracture surface of the second layer is stepped. The steps on the surface arise

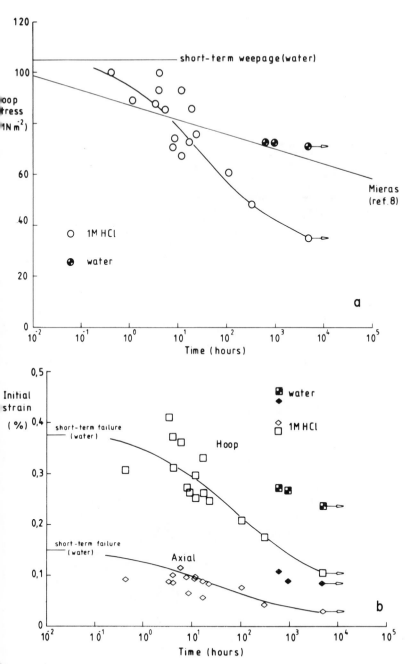

FIG. 6.   Degradation effects of an acid environment on filament wound GRP pipes tested in biaxial stress (Mode 2). (a) Time to weepage at constant hoop stress. (b) Time to failure against initial hoop and axial strains.

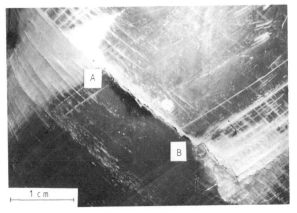

FIG. 7. Optical micrograph of a stress corrosion crack in a ± 55° pipe failed in Mode 2 using 1M HCl.

because in a ± 55° wound pipe the plane of the transverse cracks in the first layer is not normal to the fibre direction in the second layer (see reference 4 for details). The fracture markings at the end of the stress corrosion crack in the second layer extend towards the third layer which failed by transverse cracking (Fig. 9(c)). Thus, the transverse cracks appear to have been nucleated by the stress corrosion cracks in the second layer. The fourth layer has failed by stress corrosion cracking, coplanar with the third layer

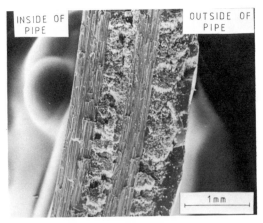

FIG. 8. Scanning electron micrograph of fracture surface of crack, along line AB in Fig. 7.

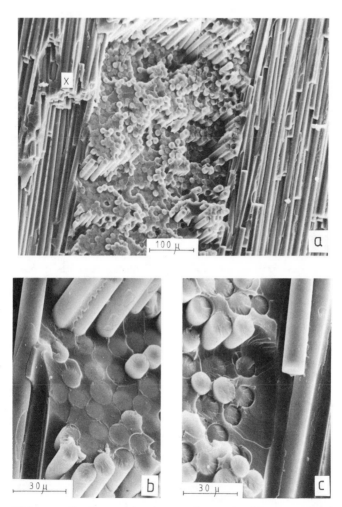

FIG. 9. Higher magnification scanning electron micrographs of fracture surface in Fig. 8 showing: (a) transverse crack in first layer (left-hand side), stress corrosion crack in second layer and transverse crack in third layer; (b) nucleation site of stress corrosion crack in second layer at tip of a transverse crack in first layer; (c) fracture surface markings from stress corrosion crack extending towards a transverse crack in the third layer.

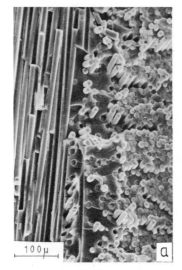 

FIG. 10. Scanning electron micrographs showing: (a) transition from transverse crack in third layer to stress corrosion crack in fourth layer; (b) nucleation of stress corrosion crack in fourth layer at tip of transverse crack in third layer.

(Fig. 10(a)). Once again the stress corrosion cracks have nucleated at the tip of the preceding transverse crack (Fig. 10(b)). The fracture process is illustrated schematically in Fig. 11.

Discussion

The relation between the times to failure obtained from internal pressure and ring compression tests is shown in Fig. 12. All the data have been expressed in terms of the effect of $\sigma_{\parallel}$ on failure times. The results of Mieras[8] from Mode 2 creep tests with water are not directly comparable with the present data because for a winding angle of $\pm 50°$ the value of $\sigma_{\perp}$ is higher than for $\pm 55°$ for a given hoop stress or $\sigma_{\parallel}$. This means that for a given hoop stress the creep rate will be higher and the time to weepage less in $\pm 50°$ pipe than in $\pm 55°$ pipe. The relatively small decay in strength shown in Fig. 12 at high hoop stresses ($\sigma_H > 90 \text{ MN m}^{-2}$) in Mode 2 tests indicates that under these conditions the life of the pipe in both acid and water is limited by transverse crack propagation which can lead to weepage without fibre fracture.[11] This is further supported by the absence of jet weepage in high hoop stress tests.

At lower stresses, stress corrosion cracking strongly affects the weepage strength as shown by the significant reduction in failure times compared

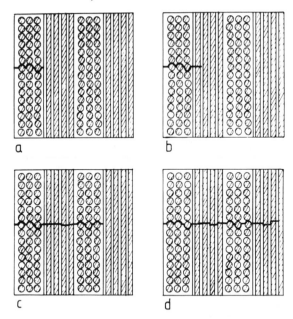

FIG. 11. Schematic representation of radial development of a stress corrosion crack, initiating from a transverse crack in innermost lamina of a ($\pm 55°$) pipe subjected to an acidic environment in Mode 2. (a) Transverse cracking in first layer; (b) stress corrosion cracks nucleated in second layer; (c) growth of stress corrosion cracks in second layer producing transverse cracking in third layer; (d) stress corrosion cracks which have nucleated and grown in fourth layer.

with a water environment. At hoop stresses below $\simeq 40\,\text{MN}\,\text{m}^{-2}$, no damage was obvious after initial loading of the pipe but quite severe damage was evident after several hundred hours. It is not clear whether or not this damage is due to corrosion-enhanced transverse cracking.[6] Tests are in hand to determine the effect of internal pressure under conditions where, according to linear elastic analysis,[10] $\sigma_\perp$ is approximately zero. In Mode 3 tests[9] this occurs in hoop wound pipes and in pipes wound at $\pm 57°$.

In the diametrical ring compression tests on hoop and $\pm 55°$ pipe sections failure is due to crack propagation by fibre fracture. Much higher values of $\sigma_\parallel$ are possible than in Mode 2 tests because transverse cracking does not dominate the failure process. In other words failure is controlled by $\sigma_\parallel$ and the slope of the failure line in Fig. 12 is determined solely by the stress or strain corrosion effect. However, the relative positions of the hoop and $\pm 55°$ lines is affected by transverse cracking. In $\pm 55°$ wound sections transverse cracking is extensive in ring compression[4] whereas in hoop

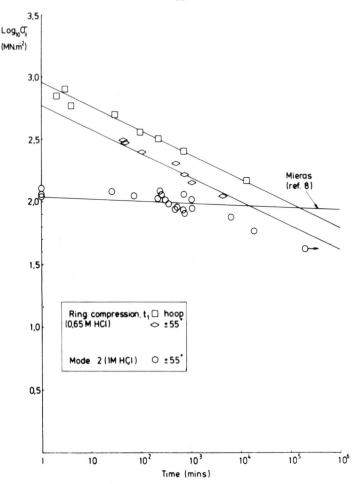

FIG. 12. Comparison of time to failure t_1 in diametrical ring compression with time to weepage in Mode 2 on basis of longitudinal stress, $\sigma_{\parallel}$, showing transition from conventional weepage behaviour to corrosion-induced failure in conditions of biaxial stress.

sections it is minimal.[2] The effect of transverse cracking is also minimised by the non-uniform stress state which exists through the wall of the pipe when it is subjected to diametrical compression. This tends to restrict cracking to the inner layers where the tensile stresses are higher, at those positions in contact with the corrosive environment.

In the long-term Mode 2 tests weepage with acid occurs before weepage with water. The weepage times depart from the water results, which are supported by the results of Mieras,[8] and follow a line parallel to the ring compression data. This similarity in stress dependence at longer times reflects the emerging dominance of a failure criterion based on $\sigma_\parallel$. The displacement of the line to shorter times compared with the ring compression data on $\pm 55°$ pipe sections is probably due to the effect of transverse cracks. Thus fractographic studies (see Fig. 10) show that the stress corrosion cracks initiated in the second layer at the tip of a transverse crack in the first layer.

These results show that stress corrosion of composite materials is determined largely by tensile stresses parallel to the fibres and that the processes which occur are affected by transverse cracking. The relative importance of these processes will depend on the fibre arrangement and loading conditions. Transverse cracking occurs at low stresses in the absence of any corrosive environment but there is evidence from long-term Mode 2 tests at low hoop stresses that the transverse cracking threshold is also affected by the environment.

In CSM laminates the relative magnitude of $\sigma_\parallel$ and $\sigma_\perp$ stresses will not be the same as in aligned laminates. This will affect the fracture path and fracture morphology. However, the examination of service failures from tanks made from CSM shows that flat fracture occurs and all the evidence indicates that stress corrosion failure is involved.

ACKNOWLEDGEMENTS

The authors are grateful to the Science Research Council for financial support, and to their colleagues, without whose help this work could not have been done.

REFERENCES

1. AVESTON, J., KELLY, A. and SILLWOOD, J. M., Long term strength of glass reinforced plastics in wet environments. In: *Advances in composite materials*, I.C.C.M. 3, Bunsell, A. R. *et al.* (eds), Oxford, Pergamon Press, 1980, pp. 556–68.
2. HOGG, P. J. and HULL, D., Micromechanisms of crack growth in composite materials under corrosive environments, *Metal Science*, **14** (1980) 441–9.
3. HOGG, P. J., HULL, D. and SPENCER, B., Stress and strain corrosion of glass reinforced plastics. To be published in *Composites* (1981).

4. HULL, D. and HOGG, P. J., Nucleation and propagation of cracks during strain corrosion of GRP. In: *Advances in composite materials*, I.C.C.M. 3, Bunsell, A. R. *et al.* (eds), Oxford, Pergamon Press, 1980, pp. 543–55.

5. ROBERTS, R. C., Design stress for glass fibre reinforced polyester and long term environmental stress failure mechanisms, BPF Reinforced Plastics Congress, Paper No. 19, Brighton, 1978.

6. BAILEY, J. E. and JONES, F., Environmental stress—corrosion edge-cracking of glass reinforced polyesters. In: *Advances in composite materials*, I.C.C.M. 3, Bunsell, A. R. *et al.* (eds), Oxford, Pergamon Press, 1980, pp. 514–28.

7. SPENCER, B. and HULL, D., Effect of winding angle on the failure of filament wound pipe, *Composites*, **9** (1978) 263–71.

8. MIERAS, H. J. M. A., Irreversible creep of filament wound glass reinforced resin pipes, *Plastics and Polymers*, **41** (1973) 84–9.

9. HULL, D., LEGG, M. J. and SPENCER, B., Failure of glass/polyester filament wound pipe, *Composites*, **9** (1978) 17–24.

10. LEGG, M. J., Ph.D. thesis, Liverpool University, 1980.

11. JONES, M. L. C. and HULL, D., Microscopy of failure mechanisms in filament wound pipe, *J. Mat. Sci.*, **14** (1979) 165–74.

9

Large Deflection Analysis of Bimodular Cross-Ply Strips

G. J. TURVEY

*Department of Engineering, University of Lancaster,
Bailrigg, Lancaster LA1 4YR, England*

ABSTRACT

The behaviour of cross-ply laminated bimodular strips in the elastic large deflection regime is studied. Deflections, etc. have been computed for uniformly loaded strips with opposite edges either simply supported or clamped, using the Dynamic Relaxation (DR) method. The bimodular strip results are compared with corresponding results for strips made of conventional *composite material, i.e. composite material with elastic constants identical with the bimodular tension values. It is shown that for strips made of highly bimodular composite material the* conventional *composite strip results provide a good approximation (except at very low pressures) to the bimodular strip results when the strip edges are simply supported; however, when they are clamped the agreement between the* conventional *and bimodular values is usually poor.*

NOTATION

a	strip width
A_{11}, A_{12}	in-plane stiffnesses
B_{11}, B_{12}	coupling stiffnesses
D_{11}, D_{12}	flexural stiffnesses
e_x	strain in the x-direction
E_L, E_T	longitudinal and transverse elastic moduli
h_0	strip thickness

k_x curvature in the x-direction

M_x, M_y stress couples

$\bar{M}_x$ $(=E_T^{t-1}h_0^{-4}a^2M_x)$ dimensionless stress couple

N_L number of laminae in the strip

N_x, N_y stress resultants

$\bar{N}_x$ $(=E_T^{t-1}h_0^{-3}a^2N_x)$ dimensionless stress resultant

q lateral pressure

$\bar{q}$ $(=E_T^{t-1}h_0^{-4}a^4q)$ dimensionless lateral pressure

$\bar{Q}_{ij}$ $(i, j = 1, 2)$ reduced stiffnesses

u displacement in the x-direction

w strip deflection

$\bar{w}$ $(=h_0^{-1}w)$ dimensionless strip deflection

x, y, z Cartesian strip co-ordinates

z_k, z_{k-1} lamina upper and lower surfaces

v_{LT}, v_{TL} Poisson's ratios

Superscripts

c   compressive value (for elastic moduli); value at strip centre

e   value at strip edge

t   tensile value

1. INTRODUCTION

The elastic compressive and tensile moduli of several unidirectional, fibre-reinforced, composite material systems are known to differ significantly in magnitude. Composite materials exhibiting this phenomenon are commonly referred to as bimodular composites. The analysis of structures fabricated from these materials is usually more difficult than that of *conventional* composites, i.e. composites fabricated from laminae possessing equal valued tensile and compressive moduli, since it is not known *a priori* which parts of the structure are in tension and which are in compression. Therefore, the analysis must be accomplished iteratively rather than directly.

At the present time the theory of bimodular materials is still under development. Several theories have been proposed[1-3] and a number of exact solutions have been obtained for some simple plate and strip problems.[4-7] In addition, Reddy[8] has used the finite element method in conjunction with the bimodular theory proposed by Bert[3] to obtain a number of approximate solutions to plate problems.

All of the plate and strip solutions referred to above relate to small deflection behaviour. Up to the present time, the only studies of large deflection bimodular behaviour appear to be due to Kamiya,[9,10] who considered only isotropic bimodular materials. To the best of the author's knowledge no large deflection analyses of plates, in which the individual laminae exhibit orthotropic properties, have yet been reported. The present paper seeks to explore—if only to a limited degree—the large deflection response of orthotropic bimodular structures. The particular problem selected for study is the uniformly loaded, cross-ply laminated, bimodular strip in a state of cylindrical bending. The principal reason for selecting this problem is that it is one-dimensional and, hence, analytical and computational effort may be minimised.

A finite-difference implementation of the Dynamic Relaxation (DR) method is used to solve the strip large deflection equations, since the method is not only relatively simple to program but also has a direct physical interpretation.

The present study concentrates on cross-ply strips built up from two or three, equal thickness, unidirectional laminae. Furthermore, only uniform loading is considered and the strip edges are assumed to be either simply supported or clamped with full in-plane restraint. Numerical results are obtained for deflections, etc. These are presented in dimensionless form and compared with similar, *conventional* composite strip results (tension modulus solution) in order to demonstrate the influence of bimodular behaviour on the large deflection response.

2. STRIP GEOMETRY AND LAMINA PROPERTIES

The strip geometry and the positive co-ordinate system are shown in Fig. 1. Each strip is assumed to be fabricated from a number of equal thickness, unidirectional, bimodular orthotropic laminae. These are stacked in a cross-ply sequence in the z co-ordinate direction, i.e. the fibre-directions of successive laminae are alternately parallel to the x and y co-ordinate directions.

The type of unidirectional lamina considered here does not correspond to any specific material system, but its properties have been chosen as being representative of a material which is strongly bimodular in character. The elastic constants of this material are listed in Table 1.

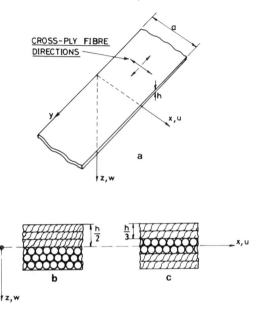

FIG. 1.   Strip and lamina details. (a) Strip geometry and positive co-ordinate system. (b) Part-section through a two-layer, cross-ply strip. (c) Part-section through a three-layer, cross-ply strip.

TABLE 1
Elastic constant ratios of a highly bimodular, unidirectional lamina

E_L^t/E_T^t	E_L^c/E_T^t	E_T^c/E_T^t	v_{LT}^t	v_{LT}^c
2·00	1·00	0·50	0·40	0·20

3.   BIMODULAR STRIP STIFFNESSES

It is possible to classify laminates fabricated from *conventional* composite materials according to their through-thickness elastic symmetry. This type of classification system breaks down for bimodular composite materials. The reason for this is that the position of the neutral surface rather than the strip mid-plane is of relevance in defining the elastic properties of each lamina, and as the strip mid-plane and the neutral surface are not generally coincident no through-thickness elastic symmetry exists. Thus, for a given type of lay-up, e.g. cross-ply, fewer stiffnesses will be zero for bimodular materials than for *conventional* composite materials. The stiffnesses required for the analysis and the expressions for their evaluation are given below:

$$A_{11} = \sum_{1}^{n} \bar{Q}_{11}(z_k - z_{k-1})$$

$$A_{12} = \sum_{1}^{n} \bar{Q}_{12}(z_k - z_{k-1})$$

$$B_{11} = \frac{1}{2} \sum_{1}^{n} \bar{Q}_{11}(z_k^2 - z_{k-1}^2)$$

$$B_{12} = \frac{1}{2} \sum_{1}^{n} \bar{Q}_{12}(z_k^2 - z_{k-1}^2) \tag{1}$$

$$D_{11} = \frac{1}{3} \sum_{1}^{n} \bar{Q}_{11}(z_k^3 - z_{k-1}^3)$$

$$D_{12} = \frac{1}{3} \sum_{1}^{n} \bar{Q}_{12}(z_k^3 - z_{k-1}^3)$$

in which

$$n = \begin{cases} 2 \text{ or } 3 \text{ for two-layer strips} \\ 3 \text{ or } 4 \text{ for three-layer strips} \end{cases} \text{ according to}$$

whether the neutral surface lies without or within the strip thickness, and the $\bar{Q}_{11}$, etc. terms are defined as follows:

(1)   Lamina fibre-direction parallel to x-axis

$$\bar{Q}_{11} = \begin{cases} E_L^t (1 - v_{LT}^t v_{TL}^t)^{-1} \text{ when the lamina is in tension.} \\ E_L^c (1 - v_{LT}^c v_{TL}^c)^{-1} \text{ when the lamina is in compression.} \end{cases}$$

$$\bar{Q}_{12} = \begin{cases} v_{LT}^t E_T^t (1 - v_{LT}^t v_{TL}^t)^{-1} \text{ when the lamina is in tension.} \\ v_{LT}^c E_T^c (1 - v_{LT}^c v_{TL}^c)^{-1} \text{ when the lamina is in compression.} \end{cases}$$

(2) Lamina fibre-direction parallel to y-axis

$$\bar{Q}_{11} = \begin{cases} E_{\mathrm{T}}^{\mathrm{t}}(1 - v_{\mathrm{LT}}^{\mathrm{t}} v_{\mathrm{TL}}^{\mathrm{t}})^{-1} & \text{when the lamina is in tension.} \\ E_{\mathrm{T}}^{\mathrm{c}}(1 - v_{\mathrm{LT}}^{\mathrm{c}} v_{\mathrm{TL}}^{\mathrm{c}})^{-1} & \text{when the lamina is in compression.} \end{cases}$$

N.B. The $\bar{Q}_{12}$ terms are identical with those given under (1) above.

4. BASIC EQUATIONS OF THE ANALYSIS

The basic equations of structural analysis consist of: equilibrium, compatibility and constitutive equations plus an appropriate set of boundary conditions. These equations, as they apply to bimodular strips, are set out below:

4.1. Strip Equilibrium Equations

The following two ordinary differential equations govern the large deflection equilibrium of a strip in cylindrical bending:

$$N_x^{\cdot} = 0$$
$$M_x^{\cdot\cdot} + N_x w^{\cdot\cdot} + q = 0 \tag{2}$$

4.2. Compatibility Equations

The strip compatibility equations are as follows:

$$e_x = u^{\cdot} + \tfrac{1}{2} w^{\cdot 2}$$
$$k_x = - w^{\cdot\cdot} \tag{3}$$

in which the geometric nonlinearity arises from the second term in the first of eqns (2).

4.3. Constitutive Equations

The constitutive equations for a bimodular cross-ply strip may be expressed as:

$$\begin{aligned} N_x &= A_{11}e_x + B_{11}k_x \\ N_y &= A_{12}e_x + B_{12}k_x \\ M_x &= B_{11}e_x + D_{11}k_x \\ M_y &= B_{12}e_x + D_{12}k_x \end{aligned} \tag{4}$$

N.B. For *conventional* cross-ply strips the B_{12}-terms of eqns (4) are zero.

4.4. Boundary Conditions

Two types of symmetric, flexural, strip edge boundary conditions are considered together with a single in-plane edge condition. These conditions are as follows:

(1)   Simply supported edges $(x = 0, a)$

$$u = w = M_x = 0 \tag{5a}$$

(2)   Clamped edges $(x = 0, a)$

$$u = w = w^{\cdot} \tag{5b}$$

5.   SOME REMARKS ON THE DR SOLUTION PROCEDURE

The DR method has been adopted for the solution of eqns (2)–(5), since it is known to be an effective method for the solution of highly nonlinear structural problems. In the present application, the method has been implemented as an iterative finite-difference procedure in which the independent variables have been specified at the nodes of two uniform interlacing meshes.

It was anticipated that the indeterminate nature of the bimodular strip stiffnesses might lead to instability of the DR iterative procedure. In practice, no instability problems arose, though there was an increase in the number of iterations required to achieve solution convergence compared with that required for a similar strip made of *conventional* composite material.

6.   DISCUSSION OF THE COMPUTED BIMODULAR STRIP RESULTS

Alternative large deflection results of bimodular composite strips are not known to the author and hence the program could only be partially verified against existing small deflection results. A typical example of such a comparison is shown in Table 2. The results for different mesh sizes shown in this table demonstrate that reasonable accuracy may be achieved using a 16-mesh analysis. This mesh size was used to obtain all of the bimodular strip results of this study.

The computer program has been used to derive results for a bimodular material with the elastic properties given in Table 1. Only a few of the results

TABLE 2
Exact and DR small deflection results for uniformly loaded ($\bar{q} = 1 \cdot 0$), highly bimodular, cross-ply strips with opposite edges clamped

Mesh intervals	N_L	$\bar{w}_c (\times 10)$	$\bar{M}_x^c (\times 10)$	$\bar{M}_x^e (\times 10)$
10	3	0·241 45	0·425 00	−0·825 00
12	3	0·235 98	0·422 44	−0·827 54
16	3	0·230 55	0·419 91	−0·830 08
(∞)	(3)	(0·223 56)	(0·416 67)	(−0·833 33)

Note: The bracketed values are exact.

obtained will be presented and briefly discussed (it is hoped that a more detailed study will be the subject of a future paper). A dimensionless graphical form has been adopted as a convenient mode for the presentation of the results (see Figs 2–5).

The central deflections of two- and three-layer, simply supported and clamped strips are shown in Fig. 2 as a function of the lateral pressure. On

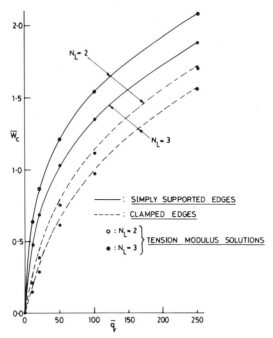

FIG. 2. Strip-centre deflection—lateral pressure curves for simply supported and clamped, two- and three-layer, cross-ply, bimodular strips.

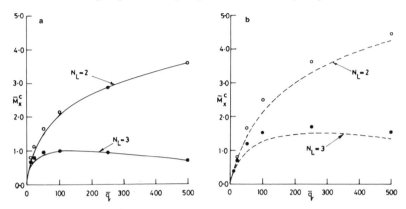

FIG. 3.   Central stress couple—lateral pressure curves for two- and three-layer, cross-ply, bimodular strips. Tension modulus solutions: $\bigcirc$, $N_L = 2$; $\bullet$, $N_L = 3$. (a) $\bar{M}_x^c$ versus $\bar{q}$ (simply supported edges). (b) $\bar{M}_x^c$ versus $\bar{q}$ (clamped edges).

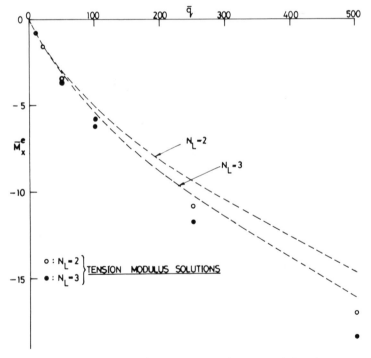

FIG. 4.   Edge stress couple—lateral pressure curves for two- and three-layer, cross-ply, bimodular strips with opposite edges clamped.

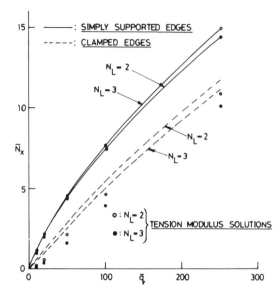

FIG. 5. Stress resultant—lateral pressure curves for simply supported and clamped, two- and three-layer, cross-ply, bimodular strips.

the same figure selected tension modulus strip results are also shown. Throughout the pressure range considered, it is evident that for both two- and three-layer, simply supported strips the bimodular and tension modulus strip centre deflections do not differ significantly. However, the difference between the deflections is significant when the strip edges are clamped, though at higher lateral pressures they tend to converge to the same value.

The stress couple, $\bar{M}_x^c$, at the strip centre is shown in Figs 3(a) and 3(b) for simply supported and clamped strips respectively. From Fig. 3(a) it is evident that, at low pressures, the bimodular values are rather smaller than the tension modulus values. Furthermore, the difference between the two stress couple values is greater in three-layer than in two-layer strips. At higher lateral pressures ($\bar{q} \geq 250$) the bimodular and tension modulus stress couples are virtually indistinguishable. For clamped conditions along the strip edges it is apparent from Fig. 3(b) that the tension modulus values also exceed the bimodular values, but by a rather greater amount, i.e. changing the support condition from simply supported to clamped accentuates the difference between the two values. In contrast to the situation in simply supported strips, the bimodular and tension modulus stress couples at the centre of clamped strips differ significantly even at high lateral pressures.

The edge stress couple, $\bar{M}_x^e$, is shown plotted against lateral pressure in Fig. 4 for the clamped strip (this quantity is, of course, zero in simply supported strips). Again, the bimodular and tension modulus values differ significantly throughout the lateral pressure range. It is of interest to observe that the magnitude of the edge stress couple is very much greater than the corresponding stress couple at the strip centre (cf. Figs 4 and 3(b)).

The final set of results presented are for the stress resultant, $\bar{N}_x$, which, of course, does not vary over the width of the strip. It is plotted as a function of the lateral pressure for two- and three-layer simply supported and clamped strips in Fig. 5. From the figure it is immediately apparent that when the strip edges are simply supported the bimodular and tension modulus values for $\bar{N}_x$ are almost identical, but that when the edges are clamped the tension modulus values are substantially less than the bimodular values.

7. CONCLUDING REMARKS

An elastic large deflection analysis of simply supported and clamped bimodular strips subjected to a uniform lateral pressure has been carried out using a finite-difference implementation of the DR method. Large deflection results have been computed for strips fabricated from unidirectional laminae possessing highly bimodular elastic properties. These results have been compared with similar results for strips fabricated from *conventional* composite material, the elastic properties of which are identical with the tensile moduli of the bimodular composite material. It has been demonstrated that the tension modulus solution often provides a good approximation to the bimodular solution when the strip edges are simply supported, but that the approximation is often poor when the edges are clamped.

The results presented lend some support to the contention that in the large deflection regime the influence of bimodular material behaviour is, from the standpoint of stiffness, rather less important than might be inferred from a knowledge of only small deflection behaviour; though, of course, from the standpoint of strength such a contention remains to be confirmed.

8. ACKNOWLEDGEMENTS

The author wishes to record his gratitude to the Department of Engineering for providing computing facilities, and to his father, Mr George Turvey, for preparing the tracings of the figures.

9. REFERENCES

1. AMBARTSUMYAN, S. A. and KHACHATRYAN, A. A. Basic equations in the theory of elasticity for materials with different stiffness in tension and compression, *Inzh. Zhur. MTT.*, **2** (1966) 44–53 (in Russian).
2. JONES, R. M. Stress–strain relations for materials with different moduli in tension and compression, *AIAA J.*, **15** (1977) 16–23.
3. BERT, C. W. *Mathematical modeling and micromechanics of fiber-reinforced bimodulus composite materials*, Report OU-AMNE-79-7, University of Oklahoma, 1979.
4. JONES, R. M. and MORGAN, H. S. Bending and extension of cross-ply laminates with different moduli in tension and compression, *Computers & Structures*, **11** (1980) 181–90.
5. BERT, C. W. *Classical analyses of laminated bimodulus composite-material plates*, Report OU-AMNE-79-10A, University of Oklahoma, 1979.
6. REDDY, J. N. and BERT, C. W. *Analysis of plates constructed of fibre-reinforced bimodulus materials*, Report OU-AMNE-79-8, University of Oklahoma, 1979.
7. BERT, C. W., REDDY, J. N., REDDY, V. S. and CHAO, W. C. *Analysis of thick rectangular plates laminated of bimodulus composite materials*, Report OU-AMNE-80-2, University of Oklahoma, 1980.
8. REDDY, J. N. and CHAO, W. C. Finite-element analysis of laminated bimodulus composite-material plates, *Computers & Structures*, **12** (1980) 245–51.
9. KAMIYA, N. Large deflection of a different modulus circular plate, *ASME Trans., J. Engng. Mat. Tech.*, **97** (1975) 52–6.
10. KAMIYA, N. An energy method applied to large elastic deflections of a thin plate of bimodulus material, *J. Struct. Mech.*, **3** (1975) 317–29.

10

Analysis of Thermally Stressed Variable Thickness Composite Discs—A CAD Technique

D. G. GORMAN

College of Engineering and Science,
National Institute for Higher Education, Limerick, Ireland

and

J. P. HUISSOON

Engineering Science School, Trinity College Dublin,
Dublin, Ireland

ABSTRACT

This chapter details a finite element technique for determining the state of in-plane stressing in a variable thickness composite disc when subjected to thermal loading and boundary restraints. By utilising two degrees of freedom annular finite elements of linearly varying thickness form, the numerical convergence rate is such that minimum computer storage is required, hence lending itself as the basis of a computer-aided design optimisation package. Laminated discs of either isotropic or polar orthotropic materials can be examined.

INTRODUCTION

Due to the wide range of engineering applications, the in-plane stress analysis of variable thickness discs has been the subject of much research this century. With the advent of high speed digital computers and advanced numerical techniques, Computer Aid Design (CAD) techniques with respect to this study have been extensively used in both industry and research establishments; for example, the technique developed by Seireg and Surana[1] was used to compute the state of stressing over a turbine disc subjected to centrifugal loading only and consequently to establish an

135

optimum disc shape (axial thickness profile) which would ensure that the combination of the stress at any point of the disc would not exceed the yield condition. In the course of research into the free transverse vibration of thermally stressed discs, Gorman[2] and Kennedy and Gorman[3] extended the above technique to include the additional effect of thermal loading. In both these studies, however, only 'solid', isotropic specimens were considered, by means of subdividing the structure into a series of uniform annular rings and applying the Lamé theory to each ring, ensuring equilibrium and compatibility between adjoining rings. Using this technique for isotropic discs, an excellent degree of accuracy was obtained and, more important from the point of view of being a CAD technique, convergence was found to be rapid.

In the general analysis of composite discs, however, and in particular the analyses of polar orthotropic laminated discs, utilisation of the Lamé expressions, which greatly contribute to the fast convergence in the isotropic case, is no longer viable. In such cases a more generalised approach is required and in order to maintain a high rate of convergence it is necessary to pursue the analysis with a series of elements which render a closer approximation between the modelled and the actual structures than in the isotropic case, where elements of uniform thickness were used.

The aim of this chapter is therefore to demonstrate the technique whereby the in-plane stress distribution can be computed for a variable thickness laminated disc composed of either isotropic or polar orthotropic materials when subjected to any proposed temperature distribution. Consequently, for any set of limiting parameters (i.e. maximum yield stress of the materials), an optimum disc configuration may be computed. Additionally, since it is now well established that thermally induced in-plane stressing can dramatically change the vibratory and stability characteristics of continuous systems,[4–6] in order to predict these changes it is necessary to establish the form of the in-plane stressing for any specified temperature distribution acting over the surface of the system.

ANALYSIS

(a) Model Structure

As shown in Figs 1(a) and (b), a variable thickness laminated disc may be modelled by a series of annular finite elements of uniform and linearly varying thickness form. Furthermore, for the case considered the two 'outer' laminates (material A) may be combined to form one structure with a common node at each extremity.

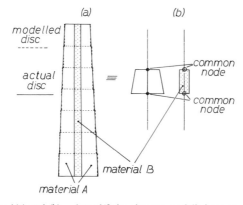

FIG. 1(a) and (b). Actual/finite element modelled structures.

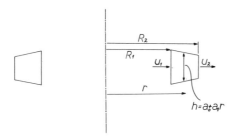

FIG. 2. Annular finite element.

(b) Finite Element Analysis

Consider the annular finite elements possessing two degrees of freedom (u_1 and u_2) as shown in Fig. 2. Assuming a radial displacement function of the form:

$$u = \left\{ 1 + \frac{R_1}{R_{21}} - \frac{r}{R_{21}}, \; \frac{r}{R_{21}} - \frac{R_1}{R_{21}} \right\} \left\{ \begin{array}{c} u_1 \\ u_2 \end{array} \right\} \tag{1}$$

and defining:

$$\varepsilon_r = \frac{\partial u}{\partial r}$$

and:

$$\varepsilon_\theta = \frac{u}{r}$$

we can write:

$$\{\varepsilon\} = [f]\{U\} \tag{2}$$

where:

$$[f] = \left[\begin{array}{cc} -1/R_{21} & \vdots & 1/R_{21} \\ \hline 1/r + R_1/rR_{21} - 1/R_{21} & \vdots & 1/R_{21} - R_1/rR_{21} \end{array}\right] \tag{3}$$

where:

$$\{\varepsilon\} = \left|\{\varepsilon_r \varepsilon_\theta\}^T \quad \text{and} \quad \{U\} = \{u_1 u_2\}^T\right.$$

Furthermore, from the constitutive relationships we can write:

$$\{\sigma\} = [D]\{\varepsilon\} \tag{4}$$

where:

$$\{\sigma\} = \{\sigma_r \sigma_\theta\}^T$$

and:

$$[D] = \frac{E}{(1-\mu^2)} \begin{bmatrix} 1 & \mu \\ \mu & 1 \end{bmatrix} \quad \text{for isotropic analysis}$$

$$[D] = \frac{E_r}{(1-n\mu_{\theta r}^2)} \begin{bmatrix} 1 & \mu_{\theta r} \\ \mu_{\theta r} & 1/n \end{bmatrix} \quad \text{for polar orthotropic analysis}$$

Hence the stiffness matrix, as defined by the relationship:

$$\{F\} = [k]\{U\} = \{F_1 F_2\}^T$$

We can obtain the general expression:

$$[k] = \int_{\text{VOL}} [f]^T [D][f] \, dv \tag{5}$$

In the case of elements exhibiting varying thickness form, the above equation reduces to:

$$[k] = \int_0^{2\pi} \int_{R_1}^{R_2} [f]^T [D][f](a_0 r + a_1 r^2) \, dr \, d\theta \tag{6}$$

Hence solving, we obtain each component of the $[k]$ matrix as listed in the Appendix.

(c) Thermal Load Vector

For the case of polar orthogonality the initial thermal strain vector $\{\varepsilon_0\}$, may be written as:

$$\{\varepsilon_0\} = T\{\alpha_r \alpha_\theta\}^T = T\{\alpha_0\} \tag{7}$$

and in the case of isotropic plates: $\alpha_r = \alpha_\theta = \alpha$.

Therefore the total strain vector $\{\varepsilon\}$ for the annular finite element may be expressed as:

$$\{\varepsilon\} = \{\varepsilon_0\} + [D]^{-1}\{\sigma\} \tag{8}$$

Or:

$$\{\sigma\} = D[\{\varepsilon\} - \{\varepsilon_0\}] \tag{9}$$

Or:

$$\{F\} = \int_{\text{vol}} [f]^T[D][f]\,dv\{U\} - \int_{\text{vol}} [f]^T[D]\,dv\{\varepsilon_0\} \tag{10}$$

Or:

$$\{F\} = [k]\{U\} - \{\mathbf{P_t}\} \tag{11}$$

where $\{\mathbf{P_t}\}$ is the thermal load vector, defined in the Appendix.

Upon computing the elemental stiffness matrix and thermal load vector for each element in the subdivided structure, the total system stiffness matrix $[K]$ and the thermal load vector $\{\mathbf{P_t}\}$ are generated using standard finite element techniques which ensure equilibrium and compability at each of the nodes. Subsequently, structural displacement vector $\{\Delta\}$ is computed from the expression:

$$\{P\} + \{\mathbf{P_t}\} = [K]\{\Delta\} \tag{12}$$

Upon computing $\{\Delta\}$ the strains and stresses within each element of the structure are obtained from eqns (2) and (4), respectively.

RESULTS

A laminated, isotropic linearly varying thickness plate with the dimensions shown in Fig. 3 is subjected to a relative temperature distribution of the form:

$$T(r) = T_0(b/r) + T_1$$

where: $T_0 = 0 \cdot 1\,°\text{C}$, $T_1 = 2\,°\text{C}$.

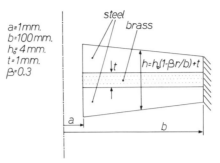

FIG. 3.   Linearly varying thickness composite disc.

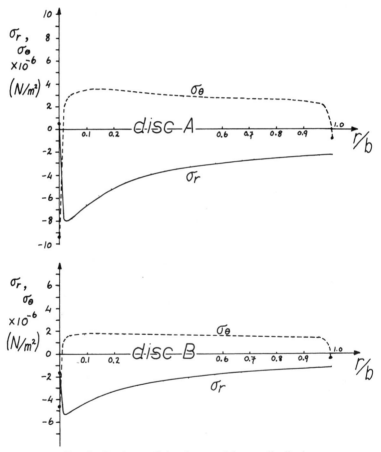

FIG. 4.   In-plane radial and tangential stress distributions.

The outer laminate is of mild steel having the following physical properties: $\alpha = 13 \times 10^{-6}$ per °C, $E = 210 \times 10^{9}$ N/m^2, $\mu = 0.3$. The inner laminate is of brass having the following physical properties: $\alpha = 18.6 \times 10^{-6}$ per °C, $E = 875 \times 10^{9}$ N/m^2, $\mu = 0.3$.

Figure 4 illustrates the resultant radial and tangential stress distributions, σ_r and σ_θ, respectively, across each laminate when the disc is free at $r = a$, and totally restrained ($u = 0$) at $r = b$.

DISCUSSION AND CONCLUSION

A finite element technique has been developed to analyse thermally induced in-plane stressing in a variable thickness disc composed of either isotropic or polar orthotropic laminates. Although the analysis has been extended to include the case where the laminates are of polar orthotropic materials, the general case of anisotropic laminates cannot be examined using the finite elements described since it would be impossible to maintain axial symmetry under such general conditions.

The effect of centrifugal loadings has not been included in the analysis, although it could readily be included by way of the technique suggested in reference 7. If, however, it was required that centrifugal and thermal loading be analysed simultaneously, the major difficulty would be in describing the effect of rotation upon the resulting temperature distribution formed over the disc. A technique analysing this problem has been described in reference 2 and at present investigations are being carried out as to how it may be applied to the analysis of laminated discs.

REFERENCES

1. SEIREG, A. and SURANA, K. S., Optimum design of rotating discs, *Trans. ASME*, **92** (1970), 1–9.
2. GORMAN, D. G., *Transverse vibration of thermally stressed stationary and rotating discs*, Ph.D. Thesis, University of Strathclyde, 1977.
3. KENNEDY, W. and GORMAN, D. G., Vibration analysis of variable thickness discs subjected to centrifugal and thermal stresses, *J. of Sound and Vibration*, **53**(1), (1977), 83–101.
4. NIEH, L. T. and MOTE, C. D., Vibration and stability in thermally stressed rotating discs, *Experimental Mechanics* (July, 1975), 258–64.
5. GORMAN, D. G. and KENNEDY, W., Membrane effects upon the transverse vibration of linearly varying thickness discs, *J. of Sound and Vibration*, **62**(1979), 51–64.

6. GORMAN, D. G., Initiation of transverse vibration in rotating discs. *J. of Sound and Vibration*, **62** (1979), 467–70.

7. ZIENKIEWICZ, O. C., *The finite element method in engineering science*, London, McGraw-Hill, 1971.

APPENDIX: STIFFNESS MATRIX AND THERMAL LOAD VECTOR

For the annular finite element as described in Fig. 2, the in-plane stiffness matrix $[k]$ can be written as:

$$[k] = \frac{2\pi}{R_{21}^2} \begin{bmatrix} k_{11} & k_{12} \\ k_{21} & k_{22} \end{bmatrix}$$

where:

$$R_{21} = R_2 - R_1$$

and:

$$
\begin{aligned}
k_{11} = \{ &\ln(R_2/R_1)[D_{11}R_2^2 a_0] + R_{21}[D_{22}R_2^2 a_1 - 2D_{12}R_2 a_0 - 2D_{22}R_2 a_0] \\
&+ \tfrac{1}{2}(R_2^2 - R_1^2)[D_{11}a_0 - 2D_{12}R_2 a_1 + 2D_{12}a_0 - 2D_{22}R_2 a_1] \\
&+ \tfrac{1}{3}(R_2^3 - R_1^3)[D_{11} - 2D_{12}]a_1 \}
\end{aligned}
$$

$$
\begin{aligned}
k_{12} = k_{21} = \{ &-D_{22}R_1 R_2 a_0 \ln(R_2/R_1) \\
&+ R_{21}[(R_1 + R_2)a_0(D_{22} + D_{12}) - D_{22}R_1 R_2 a_1] \\
&+ \tfrac{1}{2}(R_2^2 - R_1^2)[(R_1 + R_2)a_1(D_{22} + D_{12}) - a_0(D_{11} + 2D_{12} + D_{22})] \\
&- \tfrac{1}{3}(R_2^3 - R_1^3)[a_1(D_{11} + 2D_{12} + D_{22})]\}
\end{aligned}
$$

$$
\begin{aligned}
k_{22} = \{ &\ln(R_2/R_1)[D_{22}R_1^2 a_0] + (R_{21})[D_{22}R_1^2 a_1 - 2R_1 a_0(D_{22} + D_{12})] \\
&+ \tfrac{1}{2}(R_2^2 - R_1^2)[a_0(D_{11} + D_{22} + 2D_{21}) - 2a_1 R_1(D_{12} + D_{22})] \\
&+ \tfrac{1}{3}(R_2^3 - R_1^3)[a_1(D_{11} + 2D_{12} + D_{22})]\}
\end{aligned}
$$

Where:

$$
[D] = \begin{bmatrix} D_{11} & D_{12} \\ D_{12} & D_{22} \end{bmatrix}
$$

$$
= E/(1 - \mu^2) \begin{bmatrix} 1 & \mu \\ \mu & 1 \end{bmatrix} \quad \text{For isotropic analysis}
$$

$$
= E_r/(1 - n\mu_{\theta r}^2) \begin{bmatrix} 1 & \mu_{\theta r} \\ \mu_{\theta r} & 1/n \end{bmatrix} \quad \text{For polar orthotropic analysis}
$$

Where: $n = E_r/E_\theta$ and: $\mu_{\theta r}E_r = \mu_{r\theta}E_\theta$.

The thermal load vector $\{\mathbf{P_t}\}$ for each element is defined as:

$$\{\mathbf{P_t}\} = \frac{2\pi T}{R_{21}} [P][D]\{\alpha_0\}$$

where:

$$[P] = \begin{bmatrix} p_{11} & p_{12} \\ p_{21} & p_{22} \end{bmatrix}$$

where:

$$p_{11} = -\{a_{0/2}(R_2^2 - R_1^2) + a_{1/3}(R_2^3 - R_1^3)\}$$

$$p_{12} = \{R_2[a_0(R_{21}) + a_{1/2}(R_2^2 - R_1^2)]\} + p_{11}$$

$$p_{21} = -p_{11}$$

$$p_{22} = p_{21} - R_1\{a_0 R_{21} + a_{1/2}(R_2^2 - R_1^2)\}$$

11

Optimization of Laminated Shells with Multiple Loading Conditions and Fabrication Constraints*

R. T. BROWN

Atlantic Research Corporation, 5390 Cherokee Avenue, Alexandria, Virginia 22314, USA

AND

J. A. NACHLAS

Department of Industrial Engineering and Operations Research, Virginia Polytechnic Institute and State University, Blacksburg, Virginia 24061, USA

ABSTRACT

A methodology for design of laminated conical shells subject to multiple loading conditions is described. The most significant feature of the technique is the ability to tailor the composite structure to specific envelope specifications and loading conditions within the framework of restrictive fabrication constraints. The methodology combines conventional field analysis, a field analysis sampling scheme, and a dynamic program. The methodology is general with respect to loading conditions, geometry, and field theory. Application of the method yields a laminated composite shell with equivalent strength and stiffness to a baseline design with a significant reduction in weight.

INTRODUCTION

Fabrication of laminated conical shells using composite materials is a promising approach to construction of conical structures because of the resulting opportunity to obtain performance behavior equivalent to that of

*This work was sponsored by the US Air Force Materials Laboratory, Wright Patterson AFB, Ohio, USA.

144

previous designs at reduced weight. An optimized design for a laminated conical shell is defined as a composite material reinforcement architecture which most effectively resists both the magnitude and directional nature of applied loads without over design in either respect. The conical structures addressed here are subject to multiple time-dependent loading conditions including a severe thermal–chemical environment, internal pressure, lateral mechanical loads, and significant bending forces. Thus, there are a variety of failure modes and the governing modes are known *a priori.*

Fabrication methods employed for lamination of composite materials for conical structures (i.e. filament winding, braiding, tape wrapping, etc.) require that the orientation of each lamina changes as a function of meridional location. Therefore, the conical geometry dictates the reinforcement trace attainable with the composite material for each fabrication method.

The methodology developed combines conventional field analysis, a structural analysis sampling scheme, and a dynamic program. The resulting technique permits tailoring of shell architecture to performance requirements within the framework of the constraints imposed by available fabrication methods. This ability to tailor the design to performance and geometry criteria is the most significant feature of the technique developed. The most serious weakness is the computational difficulty in extending it from a discrete to a continuous design variable space. Analytical experience with both discrete and continuous design variables is described.

PROBLEM AND APPROACH

The objective of the analysis is to design a laminated composite conical shell for a defined geometry and load environment. A generalized cone is illustrated in Fig. 1. The cone envelope is defined by the throat diameter, half angle, and exit plane diameter. The task is to evolve a weight efficient structure meeting these specifications and capable of withstanding anticipated loads.

The methodology developed includes two components, one to provide an estimated design performance function and one to perform optimization. The actual steps of the procedure are (1) to select a material system based on acceptable invariant stiffness, (2) to partition the structure into sectors for evaluation of loading conditions and imposition of fabrication constraints, (3) to estimate the structural response for all designs using a field theory analysis of a sampled subset of designs, and (4) to utilize a dynamic

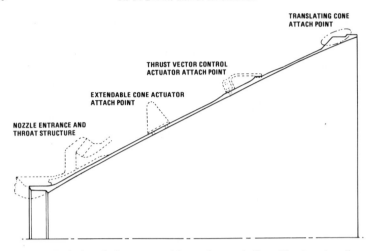

FIG. 1.   Typical conical structure with attachments indicated by dotted outline.

program to solve for the optimum design using a stagewise synthesis of
sector by sector performance estimates subject to fabrication constraints.

The first analytical step is the estimation of composite structure
performance as a function of the choice of design variables. Examination of
the governing constitutive equations for conical shells shows that while
lamina stress is linear with respect to thickness, it is nonlinear and
nonconvex with respect to orientation angle. Using well established
statistical techniques and classical laminate field theory, an estimate of the
nonlinear performance response function is constructed using a minimum
sample size. Subsequent use of this estimated function permits the
decoupling of the stress analysis from the optimization.

The remaining analytical step is the determination of the optimal design
relative to an aggregate performance measure based upon sector specific
nonlinear response functions. Previously employed methods[1,2,3] for
seeking solutions to nonlinear problems combine a constrained minimi-
zation routine with a search procedure. Search techniques are either
derivative based or derivative free. Derivative based methods have a high
probability of terminating at local optima when applied to nonconvex
functions while derivative free methods often have a reduced propensity for
termination at local optima which is achieved at the expense of sample size.

The composite shell is fabricated by lamination of unidirectional fiber
layers. Specification of layer orientation angles implies material and

structure performance characteristics. This relationship between orientation angles and performance is defined by the field theory and is the focus of the statistical sampling approach to construction of the estimated response function. Aggregation of the sector specific nonlinear response functions and determination of an optimum design is then pursued in a stagewise manner using a dynamic program.

FIELD THEORY

Design optimization is concerned with tailoring the stiffness and thermal expansion characteristics of the composite to achieve maximum strength consistent with environmental, performance, and fabrication restrictions. Classical laminate theory provides a convenient method for manipulating the stiffness and thermal strain tensors. Classical laminate theory assumes the thin shell approximation which reduces the 6×6 stiffness matrix to the symmetric 3×3 matrix $[Q]$ which consists of four constants for orthotropic lamina and six constants for an angle ply lamina. That is

$$
Q = \begin{bmatrix} Q_{11} & Q_{12} & 0 \\ Q_{12} & Q_{12} & 0 \\ 0 & 0 & Q_{66} \end{bmatrix} \quad \text{where} \quad \begin{aligned} Q_{11} &= E_1/(1 - v_{12}v_{21}) \\ Q_{12} &= v_{12}Q_{22} = v_{21}Q_{11} \\ Q_{22} &= E_2/(1 - v_{12}v_{21}) \\ Q_{66} &= G_{12} \end{aligned} \quad (1)
$$

for an orthotropic shell and

$$
[\bar{Q}] = [T][Q][T]^T \quad [T] = \begin{bmatrix} \cos^2 \theta & \sin^2 \theta & 2\sin\theta\cos\theta \\ \sin^2 \theta & \cos^2 \theta & -2\sin\theta\cos\theta \\ -\sin\theta\cos\theta & \sin\theta\cos\theta & \cos^2\theta - \sin^2\theta \end{bmatrix} \quad (2)
$$

for an angle ply lamina. Then, for a balanced ply lamina

$$
\begin{bmatrix} \sigma_1 \\ \sigma_2 \\ \tau_{12} \end{bmatrix} = Q \begin{bmatrix} \varepsilon_1 \\ \varepsilon_2 \\ \gamma_{12} \end{bmatrix} \quad \text{and} \quad \begin{bmatrix} \sigma_z \\ \sigma_\theta \\ \tau_{z\theta} \end{bmatrix} = \bar{Q} \begin{bmatrix} \varepsilon_z \\ \varepsilon_\theta \\ \gamma_{z\theta} \end{bmatrix} \quad (3)
$$

or an angle ply lamina define the composite architecture in terms of fiber properties, fiber volume, matrix properties, and lamina fiber orientation.

Laminate responses to applied loads and moments are computed for extension, coupling, and bending in both the structure body and the material directions. The BOSOR 4[4] finite difference code is used for the majority of the cases and the SAAS III[5] finite element code is used for the thick attachment sector which is the forward part of the cone. Both codes employ the above described elasticity based field theory and both are limited by the material constitutive laws employed.

STRUCTURAL RESPONSE SURFACES

Any optimization approach for design of a laminated composite structure will require the evaluation of structural response for some of the candidate design configurations. In the absence of a closed form definition of structural performance, an efficient approach to candidate design evaluation is required.

The design evaluation must provide a measure of performance as a function of the design variables and specified problem parameters. The assumed problem parameters for the laminated composite cone are shown in Fig. 2. Laminate balance is assumed in order to conform to the restrictions of the field theory. Design variables are the ply orientations and the stacking sequence. For the discrete case, angle ply orientations are varied by 15° increments and a baseline eight layer stack of balanced angle

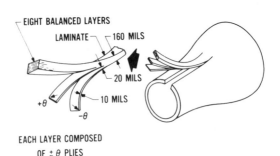

ORIENTATION INCREMENT - 15°
LAYER THICKNESS - 20 MILS
NUMBER OF LAYERS - (2 ± θ PLIES PER LAYER)
STACKING SEQUENCE - SYMMETRIC ABOUT CENTERLINE

EIGHT BALANCED LAYERS
LAMINATE — 160 MILS
20 MILS
+θ   10 MILS
−θ

EACH LAYER COMPOSED
OF ± θ PLIES

FIG. 2.   Problem parameters.

plies is optimized. The laminate is symmetric about the midplane. Consequently, four fiber orientation angles are sufficient to specify each laminate for each sector.

The measure of performance is taken to be strength as expressed by safety factor. The optimization objective is to maximize the strength for the most critical failure modes. Composite structure strength is determined using a three part partitioning of the shell geometry and the corresponding loading conditions. This approach allows selective tailoring of the laminate for sections of the cone subjected to different critical loads.

Since four fiber orientation angles specify the laminate in each sector and since each of the fiber angles may take one of seven values—0° through 90° in 15° increments—there are $7^4 = 2401$ candidate designs for each sector. Evaluation of even a modest fraction of the total number of candidate designs is computationally prohibitive. Given the problem parameters, there exists a composite structure response surface for each sector. In order to obtain an estimate for this performance function, an efficient method of sampling from the true response surface is employed.

A fractional factorial experimental design is used to define candidate design cases to be evaluated using the field analysis codes. Each of the selected cases provides one strategically located data point for use in estimating the response function. As shown in Fig. 3, fewer than 30 such cases are analyzed for each sector of the cone. Resulting data are included in a stepwise multiple regression analysis to construct a second order polynomial function which approximates the true laminate performance function.

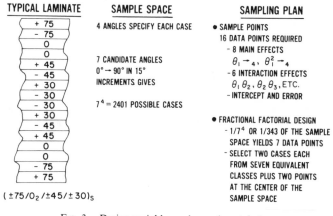

FIG. 3. Design variables and experimental plan.

The field analysis results for the sample points for each sector are used to estimate response functions for longitudinal, transverse, and shear factors of safety in the principal material directions. Two second order functions are constructed. These functions have the form

$$\hat{y}^{(1)} = b_0 + \sum_{j=1}^{4} b_j \cos \theta_j + \sum_{j=1}^{4} \sum_{k=j}^{4} b_{jk} \cos \theta_j \cos \theta_k \qquad (4)$$

and

$$\hat{y}^{(2)} = b_0 + \sum_{j=1}^{4} b_j x_j + \sum_{j=1}^{4} \sum_{k=j}^{4} b_{jk} x_j x_k \qquad (5)$$

where $\hat{y}^{(1)}$ and $\hat{y}^{(2)}$ are estimates of the true response, θ_j is the fiber orientation angle in lamina j, $x_j = \theta_j/15$ is a surrogate variable, and b_j and b_{jk} are the functional coefficients determined in the regression analysis.

For each of the hypothesized functions, two descriptors of validity are computed. The coefficient of correlation is a measure of the percentage of the variation in the true response function that is explained by the regression model and the F statistic is an indicator of the statistical significance of the constructed function. The results obtained for the regression models show that of the 18 models constructed, 15 are statistically significant. For the significant models, the correlation coefficients are generally greater than 70%.

The regression models found to be statistically significant are aggregated to define a conservative estimate of the structural response function by taking advantage of the meaningful information provided by each regression model. For each sector, $y^{*(1)}$ and $y^{*(2)}$ are defined as

$$y^{*(1)} = \min_k \{\hat{y}_k^{(1)}\} \quad \text{and} \quad y^{*(2)} = \min_k \{\hat{y}_k^{(2)}\} \qquad (6)$$

at each candidate design point. That is, for each valid functional form, the minimum of the longitudinal, transverse, and shear factors of safety is taken as the performance estimate. Then, the estimated response, z, is taken to be

$$z = \min \{y^{*(1)}, y^{*(2)}\} \qquad (7)$$

The values for z are tabulated for each candidate design and serve as the laminate performance estimates for use in the optimization model.

A comparison of the resulting estimates of structure response and the true calculated values shows strong agreement. The approximating response functions appear to preserve most of the relative ordering of the designs and are therefore appropriate surrogates for the true surfaces.

OPTIMIZATION

The optimization model is a dynamic program.[6] Structural synthesis is accomplished within the dynamic program through sector by sector design variable selection subject to variable choices in other sectors. For each sector, the decision variables, x, are the angles of fiber orientation in each layer. The state variables, u, are defined to be the angles of fiber orientation in the adjacent more forward sectors. This state variable definition is employed in order to provide a means for imposing fabrication constraints upon the selection of decision variable values. As shown in Fig. 4, the available fabrication methods dictate the feasible fiber orientation angle changes from sector to sector. The state variable definition employed assures the selection of design variables consistent with the angle change constraints implied by the fabrication methods.

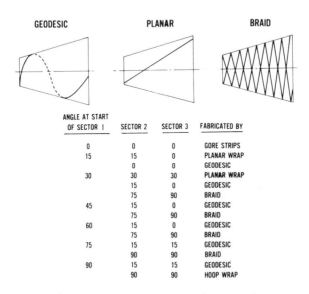

ANGLE AT START OF SECTOR I	SECTOR 2	SECTOR 3	FABRICATED BY
0	0	0	GORE STRIPS
15	15	0	PLANAR WRAP
	0	0	GEODESIC
30	30	30	PLANAR WRAP
	15	0	GEODESIC
	75	90	BRAID
45	15	0	GEODESIC
	75	90	BRAID
60	15	0	GEODESIC
	75	90	BRAID
75	15	15	GEODESIC
	90	90	BRAID
90	15	15	GEODESIC
	90	90	HOOP WRAP

FIG. 4. Fabrication method imposed design constraints.

The variables and functions that define the dynamic program are:

$\mathbf{x}^{(i)} = $ the vector of fiber orientation angles selected in sector i;

$\mathbf{u}^{(i-1)} = $ the vector of fiber orientation angles for the fibers leaving sector $i-1$ and entering sector i;

$t_i(\mathbf{u}^{(i-1)}, \mathbf{x}^{(i)}) = $ the state variable transformation function which defines $\mathbf{u}^{(i)}$ in terms of $\mathbf{u}^{(i-1)}$ and $\mathbf{x}^{(i)}$;

$r_i(\mathbf{u}^{(i-1)}, \mathbf{x}^{(i)}) = $ the design response attained in sector i when decision vector $\mathbf{x}^{(i)}$ is selected given input state $\mathbf{u}^{(i-1)}$. Equals $z(\mathbf{x}^{(i)})$ for this analysis;

$f_i(\mathbf{u}^{(i-1)}) = $ the optimal structural response obtainable in sectors i through N given input state $\mathbf{u}^{(i-1)}$;

$g_i(\mathbf{x}^{(i)}) = $ any applicable further constraint functions which limit the selection of a design vector.

Using these definitions, the dynamic program is a set of recursion equations:

$$f_3(\mathbf{u}^{(2)}) = \text{opt } \{r_3(\mathbf{u}^{(2)}, \mathbf{x}^{(3)})\}$$
$$\text{subject to } g_3(\mathbf{x}^{(3)}) \geq 0$$
$$f_2(\mathbf{u}^{(i)}) = \text{opt } \{r_2(\mathbf{u}^{(1)}, \mathbf{x}^{(2)}) \, \Delta f_3(t_2(\mathbf{u}^{(1)}, \mathbf{x}^{(2)}))\} \qquad (8)$$
$$\text{subject to } g_2(\mathbf{x}^{(2)}) \geq 0$$
$$f_1(\mathbf{u}^{(0)}) = \text{opt } \{r_1(\mathbf{u}^{(0)}, \mathbf{x}^{(1)}) \, \Delta f_2(t_1(\mathbf{u}^{(0)}, \mathbf{x}^{(1)}))\}$$
$$\text{subject to } g_1(\mathbf{x}^{(1)}) \geq 0$$

As is implied by their form, the equations are solved in a reverse order from sector three to sector one in a stepwise fashion. An optimal choice of $\mathbf{x}^{(3)}$ is identified for all possible input state vectors $\mathbf{u}^{(2)}$. Then an optimal choice of $\mathbf{x}^{(2)}$ is determined to optimize the response for sectors two and three for any given input state vector $\mathbf{u}^{(1)}$. Similar arguments are applied successively to create a nested solution that ultimately depends only upon the initial state variable $\mathbf{u}^{(0)}$. The initial state variable may be defined in terms of design requirements or may be selected on the basis of an optimization criteria. This second approach is employed in analyzing the composite cone.

An important feature of the employed form of the generic recursion equations is that the composed return functions are maximized while the composition operator, denoted by Δ, is the selection of the minimum of the quantities composed. Thus, the general form of the recursion equations is

$$f_i(\mathbf{u}^{(i-1)}) = \max \{\min \{r_i(\mathbf{u}^{(i-1)}, \mathbf{x}^{(i)}), f_{i+1}(t_i(\mathbf{u}^{(i-1)}, \mathbf{x}^{(i)}))\}\} \qquad (9)$$

which corresponds to the selection of a decision vector in sector i which maximizes the minimum structural factor of safety over the shell from that sector to the aft end. This form of the recursion is particularly appropriate for the laminated composite structure because it represents the strength of a chain of sectors of the cone by the strength of the weakest sector.

The estimated structural response associated with the selection of the vector of fiber orientation angles $\mathbf{x}^{(i)}$ in sector i is $z_i(\mathbf{x}^{(i)})$. For the laminated cone application, the definition of a constraint function $g_i(\mathbf{x}^{(i)})$ is necessary in sector two only. This function reflects the requirement that the structure withstand specific mechanical loads associated with the activation of the attached control levers.

Define the sets $h_i(\mathbf{u}^{(i-1)})$ to be the design vectors in sector i having a feasible fiber trace for the state vector $\mathbf{u}^{(i-1)}$. Using this form of the fabrication constraints, the implemented form of the dynamic program is

$$f_3(\mathbf{u}^{(2)}) = \max \{z_3(\mathbf{x}^{(3)})\}$$
$$\mathbf{x}^{(3)} \varepsilon h_3(\mathbf{u}^{(2)})$$
$$f_2(\mathbf{u}^{(1)}) = \max \{\min \{z_2(\mathbf{x}^{(2)}), f_3(\mathbf{u}^{(2)})\}\}$$
$$\mathbf{x}^{(2)} \varepsilon h_2(\mathbf{u}^{(1)}) \qquad (10)$$
$$\text{subject to } g_2(\mathbf{x}^{(2)}) \geq 0$$
$$f_1(\mathbf{u}^{(0)}) = \max \{\min \{z_1(\mathbf{x}^{(1)}), f_2(\mathbf{u}^{(1)})\}\}$$
$$\mathbf{x}^{(1)} \varepsilon h_1(\mathbf{u}^{(0)})$$
$$\max \{f_1(\mathbf{u}^{(0)})\}$$
$$\mathbf{u}^{(0)}$$

The five best solutions obtained using this model are listed in Table 1. The estimated factors of safety indicated are based upon the estimated response functions and are therefore the criteria on which the solutions are selected. These solutions are next verified by analysis using the field theory codes. The resulting actual factors of safety are indicated in the table. The fourth solution is taken as the selected design concept. This design is illustrated in Fig. 5.

TABLE 1
Families of optimal laminations

SECTOR	ARCHITECTURE	ESTIMATED SAFETY FACTOR	COMPUTED SAFETY FACTOR
1:	(90/0/90/75)		
2:	$(90/0/90_2)_s$	2.08	1.1
3:	$(90/0/90_2)_s$		
1:	$(90/15/90/75)_s$		
2:	$(90/0/90_2)_s$	2.0	1.1
3	$(90/0/90_2)_s$		
1:	$(30/90/0/90)_s$		
2:	$(30/90/0/90)_s$	1.7	1.2
3:	$(30/90/0/90)_s$		
1:	$(30/90/0/90)_s$		
2:	$(15/90/0/90)_s$	1.7	1.5
3:	$(0/90/0/90)_s$		
1:	$(60/90/0/75)_s$		
2:	$(15/90/0/90)_s$	1.7	1.2
3:	$(0/90/0/90)_s$		

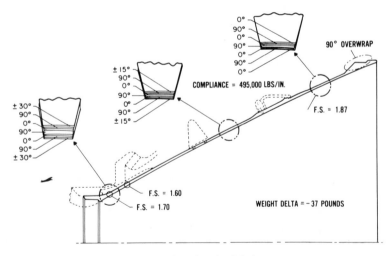

FIG. 5.   Selected optimal design.

SENSITIVITY

It is appropriate to note that the solutions obtained form a family of similar designs and are reasonable when considered qualitatively. The frequent occurrence of hoop fibers (90°) assures resistance to internal pressure and thermal expansion while the axial fibers (0°) provide stiffness in the direction of the applied mechanical loads. The absence of 45° fibers is reasonable in view of their propensity for shear failure. Thus, the analytical solutions conform well to engineering judgements concerning the design of the laminated composite cone.

Relaxation of the fabrication constraints significantly affects the model solutions. The fabrication constraints dictate which fiber orientations may be used in the same ply over more than one sector. Relaxing these constraints implies that the fibers may be wrapped in any pattern. For this case, several solutions with higher minimum safety factors than those of the constrained solutions are found. This suggests that fabrication constraints strongly influence the problem solution and that improvements in fabrication methods that result in greater wrapping flexibility may lead to stronger designs. However, it should be noted that the optimal estimated safety factor for the unconstrained analysis is only 20% greater than that for the constrained solution and that increases in minimum safety factor beyond 1·5 are of questionable utility. Thus, the constrained solution is a relatively good design and the value of pursuing improvements in the solution through the development of new fabrication techniques is not obvious.

EXTENSION TO CONTINUOUS DESIGN VARIABLES

Intuitively, it is appealing to represent the design problem in terms of continuous design variables. The generic dynamic programming formulation is unchanged and optimization can presumably be accomplished in closed form using calculus rather than numerically. However, maximization of a minimum response is discrete and the choice of wrapping method is also discrete.

The discrete selection of wrapping method cannot be modified to take a continuous form. Given/ this restriction, a continuous choice of design vectors can be formulated separately for each combination of wrapping methods. Gore strip and hoop fiber wrapping can be viewed as specific planar wrap cases. There are then three wrapping patterns available for each of four laminae. As a result there are $3^4 = 81$ feasible design wrapping patterns each of which may be optimized in terms of fiber orientation angles.

For each of the resulting 81 subproblems, an overall structural response function must be constructed. This is accomplished by using the estimated response functions to obtain a single aggregate performance estimate over the entire structure which is defined as the minimum safety factor for each candidate design. Regression analysis is again applied to construct the 81 response functions each of which has the form given in eqn. (4). State variable transformation equations are the equations for the fiber traces under the given wrapping methods. Partial differentiation of the response function yields

$$\frac{\delta \hat{z}}{\delta \theta_j} = -\sin \theta_j \left[b_j + b_{jj} \cos \theta_j + \sum_{k=1}^{4} b_{jk} \cos \theta_k \right]$$

$$\frac{\delta^2 \hat{z}}{\delta \theta_j^2} = 2 b_{ii} \sin^2 \theta_j \tag{11}$$

$$\frac{\delta^2 \hat{z}}{\delta \theta_k \delta \theta_j} = b_{jk} \sin \theta_j \sin \theta_k$$

Setting the first partial derivatives equal to zero yields a set of four simultaneous nonlinear equations in four unknowns which must be solved for stationary points. Then the Hessian matrix of second partial derivatives must be analyzed at each stationary point to select the maxima which are compared to determine the global maximum. The resulting solutions for each of the 81 cases are then compared to identify the optimal design. Obviously, solution of the design problem for continuous design variables requires considerable computational effort. This work is in progress. It is significant that for each of the (a) planar, geodesic, planar, braid, (b) geodesic and three planars, (c) geodesic, two planars and braid, and (d) three planars and braid, the solutions match those found for the discrete analysis. This is attributed largely to the aggregate response functions obtained. Further sampling to construct better aggregate response functions is appropriate and is part of the continuing analysis of the problem.

CONCLUSIONS

An analytical approach to the design of a laminated composite conical shell has been demonstrated. For the imposed limitations of laminate symmetry and state of the art fabrication methods, the method efficiently locates the optimal reinforcement pattern. The method permits decoupling of the field analysis codes and the optimization activity with a modest sacrifice in accuracy. The solution identified yields a weight saving of 37 pounds when compared to the baseline design.

Several advantages of the approach are apparent. The procedure is completely general with respect to the type of field analysis, loading conditions, and structure geometry considered. Measures of performance employed are conservative assuring an acceptable design. The stepwise approach to solution permits the designer complete visibility and control of the procedure. Finally, accuracy versus expense decisions can be made for any part of the analysis with recognizable consequences.

The optimized composite design identified is being fabricated to permit testing of material and structure behavior. The continuous variable model is being analyzed further and is being extended to consider constraints based upon wrapping tension and surface friction resulting in nongeodesic patterns and a greater degree of design flexibility.

REFERENCES

1. POPE, G. G. and SCHMIT, L. A. (eds), Structural design applications of mathematical programming techniques, *AGARDograph*, **149** (1971).
2. SCHMIT, L. A., The structural synthesis concept and its potential role in design with composites, *Mechanics of Composite Materials*, ONR, May, 1967.
3. BALDUR, R., Structural optimization by inscribed hyperspheres, *J. Engng Mech.*, **98** (1972).
4. BUSHNELL, D., *Stress, stability, and vibration of complex branched shells of revolution: Analysis and user's manual for BOSOR4*, Lockheed Missiles and Space Company, Inc., AD748639, March 1972.
5. CROSE, J. G. and JONES, R. M., *Finite element stress analysis of axisymmetric and plane solids with different orthotropic, temperature-dependent material properties in tension and compression*, SAMSO-TR-71-103, 1971.
6. SASIENI, M., YASPEN, A. and FRIEDMAN, L., *Operations research*, New York, John Wiley and Sons Inc., 1959.

12

Recent Developments in Polyester Matrices and Reinforcements for Marine Applications, in Particular Polyester/Kevlar* Composites

L. S. NORWOOD

*Scott Bader Co. Ltd, Wollaston, Wellingborough,
Northamptonshire NN9 7RL, England*

AND

A. MARCHANT

*Anthony Marchant and Associates, Bell House,
32 Bell Street, Romsey SO5 8GW, Hampshire, England*

ABSTRACT

Since their acceptance into the marine world, reinforced plastics—and, in particular, glass-reinforced polyester resins—have been used for the fabrication of boats and other marine structures.

In the early learning days overdesign made it possible to use inexpensive systems, which were not necessarily the best for the job. However, in recent years, as composite marine structures have increased in size (the largest, at present, being the 55 m 'Brecon' class of mine countermeasure vessel) or where performance is the major criterion (as in power boats and racing yachts) a move has been made towards the use of improved material for construction.

On the resin side, isophthalic acid and isophthalic-neopentyl glycol based unsaturated polyester resins have been developed to give improved water resistance, blister resistance, toughness and long-term retention of mechanical properties. These systems also form the basis of many successful gelcoats, which afford the first barrier to protect the structural laminate. In

* Kevlar is Du Pont's registered trade mark.

recent years it has been recognised that sensible practice would be to use these types of resin throughout the structure to provide a material with the best possible performance.

On the reinforcement side there have been developments in size technology for E-glass to provide improved fibre/resin bonding. Also, many types of glass mat have been developed to provide maximum reinforcing action in any direction. However, glass is rather dense compared with polyester resin and, for applications where weight and performance are paramount, lightweight reinforcements are being developed which are currently receiving careful consideration as alternatives to glass fibre.

In particular, polyaramid fibres are strong contenders to partially or completely replace glass in such applications because of their low density, high strength and high stiffness.

Data obtained from composites constructed using Crystic* polyester resins and Kevlar polyaramid fibres are discussed in detail in this chapter and the possibility of combining glass and Kevlar fibres to give optimum laminate and structural performance is investigated.

INTRODUCTION

For more than thirty years thermosetting resins have been developed for combining with reinforcements to form composite materials. In the early days the demand was for inexpensive water-resistant materials and the first generation of polyester resins became available. Since that time, intensive research and improved processing facilities have resulted in other types of polyester resin being produced with improved long-term water resistance, flexibility and heat distortion temperature. This means improved long-term performance but, as always, resistance is met on price.

On the reinforcement side, E-glass fibre was developed and various size and coating treatments enabled good bonding to be obtained between glass and polyester resin matrices. Hence, weak matrices could be transformed into stronger, stiffer materials, and strong fibres could be given rigidity— the resulting combination being useful composite materials. Development in glass technology has mainly been restricted to the surface treatment rather than fundamental changes to the properties of the glass fibres themselves. The changes and improvements in surface technology have

---

* Crystic is Scott Bader's registered trade mark.

provided products which wet-out in a variety of ways, depending on the fabricator preference, but do not necessarily give any improvements in performance. However, the introduction of directional fabrics enables large GRP structures to be manufactured, by allowing the reinforcement in critical areas to act in the directions of the maximum stresses.

All this is acceptable, provided the lightest possible structure is not required. Even then, the use of foam sandwich construction does allow useful reductions to be made in weight, whilst still retaining rigidity, but the skin-to-core interfacial bond strength and the low core shear stress can present problems.

In the last few years the introduction of new lightweight fibres has paved the way for the fabrication of strong, stiff composites without resorting to sandwich structures. Carbon fibre has never been a contender for marine use with polyester resins because of its very high cost and low elongation to failure. However, the use of polyaramid fibres is now receiving considerable attention, enabling strong, stiff, lightweight reinforced polyester resin laminates to be produced.

By combining polyaramid and glass fibre fabrics, optimum mechanical performance can be economically obtained to provide competitive materials for marine applications.

Glass, polaramid (Kevlar 49) and glass/polyaramid hybrid isophthalic acid based resin (Crystic 272) systems are discussed in this paper and their technical performance is examined.

Polyester/polyaramid reinforced composites are shown to have equivalent mechanical performance to vinyl ester/polyaramid reinforced composites. Long-term retention of properties after immersion in water has also been examined.

CHOICE OF CONSTRUCTION MATERIALS FOR MARINE APPLICATIONS

Resin System

Careful consideration should be given to the choice of resin system for a given application, taking into account factors such as location of use, type of structure and the use to which it is to be put.

Where continuous immersion occurs, careful thought must be given to the effect of water absorption on strength retention and appearance.[1] It is important for applications in warmer water to use better quality resins, since heat distortion temperature, toughness and water pick-up properties all become more significant in demanding environments. Resins based on isophthalic acid, such as Crystic 272, Crystic 491 PA, Crystic 625, etc., offer improved performance, showing lower water pick-up in the cast form, than resins based on orthophthalic acid (Fig. 1) and better property retention at higher temperatures (Fig. 2).

It is, however, difficult to assess composite performance with reference to cast resin properties alone. It is better practice to consider the performance of matrix/reinforcement systems but bearing in mind that important resin

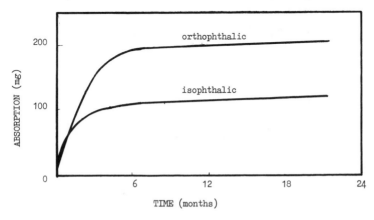

FIG. 1.   Water absorption of cast resin at 25 °C. (Tested to B.S.2782, Method 502).

characteristics, such as rate of cure, degree of cure, rheology and handleability, are largely independent of the reinforcement. Such properties should receive careful consideration to ensure ease of fabrication and good fibre wet-out and it is worth remembering that:

(i)   Rate of cure is a function of resin reactivity and the cure system used, and can generally be tailored to suit individual requirements whilst still retaining an acceptable level of cure.

(ii)   Degree of cure is very much dependent on the initial cure and, although room temperature cure is acceptable for many applications, full cure—and hence optimum environmental resistance—is only obtainable by using high temperature post-cure treatments.

Barcol hardness measurements give an indication of the degree of the initial cure; hardnesses of around 35 within a few days at room temperature are indicative of good initial cure. For *Lloyd's*

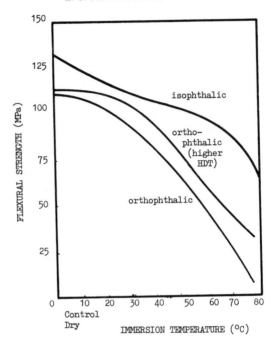

FIG. 2. Flexural strength retention of cast resin after immersion in tap water for 7 days at various temperatures.

Register of Shipping approval purposes a post cure of 24 h at room temperature, followed by 16 h at 40 °C, is allowed for test specimens; this is based on empirical evidence that such a cure equates to a 28-day room temperature cure. Data obtained on cast Crystic resins, from an independent test house, for *Lloyds Register* approval purposes are shown in Table 1. All specimens were post cured for 24/16 h RT/40 °C.

Considerable changes in mechanical properties occur when using higher temperature post cure, giving improvement in some properties and loss of others (see Table 2). A compromise has to be struck, when considering post-cure conditions, on the need for best environmental resistance versus the need for toughness.

(iii) Rheology and handleability have a bearing on the ease with which reinforcement is wetted, the time of wet-out and resin drainage. The use of thixotropic resins with the correct base viscosity can solve these problems.

TABLE 1

Typical cast resin properties of Crystic resins with a post-cure of 24/16 h RT/40°C. MEKP catalyst and cobalt accelerator

Resin type	Orthophthalic acid based						Isophthalic acid based				
Property Resin	Crystic 189LV	Crystic 196	Crystic 405PA	Crystic 406PA	Crystic 2-406PA*	Crystic 272	Crystic 489PA	Crystic 2-489PA*	Crystic 491PA	Crystic 2-491PA*	Crystic 625MV
Heat distortion temperature (°C)	55	56	56	55	56	56	55	55	56	56	59
Barcol hardness	40	40	41	39	42	37	34	34	37	37	40
Tensile strain to failure (%)	5·1	4·1	3·2	3·3	3·0	4·2	5·6	5·6	5·0	5·0	3·5
7 days' water (mg) absorption at 23°C (%)	40	45	45	35	35	39	32	32	40	40	45
	0·40	0·46	0·46	0·35	0·34	0·39	0·35	0·35	0·39	0·39	0·44
Tensile strength (N/mm²)	55	57	60	60	58	55	54	54	55	55	62
Tensile modulus (N/mm²)	2 500	2 800	3 200	3 000	3 000	3 000	3 000	3 000	3 000	3 000	3 000

* Low styrene emission versions of base resin system.

TABLE 2

Typical cast resin properties of Crystic resins with a post-cure of 24/3h RT/80°C. MEKP catalyst and cobalt accelerator

Resin type	Orthophthalic acid based					Isophthalic acid based					
Property	Crystic 189LV	Crystic 196	Crystic 405PA	Crystic 406PA	Crystic 2-406PA*	Crystic 272	Crystic 489PA	Crystic 2-489PA*	Crystic 491PA*	Crystic 2-491PA	Crystic 625MV
Heat distortion temperature (°C)	66	76	71	65	62	78	75	75	75	75	93
Barcol hardness	45	46	49	46	45	44	43	42	43	43	46
Tensile strain to failure (%)	4·0	2·5	2·2	2·3	2·5	3·8	3·5	3·5	3·0	3·0	2·5
7 days' water absorption at 23°C (mg)	45	50	50	42	41	45	46	46	45	45	53
(%)	0·46	0·51	0·51	0·42	0·41	0·46	0·46	0·46	0·46	0·46	0·52
Tensile strength (N/mm²)	72	71	66	69	69	77	75	75	75	75	70
Tensile modulus (N/mm²)	3 400	3 800	3 800	3 000	3 600	3 500	3 500	3 500	3 500	3 500	3 700

* Low styrene emission versions of base resin system.

The Reinforcement System
Glass fibre

Ease of laminate construction is just as much a function of glass parameters, such as filament diameter, surface coatings and type of fabric, as it is of resin parameters. In fact, the wetting-out of glass is very much a surface phenomenon, more under the control of the glass suppliers than the resin supplier.

Laminate properties can be tailored to meet specific requirements by correct choice of glass reinforcement, ranging from random mat to uni-directional rovings. Bi-directional fabrics of various weights are commonly in use. By careful combination of the numerous types and weights of glass fabrics available, composites with the desired properties can be produced.

During design, reference can be made to minimum strength and stiffness values, often found in the standards:[2] typical values are shown in Table 3.

However, such values are of little use unless property predictions can be made with confidence. The data[3] given in Table 4 for the various glass fibre constructions shown in Fig. 3, emphasise the good correlation that can be obtained from experimental and predicted data. Simple rules, such as adding the expected values of strength and stiffness for all the individual layers, according to their nominal weights, can be used. The agreement between predicted and experimental data is expected to be good for stiffness since these ought to be additive. However, strengths are only strictly additive if all layers break at the same strain; this is not the case here but, even so, the calculated data for the constructions under consideration (Fig. 3) are comparable with the measured data, providing a useful guide to potential performance.

TABLE 3
Minimum strength and stiffness data for glass reinforcements

Glass type	Unit tensile strength (N/mm per kg/m² of glass)	Unit tensile stiffness (kN/mm per kg/m² of glass)
Chopped strand mat at all angles	200	12·7
Woven rovings at 0° and 90°	300	16·2
Uni-directional		
0°	600	32·4
90°	0	10·8

TABLE 4
Predicted and measured laminate properties for the laminates shown in Fig. 3

Laminate No.	1		2		3		4		5	
Direction of test	0°	90°	0°	90°	0°	90°	0°	90°	0°	90°
Measured glass content (% by wt) (kg/m²)	41·2 2·96		41·0 3·74		40·2 3·97		35·5 1·08		41·3 1·30	
Property										
Predicted tensile strength (N/mm width)	1230	510	1260	900	1140	1140	270	270	540	180
Measured tensile strength (N/mm width)	1270	465	1350	874	1240	1100	269	237	501	141
Measured tensile stress (N/mm²)	256	94	231	144	206	184	105	101	232	66
Predicted tensile stiffness (kN/mm width)	68·7	42·8	77·1	64·1	76·8	76·8	16·3	16·3	30·0	17·1
Measured tensile stiffness (kN/mm width)	70·7	45·7	72·2	65·5	72·8	68·4	21·6	19·5	31·1	19·8
Measured tensile modulus (N/mm²)	14 200	9 200	12 300	10 800	12 100	11 400	8 400	8 300	14 300	9 200

LAMINATE

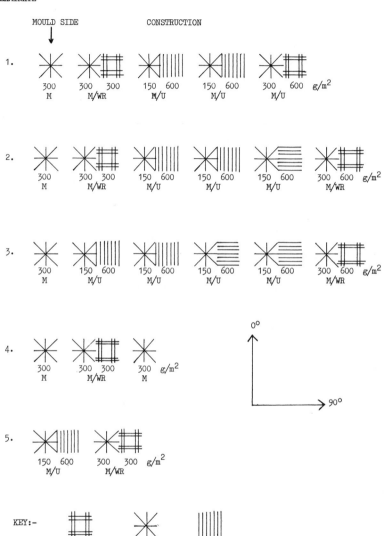

FIG. 3.   Laminate constructions. M/WR = Combination csm/woven roving glass reinforcement. M/U = Combination csm/uni-directional glass reinforcement. M = Mat (csm).

What is undoubtedly more important for composite materials than ultimate properties is the strain-to-first-noise and/or the limit of proportionality on the stress and strain.

Invariably, resins are chosen on the basis of their toughness or extensibility in the cast form. In general, it follows that the tougher or more flexible the resin, the higher the strain limitations of the laminate[1,4] before initial damage occurs. It must be noted, however, that very flexible resins, although desirable because of their high elongation to failure, suffer in other ways, usually with unacceptably low heat distortion temperatures and high water pick-up.

Strain-to-first-noise is also dependent on glass fibre type, and the recent work carried out on the laminates shown in Fig. 3 exhibits a range of strains-to-first-noise and limit of proportionalities depending on the construction and direction of applied load—see Table 5. It is well known that uni-directional material is very weak in the 90° direction and elongation at break is believed to be no more than 0·3–0·4%;[5] this explains the reduction in strain-to-first-noise in the 90° direction for the laminates containing uni-directional glass. Where the uni-directional material has been cross-plied, low values of strain-to-first-noise are observed in both directions examined as a result of the influence of the transverse properties of each uni-directional ply.

Polyaramid fibre

Advances in fibre technology in recent years have seen the introduction of polyaramid fibre in the form of Kevlar 49. Polyaramid fibre is

TABLE 5
Initial damage to laminates shown in Fig. 3

Laminate		1		2		3		4		5	
Property	Direction	0°	90°	0°	90°	0°	90°	0°	90°	0°	90°
Strain-to-first-noise (%)		0·52	0·34	0·50	0·38	0·45	0·42	0·76	0·70	0·77	0·41
Strain to limit of proportionality on the stress–strain curve (%)·		0·73	0·44	0·75	0·48	0·51	0·69	0·58	0·64	0·77	0·38
Ultimate tensile strain to failure (%)		2·0	1·7	2·2	1·7	2·1	2·0	1·5	1·4	1·7	1·0

approximately 55 % the density of glass fibre and is inherently stronger and stiffer, but its potential in marine applications has barely been realised, for several reasons:

(i)   Its use has predominantly been linked with epoxy and vinyl ester resins.
(ii) It is reputed to have inferior water resistance, although no evidence supporting this view has been put forward.
(iii) It is more expensive than glass fibre, although its cost must not be looked at in isolation as reduced structural weight has obvious economic benefits.
(iv) Its compression strength is low but careful combination with glass helps to compensate for this deficiency.

In fact, the most cost-effective polyaramid reinforced laminates are those incorporating glass fibre reinforcement and using polyester resin matrices.

Over the past three years an intensive programme of research has been carried out to assess the performance of Kevlar 49/Crystic 272 and Kevlar 49/glass/Crystic 272 systems. Their light weight, high strength and high stiffness advantages are clearly demonstrated.

The effects of using a fibre treatment incorporating an aqueous epoxy size on the mechanical performance of the polyester polyaramid composite have been investigated. The data in Table 6 show improvement in stress-to-first-noise for the sized Kevlar systems, clearly indicating improvements in the fibre–resin interfacial bond.

TABLE 6

Mechanical properties of woven Kevlar 49/csm/polyester ('Finished'—An aqueous epoxy size has been applied to the Kevlar woven roving)

All properties in N/mm^2	Unfinished Kevlar 49/csm composites	Finished Kevlar 49/csm composites
Ultimate tensile strength	355	380
Ratio of stress-to-first-noise to ultimate tensile stress	0·47	0·72
Tensile modulus	20 000	18 800
Ultimate compressive strength	120	150
Compressive modulus	14 600	18 600
Flexural strength	230	300
Flexural modulus	17 200	16 100
Lap shear strength	7·7	9·4

TABLE 7

Mechanical properties of woven Kevlar-reinforced/vinyl ester and polyester composites. (The Kevlar reinforcement is not coated (sized).) (VE = vinyl ester, P = polyester)

All properties in N/mm²	Kevlar composites		Kevlar/csm composites	
	VE	P	VE	P
Ultimate tensile strength	430	430	300	300
Tensile modulus	26 400	26 800	17 100	17 700
Ultimate compressive strength	90	80	125	115
Compressive modulus	23 600	25 400	17 800	16 400
Flexural strength	195	200	290	250
Flexural modulus	22 000	23 000	16 600	16 100
Lap shear strength	7·2	8·4	8·6	7·5

Kevlar fibre can currently be obtained in uni-directional fabric, bi-directional woven roving fabric and cloth forms. Hybrid Kevlar/glass bi-directional products are being developed and uni-directional Kevlar/glass hybrids are currently being researched. Recent test work has shown that there is no justification for using expensive vinyl ester matrices instead of polyester resin matrices with Kevlar reinforcement. Data obtained from identically reinforced polyester and vinyl ester resins are shown in Table 7.

MECHANICAL PROPERTIES OF FIBRE-REINFORCED COMPOSITES

Short-term Properties

Improvements in composite weight, stiffness and strength can be achieved by increasing: (a) the reinforcing fibre properties; (b) the efficiency of the resin to fibre interface and (c) the resin strength; and by decreasing the weight of the matrix and reinforcement.

Glass-reinforced composites improve in efficiency in going from random glass mats through bi-directional or woven materials to uni-directional products. Combinations of all three have provided a means of optimising the mechanical properties relative to the production method.

Table 8 shows the mechanical properties of the more commonly used glass reinforcement systems in a polyester resin matrix.[6,7]

As expected, the composite strength and stiffness increase as the reinforcing material changes from random mat to uni-directional, i.e. more

TABLE 8

Ultimate design properties of polyester glass-reinforced composites

All properties in N/mm^2	A csm	B wr/csm	C wr	D uni-d/csm
Ultimate tensile strength	108	186	250	460
Tensile modulus	8 100	11 720	15 800	21 000
Ultimate compressive strength	150	145	190	240
Compressive modulus	8 100	11 930	18 000	—
Lap shear strength	7	10	9	12
Flexural strength	190	280	285	—
Flexural modulus	6 700	9 500	13 400	—

A. Chopped strand mat (nominal resin-to-glass ratio = 2:1 by weight).
B. Combination product of woven rovings and chopped strand mat (nominal resin-to-glass ratio = 1·3:1 by weight).
C. Woven rovings (nominal resin-to-glass ratio = 1:1 by weight).
D. Combination product of uni-directional reinforcement and chopped strand mat (nominal resin-to-glass ratio = 1·2:1 by weight).

reinforcement is acting in the direction of the principal stress. What is important about the change in mechanical properties, as a function of reinforcement type, is the freedom it allows designers to vary composite design for given stress situations. The optimum properties of glass-reinforced polyester resins, still the most commonly used composite in the marine industry, are obtained by using uni-directional fibres. However, incorporation of polyaramid fibre, such as Kevlar 49, results in improved mechanical performance, with the exception of the compressive strength,[8] as shown in Table 9.

The addition of glass chopped strand mat reinforcement, as an interlayer between Kevlar layers, increases the flexural strength, the compressive strength and the compressive modulus but causes reductions in flexural modulus, tensile strength and tensile modulus.

Uni-directional Kevlar/csm has superior tensile strength and modulus properties compared with uni-directional glass/csm reinforcement (see Tables 8 and 9).

Immersion Properties

Laboratory immersion tests are usually carried out under accelerated conditions involving double-sided immersion in distilled or tap water, without gelcoat protection and at elevated temperatures.

In general, such tests only give an indication of relative performance but

TABLE 9

Ultimate design properties of Kevlar/polyester composites and Kevlar/glass/ polyester composites

All properties given in N/mm^2	E Woven Kevlar/csm	F Woven Kevlar	G Uni-d Kevlar/csm
Tensile strength	380	420	610
Tensile modulus	18 800	26 000	31 000
Compressive strength	150	115	135
Compressive modulus	18 600	16 300	26 800
Lap shear strength	9·4	12·8	7·1
Flexural strength	300	255	320
Flexural modulus	16 100	23 400	22 700

E. Kevlar 49 (woven and treated with aqueous epoxy finish) and glass chopped strand mat at a resin-to-fibre ratio of 1·04:1 by weight.
F. Kevlar 49 (woven and treated with aqueous epoxy finish) at a resin-to-fibre ratio of 0·82:1 by weight.
G. A combination product of Kevlar 49 (untreated) uni-directional fibres and glass chopped strand mat at a resin-to-fibre ratio of 1·20:1.

it is reasonable to assume that, under the test conditions, if one system out-performs another it will do so under less severe conditions. For marine applications where post curing is rarely applied and is generally unnecessary, accelerated testing is best restricted to a maximum temperature of 40 °C[1] in order not to change the mechanism of any degradation process that might occur. Test periods in excess of two years may therefore be necessary. van der Beek,[9] from his studies of GRP, has assessed the acceleration factor for single-sided exposure to distilled water at 40 °C as 5–6 times that at ambient temperature. This factor could be higher for contact with seawater at ambient temperatures. Tests conducted under double-sided immersion conditions are more accelerated. However, even after ten years of double-sided immersion testing[1] property retention more than satisfied the working stress level requirements for normal boat hull construction, including those constructed using general purpose resins. Laminates constructed with more resistant systems, such as isophthalic acid resins and glass fibre containing a minimum of hydrolysable binder, generally perform better.[10,11]

Table 10 contains data comparing the performance of two marine Crystic polyester resins used to construct chopped strand mat, woven roving/ chopped strand mat and uni-directional/chopped strand mat laminates,

TABLE 10

Mechanical properties of GRP after long-term total immersion in tap water at 30°C

Property N/mm^2	Resin Immersion time	Marine orthophthalic resin			Marine isophthalic resin		
		Control	100 days	4 years	Control	100 days	4 years
	Construction						
Ultimate tensile strength	1	110	95	70	105	85	75
	2	209	175	126	203	181	162
	3	210	171	138	194	174	166
	4(0°)	273	230	174	267	243	215
	4(90°)	77	—	—	72	—	—
Tensile modulus	1	7000	6700	6200	7400	6800	6400
	2	13000	12400	11000	13700	12900	11900
	3	12800	11900	9100	12500	12200	11800
	4(0°)	14100	13800	13200	14400	14200	13800
	4(90°)	9900	—	—	9000	—	—
Flexural strength	1	187	162	112	175	160	128
	2	281	245	141	278	254	190
	3	314	273	160	321	286	245
	4(0°)	386	341	234	386	345	298
	4(90°)	166	142	88	167	151	122
Flexural modulus	1	6600	5800	4400	6400	6100	4800
	2	9900	9500	8000	9900	9300	8200
	3	10100	9500	7800	10300	9900	9500
	4(0°)	10200	9800	9500	10700	10500	9900
	4(90°)	6800	6300	4800	7500	7100	6200

Cure: MEKP + cobalt. Post-cure: 24 h RT, then 16 h at 40°C.
Construction:
1. Four layers of csm (450 g/m^2 per layer)—30% glass by wt.
2. csm/wr/csm/wr/csm/wr/csm (csm = 300 g/m^2, wr = 800 g/m^2)—44% glass by wt.
3. csm, three layers wr/csm combination mat (csm = 300 g/m^2, wr = 800 g/m^2)—43% glass by wt.
4. csm, three layers uni-directional/csm combination mat (csm = 300 g/m^2, uni-directional = 600 g/m^2)—42% glass by wt.

after 100 days' and 4 years' double-sided immersion in tap water at 30°C without gelcoat or paint protection. Retention of properties is very good under these harsh test conditions but, as expected, the isophthalic acid based system shows superior performance. Even so, orthophthalic acid based systems are perfectly acceptable for many marine applications, as proved by the countless examples of successful marine vessels and

TABLE 11

Mechanical properties of Kevlar and Kevlar/csm composites after immersion in tap water at 30°C

All properties given in N/mm^2	Woven Kevlar/csm composites		Woven Kevlar composites		
	Control	100 days	Control	100 days	1 year
Ultimate tensile strength	380	380	420	380	400
Tensile modulus	18 800	22 200	26 000	25 400	24 700
Flexural strength	300	310	255	270	260
Flexural modulus	16 100	18 400	23 400	21 500	23 900
Lap shear strength	9·4	7·3	12·8	11·2	14·3

structures made using them over the past 30 years. However, when optimum performance is required, isophthalic acid based systems give improved long-term properties as a result of their chemical make-up, rendering them less susceptible to chemical degradation.

Immersion tests have been carried out on Kevlar and Kevlar/chopped strand mat isophthalic (Crystic 272) polyester resin laminates (see Table 11).

Although the immersion programme on Kevlar composites is only at an interim stage it is clear that Kevlar-reinforced isophthalic polyester resin composites retain their properties equally as well as glass-reinforced isophthalic polyester composites. In fact, there is no significant loss of properties after the one-year immersion period.

DESIGN CHARACTERISTICS OF COMPOSITES RELATIVE TO MARINE STRUCTURES

The advantages of lightweight, high strength, high modulus reinforcements like polyaramid fibres are summarised in Figs 4 to 7 by comparing the specific properties of composites made using different reinforcement systems.

Figures 4 and 5 show that the specific tensile strengths and moduli of composites are functions of the reinforcement type—increasing when bi-directional and directional reinforcement are used and exhibiting significant increases in properties whenever Kevlar reinforcement is incorporated. Figures 6 and 7 show the specific compressive properties of composites. The specific compressive modulus of Kevlar-reinforced

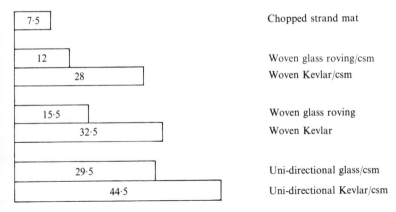

Fɪɢ. 4.   Specific tensile strength of fibre-reinforced polyester composites.

material is improved by incorporating chopped strand mat into the construction. The advantage of using high strength, high stiffness (or modulus) reinforcements is a reduction in the weight of a structure for a given set of loading conditions. This can be illustrated by considering the simple bend conditions of a laminate. By using different reinforcements having different moduli, retention of stiffness can be achieved with considerable weight saving (see Fig. 8).

If, however, the laminate deflection or stiffness is not the criterion but strength is, then the weight comparison, shown in Fig. 9, applies for a common bending condition.

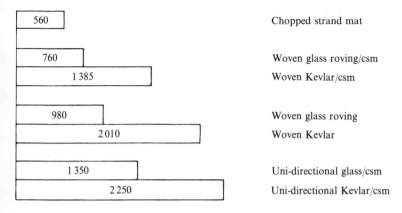

Fɪɢ. 5.   Specific tensile moduli of fibre-reinforced polyester composites.

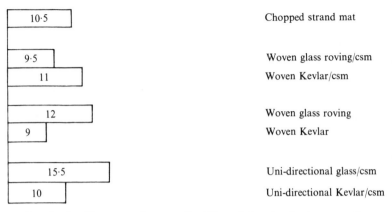

FIG. 6. Specific compressive strength of fibre-reinforced polyester composites.

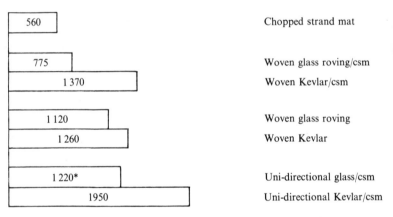

FIG. 7. Specific compressive moduli of fibre-reinforced polyester composites. (* Calculated value.)

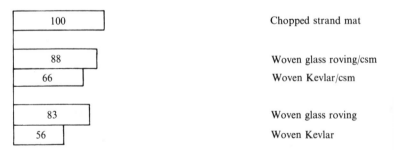

FIG. 8. Relative weights of fibre-reinforced polyester composites for equal stiffness in bend

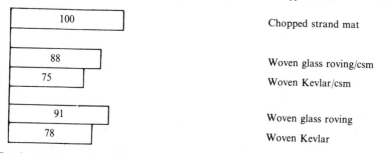

100	Chopped strand mat
88	Woven glass roving/csm
75	Woven Kevlar/csm
91	Woven glass roving
78	Woven Kevlar

Fig. 9. Relative weights of fibre-reinforced polyester composites for equal flexural strength.

The comparisons made in Figs 8 and 9 only apply if the laminate behaves according to normal bending theory, which breaks down if the laminate deflection exceeds half its thickness. Beyond this point the laminate will be subjected to membrane stresses. This is particularly applicable to high speed vessels where the slamming pressures are such that the external skin is undergoing large deflections.

The relationship between membrane effects and normal bending theory are shown in Fig. 10 for a panel subjected to a pressure normal to the surface, for both membrane and bending conditions.

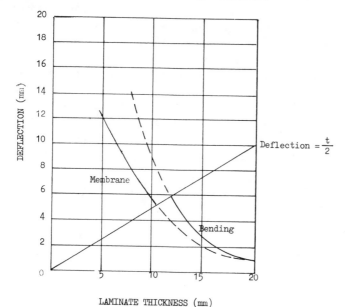

Fig. 10. Deflection versus thickness for a given loading condition (calculated using membrane and bending theories).

The curves show the effect of a pressure of 0·5 kg/cm² on a panel, made of material with tensile |modulus| 11 720 N/mm², with edges fully restrained against rotation and translation.

Material properties which contribute to the membrane effect are strength and modulus. Hence, lighter Kevlar-reinforced composite panels perform equally as well as heavier glass-reinforced panels for a given loading condition (see Fig. 11).

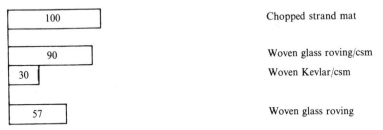

FIG. 11. Relative weights of fibre-reinforced polyester panels, designed using membrane theory, to the ultimate stress, for a pressure of 2·0 kg/cm².

If deflection is the criterion, then again Kevlar-reinforced composites show improvements compared with glass-reinforced material (see Fig. 12)

All the examples discussed so far give an indication of the advantages to be gained by using an engineered composite, incorporating high strength high modulus reinforcements, such as Kevlar-reinforced polyester composites. The significance of using this type of material can best be put into perspective by considering an actual structure. A 30 % weight saving can be achieved in the manufacture of a 13 m, 30 knot, patrol vessel by using a glass/Kevlar–polyester resin construction instead of the conventional glass–polyester construction, for identical loading conditions and structural analysis (see Table 12).

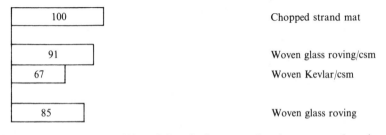

FIG. 12. Relative weights of fibre-reinforced polyester panels under pressure and membrane

TABLE 12

Comparative weights for the hull and deck structure of a 13 m patrol vessel constructed in: (i) Kevlar/csm/polyester resin. (ii) Glass (wr/csm)/polyester resin

Composite	Material weight (kg): Hull and deck structure			
	Glass	Kevlar	Resin inc. gelcoat	Total
Kevlar/csm (480 g/m² K49) (300 g/m² csm)	440	550	1 430	2 420
Glass/wr/csm (800 g/m² wr)	1 510	—	1 940	3 450

The weight saving achieved in the Kevlar-reinforced structure means greater vessel speed and/or greater range for a given power.

The exercise carried out on a 'Spear' class patrol boat by Fairey Allday Marine Limited enabled a direct performance comparison to be made between identical 'Spear' patrol vessels constructed in glass-reinforced polyester resin and Kevlar/csm reinforced polyester resin. The measured weight savings for each of three mouldings are shown in Table 13.

The Kevlar specification was calculated on an equal strength and stiffness basis. Chopped strand mat was incorporated in the construction between layers of Kevlar woven roving. The resultant structure was 20 % lighter than the all glass-reinforced version, giving a total weight-saving, on the all-up weight of the boat, of 9·1 %. The boats were tested in the Solent—the Kevlar version, using the same horsepower as the glass boat, showed a 1·7 knot

TABLE 13

Weight saving achieved for the Kevlar/csm reinforced 9 m 'Spear' patrol boat compared with the equivalent all glass-reinforced structure

Mouldings	Weight saved (kg)
Hull, with all internal stiffening, girders and bulkheads	304
Deck structure, including cabin top and part of the cockpit	101
Wheelhouse and aft bulkhead	25
Total weight saved	430

speed increase and a fuel consumption of 5 gallons per hour less when at full throttle, which means reduced running costs.

Sound level readings were taken within the boats—the noise level in the Kevlar version was two points less on the decibel meter.

CONCLUSIONS

(a) Resin systems must be chosen with careful reference to the proposed application, but for the best long-term property retention, isophthalic acid based polyesters are a natural choice because of their improved water resistance and heat resistance properties.

(b) Polyester resins can be used in conjunction with polyaramid (Kevlar) reinforcement to produce composites with excellent mechanical properties.

(c) The mechanical properties of glass-reinforced polyester composites can, in general, be exceeded by the use of the lighter weight reinforcements.

(d) Considerable weight saving can be achieved in marine structures by using polyaramid reinforcements, resulting in improved performance.

(e) Structural and economic optimisation is best achieved by combining Kevlar and glass reinforcements which, when used in conjunction with an isophthalic polyester (Crystic 272) resin, provide composites with very suitable properties for marine application.

ACKNOWLEDGEMENTS

The authors wish to thank Scott Bader Company Limited and Du Pont de Nemours International SA for their permission to publish the information used in this paper, and Fairey Allday Marine Limited for their permission to publish data on the 'Spear' patrol boat.

REFERENCES

1. CLARKE, G. M. and NORWOOD, L. S., *Reinforced plastics* (November, 1978) 370.
2. B.S.4994: 'Vessels and Tanks in Reinforced Plastics'.

3. FAREBROTHER, T., *Crystic 272 Laminate Properties*, Scott Bader Materials Science Internal Report, December, 1979.
4. NORWOOD, L. S. and MILLMAN, A. F., *Composites* (January, 1980) 39.
5. *Military Handbook—Plastics for Aerospace Vehicles, Part 1. Reinforced Plastics*, MIL-HDBK-17A, January, 1971.
6. JOHNSON, A. F., *Engineering Design Properties of GRP BPF/NPL*.
7. MARCHANT, A., Southampton Boat Show Symposium, 1978, PRI Meeting: Alternative Materials Reviews.
8. MARCHANT, A. and Associates, Internal Test Reports.
9. VAN DER BEEK, M. H. B., Scott Bader Symposium, March 1979, Creaton Hall, Northamptonshire.
10. Scott Bader literature—Crystic 489A leaflet.
11. NORWOOD, L. S., EDGELL, D. W. and HANKIN, A. G., 12th Reinforced Plastics Conference of the BPF, Brighton, November, 1980. Paper 39, p. 185.

13

The Testing and Analysis of Novel Top-Hat Stiffener Fabrication Methods for Use in GRP Ships

A. K. Green and W. H. Bowyer

*Fulmer Research Laboratories Limited,
Stoke Poges, Slough SL2 4QD, England*

ABSTRACT

Reinforced and unreinforced specimens representative of GRP ship hull/ frame structures have been tested by a slow pull-off method. Changes in bondline stress distribution and overall test system compliance produced by the three loading modes used affected bondline crack initiation loads and the crack propagation behaviour. The loading mode did not influence the ranking order of the various fabrication techniques. As a simulation of the behaviour of the structure under shock loading, the use of the most severe, centre clamp, loading method is recommended.

The beneficial effects of incorporating a low modulus acrylic matrix in the bondline stress concentration under the heel of the stiffener web are described. The crucial role of the stiffness of the stiffener frame web/flange corner in determining the failure pattern is discussed. It is suggested that local reductions in stiffness in the web/flange corner may have a beneficial effect in inhibiting damage to the frame/hull connection during overloads.

INTRODUCTION

The use of the familiar hand laid-up glass-fibre/polyester resin GRP for ship construction is commonplace. The usual reasons for its choice involve lower tooling and fabrication costs for limited production runs compared with other materials, the possibility of construction by relatively unskilled personnel and the ready applicability of laminated products to the fabrication of the complex curved shapes that occur in hulls. It is rarely the properties of GRP *per se* that cause it to be selected. An exception to this is its use in the 'Hunt' class of naval MCMV's that are just entering service with the Royal Navy, where the non-magnetic properties of GRP were of prime importance in dictating its selection. This is a demanding application and much background work was carried out on the GRP materials, their fabrication methods and quality control systems, before the 60 m long vessels were constructed, as described in references 1 and 2.

A fabrication method was chosen that incorporates a single skin hull constructed with laminations of woven glass roving fabric. Due to the inherently low elastic modulus of GRP, top-hat stiffeners are used on the hull, decks and bulkheads to achieve the required overall stiffness in the hull structure. Top-hat stiffeners are in widespread use in the GRP shipbuilding industry since they can be tailored readily to the complex curvature of hulls and provide built-in buoyancy by the fabrication method of laminating over rigid polymeric foam cores. The conventional fabrication method involves the lamination of the hull shell and flat or gently curved deck and bulkhead panels. Rigid foam cores are bonded to these unstiffened structures where stiffness is required and GRP laminations are built up around the cores. When constructing a large hull, it is not uncommon for a substantial delay to occur between shell lamination and the addition of the stiffeners. The early exploratory work[1] established that a delay of greater than 7 days prior to stiffener lamination led to an excessively weak secondary bond between hull shell and the flanges of the stiffener, if no special precautions were taken. Surface treatments for the hull immediately prior to stiffener lamination were evolved, these involving abrasion, wiping with solvents and the use of peel plies. These precautions enable stiffeners to be fabricated that perform satisfactorily in most circumstances, with a secondary bond whose transverse tensile strength equals the interlaminar tensile strength of the main hull laminate.[3]

However, the service requirements for naval minesweepers and minehunters include resistance to the effects of underwater explosions close to the hull. Shock testing of stiffened panels representative of hull designs revealed a tendency for the bond between the stiffener and the underlying panel to fail,[1] producing a significant stiffness reduction in the structure. This stiffness loss would be critical to the ship's performance and mechanical fasteners were introduced into the structure to prevent stiffener separation and maintain structural integrity. The method adopted was through-bolting of the stiffener flanges to the hull shell.[2,3] The non-magnetic hull specification, the corrosive marine environment and the fatigue and vibration service requirements necessitated the use of, initially, aluminium–silicon–bronze and, currently, titanium nuts, bolts and washers. The additional obvious requirement of maintaining the water tightness of the hull led to an insertion scheme involving boring and counterboring the hull laminate, the use of sealants and manually torque tightening the bolts and nuts from both inside and outside the hull following demoulding. The high costs associated with both the fasteners and the insertion procedure led to the study of other fabrication methods that could resist the effects of shock loading and could be implemented simply and cheaply in the shipyard.

This programme investigated the possibility of a direct replacement by a relatively inexpensive commercially available mechanical fastener and insertion scheme for the titanium through bolts. However, all mechanical fasteners are only a partial solution to the problem, since the bond failure initiates at the stiffener web/flange corner, remote from the fastener, and all the fastener can do is act as a crack arrester. To achieve a fundamental improvement in performance, it is necessary to inhibit the crack initiation process. Efforts were made to achieve this by altering the lamination procedure and/or incorporating in selected regions a higher toughness matrix than the conventional polyester resin. These modified lamination procedures were aimed at utilising the available potential bond area of the currently redundant region under the foam core, and reducing the stress concentration at the stiffener web/flange corner. In addition to testing the performance of these reinforcement methods by slow pull-off testing of representative top-hat section specimens, the production practicality of all methods was assessed. Unreinforced and titanium bolt reinforced specimens were also tested to provide a basis for comparison of the suggested substitute reinforcement methods.

FABRICATION METHODS

The fabrication methods for the test specimens used in this study are described in detail in reference 4. Briefly, five specimen types were fabricated as stiffened panels, from which 150 mm wide test specimens were cut, as follows.

(a) Unreinforced

Representative, at approximately $\frac{2}{3}$ scale, of a typical below water-line hull stiffener with a $7\frac{1}{2}°$ inclined top-hat side web. The fabrication used

Fothergill and Harvey Y920 woven roving fabric of 814 g/m^2 and BP Cellobond A2785CV polyester resin for both base panel and stiffener, with additional 630 g/m^2 unidirectional glass roving tape plies in the stiffener top and glass roving bundles as an infill at the stiffener flange/web corner.

(b) Titanium Bolt Reinforced

As (a) with an M10 titanium bolt inserted through the flange and base panel, torqued to 27·1 Nm, using Bostik 2115.5 polysulphide sealant as an interlay. The bolt head was counterbored flush to the base panel and bolts and holes were degreased with styrene monomer prior to assembly. This bolt size and procedure were adopted, following consultation with Vosper Thorneycroft, as representative of shipyard production practice.

(c) Stainless Steel Screw Reinforced

As (a) with three 25·4 mm No. 8 Type B self-tapping screws inserted into each flange. The screws were standard commercial 'Supadriv' screws made of 18Cr/9Ni/3Cu austenitic stainless steel. The screws were pneumatically driven into holes filled with liquid A2785CV resin, which was then allowed to cure. Driving torque and speed and pilot hole diameter were optimised as described in reference 4. The strength of three screws is approximately equivalent to that of one M10 titanium bolt.

(d) Complex Stitched Cloth

This involved the use of a lightweight 220 g/m^2 plain weave glass cloth stitched to the Y920 cloth with 1420 denier Kevlar yarn along two parallel lines, 200 mm apart, this being the width of the stiffener foam core base. During lamination, the Y920 cloth formed the topmost layer of the base panel and the lightweight cloth was wrapped completely around the foam core. This fabrication method is an attempt to utilise the redundant stiffener base area and reduce the stress concentration in the stiffener web/flange corner by bridging the acute angle between the stiffener web and the base panel with Kevlar fibre stitching.

(e) Complex Stitched Cloth/High Toughness Matrix

This method was as described in (d) except that lamination across the stiffener base and in the critical stiffener web/flange corner was carried out using a high toughness matrix. A flexibilised acrylic system, Flexon 241, marketed as a high peel strength adhesive, was used. The cloth was impregnated by the adhesive initiator dispersed in a solvent, the solvent was allowed to evaporate and lamination was accomplished by hand-working in the gelatinous adhesive. Specimen lamination was completed using the conventional A2785CV polyester resin, with no problems of incompatibility between the two matrix systems being encountered.

SPECIMEN TESTING

The specimens were assessed by slow pull-off testing of the top-hat stiffener, at a displacement rate of 1 mm/min, on a 570 kN Mand servo-hydraulic testing machine. Three support systems for the specimen base panel were used. All specimen variants were tested as shown in Fig. 1, with the base panel clamped to the bed of the testing machine using three clamps and a 125 mm wide loading shackle that distributes the load over the stiffener top area. Unreinforced specimens were similarly tested, but with the omission of the central clamp on the specimen base, i.e. only the two outer clamps were used. All specimen variants were further tested using only a single central base panel clamp, as shown schematically in Fig. 2, comprising a 50 mm square section mild steel bar. For these tests, an additional 38 mm wide by 6·3 mm deep steel strip was placed centrally within the 125 mm wide top-hat loading shackle to concentrate the loading at the top-hat centre. Triplicate testing was performed in most cases. The stitched cloth panel using only polyester resin lamination had been manufactured imperfectly, as described in detail in reference 4, such that the position of one of the lines

Fig. 1.   Three clamp loading arrangement. Crown copyright.

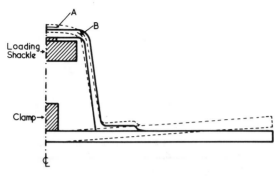

FIG. 2. Schematic loading arrangement, centre clamped.

of stitching did not coincide perfectly with the stiffener web/flange corner. Accordingly, four specimens of this type were tested by three-clamp loading, one of which was nominally perfect, the other three being variably faulty. The three such specimens tested in single clamp loading were all nominally perfect.

EXPERIMENTAL RESULTS

The experimental results for unreinforced specimens tested by one, two and three clamp loading are presented in Table 1. The data for the various reinforced specimens tested by one and three clamp loading are presented in Table 2. Figures for work done are calculated from planimeter

TABLE 1
Pull-off test results for unreinforced top-hat specimens
(Displacement rate, 1 mm/min)

Clamping system	Maximum load (kN)	Secant stiffness (kN mm^{-1})	Total work done (J)	Work done to first load drop (J)
Centre clamp	14·6	2·6 (2–12 kN)	47·0	47·0
	21·5	3·0 (5–20 kN)	80·3	80·3
	23·4	2·9 (5–20 kN)	97·2	97·2
Two clamp	33·1	2·41 (5–15 kN)	277	—
	32·1	2·35 (5–15 kN)	261	—
Three clamp	32·2	8·33(10–25 kN)	176	129
	29·2	7·14(10–25 kN)	261	115
	31·5	7·50(10–25 kN)	224	139

TABLE 2
Pull-off test results for reinforced top-hat specimens
(Displacement rate, 1 mm/min)

Specimen reinforcement	Maximum load (kN)		Secant stiffness (kN mm^{-1})		Total work done (J)		Work done to first load drop (J)	
	Three clamp	One clamp	Three clamp (10–25 kN)	One clamp (5–20 kN)	Three clamp	One clamp	Three clamp	One clamp
Ti bolts	39·3	36·0	11·5	3·2	1 230	640·5	117	68·9
	40·4	38·9	10·7	3·2	1 150	839·2	81·2	99·1
	39·8	38·1	13·1	3·1	1 040	>1 027·0†	86·4	73·5
Self-tapping screws	41·0	33·5	10·7	3·1	281	308·1	104	86·7
	43·0	34·2	10·7	3·1	331	375·3	86·0	97·7
	40·4	34·4	9·67	3·1	336	345·5	130	72·4
Stitched cloth/ Polyester resin	34·8	24·3	9·4	3·3	158	92·8	59·9	92·8
	32·3	19·9	11·5	3·4*	201	72·6	53·7	57·9
	26·3	27·5	10·0	3·2	228	121·0	41·3	121·0
	24·0		9·0		121		33·6	
Stitched cloth/ Acrylic bonded	43·4	31·9	10·00	2·9	237	423·5	116	214·0
	41·9	32·9	9·4	2·9	239	303·6	120	230·5
	44·5	31·2	10·7	2·9	264	296·8	131	205·3

* Secant extrapolated to 20 kN.
† Test terminated when bolt head continued to pull through at constant load.

measurements of the specimen load/displacement records. Representative load/displacement records for the five specimen variants tested by centre clamp loading are given in Fig. 3, and by three clamp loading in Fig. 4. The records for the unreinforced specimens are repeated in Fig. 5, together with a representative record of an unreinforced specimen tested by two clamp loading. In all cases, audible evidence of damage occurred prior to any visible damage to the specimen during testing.

Unreinforced Specimens

It is obvious from Fig. 5 that the mechanical response of the specimens during slow pull-off testing is influenced strongly by the manner in which they are loaded. The loading arrangement also affected the sequence of failure. It is obvious that the load/displacement history for centre clamped specimens is essentially linear to failure. During testing, the specimens assumed the position indicated by the dotted lines in Fig. 2. Noise generated was associated with the development of a network of fine cracks in the region A in Fig. 2. The top of the top-hat deformed as indicated, but the side webs of the top-hat remained essentially straight under load. No damage was visible at, or close to, the secondary bond line prior to catastrophic failure, which was by complete separation of one stiffener flange from the base panel along the secondary bond line.

With the two clamp loaded specimen, no visible—and little audible—damage occurred until the two small load drops at about 15 kN. The stiffener flange/web corner cracked internally through most of its thickness along interlaminar planes, first on one side of the top-hat, then the other, at these load drops, but no cracks were produced either in the secondary bond line or in the base panel. The load increased evenly until catastrophic failure of the specimen at its maximum load by complete separation of one of the stiffener flanges from the base panel. No cracks were observed in the secondary bond line prior to final failure. A small amount of delamination was apparent in the flange area of the base panel after test, between the topmost and the adjacent roving ply.

The failure sequence for three clamp loaded specimens was described in detail in reference 4. Audible evidence of damage occurred prior to any visible damage to the specimen. This noise was much louder than that associated with the centre and two clamp tests. The first crack appeared at, or just before, the point of maximum load, under the heel of the stiffener side web, in the secondary bond line, and arrested after only a few millimetres of growth. As displacement increased, the stiffener cracked internally along interlaminar planes in the web/flange corner and the base

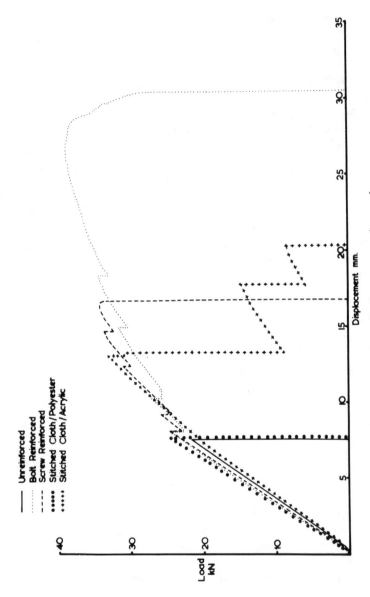

FIG. 3.   Top-hat pull-off tests, centre clamp only.

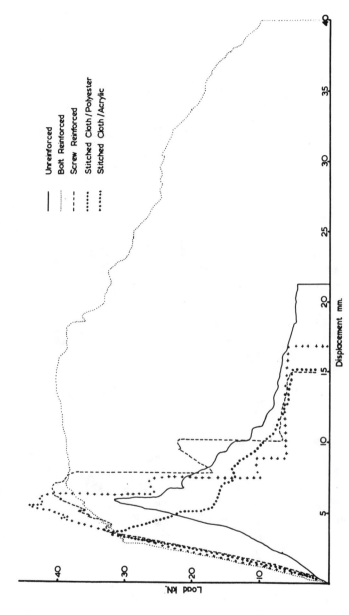

FIG. 4. Top-hat pull-off tests, three clamp loading. Crown copyright.

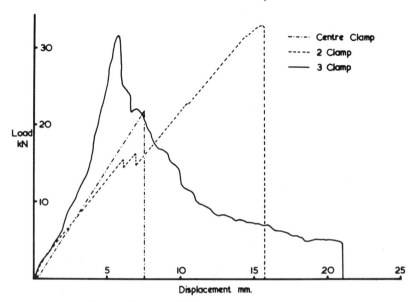

FIG. 5.   Top-hat pull-off tests, unreinforced specimens.

panel started to delaminate under the crack in the secondary bond line. This base panel delamination grew along the region adjacent to the flange and extended down three of four plies into the base panel. Final failure was by rapid crack growth along the secondary bond line leading to separation of a stiffener flange from the base panel. Following growth of the initial crack, crack growth in the secondary bond line was minimal prior to final failure.

Reinforced Specimens—Centre Clamp Loaded

The load displacement records were essentially linear in all cases up to a load in the range 21–24 kN with noise generation associated with the growth of a crack network as indicated at A in Fig. 2. The stitched cloth all polyester resin construction specimens failed at, or shortly after, this point by complete separation of one stiffener flange from the base panel along the secondary bond line, with no damage visible at, or close to, the secondary bond line prior to final failure.

In the mechanically fastened specimens, the small load drops evident in Fig. 3 at about 24 kN were associated with the formation of a crack in the secondary bond line under the heel of the stiffener side web that was arrested at the bolt or just before the line of screws. The second load drop at

~ 26 kN was associated with a similar crack under the other stiffener flange of the specimen. The load increased as the crack either grew around the bolt and the nut or bolt head started to pull through the laminate, or the crack grew up to, and a few millimetres past, the line of screws. The small load drops at ~ 32 kN were associated with the formation of gross interlaminar cracks within the corners of the top-hat, in the position indicated by B in Fig. 2. Catastrophic failure occurred by pull-through of either the bolt head or the nut and washer, or by snapping of the screws with separation along the secondary bond line under one flange.

For the stitched cloth specimen incorporating the acrylic matrix, the loss of linearity in the load displacement record at ~ 20 kN was associated with the appearance of a small interlaminar crack, under the heel of the stiffener side web between the first and second plies of the base panel, i.e. below the secondary bond line. No load drop occurred and the specimens bore increasing loads with a marginally reduced stiffness. As the load increased, the crack grew only slightly until failure occurred by rapid crack growth of the pre-existing crack in both directions within the base panel and by partial failure of the secondary bond line under the toe of the stiffener flange, i.e. remote from the acrylic matrix region, with the load drop to ~ 10 kN. The further load increases were associated with progressive delamination of the top ply from the base panel. Final failure was by tearing of both the heavyweight and lightweight glass cloths. No cracks grew into, or within, the acrylic matrix region prior to final failure.

Reinforced Specimens—Three-clamp Loaded

The test results obtained with these specimens are described in more detail in reference 4 and are briefly summarised here. In the specimens reinforced with mechanical fasteners the first crack always appeared in the range 29–34 kN, corresponding to the peak loads for three-clamp loaded unreinforced specimens, and a discontinuity was apparent in the load/displacement records. Interlaminar cracks growing in the base panel arrested temporarily at the fasteners while the load increased. In bolted specimens, the cracks grew around the bolt until, at maximum load, the titanium nuts and washers started to pull through the stiffener flange. The large amount of subsequent deformation increased the degree of pull through until the flange failed, extensively damaged. In screw-reinforced specimens, local yielding of the screws near the secondary bond line apparently occurred at the peak load, and final failure was by a combination of screw snapping and screw pull out from the specimen base.

For stitched cloth specimens, audible damage usually occurred in the

20–25 kN load range. The initial visible failure event was delamination cracking within the base panel, immediately below the stitched ply, under the heel of the stiffener side web. The four results in Table 2, for specimens with only polyester resin matrix, show a dependence of peak load on the stitching position. The highest peak load occurred in the nominally perfect specimen and the lowest in the most defective specimen. Following maximum load in these specimens, the load supported fell gradually and a progressive pattern of interlaminar cracks grew along and below the secondary bond line. For the specimens incorporating the acrylic matrix, little visible damage was apparent prior to the peak load at 40–45 kN. Ultimate failure occurred by snapping of the Kevlar stitching and tearing of the lightweight cloth across the specimen width. The acrylic impregnated roving bundle at the stiffener web/flange corner then separated with increasing displacement, followed by interlaminar cracking along the secondary bond line and the adjacent base panel ply. At final failure, the separated surfaces of the stiffener flange and the base panel were still linked by individual fibres and fibre clumps, these being the remains of the acrylic impregnated roving bundle.

DISCUSSION OF RESULTS

It is obvious that significant differences in performance have been produced both by the specimen fabrication method and testing mode. These differences may be summarised as follows.

(1) For centre clamp loading, failure or the first visible sign of damage at, or close to, the secondary bond line occurs in the load range 20–25 kN: For three clamp loading, the comparable damage occurs in the load range 30–34 kN.

(2) For specimens with an acrylic matrix region, the load/displacement histories are atypical for each loading mode. When centre clamp loaded, no load drops occur prior to peak load, even though damage is visible prior to peak load. When three clamp loaded, no damage is visible in these specimens prior to failure.

(3) For unreinforced specimens, the peak loads attained and the work done to failure are significantly lower for centre clamp loading than for either two or three clamp loading.

(4) Similar maximum loads were achieved by the bolted specimens, whether centre clamp or three clamp loaded. These peak loads were associated with large test machine displacements, when either the bolt head

or the nut started to pull through the GRP, and resulted in the high total work done to fail bolt-reinforced specimens.

(5) Similar peak loads were achieved by the screw reinforced and acrylic matrix specimens for both loading modes. For both modes, substantial damage was visible in the screw reinforced specimens prior to the peak load, whereas little or no visible damage was apparent in the acrylic matrix specimens.

(6) When considering total work to failure, the comparative merits of the reinforcement methods are similar for both clamping modes.

(7) An interesting effect is observed when comparing work done to first load drop. When three clamp loading is used, the stitched cloth/polyester resin specimens are markedly inferior to the other four specimen types, these four types exhibiting similar performance. For centre clamp loading, the stitched cloth/acrylic matrix specimens are vastly superior to all other specimen types, the performances of which are similar. However, this latter distinction does not give a complete picture, since small cracks were visible in these specimens and this damage occurred without an associated load drop.

THE EFFECT OF CLAMPING MODE ON STRESS DISTRIBUTION

1. Centre Clamp Loading

This loading mode is relatively simple, in that the base panel can be considered as two cantilevers, each rigidly fixed by the centre clamp, and loaded via the top-hat flange. The energy stored in the system is contained within the two unclamped regions of the base panel under the top-hat, and within the side webs and top of the top-hat. The load was concentrated at the centre of the top-hat, as illustrated in Fig. 2, and caused significant deformation of the top of the top-hat and little, if any, bending of the side webs. This is in contrast to the result predicted by finite element analysis[5] of significant side web bending for this loading condition, when the loading is distributed uniformly across the stiffener top (Loadcase 5, reference 5), as shown in Fig. 6(a). This analysis further predicts a very high tensile stress across the secondary bond line under the heel of the side web and a smaller compressive stress under the root of the stiffener flange. This compressive stress is associated with bending of the side web as predicted by the analysis, and both are indicated in Fig. 6. It is likely that in this work, where little side

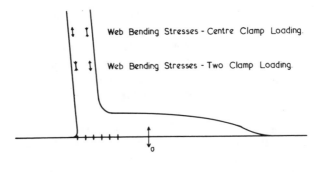

FIG. 6(a).   Stress distribution in stiffener flange region.

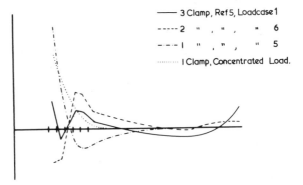

FIG. 6(b).   Stress distribution in stiffener flange region.

web bending is induced as a result of the concentrated loading of the top-hat, the predicted compressive stress across the secondary bond line will be much reduced, or even absent. A suggested stress distribution is shown by the dotted line in Fig. 6(b), indicating a high stress concentration under the heel of the side web, initiating failure at a low load. As the crack grows, the system compliance increases simultaneously with an overall reduction in load, due to the effective fixed displacement condition, and stored elastic energy is released to propagate the crack. The balance of these factors will determine whether the crack propagates to failure or arrests, and rapid crack propagation has occurred in this loading mode.

2. Two Clamp Loading

In this loading mode, the length of the base panel between the clamps is free to flex, and energy is stored within this flexed portion of the base panel, the stiffener side webs and the flanges that bend in response to the base panel bending. The loading was distributed evenly across the stiffener top, bending of which is thus small. The finite element analysis for this mode (Loadcase 6, reference 5) shown in Fig. 6, predicts a compressive stress under the heel of the stiffener side web, a similar magnitude tensile stress under the flange root and pronounced side web bending. This bending is in the opposite sense to that predicted for centre clamp loading and produces close to mirror image stress distributions across the bond line for the centre clamp and two clamp loading modes. The predicted side web bending is borne out by the observed initial failure event in these specimens, of cracking within the side web/flange corner, relieving the side web bending stresses. Relief of this bending stress presumably also reduces the predicted compressive stress across the secondary bond line under the heel of the side web, allowing a tensile stress to be built up as displacement increases. If so, this analysis then predicts the ultimate rapid failure observed, since the cracked side web/flange corner can effectively act as a compliant hinge and produce rapid catastrophic failure by a cleavage mechanism, once fracture initiates under the heel of the side web. However, the displacement required will be large, as observed, since any remaining tendency for compressive stress generation under the side web heel must be overcome before the necessary tensile stress can be built up at this point to initiate secondary bond line fracture.

3. Three Clamp Loading

This complex loading mode limits bending of the base panel to the region close to the flange. This bending can be substantial, as shown by Fig. 9 of reference 4, even though the overall system compliance is relatively low. Stored energy is contained in a relatively confined region of the specimen and flexing of the stiffener top is slight due to the distributed loading used. The stress distribution suggested by the finite element analysis for this loading mode is shown in Fig. 6(b) (Loadcase 1, reference 5). The analysis of web bending unfortunately does not consider this loading mode explicitly, but by comparison with the analyses of Loadcases 5 and 6 one can predict that both faces of the side web will be in tension, i.e. the clamping mode restricts the tendency for side web bending. The relatively small tensile stress concentration under the heel of the side web necessitated a high applied load for crack initiation compared with that for centre clamp loading. This loading mode has produced a system with an overall low compliance, and a consequent rapid load drop for an increment of crack growth. This rapidly falling load becomes insufficient to maintain the high force required for crack propagation, due to the relatively small crack tip stress concentration factor in this case, and the crack arrests. By contrast, when single clamp loaded, the load drops at a lower rate as a function of crack length, due to the overall higher initial system compliance. This behaviour, coupled with the very high geometric crack tip stress concentration factor when single clamp loaded, results in the relatively low force necessary for crack propagation to be maintained. Thus, crack arrest occurs in three clamp loading, and catastrophic failure occurs in single clamp loading, even though the stored elastic energy at crack initiation (see Table 1) is higher in the former case than the latter. This crack initiation and arrest is a function of system stiffness, and the presence of a mechanical fastener some distance from the crack initiation site should have little effect on crack initiation behaviour, as observed. Similarly, the small local change in stiffness produced by the fibres bridging the secondary bond line in stitched cloth specimens can be expected to have only a slight effect on crack initiation. However, the relatively low modulus acrylic matrix region in this critical area can be expected to produce the dramatic effect observed, since the relative increase in compliance it confers to this region has the function of redistributing stresses and reducing the effect of the geometric stress concentration. The observation of cracking within the base panel following crack initiation in the secondary bond line suggests that stress redistribution is occurring with the stiffener web/flange corner acting as a hinge.

4. Comparison of Loading Modes

In this work we are attempting to gain an insight into the behaviour of shock loaded stiffeners and which of the loading modes used most closely represents the service condition is a moot point. It was shown in reference 5 that the magnitude and distribution of tensile stresses across the secondary bond line are very sensitive to the form of the applied load. In particular, they are dependent on bending of the stiffener side web, as has been apparently confirmed in this work. Shock loading in service could produce complex hull flexure such that it is possible that no two stiffener frames will experience the same deflection history. Consequently, it seems sensible to assess candidate fabrication methods by the most severe slow pull-off testing system, so that the worst service case may perhaps be simulated.

The centre clamp loading mode is apparently the most severe, as demonstrated by the relatively low crack initiation loads and the catastrophic ultimate failure along the secondary bond line. This behaviour is a consequence of the high tensile stress concentration under the heel of the stiffener web and the readily available stored energy reservoir as a result of the relatively compliant test set up. By contrast, the three clamp loading mode requires higher crack initiation loads resulting from the relatively small tensile stress concentration under the heel of the side web, and crack arrest is produced as a result of the relatively stiff test set up providing only a limited stored energy reservoir available for crack driving.

Further, it is known that shock loading can cause complete separation of unreinforced stiffener frames from hulls and cracks in the secondary bond line arrested at through bolts. Both these phenomena occur in the centre clamped slow pull-off test, suggesting this loading mode to be the most representative of shock loading. By contrast, cracks in unreinforced secondary bond lines that arrest after a short propagation length, as observed in three clamp loading, are not observed in shock tests. However, it is known that cracks can occur below the secondary bond line in shock testing, and this phenomenon was mostly observed in this work in three clamp loaded specimens. Consequently, although centre clamping is suggested as the most appropriate simulation for the shock loading condition, it is evidently not completely satisfactory.

It is noteworthy that the stitched cloth/acrylic matrix specimens behaved in an atypical manner in both loading modes. When specimens are three clamp loaded, the higher failure strain, lower modulus (compared to polyester resin) matrix at the critical stress concentration redistributes stresses away from the stress concentration. This delays crack initiation until a higher applied load. When centre clamp loading is used, crack initiation at the secondary bond line under the stiffener side web heel, as occurs in all other specimens, is suppressed in specimens incorporating the acrylic matrix. The small performance improvement of the stitched cloth all polyester resin specimens compared with those unreinforced suggests that these improvements result from the use of the acrylic matrix rather than the stitching.

Throughout this work it has become apparent that in the standard stiffener construction technique, bending of the side web has an important influence on the stress condition at the secondary bond line. Further, failure has often involved cracking within the web/flange corner, apparently to relieve web bending stresses. Hence, the stiffness of the web/flange corner must be important in determining the bond line stress profile. The overall results suggest that it may be advantageous to make the web/flange corner more compliant, so that web bending stresses may be more readily accommodated by the structure as a whole, without damaging the stiffener to hull connection. Further, the relocation of the damage zone by the acrylic matrix into regions laminated using polyester resin suggests that it could be advantageous to use a more compliant matrix for hull lamination in the highly stressed regions under the frame flanges. By using a more compliant—and also inherently tough—matrix, such as the acrylic, in these regions, it should be possible to both redistribute the concentrated stresses under the web heel over a wider region of the structure and inhibit crack initiation, and so improve resistance to damage caused by overloads.

CONCLUSIONS

The loading mode for slow pull-off testing of top-hat stiffened specimens determines both the crack initiation load and whether the failure is progressive or catastrophic. The centre clamped loading mode is apparently the most representative of shock loading service conditions, by consideration of both the failure mode of, and the damage pattern in, specimens. Specimens reinforced with mechanical fasteners do not show any inhibition of crack initiation, but crack propagation from the initiation site is hindered by the presence of fasteners through the bond line. The use of the acrylic matrix around the geometric stress concentration at the heel of the stiffener web increases the load required for crack initiation. This is a result of the relatively low modulus of the acrylic matrix, that redistributes stresses away from the stress concentrator, and the higher failure strain of the acrylic matrix compared with polyester resin. The stiffener web/flange corner plays an important role in determining the stress distribution across the secondary bond line and in defining the failure sequence. Local increases in compliance of both the stiffener and hull laminate in this region could have a beneficial effect on performance.

ACKNOWLEDGEMENT

The work described is part of a programme carried out with the support of the Procurement Executive, Ministry of Defence.

REFERENCES

1. Dixon, R. H., Ramsey, B. W. and Usher, P. J., Design and build of the GRP hull of HMS Wilton, *Proc. Symp. on GRP Ship Construction*, London, RINA, 1973, 1–32.

2. SMITH, C. S., Structural problems in the design of GRP ships, *Proc. Symp. on GRP Ship Construction*, London, RINA, 1973, 33–56.
3. SMITH, C. S. and PATTISON, D., Design of structural connections in GRP ship and boat hulls, *Proc. Conf. Designing with Fibre Reinforced Materials*, London, I. Mech. E., 1977, 33–6.
4. GREEN, A. K. and BOWYER, W. H., The development of improved attachment methods for stiffening frames on large GRP panels, *Composites*, **12** (1981) 49–55.
5. Unpublished MoD data.

14

The Development of Improved FRP Laminates for Ship Hull Construction

J. BIRD AND R. C. ALLAN

Admiralty Marine Technology Establishment, Ministry of Defence, St. Leonard's Hill, Dunfermline, Fife KY11 5PW, Scotland

ABSTRACT

This paper describes recent research aimed at improving the quality and efficiency of fibre reinforced plastic laminates for use in the construction of ship hulls. Methods of improving through thickness properties, where delays in lamination occur, are reported. The results of long term seawater immersion tests are also presented. The tests, carried out in three point bending, suggest that high stress levels, 80% of failure load in bending, are necessary before significant degradation takes place with isophthalic polyester/'E' glass woven roving laminate combinations.

INTRODUCTION

It was around 1951 when the decision was taken to evaluate glass reinforced plastic for Naval service. The use of GRP, a radical departure from the more conventional materials of hull construction, was proposed for minesweeper hulls because of its unique properties.[1-4] The use of GRP for these hulls was thought most suitable because it had the following advantages:

(1) It is non-magnetic.
(2) It is resistant to induced eddy currents. Aluminium framed wooden minesweepers are susceptible to induced eddy currents in the framing, so creating a magnetic signature.

(3) It is corrosion-resistant. Both wood and steel eventually degenerate because of contact with seawater unless the protection system is extremely efficient and well maintained.
(4) It is resistant to marine organisms.
(5) It has a high strength to weight ratio.
(6) Fabrication of complex shapes in a variety of thicknesses is achievable at will.
(7) The use of a simple once-built hull mould allows many identical vessels to be built with a high degree of finish using relatively unskilled labour.
(8) It is resistant to warping and shrinkage.
(9) It has low thermal conductivity.

The laminates used in minesweeper construction were manufactured from 'E' glass woven rovings and isophthalic polyester resin. Both were chosen on their availability and the woven rovings because they produced adequate strength at reasonable cost. The labour-intensive nature of GRP hull construction, however, means the total hull cost is significantly more expensive than the material costs.

A possible means of reducing the cost of mine countermeasures vessel hull construction in future ships of the class would be to replace the original shell laminate, i.e. polyester resin reinforced with $820 \, g/m^2$ balanced woven rovings (WR) having approximately equal distribution of fibres in longitudinal and transverse directions, with a laminate reinforced by plies of $600 \, g/m^2$ chopped strand mat (CSM) alternating with plies of $820 \, g/m^2$ WR in which the fibre distribution is biased in the longitudinal direction of the hull. The advantage of biased WR reinforcement is that it should be possible to compensate in the longitudinal direction for the lower mechanical properties of CSM. The basic cost-saving potential of this configuration lies in the greater thickness per ply of CSM ($1.40 \, mm$) compared with that of WR ($0.95 \, mm$): fewer plies are required per unit thickness and a proportionate saving in labour cost (about 30 %) should result in the lay-up of shell, deck and bulkhead laminates.

In the manufacture of large surface ship hulls in GRP one area stands out as a particular point of weakness; namely the through thickness properties of the GRP laminate. The problem can be conveniently considered under three headings:

(1) The inherent through thickness properties of the laminate which may be affected by resin type and/or the form of the constituent fibre (WR or CSM).

(2) The problem of attaching framing to a hull in order to withstand shock loading.

(3) The choice of technique for conditioning surfaces on to which further lamination has to take place following a delay of some kind. In this instance the surface could be partially or even fully cured depending on the extent of the delay.

The effectiveness of any production variable on the through thickness properties of a laminate can only be judged if a reliable test method can be devised. This paper describes such a method based on some previous work[5] and uses the method to investigate various means of improving the bond strength where delay during lamination has occurred.

A number of workers have investigated the problem of the mechanical property degradation of GRP in an aqueous environment. Aveston *et al.*[6] and Steel[7] suggest that degradation can be severe in seawater under statically loaded conditions where failure is by a stress corrosion mechanism of the glass fibres. Hulls of surface ships are only lightly stressed and thus this problem may not therefore be encountered. Proposals have been made for GRP to be used in statically immersed environments, e.g. dock gates, where for efficiency of design considerably higher stresses could be used. It is sensible, however, to examine the effect of prolonged immersion of the type of GRP laminate used in surface ship hulls, under statically loaded conditions, in order to develop confidence in the use of the material for future construction. This is particularly relevant for high performance craft such as fast patrol craft where more highly stressed laminates will undoubtedly be used because of weight-saving requirements.

USE OF FIBRE REINFORCED PLASTICS IN SHIP HULLS

When designing surface vessels using conventional materials, assumptions of nominal load levels, factors of safety, and satisfactory performance of scantlings from previous design experience, are already built into standard design procedures. The first design using GRP for a surface vessel with a large hull was lacking in design experience and no service data were available.[8-11] One important factor in the hull design was deemed to be the low modulus relative to the strength of GRP. This low modulus would result in large hull deflections for a stress-based design, giving rise to buckling problems in large flat panels subject to compressive and shear loadings.

The final hull configuration of *HMS Wilton* consisted of a single skin hull composed entirely of 'balanced' 5/4, 815 g/m² woven rovings at a 50 % by weight glass/resin ratio, transversely stiffened, the stiffeners being attached by means of metallic fasteners to assist the bonded connections under explosive load. The fabrication cost of this hull design is significantly more expensive than the material of construction due to the labour-intensive nature of production. With the experience gained in the building of *HMS Wilton* and looking to future designs, it might be possible to reduce construction costs with no loss in properties with the use of directional fabrics to fully utilise the advantages of glass reinforced plastic laminates.

Laminates with biased directional properties can be used to build in strength and stiffness where it is most desirable, i.e. hull longitudinal strength. Experience has shown that the all-round strength provided by a balanced reinforcement is not always required and may lead to excess weight in parts of the finished structure. In hull design it is usual to require greater longitudinal strength in order to withstand more bending loads. In the transverse hull direction relatively low membrane stiffness is acceptable provided that adequate flexural rigidity of the framing is maintained.

The possibility of effectively using directional fabrics and chopped strand mat was first investigated using a computer program developed by Dr C. S.

TABLE 1
Selected mechanical properties for mixed fibre laminates

Mechanical properties	5/4 WR			4/1 WR			4/1 WR + 30% CSM		
	0°[a]	90°[b]	45°	0°	90°	45°	0°	90°	45°
Theoretical values									
Modulus E (MN/m² × 10⁴)	14·7	16·7	9·5	20·7	10·8	9·3	15·5	10·0	9·6
Poisson ratio μ	0·13	0·11	0·47	0·09	0·17	0·43	0·12	0·18	0·32
Shear modulus G	3·2	3·2	6·8	3·2	3·2	6·2	3·7	3·7	5·2
Measured values									
Modulus E (MN/m² × 10⁴)	18·1	18·6	10·9	21·8	14·8	9·2	15·4	9·7	10·1
Poisson ratio μ	0·09			0·20					
Shear modulus G				3·1					
Strength (MN/m²)	240	270	101	287	145	80	226	121	57
Fibre content	55%			52%			44%		

[a] 0° ≡ parallel to weft.
[b] 90° ≡ parallel to warp.

Smith[12] and based on work by Tsai,[13] Hashin and Rosen[14] and Whitney and Riley[15] for calculating the moduli of fibre reinforced plastics. The elastic analysis is based on two assumptions; first that the material is macroscopically homogeneous, its gross behaviour being governed by equations of anisotropic elasticity, and secondly that for a specific laminate configuration the elastic properties of the whole may be determined, knowing the fibre and matrix moduli, by examining the behaviour of a representative element of the composite.

Some computed data relating to mixtures of 5/4 woven roving, 4/1 woven roving and chopped strand mat +4/1 WR, are given in Figs 1, 2 and 3 for Young's modulus, shear modulus and Poisson's ratio respectively. Some experimental data points are also shown at 0, 90 and 45° to the weft of the fabric; these values are also reported in Table 1. This data is based on

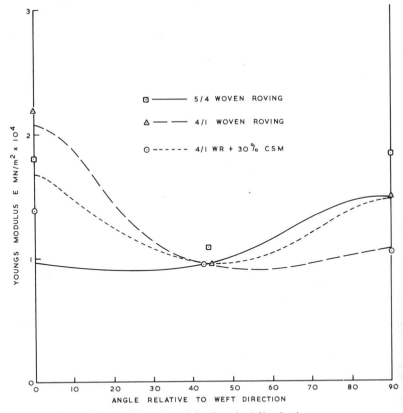

Fig. 1. Young's modulus for mixed fibre laminates.

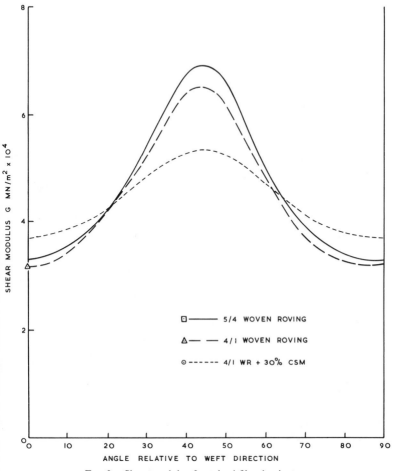

FIG. 2.   Shear modulus for mixed fibre laminates.

experimental work whereby 5/4 balanced WR was mixed with various quantities of unidirectional glass to build up a variety of biased laminates. The 4/1 ratio fabric and a number of others were eventually chosen for further study as fabrics in their own right. On the basis of this study it was concluded that the best compromise was achieved using the 4/1 biased woven roving which was still easy to handle, was stable, possessed the correct drape characteristics, readily wetted out and gave a high glass to resin ratio. Because of the structural requirements of ship building the 4/1 bias was incorporated in the weft, thus enabling the longitudinal strength to

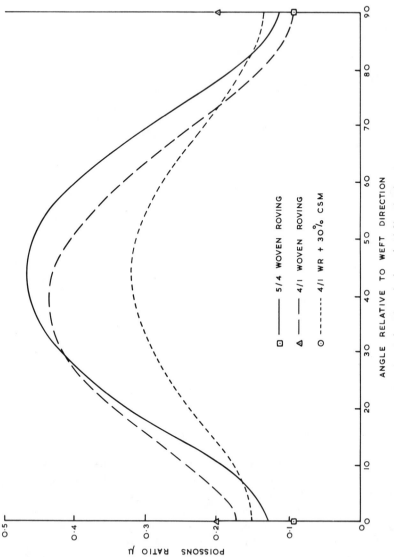

Fig. 3. Poisson's ratio for mixed fibre laminates

be built in whilst still allowing the fabric to be laid transversely with existing equipment. A considerable amount of weft oriented 4/1 fabric has now been made and used in manufacturing quite large structures, e.g. 20 ft × 20 ft stiffened panels, with the same ease as would be expected from more conventional fabric.

Comparing the theoretical and experimental results shows close agreement in some cases. However, where less agreement is apparent the cause could be simply due to the effect of the glass to resin ratio. In the case of the 4/1 laminates an 8 % change in glass content effectively increases the modulus by some 20 %. However, fibre straightness due to lamination problems could also be partly responsible for the low values observed.

SECONDARY BONDING AND THROUGH THICKNESS PROPERTIES OF GRP LAMINATES

During the production of large GRP structures, such as minesweeper hulls, delays in lay-up may occur ranging from short interruptions of typically 12 h to avoid exotherm build-up, to longer periods of days or even weeks resulting from holidays or other interruptions. When delays are long enough for a partial cure of the resin to occur, before lay-up can restart the surface must be prepared in some way. There are various methods available for preparing a cured or partially cured resin surface to achieve a bond with fresh resin and the choice of technique will depend on the delay incurred, the type of laminate, conditions of working, etc.

The method favoured by shipbuilders for many years has been the use of vacuum grit blasting, especially in the interests of ship cleanliness. Terylene tear-off cloth has also been used but confined mainly to localised areas, e.g. in the frame or bulkhead attachment. Grinding causes a dust problem and is used only sparingly to remove any residue following the removal of terylene tear-off cloth.

Recent work at AMTE(S) has investigated the effectiveness of various techniques used to prepare the surfaces of cured or partially cured laminates in order to achieve a satisfactory bond with fresh resin. The effectiveness of the secondary bond preparation can be tested by evaluating the through thickness interlaminar tensile strength of the finished, fully cured laminate. The method used at AMTE(S) to measure the interlaminar tensile strength is based on the work of Kimpara and Takehama[5] and has been found by experience at AMTE(S) to give reasonably consistent

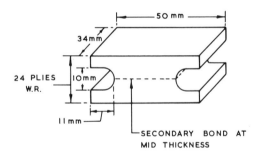

Fig. 4.   Specimen for testing through thickness tensile strength of GRP laminates.

measurements of the interlaminar tensile strength of glass reinforced composites.

Test panels were fabricated incorporating the various surface preparations and these will be described later. From these panels, all of which were 1000 mm × 500 mm × 25 mm, specimens were manufactured with 'U' notches ground in two opposite faces in accordance with the details given in Fig. 4.

Initial attempts to use the through thickness test produced an unacceptable amount of scatter as a result of premature failure of the test pieces. In most cases premature failure was associated with a very low interlaminar tensile stiffness as was evident from the pre-failure slope of the load displacement curve. Test pieces also often failed outside the gauge length and the presence of flaws in the laminate was suspected. Using dye penetrant techniques and microscopical examination, very fine cracks were detected in the resin matrix of many test pieces, Fig. 5. It was concluded that

Fig. 5.   Cracks in through thickness specimen produced by wrong machining technique.

these cracks were the cause of the premature failures experienced during testing giving rise to the unacceptable scatter in the results obtained. The initiation of the cracks was traced to the machining process used to manufacture the test pieces and by making changes in tooling and feed rate, etc. cracking was virtually eliminated.

For testing, the test pieces were axially mounted, using adhesive bonding, between 25 mm thick, parallel, steel mounting plates using the jig shown in Fig. 6. The mounted test pieces were left for 3 days at 15°–20 °C to allow the bond to cure before bolting into the testing machine using holes drilled in the mounting plates.

Various ways of preparing cured or partially cured resin surfaces prior to further lay-up have been examined using the through thickness test to measure the strength of the finished laminate. In addition to the standard shipyard techniques of surface preparation, namely grit blasting and tear-off cloth, the effectiveness of treating the surface with liquid styrene monomer was also examined. Two delay periods of 10 days and 21 days were incorporated in the programme for test panel production, thus providing resin surfaces at two stages of cure. These periods were seen as typical of shipyard production delays associated with holidays or other stoppages. Details of the lay-up programme are given in Table 2; the resin

FIG. 6.   Jig for assembling through thickness specimens and mounting plates.

TABLE 2

Panel no.	Period of cure after initial lay-up	Surface preparation	Final lay-up
A B	10 days⎫ 21 days⎭	Vacuum cleaned only	
C D	10 days⎫ 21 days⎭	Grit blasted using Dynablast 24–30 grit grade 1, followed by vacuum cleaning	
E F	10 days⎫ 21 days⎭	Terylene tear-off cloth rolled into wet resin. After delay period cloth removed and any residue ground off with No. 24 sanding disc and vacuum cleaned	After the various surface preparations all panels received a final lay-up of 12 plies
G	Fully cured by heating at 80 °C for 2 h to a maximum Barcol hardness of 55	Styrene wipe using a clean lint free cloth followed by a generous application of styrene using a mohair roller so as to flood the surface. The surface was left for a minimum of 15 min and at least until dry before recommencing lay-up	

used was Scott Bader's Crystic 625 TV isophthalic polyester and the reinforcement was Fothergill & Harvey's Y920, 815 g/m^2 woven rovings.

All the panels were laid up to a final thickness of 24 plies and post-cured at 80 °C for 2 h to a minimum Barcol hardness of 55 thus ensuring all were in a similar condition. After post-curing the panels were carefully sawn into manageable sections and through thickness test pieces manufactured.

The tests were carried out in a 100 kN servo hydraulic tensile machine using actuator displacement at a rate of 1 mm per min and a full range load of 10 kN. Load and actuator displacement were continuously recorded on a X–Y recorder so that the maximum load reached in each test was clearly seen.

A statistical examination of the results shown in Table 3 show that except for panels A and B, there are apparent differences related to delay time between the mean values of interlaminar tensile strength for the various surface treatments.

The observed effects could be regarded as being caused by improved mechanical keying. After 21 days the resin surface will be more fully cured and therefore harder than after 10 days and in panel F for example the

TABLE 3
Results of delayed lamination tests

Sample	Mean failure stress	Standard deviation	Sample size	t test value	t at 95% CL	Remarks	Analysis of variance	
							For 9 days delay	For 21 days delay
A	9·79	3·00	17				$F_{3,73}=4·55$	$F_{3,67}=13·07$
B	11·13	2·15	21	1·606	2·042 ($k=30$)	No effect of delay time		
C	8·26	2·14	18				$F_{3,60}=3·34$ at 95 % CL	$F_{3,60}=3·34$ at 95 % CL
D	10·44	2·84	19	2·619	2·042 ($k=30$)	Delay time effective	Treatments show differing effects	Treatments show significantly differing effects
E	9·62	2·75	17					
F	11·56	2·71	18	2·372	2·045 ($k=29$)	Delay time effective		
G	13·01	1·60	19					

removal of the tear-off cloth will probably result in a better mechanical key than might be expected with the lesser cured panel C. A similar argument may be applied to the grit blasted surfaces of panel D.

The best result was obtained from panel G in which a fully cured resin surface was styrene-treated before recommencing lay-up. Whilst the mechanism for the use of styrene is not clear and is not the concern of this paper, it is thought likely to be a combination of cleaning, degreasing, physical bonding and chemical bonding. The contributions of cleaning and degreasing are probably the most important, whereas any chemical cross-linking effect can only be present on partially cured resin surfaces. Styrene monomer is claimed by some to bring about an 'opening up' of the surfaces of both partially and fully cured resin leading to an improved physical bond.

Whatever the mechanisms of the use of styrene monomer the following conclusions can be reached:

(1)   There are differences between the interlaminar tensile strengths of those laminates where the intermediate surface has been subjected to the mechanical surface treatment of either grit blasting or terylene tear-off cloth at either 10 days or 21 days.

(2)   Mechanical techniques of surface preparation tend to give better results on the more fully cured and therefore harder resin. The harder resin will offer a less yielding surface to grit blasting and will therefore 'suffer' more damage and provide a better key. Similarly, terylene tear-off cloth will be more firmly held by the more fully cured resin and will promote more 'matting' of the surface.

(3)   The use of styrene to prepare partially cured resin surfaces for further lay-up was considered to give the most consistent results with a higher mean interlaminar tensile strength than the abrasive methods.

(4)   The effectiveness of styrene was found to be as good on a fully cured as on a partially cured resin surface.

THE EFFECTS OF LONG TERM IMMERSION IN SEAWATER ON THE MECHANICAL PROPERTIES OF HULL LAMINATES

Many workers have been studying the effects of various environments on the mechanical properties of reinforced plastics. Much of the evidence so far produced suggests there may be severe degradation of some reinforced plastics due to absorption of water where immersion is combined with continuous loading. Of particular interest are the results of those workers studying the effects of long term exposure to water on polyester resins reinforced with 'E' glass.[6,7]

Recently the use of GRP for a large dock gate was considered but there was sufficient lack of confidence in the performance of such a structure exposed to prolonged levels of static loading and long term immersion in seawater. With a need for reassurance that large GRP structures are not seriously at risk of degradation as a consequence of high loading in the presence of seawater, a series of tests involving typical levels of service stress coupled with seawater immersion were conducted at AMTE(S).

In order to carry out 24 tests simultaneously a large seawater filled tank was set up. Flexure specimens in three point loading were chosen for the current programme and although the interpretation of results may be complicated by the presence of both non-uniform stress and non-uniform permeation by water the tests are seen as being representative of the practical situation. During testing the specimens were supported at a span of 610 mm and loaded by a simple dead weight system acting directly on to the centre of each specimen. The tank and loading arrangements are shown in Figs 7 and 8.

Fig. 7.   Facility for environmental testing of reinforced plastics.

FIG. 8.   Detail of arrangement for testing reinforced plastics under load and immersed in
seawater.

TEST PROCEDURE

Because of the direct acting design of the loading system very heavy loads would be needed to carry out tests on full thickness hull laminates. In order to limit the size of the dead weights, half thickness laminates were used made up of 12 plies of Fothergill & Harvey's Y920, 815 g/m^2 woven rovings and Scott Bader's 625 TV isophthalic polyester resin, to a finished glass content of approximately 50 % by weight and a thickness of 12·5 mm.

All the panels were post-cured at 80 °C for 2 h to a minimum Barcol hardness of 55. From the panels were cut strips 50 mm wide and 1 m long, some being left with sawn edges and others sealed with two coats of the laminating resin.

Three strips from each panel were tested in three point bend at a span of 610 mm to establish a nominal breaking load (NBL) for the laminate and this proved to be of the order of 180 kg. Tests were set up in the tank at 15, 20, 40, 65 and 85 % of the established NBL using both sealed and unsealed strips. Several strips, some with sealed edges and others unsealed were placed in the bottom of the tank and left unloaded. These unloaded specimens were removed at various intervals and their residual strength measured by three point bend loading.

RESULTS

At the beginning of testing the deflection of each test piece was measured daily but after several days the frequency of readings was gradually extended to periods of a week. Typical deflection curves are shown in Figs 9 and 10 and these suggest that 'creep' is occurring in three stages as defined by variation in the increases in rate of deflection. In the primary stage the increase is quite rapid; a secondary stage is indicated by a marked slowing down in the rate of increase. In the case of those specimens loaded to 85 % of NBL a tertiary stage is identified by a dramatic increase in the rate of deflection leading to eventual failure. The onset of the tertiary stage was seen to be coupled with the occurrence of obvious damage to the test piece revealed as an opacity or whitening in the laminate in the area immediately surrounding the central point of loading.

Those specimens loaded to or below 65 % of NBL would appear from Fig. 9 to be passing through the secondary stage of 'creep. with no indications after 500 days to suggest the onset of the tertiary stage and eventual failure.

Those samples which were subjected to a bending load of greater than 80 % breaking load show evidence of initial failure after quite short test times, Fig. 10. It is assumed that this failure is due to some breakage of the individual filaments on the tensile side of the bend. Similar effects have been reported by Hull and Hogg[16] during the testing of GRP pipe in 0·65 N HCl. There is, however, no suggestion that after the initial failures further intermediate failures take place before the final break. Nevertheless, buckling of the surface fibres on the compressive side of the bend is seen as failure becomes more apparent. A typical example of a failed sample is shown in Fig. 11. It is not clear at present whether significant buckling occurs before the initial failure or only towards final failure. The samples showing early failure are representative of the unsealed test only. It would appear therefore that early ingress of water may be the cause, with a failure mechanism similar to that described by Aveston *et al.*,[6] but then only on the fibres adjacent to the cut surfaces. Sealing effectively slows down the degrading process until all the fibres are equally affected and failure occurs by the same mechanism but at a later date.

In all the tests carried out there has been no evidence to suggest specimens with resin sealed edges perform in the long term any better than the unsealed specimens. The use of polyester resin to seal sawn edges although a reflection of shipyard practice appears to offer little benefit and at best only delays water pick-up by the 'wicking' action of exposed fibres and

J. Bird and R. C. Allan

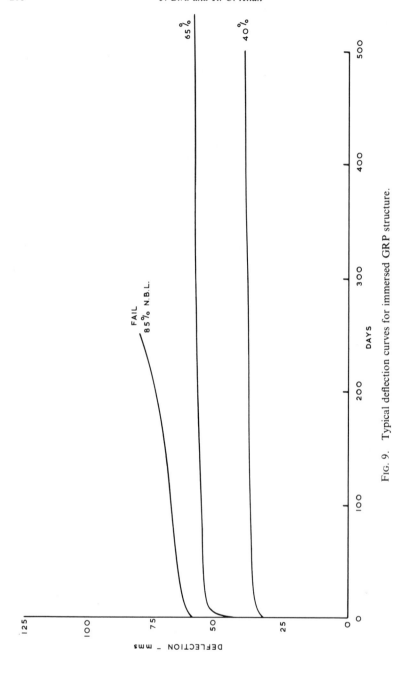

FIG. 9. Typical deflection curves for immersed GRP structure.

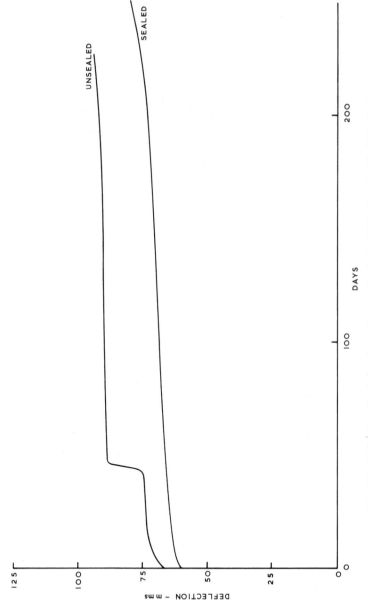

FIG. 10.   Deflection of sealed and unsealed tests loaded to 85% of nominal breaking load.

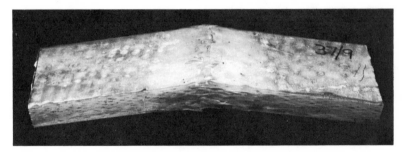

FIG. 11.   Typical failed specimen with tension face uppermost.

penetration along the fibre–resin interface. If water pick-up is essentially a resin-controlled mechanism then a more effective sealant is required if water penetration is to be prevented.

Some specimens, both sealed and unsealed, were immersed in seawater for up to 500 days but not subjected to any loading. It can be reported that none of these specimens on removal from the tank and on testing to failure in three point bend showed any significant change in modulus or departure in strength when compared with values for the original laminate. It is interesting to note that tensile tests carried out recently on material taken from the 10-year-old hull of *HMS Wilton* revealed no loss in modulus or ultimate strength when compared with the well-established figures for the hull laminate. Thus it would seem that similar measurements of the residual strength of unstressed specimens can only give a rough idea of the factor of safety for any particular laminate in a given period of immersion and such measurements are of little use for design purposes. *HMS Wilton* was designed to sustain maximum loads with a fatigue limit of 25 % of the ultimate strength, on the basis that fatigue failure will occur before creep. The immersion tests carried out so far with a '*Wilton*' type of laminate indicate that very high levels of static stress in excess of 2/3 of the value of the ultimate strength determined by three point bending, are required to bring about failure within a seawater immersion period of 500 days.

FUTURE WORK

It is planned to establish more conclusively the levels of static stress required to bring about failure in laminates exposed to a seawater environment. In addition to the current 'E' glass/polyester laminate it is intended to examine a wide range of laminates incorporating aramid and carbon fibres. A further programme is planned to study the effects of water on sandwich constructions and with these aims in view a further tank has now been set up to provide a further 24 test positions. Further tests are also planned where the loading will be purely tensile.

CONCLUDING REMARKS

The use of a mixture of biased fabric and chopped strand mat will, assuming the thicknesses of future construction remains the same as present ship design, have the following features when incorporated in deck and shell laminates:

(1) Strength and stiffness of the laminate (and hence the hull) would be maintained in the longitudinal direction.
(2) Laminate strength and stiffness in the transverse direction would be reduced by some 50 %; reduction in transverse strength and stiffness of the stiffened shell, which depends primarily on frame rigidity, should be generally acceptable.
(3) A saving of 25–30 % in fabrication time and therefore labour cost for lay-up of deck and shell laminate should result from the reduced number of plies per unit thickness.
(4) A weight saving of 5–7 % in the deck and shell laminate should result from the lower specific gravity of the WR/CSM laminate.

When manufacturing thick GRP laminates, it is clearly advantageous to prepare those surfaces which have been subjected to delays in lamination, before lamination is continued. The use of mechanical abrasion shows some immediate improvements in through thickness properties but the use of liquid styrene appears to give the best results. The correct application of styrene alone should be sufficient for all production needs where delays have occurred. By using styrene alone significant cost savings over the use of mechanical abrasion can be made.

Immersion of GRP plastics under load in seawater results in a degradation in mechanical properties. In the tests being carried out by AMTE(S) it is clear that the degradation is only severe at high load levels, i.e. at a large percentage of the breaking load.

For ship hulls where the dead loads are very low and where large superimposed transient service loads only occur infrequently, degradation leading to premature failure is not to be expected. Although this is reassuring, the results reported by AMTE(S) contradict the previously reported work of Aveston *et al.*[6] and Steel.[7] Whether the differences are related to the rate of ingress of water into the laminate (some resins being better than others in this respect) or whether fibre surface treatment is also a factor, is not known. In order to ensure that future designs can be made with confidence it is essential to investigate the reasons for the observed behavioural differences of apparently similar materials.

ACKNOWLEDGEMENTS

The authors wish to acknowledge the help of colleagues at AMTE (Dunfermline) for their assistance in the preparation of this paper.

Any views expressed are those of the authors and do not necessarily represent those of the Procurement Executive.

REFERENCES

1. HENTON, D. Glass reinforced plastics in the Royal Navy, *Trans. of RINA*, **109** (1967) 487–510.
2. CHEETHAM, M. A. Naval applications of reinforced plastics, *Plastics & Polymers*, **36** (1968) 15–20.
3. BEALE, R. F. Selection of glass reinforced plastic materials for large marine structures, *Brit. Polymer Journal*, **3** (1971) 1–8.
4. GIBBS, H. 'Materials for marine structures', *Plastics & metals: Competitors & allies*, 4th International TNO Conference (Utrecht, January 1971) pp. 3–9.
5. KIMPARA, I. and TAKEHAMA, M. *Static and dynamic interlaminar strength of glass reinforced plastics thick laminates*, Proceedings of Reinforced Plastics Congress, Brighton, 1976.
6. AVESTON, J., KELLY, A. and SELLWOOD, J. M. 'Long term strength of glass reinforced plastics in wet environments', *Advances in composite materials*, 3rd International Conference on Composite Materials, Paris, August 1980.
7. STEEL, D. J. The creep and stress rupture of reinforced plastics, *Trans. J. Plastics Institute*, (1965) 161–7.
8. SMITH, C. S. 'Structural problems in the design of GRP ships', *RINA Symposium of GRP ship construction* (London, October 1972) pp. 33–56.
9. DIXON, R. H., RAMSEY, B. W. and USHER, P. J. 'Design and build of the GRP hull of HMS Wilton', *RINA Symposium of GRP ship construction* (London, October 1972) pp. 1–32.
10. SMITH, C. S. Buckling problems in the design of fibreglass reinforced plastic ships, *J. Ship Research*, **16** (1972) 174–90.

11. LANGFORD, B. W. and ANGERER, J. F. Reinforced plastics developments for application to minesweeper construction, *Naval Engineers Journal*, **83** (1971).
12. SMITH, C. S. 'Calculation of elastic properties of GRP laminates for use in ship design', *RINA Symposium of GRP ship construction* (London, October 1972) pp. 69–84.
13. TSAI, S. W. 'Structural behaviour of composite materials', NASA CR-71 (1964).
14. HASHIN, Z. and ROSEN, B. W. The elastic modulii of fibre reinforced materials, *J. Applied Mechanics*, **31** (1964) 223–32.
15. WHITNEY, J. M. and RILEY, M. B. Elastic properties of fibre reinforced composite materials, *AIAA Journal*, **4** (1966) 1537–42.
16. HULL, D. and HOGG, P. J. 'Nucleation and propagation of cracks during strain corrosion of GRP', *Advances in composite materials*, 3rd International Conference on Composite Materials, Paris, August 1980.

15

Development of Cylindrically Orthotropic Model Material for Transmission Photoelasticity

P. K. SINHA

Department of Mechanical Engineering,
Bihar College of Engineering, Patna 800005, India

AND

B. L. DHOOPAR

Department of Mechanical Engineering,
Indian Institute of Technology Kanpur, Post Office I.I.T.,
Kanpur 208 016, India

ABSTRACT

In the present work, a cylindrically orthotropic model material for transmission photoelasticity has been developed. Many investigators have reported some unusual optical and photoelastic properties in the rectilinearly orthotropic model materials developed by them which are not observed in an isotropic photoelastic model material. The model material developed in the present investigation has also been found to possess similar properties. These unusual properties seem to be due to a slight mismatch of the refractive indices of its constituents. It is interesting to note that the above-mentioned cause also affects its photoelastic response. Simple experiments have been devised for detecting the above-mentioned defects. Further, a criterion has been developed for quantifying these effects which may be used for rating the relative qualities of available model materials.

INTRODUCTION

The basic requirements of a model material for transmission photo-orthoelasticity are that it should be transparent, optically birefringent and mechanically anisotropic. Pih and Knight,[1] Dally and Prabhakaran[2] and

Agarwal and Chaturvedi[3] have reported the development of rectilinearly orthotropic unidirectionally and bidirectionally reinforced photoelastic model materials, whilst Knight[4] has reported the development of a cylindrically orthotropic model material. However, as Knight himself reports, his model was not sufficiently transparent and thus he used the techniques of photography and microdensitometer scanning of photographic negatives for the collection of photoelastic data. In the present investigation a cylindrically orthotropic model material with a superior level of transparency has been developed.

Agarwal and Chaturvedi[3] and Prabhakaran[5] have observed some unusual optical and photoelastic properties in the model material they developed. Similar peculiarities have been also observed in the model material developed in the present investigation. Simple experiments have been devised for the measurement of transparency and for observing the phenomena of light beam distortion and smearing of sharp boundaries. Further, the distinguishability of isochromatic fringe patterns at different orientations to the radial direction has been examined and for this purpose a model similar to that reported on by Prabhakaran[5] has been used. The fringe resolution capability has been found to be affected along the radial direction whereas it is insignificant in the circumferential direction. Further, an attempt has been made to evolve a quantitative criterion on the basis of which a model material with a minimum of the undesirable properties mentioned above may be selected. It has been observed that, when two light beams, separated by a given centre to centre distance, are viewed through the model material, the distortion of light increases with the increase in the gap between the sources and the model and, at a particular distance, they merge into each other so that it becomes difficult to differentiate between the two sources. This limiting distance between the source of light and the model has been used as an index for rating the quality of the material.

A method has also been developed for the determination of the material fringe values and has been reported on by the present authors elsewhere.[6] It was also established in reference 6 that the stress optic law suggested by Sampson[7] is valid for cylindrically orthotropic model materials.

PREPARATION OF MODEL MATERIAL

The refractive index of E-glass $(1 \cdot 548 + 0 \cdot 003)$ is quite close to that of polyester resin $(1 \cdot 520)$. This resin can be blended with styrene monomer to make its refractive index very close to that of E-glass. The matrix thus obtained can be cured at room temperature by adding a small amount of methyl ethyl ketone peroxide and cobalt octate. The transparency of the matrix material appears to deteriorate as the matrix solidifies at the time of curing. In the present work, the polyester resin was blended with styrene monomer with different percentages (10 % to 50 %) and then 0·5 % each of methyl ethyl ketone peroxide and cobalt octate were mixed with it. The matrix thus prepared was used for making orthotropic discs of different compositions. After curing, their transmission ratios (defined by the ratio of light transmitted through the model to the light incident on the model) were determined. The technique for the determination of the transmission ratio is described later. A maximum transmission ratio of 0·512 was found in a disc with the matrix composition shown in Table 1. With this

TABLE 1
Composition by weight of matrix material used for the preparation of the photoelastic model

Polyester resin	100 g
Styrene monomer	30 g
Methyl ethyl ketone peroxide	0·5 g
Cobalt octate	0·5 g

composition, discs of 19 cm diameter were fabricated. While making the disc, the E-glass fibre, in roving form, was first wound dry in a reel form in between two glass plates, the gap between which was maintained according to the thickness of cast sheet desired. The last few turns of the glass fibres were wound after wetting with CY-230 resin mixed with 9 % of HY-951 hardener. After curing, these served as a retainer ring when the polyester resin based matrix was poured. After the removal of entrapped air bubbles by squeezing, the fabricated disc was cured at room temperature under a constant pressure. The arrangements used for observing and measuring some unusual properties in the above-mentioned fabricated disc are discussed in the following section.

ARRANGEMENTS FOR MEASURING AND OBSERVING SOME PROPERTIES OF THE MODEL MATERIAL

The arrangements used for the measurement of transparency and for observing the phenomena of light beam distortion and smearing of sharp boundaries are as follows.

(a) Transparency

The determination of the transmission ratio requires the measurement of the incident light on the model and the transmitted light through it. The incident light was measured by keeping a screen with a photocell behind the lens of the transmission polariscope, as shown in Fig. 1. The density of light falling on the photocell was measured with a photometer. The transmitted light was measured by keeping the model in contact with the screen over the photocell. The transmission ratio was calculated by the ratio of transmitted light to incident light.

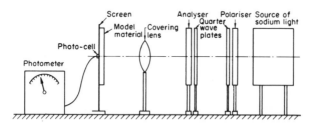

FIG. 1. Schematic diagram of the instruments with a polariscope used in the measurement of the transmission ratio of the model.

(b) Light Beam Distortion

It has been observed that a light beam becomes distorted when it passes through the model. This was studied by making a simple arrangement, as shown in Fig. 2. An aluminium disc with a number of holes, 1·75 mm in diameter, was kept over the source of light, providing a number of light beams. The model material was then kept at different distances from the source beam and the distortion of light was observed.

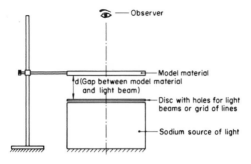

FIG. 2. Arrangement used for studying light beam distortion and smearing of sharp boundaries through the photoelastic model.

(c) Smearing of Sharp Boundaries

This has been observed through the model by putting a grid of equally spaced concentric circular and radial lines below it, as shown in Fig. 2. The distance of the grid from the model was adjustable.

DISCUSSION OF RESULTS AND A CRITERION FOR THE SELECTION OF A SUITABLE MODEL MATERIAL

Of the various factors which control the selection of a model material, transparency is one of the basic requirements. In the present work, the transparency has been quantified by transmission ratio, as described earlier. The transmission ratio (in the material prepared, Fig. 3) has been found to vary with radius. The variation, presented in Fig. 4, shows that the level of transparency decreases towards the centre. It is to be noted that entrapped air bubbles have been removed at the time of fabrication by the process of squeezing, using a conical aluminium roller. While rolling, the velocity of a point on the roller is linearly proportional to its radius. Thus, at larger radii, the squeezing of entrapped air bubbles is better than at smaller radii. Uniform velocity is possible with a cylindrical roller, but fibre

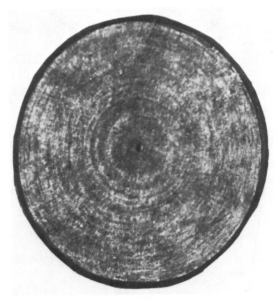

FIG. 3.   Cured glass polyester disc with retainer ring.

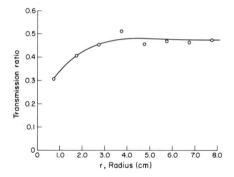

FIG. 4. Variation of transmission ratio along the radius in a cylindrically reinforced photoelastic model.

alignment is disturbed when it is used. Although the transparency level in terms of transmission ratio is lower towards the centre, it has been found to be sufficient to observe lines drawn on a piece of paper through this material.

The transparency also becomes affected due to mismatching of the refractive indices of the two constituents. Although the refractive indices have been very carefully matched, some slight mismatch has been found. Before discussing any basis for quantifying this mismatch, some of its probable effects are discussed.

A beam of sodium light coming through a circular hole, 1·75 mm in diameter, is distorted when it passes through the model material. A number of such light beams are shown in Figs 5(a) and 6(a). When the model material was kept on the light source, Fig. 5(a), the beams appeared to be slightly elliptical with the major axis along radial lines, as shown in Fig. 5(b). This distortion increased when the distance between the source of light and the model was increased. Figure 5(c) shows the source of light at a distance of 9 mm from the model material. It should be pointed out that the distortion of the beam is only along the radial direction whereas, in the circumferential direction, such an effect is absent. Similar effects have also been observed in the case of the unidirectionally reinforced composite model and these are shown in Figs 6(b) and 6(c). Here, also, the distortion is along the transverse direction and the beams appear as parallel thick lines.

Another optical effect is the smearing of sharp boundaries. A photograph of equally spaced concentric semicircles and radial lines is shown in Fig. 7(a). Both these lines were clearly visible when the model was kept in contact with them (Fig. 7(b)). When the grid of lines and the model were separated

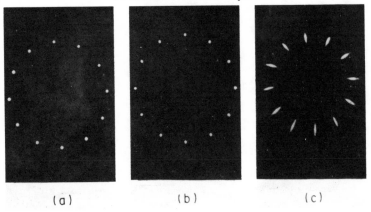

FIG. 5. Distortion of a beam of light through cylindrically orthotropic model material and views of: (a) the source of light, (b) the source of light in contact with the model material and (c) the source of light kept at a distance of 9 mm from the model material.

by a distance, the semicircular lines became less distinguishable from each other, whereas the radial lines were clearly distinguishable. This effect is shown in Fig. 7(c). Similar effects have also been observed by Agarwal and Chaturvedi[3] for the rectilinearly orthotropic model material prepared by them.

The distinguishability of isochromatic fringe pattern was also studied. This required the development of a model so that the fringes could be

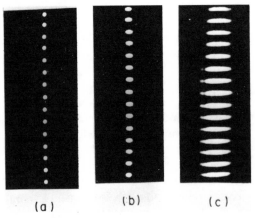

FIG. 6. Distortion of a beam of light through unidirectionally reinforced rectilinearly orthotropic model material and views of: (a) the source of light, (b) the source of light in contact with the model material and (c) the source of light kept at a distance of 9 mm from the model material.

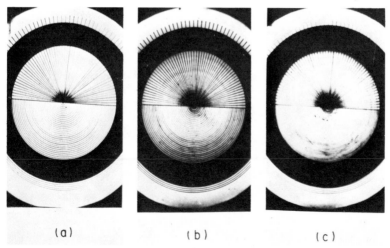

(a)   (b)   (c)

FIG. 7.   Line resolution through cylindrically orthotropic model material and views of: (a) the grid of circumferential and radial lines, (b) the grid when the model material is kept in contact, (c) the grid when the model material is kept at a distance of 9 mm.

obtained very close to each other. Some of the simple models—a disc, a disc with a hole and semicircular beams—were examined but it was not possible to obtain very close fringes when they were loaded on and viewed through a polariscope. Ultimately, a shape of model, shown in Figs 8(a) and (b) was found to be suitable for this purpose. When the fibre orientation was parallel to the load axis, the isochromatic fringe patterns obtained were not very clear and it was almost impossible to count their fringe orders (see Fig. 8(a)) whereas, for a load perpendicular to the fibre direction, even very closely spaced fringes were easily distinguishable. In zone 1 of Fig. 8(b) the fringes are perpendicular to the fibre direction and are thus clearly distinguishable whereas, in zone 2 of the same Figure, they are inclined at approximately 80° to the fibres. It is to be noted that this slight change of inclination has led to a deterioration in the fringe resolution capability. This variation in fringe resolution capability with the fibre direction has also been observed by Prabhakaran[5] for the case of the rectilinearly orthotropic composite model.

It is thus observed that the above-mentioned effects are prominent along the radial direction whereas they are insignificant in the circumferential direction. These effects are due to the lens effects caused by the mismatch of the refractive indices of the constituents and the cylindrical shape of the fibres.

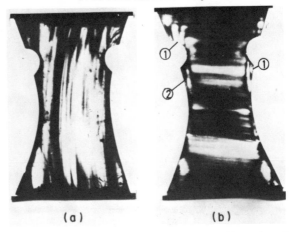

FIG. 8. Isochromatic fringe patterns showing the effect of fibre orientation on fringe resolution in cylindrically reinforced model for: (a) load approximately parallel to the fibre direction and (b) load approximately perpendicular to the fibre direction.

Now, given a number of model materials as alternative choices, one needs a quantitative criterion on the basis of which a model material with a minimum of the undesirable properties mentioned above may be selected. In an attempt to evolve one such criterion, it was observed that when two light beams, separated at a particular distance, were viewed through the model material, the distortion of light increased with the increase in the gap

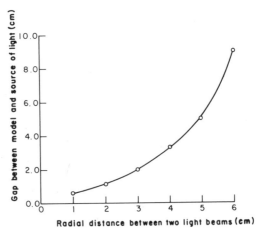

FIG. 9. The limiting gap between sources of light and the model at which images merge versus the radial distance between two light beams.

between the sources of light and the model. Finally, at a particular distance, they merged into each other and it became difficult to differentiate between the two sources of light. The variation of this limiting distance with the gap between the two beams of light is shown in Fig. 9 for the cylindrically orthotropic 0·4 cm thick model material prepared in the present investigation. This limiting distance between the source of light and the model can be used as an index for rating the quality of the material at the time of its selection.

CONCLUSIONS

The following conclusions are drawn from the present investigation.

(1) A cylindrically orthotropic photoelastic glass polyester model material has been developed.

(2) Simple experiments have been devised for studying the transparency, distortion of light and smearing of sharp boundaries. These optical properties seem to be due to the mismatch of the refractive indices of the reinforcing glass fibre and the matrix polyester.

(3) The fringe resolution capability in this material has been observed to be prominently affected along the radial direction whereas it has been found to be insignificant in the circumferential direction. Slight changes in the inclination of fringes from the radial direction greatly reduce the fringe resolution capability.

(4) A criterion has been developed for quantifying the unusual optical and photoelastic properties observed in this material and this may help in rating the quality of a given model material over other available ones when selecting it for various applications.

REFERENCES

1. PIH, H. and KNIGHT, C. E., Photoelastic analysis of anisotropic fibre reinforced composites, *J. of Composite Materials*, 3(1) (1969) 94–107.
2. DALLY, J. W. and PRABHAKARAN, R., Photo-orthotropic elasticity, Parts I and II, *Experimental Mechanics*, 11(8) (1971) 346–56.
3. AGARWAL, B. D. and CHATURVEDI, S. K., Development and characterisation of optically superior photoelastic composite materials, *Int. J. Mech. Sci.*, 20(7) (1978) 407–14.
4. KNIGHT, C. E., Orthotropic photoelastic analysis of residual stresses in filament-wound rings. Paper presented at SESA Fall Meeting, Boston, USA, 1970.

5. PRABHAKARAN, R., Model materials for photo-orthotropic elasticity, *Fibre Science and Technology*, **13**(1) (1980) 1–11.
6. SINHA, P. K. and DHOOPAR, B. L., Stress-optic law for cylindrically orthotropic composites. Paper presented at All India Seminar on Experimental Stress Analysis, Pune, April, 1979.
7. SAMPSON, R. C., A stress-optic law for photoelastic analysis of orthotropic composites, *Experimental Mechanics*, **10**(5) (1970) 210–15.

16

Photoelastic Techniques for the Complete Determination of Stresses in Composite Structures

R. Prabhakaran

Associate Professor, Mechanical Engineering and Mechanics,
Old Dominion University, Norfolk, Virginia 23508, USA

ABSTRACT

Transmission photoelastic analysis of composite models has attracted increasing attention in recent years. The interpretation of the photoelastic response in terms of the average (macroscopic) composite stresses is more involved than for isotropic models. Methods of determining the individual principal stresses (or strains) which have been suggested so far are not satisfactory. In this paper, three new methods are examined. The first method is an extension of the oblique incidence technique in which the model (or the light beam) is rotated about one of the material symmetry axes. In the second method, transmission and reflection photoelastic responses are combined. The third method requires the drilling of small holes and the determination of the fringe orders at selected points on the hole boundary. The three methods are applied to an orthotropic circular disk under diametral compression. Results are compared with strain gage data.

INTRODUCTION

A complete stress analysis and reliable failure criteria are essential for optimum utilisation of the unique properties of composite materials in structural applications. The case for micromechanics analysis of composites is very strong because the materials are heterogenous and exhibit several modes of failure. However, micromechanics analysis is complicated and the results from such analysis cannot directly be applied to design. An engineering or macromechanics analysis of composites is, therefore, needed and the results from such analysis have been found to agree well with experimental results. For the design and analysis of composite structures on a macroscopic scale, for instance in the failure theories such as the tensor polynomial theory, the individual average composite stresses are required.

When polarized light is passed through a transparent birefringent composite, the phenomenon on a microscopic scale is very complicated. But over-all fringe patterns are observed. Considerable progress has been achieved in the application of transmission photoelastic techniques to composite orthotropic models in recent years. The developments in the subject have been reviewed by the author.[1] The isochromatic fringe order is a complex function of the principal stresses (or strains), their orientations, etc. The isoclinic parameter gives the directions of the principal birefringence components according to a Mohr circle of birefringence.

For the transmission photoelastic analysis of an orthotropic birefringent model to yield useful information, methods must be developed to determine the individual values of the principal stresses or strains. Several methods have already been proposed, such as shear difference, numerical solution of the compatibility equation and holography. These methods have been reviewed by the author.[2] Some of these proposed techniques suffer from the disadvantage that they use the photoelastic response partially and rely on analytical procedures which either give rise to error or are involved. The holographic method of combining isochromatics and isopachics is not feasible for composites because of the complex nature of both families of fringes.[3] There is consequently a need for a simple and completely experimental method of determining the individual values of principal stresses or strains. Three such methods are proposed and examined in this paper.

DRILLING SMALL HOLES

In order to determine the magnitudes and directions of the principal stresses at a given point in the interior of an isotropic photoelastic model, Tesar[4] suggested making a very small circular hole at the point. Referring to Fig. 1, the stresses at the points A and C on the boundary of the hole of radius a are determined from the isochromatic fringe orders at these points.

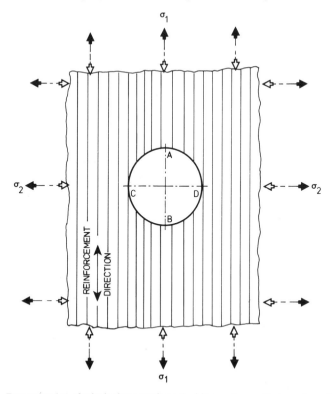

FIG. 1. Determination of principal stresses from the fringe orders on the boundary of a small hole.

It can be shown that the principal stress magnitudes, corresponding to the center of the hole and in the absence of the hole, are given by

$$\sigma_1 = \frac{\sigma_A + 3\sigma_C}{8} \qquad (1)$$

$$\sigma_2 = \frac{\sigma_C + 3\sigma_A}{8} \qquad (2)$$

This procedure has the disadvantage of depending on the precise determination of the boundary stresses at the edge of a small hole. Durelli and Murray[5] have overcome this disadvantage by determining the principal stresses corresponding to the hole center from the principal stress differences measured at interior points. If these interior points on a circle of radius $2a$ beyond A and C are designated as E and F, respectively, then it can be shown that

$$\sigma_1 = \frac{7\sigma_E + 15\sigma_F}{11} \tag{3}$$

$$\sigma_2 = \frac{7\sigma_F + 15\sigma_E}{11} \tag{4}$$

where σ_E and σ_F are the principal stress differences. Compared to Tesar's method, the improved procedure represented by eqns (3) and (4) has the disadvantage of larger errors due to stress gradients.

The author[6] has suggested the extension of Tesar's method to birefringent composites. The state of stress around a circular hole in a composite plate subjected to a biaxial loading is quite complex in the general case. Simplifications can be made if the composite plate is considered to be subjected to stresses which act along the material symmetry axes, as shown in Fig. 1.

When only the stress parallel to the reinforcement, σ_1, is acting, the tangential stress on the hole boundary is given by

$$\sigma_\theta = \sigma_1 \frac{E_\theta}{E_L} \left[-k\cos^2\theta + (n+1)\sin^2\theta \right] \tag{5}$$

where

$$k = \sqrt{\frac{E_L}{E_T}} \tag{6}$$

$$n = \sqrt{2\left(\frac{E_L}{E_T} - v_{LT}\right) + \frac{E_L}{G_{LT}}} \tag{7}$$

In the above equations E is the Young's modulus, G the shear modulus, v the Poisson's ratio, L and T the material symmetry axes and θ the angle measured from the σ_1-direction. At the points A and B ($\theta = 0, \pi$)

$$\sigma_{A,B} = -\frac{\sigma_1}{k} \tag{8}$$

and at the points C and D ($\theta = \pi/2, 3(\pi/2)$)

$$\sigma_{C,D} = \sigma_1(1 + n) \tag{9}$$

When the stress perpendicular to the reinforcement, σ_2, is acting alone, the tangential stress on the hole boundary is given by

$$\sigma_\theta = \sigma_2 \frac{E_\theta}{E_L} \left[(k+n)k\cos^2\theta - k\sin^2\theta \right] \tag{10}$$

At the points A and B

$$\sigma_{A,B} = \sigma_2 \frac{k+n}{k} \tag{11}$$

and at the points C and D

$$\sigma_{C,D} = -k\sigma_2 \tag{12}$$

Superposing the stresses σ_1 and σ_2 and solving for them,

$$\sigma_1 = \frac{k^2\sigma_A + (k+n)\sigma_D}{n(n+k+1)} \tag{13}$$

$$\sigma_2 = \frac{k(1+n)\sigma_A + \sigma_D}{n(n+k+1)} \tag{14}$$

While the measurement of the isochromatic fringe order is difficult on the hole boundary and it would be preferable to make the measurement at interior points, the analytical expressions for stresses at interior points in an orthotropic plate with a circular hole are not available in a closed form. Experiments to substantiate this proposed method are described in a later section.

OBLIQUE INCIDENCE

In orthotropic birefringent models, the isochromatic fringe order under normal incidence of light is related to the principal stresses by the equation

$$N_n = h \left\{ \left[\frac{1}{f_L} (\sigma_1\cos^2\alpha + \sigma_2\sin^2\alpha) - \frac{1}{f_T}(\sigma_1\sin^2\alpha + \sigma_2\cos^2\alpha) \right]^2 \right.$$
$$\left. + \left[\frac{1}{f_{LT}}(\sigma_1 - \sigma_2)\sin 2\alpha \right]^2 \right\}^{1/2} \tag{15}$$

where σ_1, σ_2 are the principal stresses, α the angle between σ_1 and L directions and f_L, f_T, f_{LT} are the principal stress-fringe values. As the author[7] has pointed out, rotation of the model about either of the principal stress directions is not possible because the principal stress angle, α, is not given by the optical isoclinic. It is possible to rotate the model about either of the principal strain directions, if it is assumed that the isoclinic parameter approximately gives the principal strain directions. The equations resulting from this approach are involved and the procedure is very complex. It is also necessary, in this approach, to directly or indirectly determine some out-of-plane elastic constants.

Instead of trying to determine the principal stresses or strains directly, the oblique incidence technique can be adapted to composite models by seeking the stress components σ_L, σ_T and τ_{LT}, referred to the material symmetry axes. One of the three equations required for this purpose is eqn. (15), which can be rewritten as

$$N_n = h \left\{ \left(\frac{\sigma_L}{f_L} - \frac{\sigma_T}{f_T} \right)^2 + \left(\frac{2\tau_{LT}}{f_{LT}} \right)^2 \right\}^{1/2} \tag{16}$$

According to the Mohr circle of birefringence, the optical isoclinic parameter, ϕ, is related to the stress components by

$$\tan 2\phi = \frac{2\tau_{LT}/f_{LT}}{(\sigma_L/f_L - \sigma_T/f_T)} \tag{17}$$

The third equation required can be obtained by rotating the model about the L-axis by θ, are shown in Fig. 2. The oblique incidence fringe order is given by

$$N_\theta = \frac{h}{\cos \theta} \left\{ \left(\frac{\sigma_{L^1}}{f_{L^1}} - \frac{\sigma_{T^1}}{f_{T^1}} \right)^2 + \left(\frac{2\tau_{L^1T^1}}{f_{L^1T^1}} \right)^2 \right\}^{1/2} \tag{18}$$

where σ_{L^1}, σ_{T^1}, $\tau_{L^1T^1}$ are the transformed stress components and f_{L^1}, f_{T^1}, $f_{L^1T^1}$ are the transformed stress-fringe values. While the transformed stress components are given by

$$\sigma_{L^1} = \sigma_L$$
$$\sigma_{T^1} = \sigma_T \cos^2 \theta \tag{19}$$
$$\tau_{L^1T^1} = \tau_{LT} \cos \theta$$

the transformed stress-fringe values, due to transverse isotropy, are given by

$$f_{L^1} = f_L$$
$$f_{T^1} = f_T \tag{20}$$
$$f_{L^1T^1} = f_{LT}$$

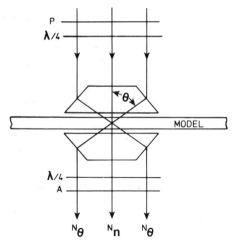

FIG. 2. Chopped prism oblique incidence arrangement.

t is therefore possible to rewrite eqn. (18) as

$$N_\theta = \frac{h}{\cos \theta} \left\{ \left(\frac{\sigma_L}{f_L} - \frac{\sigma_T \cos^2 \theta}{f_T} \right)^2 + \left(\frac{2\tau_{LT} \cos \theta}{f_{LT}} \right)^2 \right\}^{1/2} \tag{21}$$

The three stress components can be obtained by solving eqns (16), (17) and (21). Experiments verifying the proposed method are described in a later section.

COMBINED TRANSMISSION AND REFLECTION

The author[8] had proposed, without applications, combining the transmission and reflection photoelastic methods in order to determine the principal stresses or strains. If the symmetry axes for the composite model coincide with the material symmetry axes and if the loads are applied along these directions, then the principal stress and strain directions are the same. Assuming faithful strain transmission from the composite to the coating, the reflected isochromatic fringe order can be expressed as

$$N_r = \frac{2h^c}{f_\varepsilon^c} \left[\frac{\sigma_L^s}{E_L} (1 + v_{LT}) - \frac{\sigma_T^s}{E_T} (1 + v_{TL}) \right] \tag{22}$$

where the superscripts c and s refer to the coating and the composite specimen, respectively, and f_ε^c is the strain-sensitivity of the coating. The transmitted isochromatic fringe order, N_t, given by eqn. (16), simplifies to

$$N_t = \pm h^s\left(\frac{\sigma_L^s}{f_L} - \frac{\sigma_T^s}{f_T}\right) \tag{23}$$

where the positive or negative sign is chosen appropriately to keep the fringe order positive. The principal stresses σ_L^s and σ_T^s can be obtained from eqns (22) and (23) as

$$\sigma_L^s = \frac{2N_r h^c \left\{\dfrac{f_\varepsilon^c(1+v_c)}{E^c}\dfrac{f_L}{f_T}\right\} - N_t h^s \left\{\dfrac{f_L^2(1+v_{LT})}{f_T E_L} - f_L\right\}}{\dfrac{1+v_{LT}}{E_L}\dfrac{f_L}{f_T} - \dfrac{1+v_{TL}}{E_T}} \tag{24}$$

$$\sigma_T^s = \frac{2N_r h^c \dfrac{f_\varepsilon^c(1+v_c)}{E^c} - N_t h^s \dfrac{f_L(1+v_{LT})}{E_L}}{\dfrac{1+v_{LT}}{E_L}\dfrac{f_L}{f_T} - \dfrac{1+v_{TL}}{E_T}} \tag{25}$$

Experiments verifying the proposed method are described in the next section.

TESTS

To verify the proposed experimental methods, three circular disks of 7·6 cm diameter were tested in diametral compression. The disks were machined from a unidirectionally reinforced E-glass–polyester laminate. The elastic and photoelastic constants for the material were determined by standard calibration procedures and are given in Table 1.

On one of the disks, circular holes of 0·32 cm diameter were drilled on radial lines parallel and perpendicular to the reinforcement, at locations 1·27 cm and 2·54 cm from the center. Electrical resistance strain gages were mounted at similar points diametrically across from the holes. The disk was loaded parallel and perpendicular to the reinforcement and strain gage readings as well as fringe patterns were recorded. Typical isochromatic fringe patterns are shown in Fig. 3. The values of principal stresses given by eqns (13) and (14) were found to differ from the strain gage results by a maximum of 10%.

TABLE 1
Elastic and photoelastic properties of birefringent composite model

Property	Value
E_L	28·8 GPa
E_T	9·4 GPa
G_{LT}	3·2 GPa
v_{LT}	0·3
f_L	156 kPa/m/fringe
f_T	78 kPa/m/fringe
f_{LT}	69 kPa/m/fringe

On a second disk, a circular photoelastic coating of slightly smaller diameter was bonded. The disk was again loaded under diametral compression, parallel and perpendicular to the reinforcement. The isochromatic fringe patterns for the coating are shown in Fig. 4. The fringe patterns for the composite disk in transmitted light were also recorded as a reflected pattern from the back of the photoelastic coating. Fringe patterns obtained in this manner are shown in Fig. 5. As birefringent composites incorporating glass fibres as reinforcement are usually photoelastically insensitive, this procedure doubles the maximum fringe order. For comparison, isochromatic fringe patterns for a third composite disk in transmitted light are shown in Fig. 6. All the fringe patterns shown in Figs 3, 4, 5 and 6 correspond to a diametral compressive load at 1780 N. The values

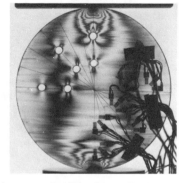

FIG. 3.   Isochromatic fringe patterns for circular composite disk with small holes under parallel and transverse diametral compression.

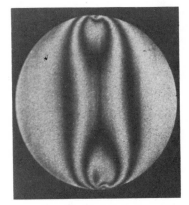
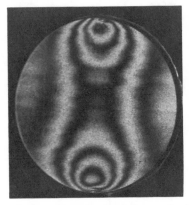

FIG. 4. Isochromatic fringe patterns for the photoelastic coating bonded to a circular composite disk under parallel and transverse diametral compression.

of principal stresses given by eqns (24) and (25) were found to differ from the strain gage results by a maximum of 5 %.

Oblique incidence measurements were conducted on the third composite disk for which the angle of oblique incidence with the chopped prism arrangement was found to be about 30°. The oblique incidence fringe order was combined with the normal incidence fringe order and the isoclinic parameter by eqns (16), (17) and (21). The results obtained in this manner differed from the strain gage results by a maximum of 7 %.

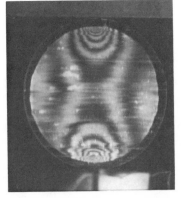

FIG. 5. Isochromatic fringe patterns for circular composite disk obtained by reflection from the back of photoelastic coating.

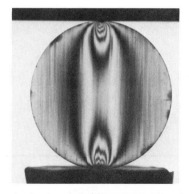

Fig. 6. Isochromatic fringe patterns for circular composite disk obtained in transmitted light.

CONCLUSIONS

Three completely experimental procedures have been proposed and compared. In one of the methods, small circular holes are drilled at the points of interest and the isochromatic fringe order at selected points on the hole boundaries is measured. In the second method a photoelastic coating is bonded to the birefringent composite model and the transmitted and reflected fringe orders are combined. In the third method, the oblique incidence fringe order, obtained by rotating the model or the light beam about the material symmetry axis, is combined with the normal incidence fringe order. The three methods have been applied to a circular disk under diametral compression, with the load parallel or transverse to the direction of reinforcement. Comparison with strain gage results indicates that the method of drilling holes is the least accurate, due to the difficulty in determining the fringe order on the boundary of a hole and the stress gradient from the hole center to the hole boundary; this method also requires extension to the more general biaxial loading where the material symmetry axes are not the principal stress directions.

The method of combining transmitted and reflected isochromatic fringe orders has the added advantage of doubling the transmitted photoelastic response if it is obtained by reflection from the back of the coating. However, the method requires several corrections due to the coating.

The oblique incidence method is easily applicable to general biaxial loading but the method depends on the isoclinic parameter and also yields the stress components referred to the material symmetry axes. Use of the chopped prism restricts the angle of oblique incidence to just one value.

ACKNOWLEDGEMENT

This research was supported by an equipment grant, CDP-80-16606, from the National Science Foundation and a cooperative agreement, NCCI-26 with NASA-LRC. The author would like to thank Dr Howard H. Hines of NSF and Dr Paul A. Cooper of NASA-LRC for their support and encouragement.

REFERENCES

1. PRABHAKARAN, R., *Developments in photo-orthotropic-elasticity*, International Union of Theoretical and Applied Mechanics Symposium on Optical Methods in Solid Mechanics, Poitiers, France, Sept. 1979.
2. PRABHAKARAN, R., Applications of transmission photoelasticity to composite materials. In: *Developments in composite materials, Vol. 2*, Holister, G. S. (ed.), Barking, England, Applied Science Publishers, in press.
3. ROWLANDS, R. E., DUDDERAR, T. D., PRABHAKARAN, R. and DANIEL, I. M., Holographically determined isopachics and isochromatics in the neighbourhood of a crack in a glass composite, *Exp. Mech.*, **20** (1980) 53–6.
4. TESAR, V., La Photoelasticimetrie et ses applications dans les constructions aeronautiques, *La Science Aerienne*, Paris (1932).
5. DURELLI, A. J. and MURRAY, W. M., *Stress distribution around a circular discontinuity in any two-dimensional system of combined stress*, 14th Semiannual Eastern Photoelasticity Conference, Yale Univ., 1941.
6. PRABHAKARAN, R., *Extension of the hole-drilling method to birefringent composites*, 36th Annual Conference of the Society of the Plastics Industry, Washington, DC, Feb. 1981.
7. PRABHAKARAN, R., *Extension of oblique incidence method to photo-orthotropic elasticity*, presented at the Spring Meeting of the Society for Experimental Stress Analysis, Dearborn, Michigan, USA, May 1981.
8. PRABHAKARAN, R., Separation of principal stresses in photo-orthotropic-elasticity, *Fibre Science and Tech.*, **13** (1980) 245–53.

17

A Boundary Layer Approach to the Calculation of Transverse Stresses along the Free Edges of a Symmetric Laminated Plate of Arbitrary Width Under In-plane Loading

D. ENGRAND

Office National d'Etudes et de Recherches Aérospatiales,
92320 Châtillon, France

ABSTRACT

In order to calculate the transverse and normal stresses along a free edge in a symmetrically laminated plate under in-plane loading, we derive the boundary layer equations, and then give an approximate solution, using the complementary energy principle. The method gives a stress field which satisfies all the equilibrium, continuity, and boundary conditions. Several examples are given, illustrating the great versatility of the method, which is applicable regardless of the number of layers and the layer orientations.

NOTATION

Standard notations are used throughout. In Cartesian axes, we denote the coordinates in the mid-plane of the plate by x, y, and the coordinate normal to the mid-plane by z ($-h \leq z \leq +h$). In order to make conspicuous the influence of the small parameter h (thickness $= 2h$), we put

$$\eta = \frac{z}{h}(-1 \leq \eta \leq +1)$$

The material is supposed to be heterogeneous in the thickness direction, so that we can define the matricial compliances δij as functions of η. In the usual case of layered plates, these functions are piecewise constant. The symmetry of the laminate imposes

$$\delta ij(-\eta) = \delta ij(\eta)$$

The non-zero δij coefficients in the 6×6 symmetric matrix S are $S_{11}, S_{12},$ $S_{13}, S_{16}, S_{22}, S_{23}, S_{26}, S_{33}, S_{36}, S_{44}, S_{45}, S_{55}, S_{66},$ and the tangential part of S is denoted by S^{t}

$$S^{\mathrm{t}} = \begin{bmatrix} S_{11} & S_{12} & S_{16} \\ S_{12} & S_{22} & S_{26} \\ S_{16} & S_{26} & S_{66} \end{bmatrix}$$

The stress tensor is denoted by σ, as usual, and the additional stress tensor in the boundary layer is denoted by F. The engineering components are used for the strains.

INTRODUCTION

In the present study, we attempt to give a reasonable approximation to the free edge shear and normal stresses in arbitrarily wide symmetrically laminated plates under in-plane loading.

We first recall the main results of the Classical Laminate Theory (CLT), which can be obtained by means of asymptotic expansions,[1,2,3] taking the thickness $2h$ as a small parameter. Then, following Tang's basic ideas,[4] we derive the boundary layer equations, as was previously done by Friedrichs and Dressler,[3] or by Reiss and Locke,[2] in terms of two coupled partial differential equations for two stress functions. It can be noted that these equations are quite different to Tang's ones. In the last part, following Horvay's method for end problems, we construct a statically admissible stress field depending on two scalar functions, that are determined by solving a system of two ordinary differential equations obtained by minimising the complementary energy functional. It results in an approximate stress field which satisfies exactly all the equilibrium, continuity, and boundary conditions. In addition, it can be seen that this stress field is not perturbed by any singularity.

THE CLT INTERIOR PROBLEM

It is well known that the stresses in the laminate, away from the edges, are given by the Classical Laminate Theory. This theory can be obtained by means of asymptotic expansions.[1,2,3] We only recall here its principal features. If we consider the asymptotic expansion of the stress tensor

$$\sigma(x, y, \eta) = \sum_{n=0}^{\infty} h^n \sigma^n(x, y, \eta) \tag{1}$$

the CLT gives for zero order terms the Generalised Plane Stresses equations, valid for thin plates

$$\left[\begin{array}{l} \sigma_{xz}^0 = \sigma_{yz}^0 = \sigma_{zz}^0 = 0 \\ S^t \Sigma^0 = \varepsilon^0 \text{ with } \Sigma^0 = \left\{ \begin{array}{l} \sigma_{xx}^0 \\ \sigma_{yy}^0 \\ \sigma_{xy}^0 \end{array} \right\} \quad \text{and} \quad \varepsilon^0 = \left\{ \begin{array}{l} \varepsilon_{xx}^0 \\ \varepsilon_{yy}^0 \\ \gamma_{xy}^0 \end{array} \right\} \end{array} \right. \tag{2}$$

where S^t and Σ^0 are functions of η, while the strain ε^0 is independent of η (i.e. constant in the whole thickness).

The associated equilibrium equations (without volume forces) are given, in terms of membrane stresses, by

$$\left[\begin{array}{l} N_{xx,x}^0 + N_{xy,y}^0 = 0 \\ N_{xy,x}^0 + N_{yy,y}^0 = 0 \end{array} \right. \tag{3}$$

with

$$N^0 = \int_{-1}^{+1} \Sigma^0(\eta)\, d\eta \tag{4}$$

and the stress–strain relation between N^0 and ε^0 is given by

$$N^0 = Q\varepsilon^0 \tag{5}$$

$$Q = \int_{-1}^{+1} (S^t)^{-1}\, d\eta \tag{6}$$

It is worth noting that the stress components σ_{xz}^1, σ_{yz}^1, σ_{zz}^2 ($\sigma_{zz}^1 = 0$), are also known through an integration of the equilibrium equations in η, together with stress free conditions for $\eta = \pm 1$.

Within the frame of CLT, it is generally not possible to satisfy exactly the boundary conditions for loaded or free edges. If we consider for instance a free edge defined by $y = 0$, the only conditions that can be satisfied are

$$N_{yy}^0 = 0 \qquad N_{xy}^0 = 0$$

It is thus necessary to analyse in more detail the stress field near the boundaries. In the present paper, we restrict ourselves to a straight free edge problem, though the loaded or curved edge problem can be treated in a similar way.[2,3]

BOUNDARY LAYER EQUATIONS

As in the study by Reiss and Locke,[2] the free edge is defined by $y = 0$. We suppose that a CLT solution is known, and we focus our attention on this free edge, where σ_{yy}^0 and σ_{xy}^0 have non-zero values and their mean values N_{yy}^0 and N_{xy}^0 are zero. This defect in the satisfaction of the boundary conditions can be removed if we superpose on the CLT solution an additional stress field $F^0(x, y, \eta)$ such that, on the boundary ($y = 0$)

$$\begin{cases} F_{yy}^0 = -\sigma_{yy}^0 \\ F_{xy}^0 = -\sigma_{xy}^0 \end{cases} \tag{8}$$

and such that it vanishes at a small distance to the edge. As pointed out by Tang,[4,6] this can be achieved by a boundary layer theory. Following Tang's study, we put

$$\xi = \frac{y}{h} \quad \text{together with} \quad \eta = \frac{z}{h} \tag{9}$$

so that the three-dimensional equilibrium equations become

$$\begin{cases} hF_{xx,x} + F_{xy,\xi} + F_{xz,\eta} = 0 \\ hF_{xy,x} + F_{yy,\xi} + F_{yz,\eta} = 0 \\ hF_{xz,x} + F_{yz,\xi} + F_{zz,\eta} = 0 \end{cases}$$

Similarly, the compatibility equations expressed in terms of the stresses through the constitutive equations may be put into the form

$$h^2 C^{(2)}(F) + h C^{(1)}(F) + C^{(0)}(F) = 0 \tag{10}$$

where $C^{(2)}$, $C^{(1)}$, $C^{(0)}$ are partial differential linear operators that are not explained here because of their lengthy nature. Now, we can observe that in eqns (9) and (10) the terms of higher order in h are precisely those involving derivatives in the x-direction. We can simply drop these terms and try to satisfy all the equations and conditions with a tress field F^0 independent of x, or alternatively consider that the total stress field $F^0 + \sigma^0$ we are looking for is given by the first term of an asymptotic expansion and then apply the matching principle of Friedrichs.[2,3] Both approaches give the same equations, but the first is simpler and does not give rise to the question of existence of an asymptotic expansion.

The boundary layer equilibrium equations are then written

$$\begin{cases} F^0_{xy,\xi} + F^0_{xy,\eta} = 0 \\ F^0_{yy,\xi} + F^0_{yz,\eta} = 0 \\ F^0_{yz,\xi} + F^0_{zz,\eta} = 0 \end{cases} \tag{11}$$

and the corresponding compatibility equation reads

$$C^{(0)}(F^0) = 0 \tag{12}$$

These equations are almost the same as those given by Tang.[6]

Classically, the equilibrium eqns (11) can be satisfied by the introduction of two stress functions, ψ and ϕ, such that

$$\begin{cases} F^0_{xy} = \psi,\eta \qquad F^0_{xz} = -\psi,\xi \\ F^0_{yy} = \phi,\eta\eta \qquad F^0_{yz} = -\phi,\eta\xi \qquad F^0_{zz} = \phi^0,\xi\xi \end{cases} \tag{13}$$

In addition to eqns (11) and (12), we have also to satisfy the boundary conditions

$$\begin{cases} F^0_{xy} = \psi,\eta = -\sigma^0_{xy}(x, o, \eta) \\ F^0_{yy} = \phi,\eta\eta = -\sigma^0_{xy}(x, o, y) \text{ (for } \xi = 0) \\ F^0_{yz} = -\phi,\eta\xi = 0 \end{cases} \tag{14}$$

and

$$\begin{cases} F^0_{xz} = -\psi,\xi = 0 \\ F^0_{yz} = -\phi,\eta\xi = 0 \quad \text{(for } \eta = \pm 1) \\ F^0_{zz} = \phi,\xi\xi = 0 \end{cases} \tag{15}$$

As we want to ensure that the influence of the stress field F^0 does not extend far away from the edge, we have also to impose

$$F^0 \to 0$$

$$\xi \to \infty \tag{16}$$

(This condition is equivalent to the 'matching condition' at zero order.)

Now it is seen that three equations among the six involved in the $C^{(0)}$ group, expressed in terms of strain, are

$$\begin{cases} \varepsilon_{xx}^0, \, \xi\xi = 0 \\ \varepsilon_{xx}^0, \, \eta\eta = 0 \\ \varepsilon_{xx}^0, \, \eta\xi = 0 \end{cases}$$

with $\varepsilon_{xx}^0 = S_{11}F_{xx}^0 + S_{12}F_{yy}^0 + S_{13}F_{zz}^0 + S_{16}F_{xy}^0$.

From eqn. (14) we immediately deduce that

$$\varepsilon_{xx}^0 = \beta\xi + \gamma\eta + \delta$$

but it can be shown that the constants β, γ, δ are all zero, for the symmetry of ε_{xx}^0 in η impose $\gamma = 0$, and eqn. (16) implies $\beta = \delta = 0$. Finally, we have $\varepsilon_{xx}^0 = 0$, and we can eliminate F_{xx}^0 from the remaining three compatibility equations in the $C^{(0)}$ group, which now take the form

$$(A_{22}F_{yy}^0), \, \eta\eta + (A_{23}F_{zz}^0), \, \eta\eta + (A_{26}F_{xy}^0), \, \eta\eta + (A_{23}F_{yy}^0), \, \xi\xi$$
$$+ (A_{33}F_{zz}^0), \, \xi\xi + (A_{36}F_{xy}^0), \, \xi\xi - (A_{44}F_{yz}^0), \, \eta\xi - (A_{45}F_{xz}^0), \, \eta\xi = 0 \quad (17)$$

$$\Delta^0, \, \xi = 0 \qquad (18(a))$$

$$\Delta^0, \, \eta = 0 \qquad (18(b))$$

with Δ^0 given by

$$(A_{26}F_{yy}^0), \, \eta + (A_{36}F_{zz}^0), \, \eta + (A_{66}F_{xy}^0), \, \eta - (A_{45}F_{yz}^0), \, \xi - (A_{55}F_{xz}^0), \, \xi \quad (19)$$

and

$$A_{ij} = S_{ij} - \frac{S_{i1}S_{1j}}{S_{11}} \qquad (20)$$

From eqns (18(a)) and (18(b)), we deduce that Δ^0 is a constant, k. But it is easy to see that the symmetry of the problem with respect to the plane $\eta = 0$ implies

$$\int_{-1}^{+1} \Delta^0(\eta)\,\mathrm{d}\eta = 0$$

for the coefficients A_{ij}, and the components F_{yy}^0, F_{zz}^0, F_{xy}^0 are even functions of η, and F_{yz}^0, F_{xz}^0 are odd functions of η. The boundary layer compatibility equations are therefore given by eqn. (17) and

$$\Delta^0 = 0 \qquad (21)$$

We are now able to express these equations in terms of stress functions

$$(A_{22}\phi, \eta\eta), \eta\eta + (A_{23}\phi, \xi\xi), \eta\eta + A_{23}\phi, \eta\eta\xi\xi + A_{33}\phi, \xi\xi\xi\xi$$
$$+ |(A_{44}\phi, \eta\xi), \eta\xi + (A_{26}\psi, \eta), \eta\eta + A_{36}\psi, \eta\xi\xi + (A_{45}\psi, \xi), \eta\xi = 0 \quad (22)$$

$$(A_{66}\psi, \eta), \eta + A_{55}\psi, \xi\xi + (A_{26}\phi, \eta\eta), \eta + (A_{36}\phi, \xi\xi), \eta + A_{45}\phi, \eta\xi\xi = 0 \quad (23)$$

Although the first equation is comparable to one of those given by Tang,[4,6,7] the second is quite different to Laplace's equation introduced by him for the determination of ψ.

APPROXIMATE SOLUTION FOR THE BOUNDARY LAYER STRESSES

One of the main difficulties in the solution of eqns (22) and (23), associated with eqns (14), (15) and (16), is that it is not possible to separate the variables both in equations and boundary conditions. However, as was done by Horvay[5] for biharmonic problems, we can look for an approximate solution by taking

$$\phi(\xi, \eta) = a(\xi)b(\eta) \qquad \psi(\xi, \eta) = c(\xi)d(\eta) \quad (24)$$

the functions b and d being determined by integration of the boundary eqns (14) and (15) in η. In fact, it is easy to see that if we take

$$d(\eta) = -\frac{1}{c(o)} \int_{-1}^{\eta} \sigma_{xy}^0(x, o, \eta') d\eta' \quad (25)$$

$$b(\eta) = -\frac{1}{a(o)} \int_{-1}^{\eta} d\eta' \int_{-1}^{\eta'} \sigma_{yy}^0(x, o, \tau) d\tau \quad (26)$$

we are able to satisfy all the boundary conditions at $\xi = 0$. Furthermore, we can verify all the remaining boundary conditions by putting

$$a(o) = c(o) = 1 \quad (27)$$

$$a, \xi(o) = 0 \quad (28)$$

and the condition of vanishing stresses as $\xi \to \infty$. It is worth noting that with such a definition of ϕ and ψ, it is possible to account for almost any kind of heterogeneity (provided that symmetry in η is ensured). In particular, for the general case of layered plates, where the S_{ij} are taken as piecewise constants in η, the CLT stresses σ_{xy}^0 and σ_{yy}^0 are also piecewise constants, and the continuity conditions at the interfaces for F_{xz}^0, F_{yz}^0, F_{zz}^0 are easily satisfied. Furthermore, it can be seen that in such a case $d(\eta)$ is piecewise linear, continuous, with zero values at $\eta = \pm 1$, and $b(\eta)$ is piecewise quadratic, continuous, with a continuous derivative, and is zero together with its derivative at $\eta = \pm 1$.

With the functions defined by eqns (25), (26), (27) and (28), we can now construct a statically admissible stress field, and obtain a system of ordinary differential equations for $c(\xi)$ and $a(\xi)$ by invoking the principle of minimum complementary energy[8] which is the weak form of the compatibility equations. After a few calculations, this system can be put into the matrix form:

$$M\dddot{X}(\xi) + N\ddot{X}(\xi) + PX(\xi) = 0 \tag{29}$$

with

$$X = \left\{ \begin{matrix} a \\ c \end{matrix} \right\} \tag{30}$$

$$M = \begin{bmatrix} \displaystyle\int_{-1}^{+1} A_{22}b^2\,\mathrm{d}\eta & 0 \\ 0 & 0 \end{bmatrix} \tag{31}$$

$$N = \begin{bmatrix} \displaystyle\int_{-1}^{+1}(2A_{23}b''b - A_{44}b'^2)\,\mathrm{d}\eta & \displaystyle\int_{-1}^{+1} A_{36}d'b - A_{45}db'\,\mathrm{d}\eta \\ \displaystyle\int_{-1}^{+1}(A_{36}d'b - A_{45}db')\,\mathrm{d}\eta & -\displaystyle\int_{-1}^{+1} A_{55}d^2\,\mathrm{d}\eta \end{bmatrix} \tag{32}$$

$$P = \begin{bmatrix} \displaystyle\int_{-1}^{+1} A_{22}b''^2\,\mathrm{d}\eta & \displaystyle\int_{-1}^{+1} A_{26}d'b''\,\mathrm{d}\eta \\ \displaystyle\int_{-1}^{+1} A_{26}d'b''\,\mathrm{d}\eta & \displaystyle\int_{-1}^{+1} A_{66}d'^2\,\mathrm{d}\eta \end{bmatrix} \tag{33}$$

together with the edge condition

$$X(o) = \left\{ \begin{matrix} 1 \\ 1 \end{matrix} \right\} \qquad \dot{a}(o) = 0 \tag{34}$$

For sake of brevity, in all these expressions the superscript dot ($\dot{}$) is used for a ξ-derivative, and the prime ($'$) for η-derivatives.

The general solution of eqn. (29) depends on the roots of the characteristic equation

$$\det(\lambda^4 M + \lambda^2 N + P) = 0 \tag{35}$$

It can be seen by elementary algebra that this is a bicubic equation which can be written

$$q_0\mu^3 + q_1\mu^2 + q_2\mu + q_3 = 0 \quad (\text{with } \mu = \lambda^2) \tag{36}$$

Our problem can be solved if $X(\xi)$ vanishes as ξ tends to infinity, i.e. if the roots of eqn. (36) are positive whenever they are real numbers, or if they have a non-zero real part whenever they are complex. In all the numerical calculations made for typical graphite–epoxy or glass–epoxy laminates, the three roots of eqn. (36) appeared to be real and positive, thus leading to a solution of the following type

$$X(\xi) = \alpha_1 X_1 e^{-\lambda_1 \xi} + \alpha_2 X_2 e^{-\lambda_2 \xi} + \alpha_3 X_3 e^{-\lambda_3 \xi} \tag{37}$$

X_1, X_2, X_3 being the eigen vectors associated with the eigen values $\lambda_1, \lambda_2, \lambda_3$, and $\alpha_1, \alpha_2, \alpha_3$ being determined in order to satisfy eqn (34) (which is always possible due to the linear independence of X_1, X_2, X_3, if $\lambda_1 \neq \lambda_2 \neq \lambda_3$).

NUMERICAL APPLICATIONS

The applications are very easy, due to the simplicity of the method. In particular, it must be emphasised that there is no limitation on the number of layers nor on their orientation. The applications shown here were made by means of a small Fortran program containing about 600 instructions, but we could have used a microcomputer (e.g. HP 85) with a Basic program. Certain applications with a small number of layers have even been made by using a TI 59 pocket calculator. It can be shown (by the calculation of $\sigma_{yy}^0(x, o, \eta)$ and $\sigma_{xy}^0(x, o, \eta)$) that all the F_{ij}^0 appear in the form

$$F_{ij}^0 = t_{ij} N_{xx}^0(x, o)$$

and depend on the CLT solution only through $N_{xx}^0(x, o)$. Therefore, the structure of the boundary layer stresses may be obtained independently of this solution, in terms of the amplification factors t_{ij}, which are functions only of the layers' elastic characteristics and stacking sequence. Though all the six components of the stress tensor can be determined, we only show here the numerical results for the transverse shear and normal stresses, t_{xz}, t_{yz}, t_{zz} for various types of laminates.

It can be seen in Figs 1–7 that these stresses are perfectly continuous. For

D. Engrand

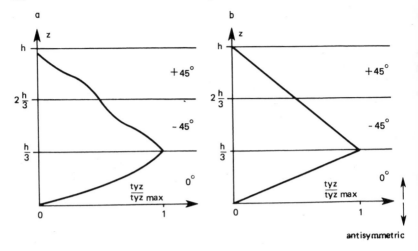

FIG. 1. Glass–epoxy $[+45°, -45°, 0°]_s$. Typical variation of t_{yz} across the thickness, near the edge. (a): finite width strip (width $= 5h$) (after Pipes[10]); (b): present study.

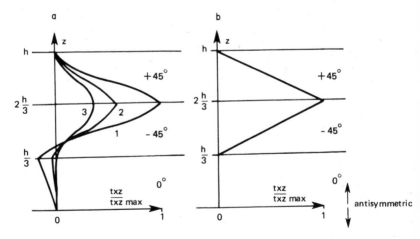

FIG. 2. Glass–epoxy $[+45°, -45°, 0°]_s$. Typical variation of t_{xz} across the thickness, near the edge. (a): finite width strip (width $= 5h$); (b): present study $y = 0$. $1 y = 0.033h$; $2 y = 0.066h$; $3 y = 0.1h$.

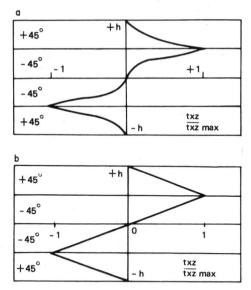

FIG. 3. Glass–epoxy $[+45°, -45°, -45°, +45°]_s$. Typical variation of t_{xz} across the thickness near the edge. (a): finite width strip (width $= 4h$) $y = 0$ (after Pipes[10]); (b): present study.

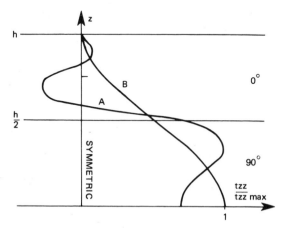

FIG. 4. Graphite–epoxy $[0°, 90°]_s$. Typical normal stress distribution across the thickness. (a): finite width strip (width $= 4h$) $y = 0$ (after Spilker and Chou[12]); (b): present study $y = 0$.

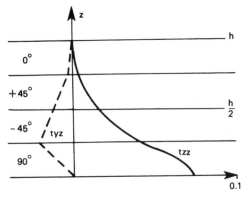

FIG. 5. Graphite–epoxy $[0°, 45°, -45°, 90°]_s$. Stress factors t_{zz} max at $y = 0$, t_{yz} max at $y = 0.534h$ (t_{zz} is symmetric, and t_{yz} antisymmetric).

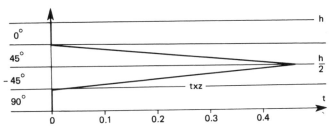

FIG. 6. Graphite–epoxy $[0°, 45°, -45°, 90°]_s$. Stress factors t_{xz} max at $y = 0$ (t_{xz} is antisymmetric).

typical graphite–epoxy laminae, the elastic characteristics used in the calculations are:

$$E_{11} = 1.3928 \times 10^5 \text{ MPa} \qquad E_{22} = 1.5238 \times 10^4 \text{ MPa} \qquad E_{33} = E_{22}$$

$$G_{12} = G_{23} = G_{13} = 5.861 \times 10^3 \text{ MPa}$$

$$v_{12} = v_{23} = v_{13} = 0.21$$

In Figs 1–3, we have plotted the results obtained with six and four layers for glass–epoxy. The stacking sequences and typical elastic characteristics are taken from Pipes.[10]

$$E_{11} = 4.137 \times 10^4 \text{ MPa} \qquad E_{22} = 1.0342 \times 10^4 \text{ MPa} \qquad E_{33} = E_{22}$$

$$G_{12} = 5.516 \times 10^3 \text{ MPa} \qquad G_{23} = 4.137 \times 10^3 \text{ MPa} \qquad G_{13} = G_{12}$$

$$v_{12} = 0.25 \qquad\qquad v_{23} = 0.30 \qquad\qquad v_{13} = v_{12}$$

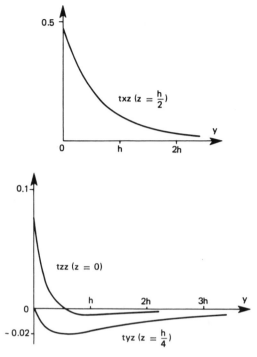

FIG. 7.   Variations of stress factors in y-direction at interfaces of maximum values. Eight layers graphite–epoxy $[0°, 45°, -45°, 90°]_s$.

Though the results of Pipes[10] are not directly comparable to ours, since he dealt with a finite width strip problem, it is interesting to see that the thickness distribution of stresses is qualitatively quite similar in the two cases.

CONCLUSION AND FURTHER IMPROVEMENTS

While it gives an exact representation of the stress functions and their η-derivatives on the free edge, our approximate solution does not guarantee such an accuracy for the stresses depending on ξ-derivatives. It is probable that the less accurate stress in our approximation is t_{zz} which is proportional to $\phi,\xi\xi$. Though the results are not quite comparable, it can be seen in Fig. 4 that the thickness distribution of t_{zz} on the edge is not the same in the present case as in the finite width case.[12]

A generalised method, that is expected to give an improved accuracy, can be outlined in the following way. As the stress functions are piecewise polynominals for $\xi = 0$, it seems natural to introduce a finite element discretisation in the η-direction, for, in such a way, it is always possible to match *exactly* the values of ϕ and ψ on the edge by choosing appropriate finite elements.

As a typical finite element, we can take

$$\phi = \sum H_i(\eta) q_i(\xi)$$
$$\psi = \sum L_i(\eta) p_i(\xi)$$

where $H_i(\eta)$ and $L_i(\eta)$ are polynomials, and q_i, p_i unknown functions. The boundary and continuity conditions, suggest taking the first degree Lagrange polynomials for L_i, and third degree Hermite polynomials for H_i.

As in the present study, the principle of complementary energy would lead to a new eigen value problem, the eigen vectors of which could be combined in order to verify exactly the conditions at $\xi = 0$. For instance, by taking one element by layer, the eigen value problem to be solved would be of dimension $3N - 2$ for a $2N$-layers laminate. However, though substantial improvements are expected from this method, the real nature of eigen values to be found does not seem mathematically evident, and the numerical solution of this problem still needs a little more work.

However, the boundary layer method, as presented in this study, can be very useful for predicting the stress distributions and risks of delamination in almost every current case of laminate and stacking sequence. Since all the six stresses are available at any point near the edge, it is easy to apply a failure criterion, as was done for example by Herakovich.[9] This can be achieved by means of a small program, and involves only minimum manual work.

Finally, it can be emphasised that this method can be used also for curved edges (e.g. a circular hole in a plate[4]) provided that the ξ-direction is taken as the direction normal to the edge, and the equations are expressed in local tangential and normal coordinates.

REFERENCES

1. DESTUYNDER, P., Sur une justification des modèles de plaques et de coques par les méthodes asymptotiques, Thèse de Doctorat d'Etat des sciences mathématiques, Université Paris VI. Paris 1980.
2. REISS, E. L. and LOCKE, S., On the theory of plane stress, *Quar. Appl. Math.*, **19**, 1961, 195.

3. FRIEDRICHS, K. O. and DRESSLER, R. F., A boundary layer theory for elastic plates, *Comm. Pure and Appl. Math.*, **14**, 1961, 1–33.
4. TANG, S., A variational approach to edge stresses of circular cutouts in composites, AIAA/ASME/ASCE/AHS 20th Structures, Structural Dynamics and Materials Conference, St Louis, April 4–6, 1979.
5. HORVAY, G., The end problem of rectangular strips, *Jal. Appl. Mech.*, **20**, 1953, 87–94.
6. TANG, S., A boundary layer theory—Part I: Laminated composites in plane stress, *Jal. Compos. Mat.*, **9**, Jan. 1975, 33–41.
7. TANG, S., Interlaminar stresses around circular cutouts in composite plates under tension, *AIAA Jal.*, **15**, 1977, 1631.
8. VALID, R., La mécanique des milieux continus et le calcul des structures, Paris Eyrolles, 1977. English version to appear March 81, North Holland Publ. Co.
9. HERAKOVICH, C. T., On failure modes in finite width angle ply laminates. In: *Advances in composite materials*. Proceedings of ICCM 3, Paris, Aug. 1980. Pergamon Press.
10. PIPES, R. B., Boundary layer effects in composite laminates, *Fibre Science and Technology*, **13**, No. 1, 1980, 49.
11. PAGANO, N. J., Stress fields in composites laminates, *Int. Jal Solids and Structures*, **14**, 1978, 385–400.
12. SPILKER, R. L. and CHOU, S. C., Edge effects in symmetric composite laminates: Importance of satisfying the traction-free edge condition, *Jal. Compos. Mat.*, **14**, 1980, 2.

18

On the Orthotropic Elastic Behaviour of a Rubber Composite

A. P. S. SELVADURAI

*Department of Civil Engineering, Carleton University,
Ottawa, Ontario K1S 5B6, Canada*

AND

N. MOUTAFIS

*Department of Civil Engineering,
The University of Aston in Birmingham, Birmingham B4 7ET, England*

ABSTRACT

This chapter outlines theoretical and experimental studies pertaining to the stress analysis of a rubber composite. In particular, the problems examined relate to the plane-strain edge loading of the rubber composite. The theoretical analysis of the edge loading problems is developed by assuming that the laminated rubber composite can be idealized as an orthotropic elastic solid. Using a Fourier integral approach, solutions are developed for the problems related to uniform loading and uniform indentation of a halfplane region. An extension of these solutions to the analysis of a quarterplane region is also briefly reviewed. The theoretical results derived for the half plane region are compared with experimental results obtained for an edge loaded rubber composite solid which consists of alternate layers of hard and soft rubber-like materials.

INTRODUCTION

Composites constructed of rubber-like materials have a variety of engineering applications which include their use in load bearing elements of structural systems, energy absorbing devices and in inflatable structures. Although the term 'composite' can refer to a variety of fabricated materials, in this chapter we are primarily concerned with the class of rubber-like laminates which is constructed by bonding together layers of rubber-like materials with differing elastic properties. In the microscale these laminates display prominent discontinuous non-homogeneous elastic characteristics. On the macroscale, which involves a large assemblage of separate layers, the composite can be visualized as an elastic material whose properties are predominantly orthotropic. The theory of orthotropic elasticity has found extensive application in the stress analysis of natural and artificial structural materials such as wood, laminated geological strata, fibre-reinforced solids and other laminated composites. Extensive accounts of the various developments in the application of orthotropic elasticity to the stress analysis of such materials are given in the texts and articles by Green and Taylor,[1] Green and Zerna,[2] Lekhnitski,[3] Holister and Thomas,[4] Wendt *et al.*,[5] Tsai *et al.*,[6] Spencer,[7] Broutman and Krock,[8] Garg *et al.*,[9] Christensen[10] and Selvadurai.[11] The orthotropic elastic behaviour of the composite is governed by the elastic characteristics of the constituent materials, their shapes, bond characteristics and volume distributions. The theoretical bulk orthotropic elastic properties of the composite can be estimated by appeal to a theory of mixtures. Such theoretical estimates are given by Hashin,[12] Hill,[13] and in references 4 to 10.

This chapter describes a programme of theoretical and experimental studies which was conducted in order to establish how closely the elastic behaviour of a laminated rubber-like composite corresponds to the orthotropic elastic idealization. The composites were constructed by bonding together, in alternate fashion, two sheets of rubber-like isotropic elastic materials with equal thickness but differing elastic moduli (modular ratio, approximately 2·5). The theoretical estimates for the bulk orthotropic elastic properties of the composite were compared with equivalent experimental results. The correlation was found to be satisfactory. The second phase of the experimental programme was devoted to the measurement of the plane-strain strain fields induced in a rubber composite block region which is subjected to a uniform stress or a uniform displacement on a part of the boundary. These strain fields are computed from the induced displacement fields which are measured by optical means. The experimental strain fields are then compared with theoretical results derived for the surface loading of an orthotropic elastic halfplane. The material constants derived previously are used in the theoretical analysis. Finally, both theoretical and experimental results derived for the edge loading of the rubber-like composite quarterplane are presented in summary form.

THEORETICAL RESULTS

We consider the plane-strain deformations of a transversely isotropic elastic material in which the principal axes coincide with the rectangular Cartesian co-ordinate system (x, y, z). The elastic material exhibits isotropic characteristics in the x–z plane. We further restrict our attention to a state of plane strain in the transversely isotropic elastic solid where u_x and u_y are independent of z and $u_z = 0$. Here, u_x, u_y and u_z are the Cartesian components of the displacement vector. The resulting elastic stress–strain relations for the transversely isotropic solid can be written as:

$$
\begin{bmatrix} \varepsilon_{xx} \\ \varepsilon_{yy} \\ \varepsilon_{xy} \end{bmatrix} = \begin{bmatrix} l_{11} & l_{12} & 0 \\ l_{21} & l_{22} & 0 \\ 0 & 0 & \dfrac{l_{66}}{2} \end{bmatrix} \begin{bmatrix} \sigma_{xx} \\ \sigma_{yy} \\ \sigma_{xy} \end{bmatrix} \tag{1}
$$

where:

$$
\begin{aligned}
l_{11} &= \frac{1 - v_{xz}^2}{E_x} & l_{12} &= -\frac{v_{yx}(1 + v_{xz})}{E_y} \\
l_{22} &= \frac{1 - v_{xy}v_{yx}}{E_y} & l_{21} &= -\frac{v_{xy}(1 + v_{xz})}{E_x} & l_{66} &= \frac{1}{G_{xy}}
\end{aligned} \tag{2}
$$

In eqn. (1) ε_{ij} follows the usual definitions of the linearized strain tensor, σ_{ij} is the Cauchy stress tensor and, by virtue of the existence of a strain energy function and the associated reciprocity relationships, $l_{ij} = l_{ji}$. By employing an Airy stress function $\Phi(x, y)$ such that:

$$
[\sigma_{xx}; \sigma_{yy}; \sigma_{xy}] = \left[\frac{\partial^2 \Phi}{\partial y^2}; \frac{\partial^2 \Phi}{\partial x^2}; -\frac{\partial^2 \Phi}{\partial x\, \partial y} \right] \tag{3}
$$

it can be shown that the differential equation governing the plane-strain problem is:

$$
\left\{ \frac{\partial^2}{\partial x^2} + k_1^2 \frac{\partial^2}{\partial y^2} \right\} \left\{ \frac{\partial^2}{\partial x^2} + k_2^2 \frac{\partial^2}{\partial y^2} \right\} \Phi(x, y) = 0 \tag{4}
$$

In eqn. (4):

$$
\left. \begin{matrix} k_1^2 \\ k_2^2 \end{matrix} \right\} = \frac{1}{2l_{22}} \{ 2l_{12} + l_{66} \pm [4l_{12}^2 + l_{66}^2 + 4l_{12}l_{66} - 4l_{11}l_{22}]^{1/2} \} \tag{5}
$$

Solutions of eqn. (4) are subject to appropriate displacement and/or traction boundary conditions applicable to the problem. The constants, k_1 and k_2, may be real or imaginary (see, for example, references 1 and 3). For the present purposes we shall assume that they are real and positive as is the case for the materials used in this investigation.

Stress Analysis of the Orthotropic Elastic Halfplane

The idealization of an elastic medium as a halfplane region serves, in general, as a useful analogue for the determination of stress distributions in finite elastic regions which are subjected to localized loads (i.e. the dimensions of the localized loaded area are small in relation to the dimensions of the elastic medium).[14,15] The earlier applications of the isotropic elastic halfplane idealization to the analysis of surface loading problems are given by Flamant,[16] Michell[17] and Carothers.[18] Similar solutions are obtained by Green and Taylor,[1] Lekhnitski,[3] Conway,[19] Brilla,[20] Akoz and Tauchert[21] for the orthotropic case. Also, Saha et al.[22] developed a generalized solution for the anisotropic case. Several results pertaining to the indentation of the surface of an orthotropic elastic halfplane by a rigid indentor are also given by Brilla,[20] Okubo,[23] Sen[24] and Conway.[25] In the following we shall list salient results pertaining to the surface loading or surface indentation of an orthotropic elastic halfplane.

It may be verified that the stress function:

$$\Phi(x, y) = \int_0^\infty \frac{1}{\alpha^2} \{C_1(\alpha) e^{-\alpha y/k_1} + C_2(\alpha) e^{-\alpha y/k_2}\} \cos \alpha x \, d\alpha \qquad (6)$$

satisfies the governing differential equation (eqn. (4)). The functions $C_1(\alpha)$ and $C_2(\alpha)$ are determined by satisfying the traction boundary conditions on the plane $y = 0$. In the particular case where the surface of the orthotropic elastic halfplane is subjected to a concentrated normal line load, P, at the origin (Fig. 1(a)) the state of stress within the medium takes the form:

$$[\sigma_{xx}; \sigma_{yy}; \sigma_{xy}] = \frac{P(k_1 + k_2)}{\pi(k_1^2 x^2 + y^2)(k_2^2 x^2 + y^2)} [x^2 y; y^3; xy^2] \qquad (7)$$

When the halfplane is subjected to a distributed normal load of stress intensity, p, and width, $2l$ (Fig. 1(b)) the state of stress in the orthotropic elastic region is given by:

$$[\sigma_{xx}; \sigma_{yy}; \sigma_{xy}] = \frac{p}{\pi(k_1 - k_2)} \left[\left(\frac{T_2}{k_2} - \frac{T_1}{k_1} \right); (T_1 k_2 - T_2 k_1); \frac{1}{2} \ln \left(\frac{t_1}{t_2} \right) \right] \qquad (8)$$

where:

$$T_i = \tan^{-1}\left\{\frac{k_i(x+l)}{y}\right\} - \tan^{-1}\left\{\frac{k_i(x-l)}{y}\right\} \quad (i=1,2)$$

$$t_i = \frac{t_i^+}{t_i^-} \qquad t_i^\pm = y^2 + k_i^2(x\pm l)^2 \qquad (i=1,2) \tag{9}$$

The problem related to the frictionless indentation of the surface of an orthotropic elastic halfplane by a plane rigid indentor (of width $2l$; see Fig. 1(c)) has been examined by Conway,[25] who reduced it to the solution of a single integral equation of the type:

$$\int_{-l}^{l} \sigma_{yy}(\zeta) \ln\left|\frac{x-\zeta}{a}\right| d\zeta = \text{constant} \tag{10}$$

where a is a constant.

Alternative formulations which involve either complex variable or integral transform techniques are given by Sneddon[26] and Gladwell.[27]

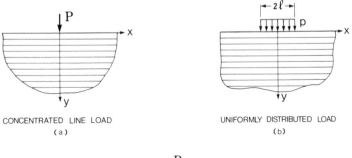

CONCENTRATED LINE LOAD
(a)

UNIFORMLY DISTRIBUTED LOAD
(b)

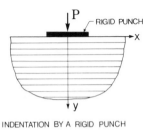

INDENTATION BY A RIGID PUNCH
(c)

FIG. 1.   Edge loading problems for the orthotropic elastic halfplane.

These analyses indicate that the contact stress distribution at the frictionless interface takes the form:

$$\sigma_{yy}(x, 0) = \frac{P}{\pi\sqrt{l^2 - x^2}} \tag{11}$$

where P is the resultant line load acting on the rigid punch. This result is uninfluenced by the degree of orthotropy of the halfplane and is in agreement with the equivalent result obtained by Sadowsky[28] for the isotropic case. With the aid of the results shown in eqns (7) and (11) it is possible to determine the distribution of stress within the halfplane region. For example, the stress, σ_{xx}, is given by:

$$\sigma_{xx} = \frac{P(k_1 + k_2)}{\pi^2} \int_{-l}^{l} \frac{y(x - \lambda)^2 \, d\lambda}{\sqrt{l^2 - \lambda^2} \, \{k_1^2[x - \lambda]^2 + y^2\}\{k_2^2[x - \lambda]^2 + y^2\}} \tag{12}$$

Stress Analysis of the Orthotropic Elastic Quarterplane

The quarterplane is a special case of a wedge shaped region. The stress analysis of wedge shaped regions which are subjected to localized or distributed surface loads has been investigated by several authors including Sternberg and Koiter,[29] Benthem,[30] Bogy[31] and Harrington and Ting[32] who make extensive use of Mellin transform techniques. References to further studies are also given by Gladwell.[27] With the stress analysis of a quarterplane region, however, it is convenient to utilize a successive superposition scheme (Schwartzian reflection principle) which makes use of a fundamental solution related to the normal loading of a halfplane region. For example, the solution to the problem of the concentrated normal surface loading of an orthotropic elastic quarterplane can be approached via a combined Flamant[16] solution which consists of an elastic halfplane subjected to concentrated forces located equidistant from the origin (see, for example, Fig. 2(a)). This is the basic state of stress $\sigma_{ij}^{(0)}$. By virtue of the symmetry of $\sigma_{ij}^{(0)}$ the shear stresses are zero on $x = 0$. The plane, $x = 0$, is subjected to only a purely normal stress, $F_0(\eta)$ (the variables $\xi = x/a$ and $\eta = y/a$ refer to distances *measured on the axes considered* and a is a length parameter). If an additional state of stress can be found such that, in relation to a quarterplane occupying the first quadrant (Fig. 2(b)): (i) the shear tractions are zero on $x = 0$; $y = 0$ and (ii) the normal tractions on the plane $x = 0$ and $y = 0$ are $- F_0(\eta)$ and zero, respectively, then this corrective state of stress, $(\sigma_{ij}^{(c)})$, together with $\sigma_{ij}^{(0)}$, constitutes a solution of the quarterplane problem. The successive superposition procedure used to

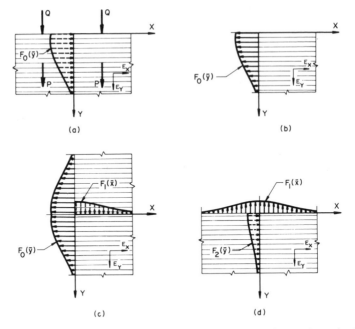

FIG. 2.   Successive superposition scheme for the stress analysis of the orthotropic elastic quarterplane.

develop $\sigma_{ij}^{(c)}$ is usually referred to as the Schwartzian reflection principle and its application to isotropic and orthotropic quarterplane problems is documented by Hetenyi[33] and Selvadurai and Moutafis,[34] respectively. The basis of the superposition scheme can be summarized as follows. For the basic state of stress $\sigma_{ij}^{(0)}$ (Fig. 2(a)) the stresses on the axis of symmetry, $x = 0$, are:

$$\sigma_{xx}^{(0)}(0, y) = F_0(\eta) \qquad \sigma_{xy}^{(0)}(0, y) = 0 \qquad (13)$$

Now consider the problem of an orthotropic elastic halfplane, occupying the region $x \geq 0$, which is subjected to a symmetric stress distribution $\sigma_{xx} = -F_0(\eta)$ on the plane $x = 0$ (step 1). The resulting stress field, $\sigma_{ij}^{(1)}$, can be obtained by superposition of the state of stress of the type shown in eqn. (7) (except for change in the frame of reference). The resulting stress field, $\sigma_{ij}^{(1)}$, can be obtained in the following form:

$$\sigma_{ij}^{(1)} = \frac{k_1 k_2 (k_1 + k_2)}{\pi} \int_0^\infty F_0(\eta) J_{ij}(\eta) \, d\eta \qquad (14)$$

where:

$$J_{ij} = J_{ij}^+ + J_{ij}^-$$

and:

$$[J_{xx}^\pm; J_{yy}^\pm; J_{xy}^\pm] = R[X^3; X(Y \pm \eta)^2; X^2(Y \pm \eta)] \tag{15}$$

$$R = \{[k_1^2 X^2 + (Y \pm \eta)^2][k_2^2 X^2 + (Y \pm \eta)^2]\}^{-1}$$

The normalized spatial variables, $X = x/a$, $Y = y/a$, refer to a general location (x, y) whereas $\xi = x/a$, $\eta = y/a$ refer to the normalized distance along the co-ordinate axes.

Thus, combining $\sigma_{ij}^{(1)}$ with $\sigma_{ij}^{(0)}$ renders the plane $x = 0$ free of traction but gives rise to a non-zero normal traction, $F_1(\xi)$, on the plane $y = 0$. To eliminate $F_1(\xi)$ we consider a symmetric external stress $- F_1(\xi)$, on $y = 0$ for the halfplane region $y \ge 0$ (Fig. 2(c)). The stresses due to this loading (step 2) are given by:

$$\sigma_{ij}^{(2)} = \frac{(k_1 + k_2)}{\pi} \int_0^\infty F_1(\xi) K_{ij}(\xi) \, d\xi \tag{16}$$

where:

$$K_{ij} = K_{ij}^+ + K_{ij}^-$$

and:

$$[K_{xx}^\pm; K_{yy}^\pm; K_{xy}^\pm] = S[(X \pm \xi)^2 Y; Y^3; (X \pm \xi)Y^2] \tag{17}$$

$$S = \{[k_1^2(X \pm \xi)^2 + Y^2][k_2^2(X \pm \xi)^2 + Y^2]\}^{-1}$$

Again, the combination of $\sigma_{ij}^{(2)}$ and $\sigma_{ij}^{(1)}$ leaves the plane $y = 0$ traction free but gives rise to a normal stress $\sigma_{xx} = F_2(\eta)$ on $x = 0$ (Fig. 2(d)). The procedures outlined in steps 1 and 2 have to be repeated to eliminate superfluous stress generated on the planes of symmetry. The successive superposition of halfplane solutions leads to a convergent corrective state of stress of the form:

$$\sigma_{ij}^{(c)} = \frac{(k_1 + k_2)}{\pi} \left(-k_1 k_2 \int_0^\infty \left\{ J_{ij}(\eta) \sum_{m=0,2,4}^\infty F_m(\eta) \, d\eta \right\} \right.$$

$$\left. + \int_0^\infty \left\{ K_{ij}(\xi) \sum_{m=1,3,5}^\infty F_m(\xi) \, d\xi \right\} \right) \tag{18}$$

where:

$$F_{m+1}(\xi) = \frac{k_1 k_2 (k_1 + k_2)}{\pi} \int_0^\infty \frac{2F_m(\eta)\xi\eta^2 \, d\eta}{T} \tag{19}$$

$$F_{m+1}(\eta) = \frac{(k_1 + k_2)}{\pi} \int_0^\infty \frac{2F_m(\xi)\xi^2\eta \, d\xi}{T}$$

and:

$$T = [k_1^2\xi^2 + \eta^2][k_2^2\xi^2 + \eta^2] \tag{20}$$

The complete solution for the orthotropic quarterplane problem is given by:

$$\sigma_{ij} = \sigma_{ij}^{(0)} + \sigma_{ij}^{(c)} \tag{21}$$

This procedure was used to examine the state of stress in the rubber-like composite block which was subjected to loads located near the corner.

THE EXPERIMENTAL PROGRAMME

The experimental study was organised in order to establish how closely the mechanical behaviour of a rubber-like composite (which was fabricated by adhesive bonding of soft and hard rubber) could be characterized by the theory of orthotropic linear elasticity. The experimental programme essentially consisted of three phases. The first phase involved the determination of the elastic properties of the individual layers constituting the composite solid. The second part of the investigation consisted of the determination of the bulk orthotropic properties of the fabricated composite. The third phase of the experimental programme involved the plane-strain testing of a composite block by edge and corner loading. Due to the lack of space, all aspects of the experimental programme cannot be given complete coverage. For further details the reader is referred to Moutafis.[35]

Elastic Properties of the Constituent Materials

The following materials were used in the fabrication of the rubber composite: a soft rubber (subscript 's') (Shotblast, 70 % natural rubber, Shore hardness (40 to 45 %), a hard rubber (subscript 'h') (Vinyl, trade name Velbex, Shore hardness 80 %) and an adhesive (Dunlop rubber adhesive S 738). (All the materials were supplied by Rubber and Plastics Industries,

Birmingham, UK). The two types of rubber were supplied in 3·5 m × 150 mm × 3 mm strips. Laboratory tests were conducted on both soft and hard rubber specimens subjected to uniaxial tension and compression. These tests indicated that the rubbers were essentially isotropic in their elastic characteristics. There were no visible creep effects in the stress ranges used in the experimental programme. Both the soft and hard rubbers exhibited linear elastic responses (in tension and compression) for strains of the order of 6%. The results of these experiments are summarized in Table 1.

TABLE 1
Properties of the constituent rubber materials

Type of rubber	Tension tests			Compression tests			v	G
	n	E	SD	n	E	SD		
Hard	6	7·5	0·11	4	7·2	0·10	0·48	2·44
Soft	6	3·2	0·08	4	2·9	0·06	0·48	1·01

n = Number of tests. SD = Standard deviation. E = Young's modulus. v = Poisson's ratio. G = Shear modulus. (All dimensional quantities are in N/mm^2.)

Orthotropic Elastic Properties of the Composite material

Four of the five elastic constants required for the complete description of the elastic behaviour of the transversely isotropic rubber block were determined experimentally. The fifth constant (G_{xy}) was estimated using the properties of the constituent materials and the theory of mixtures. Specially manufactured rubber composite samples (45 mm × 45 mm × 27 mm) identical to the large composite block used in the plane-strain test (to be described later) were tested in compression at a constant strain rate of 0·1524 mm/min along one of the principal directions. The elastic moduli (E_x, E_y) and the Poisson's ratios (v_{xy}, v_{yx}, v_{xz}) were determined by a measurement of the applied loads and the longitudinal and lateral strains. The results of these experiments are summarized in Table 2.

Theoretical estimates for the bulk orthotropic elastic properties of the laminated rubber-like composite can be established by making use of the elastic properties of the hard and soft materials (E_h, E_s, v_h, v_s) and their respective volume fractions (V_h, V_s). There are a number of theoretical estimates that can be employed for this purpose. Detailed accounts of these developments are given in references 4 to 10. For laminated materials the effective elastic constants for the orthotropic idealization can be established

TABLE 2
Orthotropic elastic constants for the transversely isotropic rubber composite

	Theory	Experiment	n	SD
E_x	5·05	4·60	8	$\pm0·06$
E_y	4·14	3·90	11	$\pm0·10$
v_{xy}	0·48	0·48	4	—
v_{yx}	0·40	0·41	11	$\pm0·01$
v_{xz}	0·48	0·48	4	$\pm0·01$
G_{xy}	1·43	1·30	—	—

n = Number of tests. SD = Standard deviation.
$l_{11} = 0·1673$, $l_{12} = -0·1544$, $l_{22} = 0·2277$, $l_{66} = 0·7692$, $k_1 = 1·332$, $k_2 = 0·6760$.
(E_x, E_y, etc., are expressed in N/mm^2; l_{11}, l_{22}, etc., are expressed in mm^2/N.)

by employing an elementary theory of mixtures. The relationships employed are as follows:

$$E_x = V_h E_h + V_s E_s \qquad E_y = \frac{E_h E_s}{E_h V_s + E_s V_h} \qquad (22)$$

$$v_{xy} = V_h v_h + V_s v_s \qquad v_{yx} = v_{xy} \frac{E_y}{E_x} \qquad G_{xy} = \frac{G_h G_s}{G_h V_s + G_s V_h}$$

A comparison between the theoretical and experimental values of E_x, E_y, etc. is given in Table 2. The constants, l_{11}, l_{22}, etc., associated with the transversely isotropic idealization are also given in Table 2.

Plane-strain Testing of the Rubber Composite

A rubber block measuring approximately 535 mm $\times$ 820 mm $\times$ 150 mm was constructed by using the soft and hard rubber strips. These strips were cleaned with trichloroethylene and neutralized with ammonia. The 'basic unit' used in the construction of the rubber block consisted of five strips of hard and soft rubber (820 mm long) glued together in alternating sequence with the rubber adhesive. These 'basic units' were later glued together in pairs and the process repeated until the thickness (150 mm) was attained. This technique of building up the block in steps was followed in order to avoid excessive self weight loading of the lower layers of the block. Such loading would induce undue straining of the soft rubber layers at the early stages when the adhesive is unhardened. In the final form the composite consisted of 83 layers of hard rubber and 82 layers of soft rubber. The large surfaces of the rubber block were machined with abrasive paper to remove irregularities and then covered with approximately 0·5 mm of Latex to obtain a smooth flat surface. A grid 10 mm square was drawn on one of the large surfaces of the block with white rubber paint (see Fig. 3).

The apparatus for plane-strain testing consisted of a steel container to accommodate the rubber block, a 20 mm thick glass plate and a set of loading devices capable of applying concentrated line loads, a distributed load and a rigid punch type load. The steel tank was constructed of 10 mm thick mild steel plates. The front face of the tank consisted of a rectangular frame containing a groove which accommodated the glass plate. The rear of the tank was stiffened with four 25-mm square steel bars to minimize the displacements which would occur in the z-direction. The inside surface of this plate was covered with a layer of formica to minimize the frictional effects.

The loads were applied to the composite block through the following loading devices: a cylinder (20 mm diameter and 150 mm long) to simulate the concentrated line loading and a rigid plate (100 mm × 150 mm × 20 mm) to simulate the rigid punch type of load. In both cases a T-beam lever (4:1 ratio) was used to apply the static loads (see Fig. 3). The uniformly distributed load was applied through a specially designed pressure cell which transmitted the load without any interference from the membrane.

Testing Procedure

The surfaces of the rubber block and the internal surfaces of the steel tank were lubricated with silicone grease to minimize the friction between the rubber composite and the glass and formica surfaces. The composite was placed inside the tank and the glass plate was slid into position and secured with the fixing screws. Two I-beams (38 mm × 76 mm) were placed at the front of the steel tank to reduce the lateral deflection of the glass plate during loading. (This is a necessary requirement to ensure that plane-strain conditions are achieved in the tests.) The actual testing under any of the three loading systems was accomplished in the following steps. (i) The appropriate loading device was fixed into position (at the centre of the rubber composite for halfplane problems or near the corner for quarterplane problems) and the thickness of the rubber block was measured with a large micrometer. (ii) Two dial gauges were positioned, one at the front and one at the rear of the steel tank, to record any lateral deformations and the initial readings were recorded. (iii) A camera (Hasselblad 500C) was fixed 500 mm from the steel tank and a photograph of the grid was taken. (iv) The load was then applied in steps and for each step a photograph of the deforming grid was taken and the dial gauge readings recorded. For further details the reader is referred to Moutafis.[35]

F<small>IG</small>. 3.   Experimental set up for the plane strain testing of the rubber composite.

Analysis of Test Data

In each test, the strains induced in the rubber block were determined from the photographic record of the grid deformation. The co-ordinates of the nodes for each load increment were determined by using a 'Universal Wild Plotter'. From the results, the lengths of each grid element were calculated and plotted against the applied load. The ratios of strain/load were determined from the slopes of these curves.

Accuracy and Errors

Several factors which could influence the accuracy of the experimental results were given due consideration. These include the measurement of loads (accuracy of the lever system, frictional effects, accuracy of measurement of the uniform pressure) and the measurement of lengths (accuracy of the Universal Wild Plotter, refraction effects of the glass plate, deformation of the reference points, displacements in the z-direction) and temperature effects. A detailed discussion of these effects is given by Moutafis.[35] In summary, the experimental data given in the paper are either uninfluenced or corrected to eliminate the potential errors.

COMPARISON OF THEORETICAL AND EXPERIMENTAL RESULTS

The series of plane-strain tests were carried out to determine the strain fields induced in a halfplane or a quarterplane by the action of externally applied loading systems with a view to providing a basis for comparison with the theoretical results. For all the tests the results are presented as a variation of the strains ε_{xx}/load, and ε_{yy}/load with the dimensionless co-ordinates $\bar{x}$ and $\bar{y}$ (where $\bar{x}$ and $\bar{y}$ are defined in terms of a characteristic length parameter of the problem). The dimensional quantities (i.e. strain/load) are expressed in mm^2/N or mm^2N^{-1}. Owing to the logarithmically divergent nature of the displacement field associated with the theoretical formulation of the two-dimensional halfplane or quarterplane problems, it is not possible to use the measured grid displacements as a basis for comparison (compressive strains are considered to be positive).

Concentrated Line Loading

Experimental investigation of concentrated force problems presents difficulties owing to the singular behaviour at the point of application of the force. Furthermore, knife edge type of loads would induce large strains in the rubber-like material which would invalidate the desired linear elastic response. To alleviate this problem, the line load was applied through a 20 mm diameter cylinder. The contact region between the cylinder and the rubber composite was measured at each stage; its average value was in the region of 12·5 mm. Figures 4 to 7 show the experimental and theoretical results obtained for ε_{xx} and ε_{yy} for both the halfplane and the quarterplane regions. As a first approximation the load is assumed to be uniformly distributed over the length, $2l$. The correlation between the theoretical and experimental results is found to be satisfactory at all locations except in the vicinity of the concentrated load.

Distributed Loads

In both cases (the halfplane and the quarterplane) the halfwidth of the uniformly distributed load ($l = 60$ mm) was assumed to be the characteristic length of the problem.

Figures 8 to 11 illustrate variations of ε_{xx} and ε_{yy} within the rubber composite. The load in the ratio 'strain/load' is the applied uniform stress, p, and for consistency of units it must be given in N/mm^2.

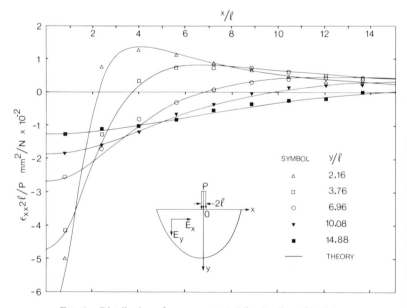

FIG. 4.   Distribution of ε_{xx}-concentrated line loading of halfplane.

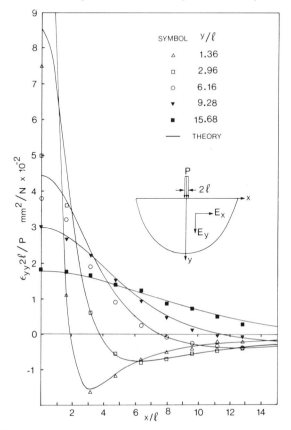

FIG. 5. Distribution of ε_{yy}-concentrated line loading of halfplane.

Rigid Punch Problem

For this particular loading attention was restricted to the experimental study of the halfplane problem. The quarterplane problem associated with the rigid punch loading cannot be analysed by using the superposition scheme since the basic solution is not available in a compact form. (It can, however, be obtained by using complex variable techniques.[27-36]) The stresses for the halfplane problem were obtained by numerical integration of results of the type of eqn. (12). Since the applied contact stress exhibits singular behaviour at $x = \pm l$, the numerical integration technique had to be modified. The applied stress was treated as a series of uniform load elements of finite width. The spacing of the elements (and their width) was

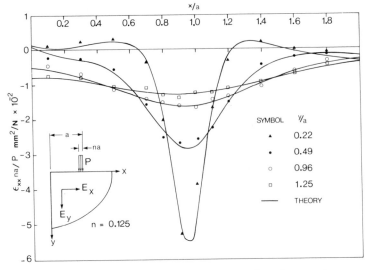

FIG. 6.   Distribution of ε_{xx}-concentrated line loading of quarterplane.

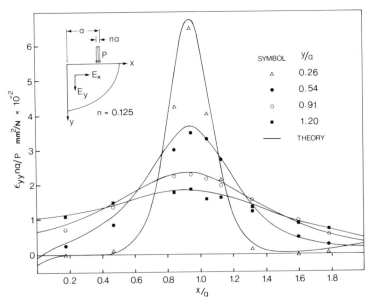

FIG. 7.   Distribution of ε_{yy}-concentrated line loading of quarterplane.

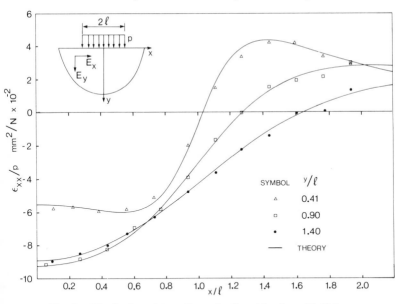

FIG. 8. Distribution of ε_{xx}-uniform distributed loading of halfplane.

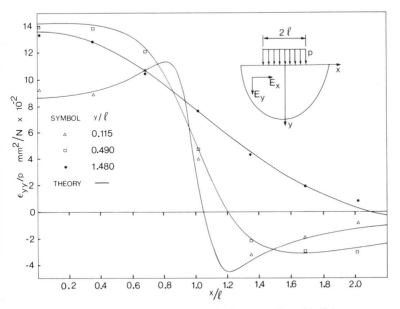

FIG. 9. Distribution of ε_{yy}-uniform distributed loading of halfplane.

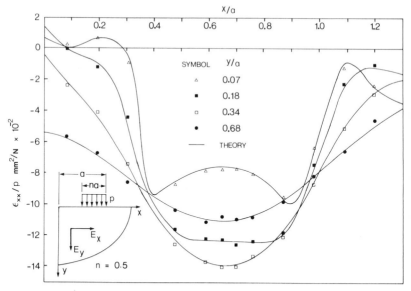

FIG. 10.   Distribution of ε_{xx}-uniform distributed loading of quarterplane.

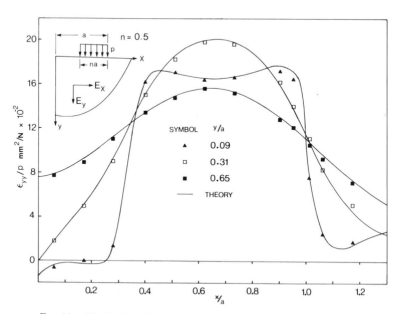

FIG. 11.   Distribution of ε_{yy}-uniform distributed loading of quarterplane.

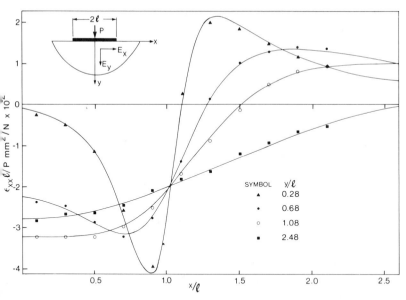

FIG. 12. Distribution of ε_{xx}-loading of the halfplane region by a rigid punch.

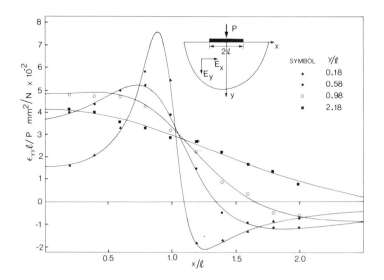

FIG. 13. Distribution of ε_{yy}-loading of the halfplane region by a rigid punch.

ordered on a logarithmic scale such that the magnitude of the stress for each load element was given by:

$$p(\bar{x}) = \frac{P}{\pi\sqrt{1 - \bar{x}^2}}$$

where $\bar{x} = 1 - \zeta$, for $-\infty < \ln \zeta \le 0$. In this way, it was possible to consider load elements as close as possible to $\bar{x} = \pm 1$ as necessary.

It was found that the best correlation between theory and experiment for the halfplane problem was obtained for a minimum of $\zeta = -2\cdot5$ and a spacing of the elements based on $d\zeta = 0\cdot1$ (i.e. 25 elements). The comparisons between theoretical and experimental results for ε_{xx} and ε_{yy} are given in Figs 12 and 13.

CONCLUSIONS

In this paper we have examined the elastic response of a bonded rubber-like composite which is composed of alternate layers of hard and soft rubber. The mechanical properties of the hard and soft rubber phases, as well as those for the composite, were determined experimentally. It was found that the orthotropic elastic properties of the composite can be accurately predicted via an elementary theory of mixtures which utilizes the elastic properties of the individual phases and their volume fractions. A series of experiments was also carried out to determine the plane-strain response of a rubber composite block which is subjected to edge and corner loads which act in a concentrated or distributed fashion. Theoretical solutions for these plane-strain problems were developed by using the theory of orthotropic elasticity. In particular, the solutions for the orthotropic elastic quarterplane were obtained via a successive superposition scheme. The results for the strain fields determined experimentally compare very favourably with equivalent theoretical results. The study concludes that the elastic behaviour of laminated materials which display periodicity in their non-homogeneity can be conveniently analysed by employing the classical theory of orthotropic elasticity. Such modelling would prove to be effective only in situations where the dimensions of the loaded area are large when compared with the dimensions of individual laminations.

REFERENCES

1. GREEN, A. E. and TAYLOR, G. I., Stress systems in aelotropic plates—I, *Proc. Roy. Soc. Ser. A*, **173** (1939) 162–84.
2. GREEN, A. E. and ZERNA, W., *Theoretical elasticity*, London, Oxford University Press, 1968.

3. LEKHNITSKI, S. G., *Theory of elasticity of an anisotropic elastic body*, San Francisco, Holden Day, 1963.

4. HOLISTER, G. S. and THOMAS, R., *Fibre reinforced materials*, Amsterdam, Elsevier, 1966.

5. WENDT, F. W., LEIBOWITZ, H. and PERRONE, N. (Eds), *Mechanics of composite materials*, Proc. 5th Symp. Naval Struct. Mech., New York, Pergamon, 1970.

6. TSAI, S. W., HALPIN, J. C. and PAGANO, N. J. (Eds), *Composite Materials Workshop*, Conn., Technomic Publ. Co., 1968.

7. SPENCER, A. J. M., *Deformations of fibre reinforced materials*, Oxford, Oxford University Press, 1972.

8. BROUTMAN, L. J. and KROCK, R. H. (Eds), *Composite materials*, Vols 1–8, New York, Academic Press, 1974.

9. GARG, S. K., SVALBONĄS, V. and GURTMAN, G. A., *Analysis of structural composite materials*, New York, Marcel Dekker, 1973.

10. CHRISTENSEN, R. M., *Mechanics of composite materials*, New York, John Wiley, 1979.

11. SELVADURAI, A. P. S. (Ed.), *Mechanics of structured media*, Vols I and II, Proc. Int. Symp., Ottawa. Amsterdam, Elsevier Scientific Publishing Co., 1981.

12. HASHIN, Z., Theory of mechanical behaviour of heterogeneous media, *Appl. Mech. Rev.*, **17** (1964) 1–9.

13. HILL, R., A self consistent mechanics of composite materials. *J. Mech. Phys. Solids*, **13** (1965) 213–22.

14. TIMOSHENKO, S. and GOODIER, J. N., *Theory of elasticity*, New York, McGraw-Hill, 1970.

15. LITTLE, R. W., *Elasticity*, New Jersey, Prentice-Hall, 1973.

16. SELVADURAI, A. P. S., *Elastic analysis of soil–foundation interaction*, Developments in Geotechnical Engineering Vol. 17, Amsterdam, Elsevier Scientific Publishing Co., 1979.

17. MICHELL, J. H., The stress distribution in an aelotropic solid with an infinte plane boundary. *Proc. Lond. Math. Soc.*, **32** (1900) 247–58.

18. CAROTHERS, S. D., Plane strain in a wedge, with application to masonry dams. *Roy. Inst. Naval Arch.*, **33** (1913) 292–306.

19. CONWAY, H. D., Some problems of orthotropic plane stress. *J. Appl. Mech.*, **20** (1953) 72–6.

20. BRILLA, J., Contact problems of an elastic anisotropic halfplane. *Rev. Mech. Appl.*, **7** (1962) 3.

21. AKOZ, A. Y. and TAUCHERT, T. R., Plane deformation of an orthotropic elastic semi-space subjected to distributed surface loads. *J. Appl. Mech.*, **95** (1973) 1135–6.

22. SAHA, S., MUKHERJEE, S. and CHAO, C. C., Concentrated forces in semi-infinite anisotropic media. *J. Comp. Mat.*, **6** (1972) 403–8.

23. OKUBO, H., Stress systems in an aelotropic rectangular plate. *Z. Angew. Math. Mech.*, **21** (1941) 383–4.

24. SEN, B., Note on two dimensional indentation problems of a non-isotropic semi-infinite elastic medium. *Z. Angew. Math. Phys.*, **5** (1954) 83–6.

25. CONWAY, H. D., The indentation of an orthotropic halfplane. *Z. Angew. Math. Phys.*, **6** (1955) 402–5.

26. SNEDDON, I. N. (Ed.), *Application of integral transforms in the theory of elasticity*, Berlin, Springer-Verlag, 1975.

27. GLADWELL, G. M. L., *Contact problems in the classical theory of elasticity*, Alphen aar den Rijn, Sijthoff and Noordhoff Int. Publ., 1980.
28. SADOWSKY, M. A., Stress concentrations caused by multiple punches and cracks. *J. Appl. Mech.*, **23** (1956) 80–4.
29. STERNBERG, E. and KOITER, W. T., The wedge under a concentrated couple: A paradox in the two-dimensional theory of elasticity. *J. Appl. Mech.*, **25** (1958) 575–81.
30. BENTHEM, J. P., On the stress distribution in anisotropic infinite wedges. *Q. Appl. Math.*, **21** (1963) 189–98.
31. BOGY, D. B., Two edge-bonded elastic wedges of different materials and wedge angles under surface tractions. *J. Appl. Mech.*, **38** (1971) 377–86.
32. HARRINGTON, J. W. and TING, T. W., The existence and uniqueness of solutions to certain wedge problems. *J. Elasticity*, **1** (1971) 65–81.
33. HETENYI, M., A method of solution of the elastic quarter-plane. *J. Appl. Mech.*, **27** (1960) 289–96.
34. SELVADURAI, A. P. S. and MOUTAFIS, N., Some generalized results for an orthotropic elastic quarterplane. *Appl. Sci. Res.*, **30** (1975) 433–52.
35. MOUTAFIS, N., *Problems in orthotropic elasticity*, Ph.D. Thesis, The University of Aston in Birmingham, 1975.
36. MUSKHELISHVILI, N. I., *Some basic problems in the mathematical theory of elasticity*, (3rd edn). English Transl. Radok, J. R. M. (ed.), Groningen, Noordhoff, 1953.

19

The Viscoelastic Response of a Graphite/Epoxy Laminate

H. F. BRINSON, D. H. MORRIS, W. I. GRIFFITH* AND D. DILLARD†

*Department of Engineering Science and Mechanics,
Virginia Polytechnic Institute and State University,
Blacksburg, Virginia 24061, USA*

ABSTRACT

An accelerated characterization method for resin matrix composites is reviewed. Methods for determining compliance master curves are given. Creep rupture analytical models are discussed as applied to polymer-matrix composites. Comparisons between creep and rupture experiments and phenomenological models are presented.

INTRODUCTION

Matrix dominated moduli and strength properties of polymer based composite laminates are time dependent or viscoelastic and are sensitive to environmental conditions such as temperature and humidity. Because of this fact, the long-term integrity of a composite structural component is an important consideration in the initial design process. Therefore, how viscoelastic matrix dominated modulus (compliance) and strength properties vary with time over the design life time is a necessary input to the initial design process. As many structural components are designed for years of service, property variations over years are often needed. Obviously, long-term testing equivalent to the lifetime of a structure is impractical and undesirable. The alternative is to develop analytical and experimental methods which can be successfully used for extrapolation.

---

* Now associated with Michelin Corp., Greenville, South Carolina, USA.
† Now Assistant Professor, University of Missouri, Rolla, USA.

An accelerated characterization procedure to satisfy the above goal for composite laminates has been proposed by Brinston *et al.*[1] The objective, to be described herein, is to discuss the accelerated characterization procedure and to provide documentation for its implementation using nonlinear viscoelastic procedures for the representation of lamina compliances and strengths.

ACCELERATED CHARACTERIZATION PLAN

The accelerated characterization procedure which was developed for polymer based composite laminates several years ago is based upon the well known time–temperature superposition principle (TTSP) for polymers and the widely used lamination theory for composite materials. A block diagram illustrating the basic details of the plan is shown in Fig. 1. The generic idea was to develop a method by which the time dependent deterioration of laminate moduli (compliances) and strength could be calculated from the results of a minimum number of tests. Hopefully, the amount of testing would only be minimally greater than that required for normal quality control and/or basic property determination procedures.[2]

Advanced composite laminates are most frequently designed using lamination theory. This theory allows the calculation of the properties of a general laminate from the knowledge of the behavior of a single lamina or ply. The stress–strain properties of a single ply may be found from constant strain-rate tests on unidirectional laminates and are normally routinely obtained when a general laminate is made. Thus, our accelerated characterization plan assumes that lamina stress–strain properties from zero load to failure are known as indicated by item A of Fig. 1.

The transformation equation for the moduli of orthotropic materials has been shown to be valid for unidirectional laminates.[1,3] Also, various orthotropic failure theories have been shown to be valid for unidirectional laminates.[4] Therefore, from item A modulus and strength properties as a function of fiber angle are known as indicated in items B and C.

Before time dependent properties of a general laminate can be predicted, the time dependent behavior of a single ply is necessary. For this reason, the constant strain-rate behavior known from routine tests as given in items A, B and C was known to be insufficient for viscoelastic predictions. Further, to perform long-term creep or relaxation tests to determine the necessary lifetime information would be impractical and would not satisfy the

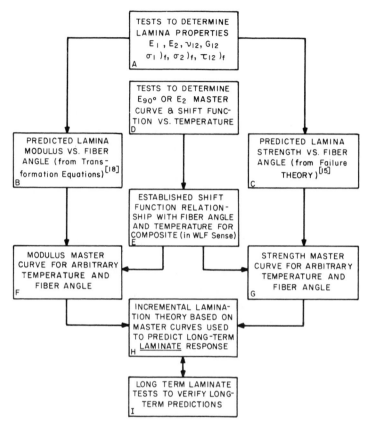

FIG. 1. Flow chart for proposed laminate accelerated characterization and failure predictions procedures.

objective of making long-term predictions from a minimum number of tests conducted in a short time.

The fundamental concept employed to overcome the above obstacle was to use the well known TTSP principle to produce a modulus (compliance) master curve for a single fiber orientation as typified by item D of Fig. 1. To do so would require the conduction of short-term creep tests on a unidirectional laminate at various temperatures. These could likely be performed in a single day. The hope was held that either an Arrhenius or WLF type equation could be modified to predict the variation of shift function with fiber angle for a single lamina without further testing as represented by item E. If such could be done, then the results of D and E combined with the information of A and B could produce the modulus (compliance) master curves of item F by simple scaling procedures without additional testing.

Of course a knowledge of time dependent strength properties was needed. Such properties often require large amounts of testing over a prolonged period of time. In our opinion, manufacturers would be reluctant to include an extensive testing program for routine quality control and property determination procedures. We therefore attempted to avoid an extensive creep rupture testing program. A means to do so was found quite simply by making the assumption that strength master curves were of the same shape as modulus (compliance) master curves for any particular fiber angle. From this assumption the lamina strength master curves as a function of fiber angle and temperature of item G could be determined from items C and F again by simple scaling procedures.

Given the master curves of F and G, an incremental lamination theory was to be developed to predict the long-term modulus and strength properties of a general laminate. The results were to be compared with experiments as specified by item I.

Our initial efforts were based upon concepts of linear viscoelasticity and a traditional TTSP procedure coupled with an incremental lamination theory. Various aspects of the plan have been documented and reported elsewhere.[1,3,5,6] However, comparisons between creep rupture predictions and measurements for $[\pm 45]_{4s}$ and $[90/\pm 60/90]_{2s}$ laminates were not very successful.[7] The use of a linear viscoelasticity method to model failure as well as the lack of a reliable anisotropic failure criteria was thought to be responsible for discrepancies between predictions and measurements. Our recent efforts center around modifications to the plan of Fig. 1 to include nonlinear viscoelastic concepts and the development of a better time dependent failure criteria.

TIME–TEMPERATURE–STRESS SUPERPOSITION PRINCIPLE (TTSSP)

In essence, the time–temperature–stress superposition principle is a simultaneous application of the well known TTSP and an analogous time–stress superposition principle (TSSP). In the former, an increase in temperature is assumed to accelerate a sequence of deformation events and in the latter an increase in stress is assumed to accelerate a sequence of deformation events. Mechanisms, of course, are assumed to remain unchanged in both cases. The combined TTSSP was first used by Daugste[7] to predict the nonlinear viscoelastic behavior of a 45° glass reinforced unidirectional composite.

A distinct advantage of this approach is that the effects of moisture can be studied through an analogous time–temperature–stress–moisture super-position principle.[8]

For illustrative purposes hypothetical transient creep compliance versus

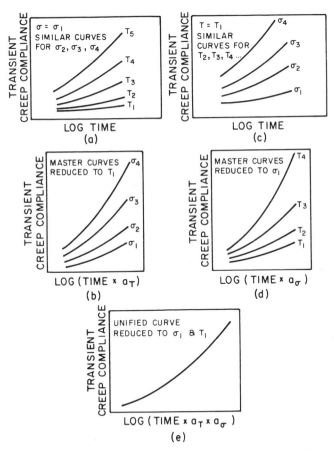

FIG. 2. Schematic diagram to illustrate the time–stress–temperature superposition principle.

log time is shown in Fig. 2 for several stress and temperature levels. The data from Fig. 2(a) for each temperature level may be shifted to obtain the σ_1 master curve shown in Fig. 2(b) using the TTSP. Similarly, master curves may be formed for stress levels σ_2, σ_3 and σ_4. An outcome of this procedure will be the temperature shift factor, $\log a_T$, and its corresponding stress dependence. The data from Fig. 2(c) for each stress level may be shifted to obtain the T_1 master curve shown in Fig. 2(d) using the TSSP. Similarly, master curves may be formed for temperature levels of T_2, T_3 and T_4. This procedure will yield the stress shift factor, $\log a_\sigma$, and its associated temperature dependence. The master curves in Fig. 2(b) or the master curves in Fig. 2(d) may now be shifted to obtain the unified master curve for a stress σ_1, and a temperature T_1, as shown in Fig. 2(e). This unified master curve can now be shifted to determine a unified master curve for any temperature and/or stress level within the range of data.

EXPERIMENTALLY DETERMINED TTSSP MASTER CURVES

S_{22} and $S_{10°}$ master curves for the creep compliances were produced by the method described in the previous section. The individual $S_{10°}$ creep compliance curves for 290 °F (143 °C), 320 °F (160 °C), 350 °F (177 °C) and 380 °F (193 °C) are given in Figs 3–6, respectively. The resulting master curve is shown in Fig. 7 and the associated temperature and stress dependent shift function surface is shown in Fig. 8. Also shown in Fig. 7 is a comparison between our $S_{10°}$ master curve and the results of a long-term creep test in excess of 150 hours. It appears that at extremely long times the master curves may tend to over predict the compliance. The reason for this may be due to additional curing or more likely some other form of aging of the material. Nevertheless, the agreement between predictions and experiment is reasonable. More details about postcuring and aging may be found in references 9 and 10.

An S_{66} master curve was generated using the orthotropic transformation equation modified to incorporate time dependence.[9] Further, the S_{22} and S_{66} stress dependent master curves were used in conjunction with the transformation equation to predict the long-term compliance of $[30°]_{8s}$ and $[60°]_{8s}$ specimens at 320 °F (160 °C). Predictions for the $[60°]_{8s}$ laminate is compared with long-term test results in Fig. 9. Agreement is seen to be fair to good.

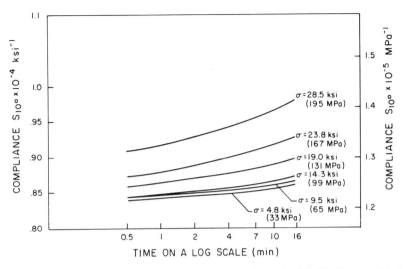

FIG. 3. S_{10° compliance at 290 °F (143 °C) as a function of stress level for T300/934 graphite/epoxy laminate.

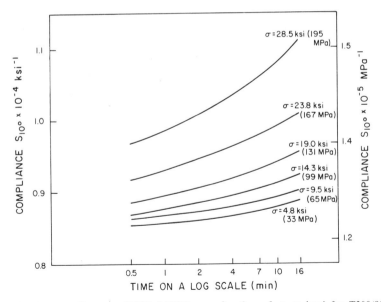

FIG. 4. S_{10° compliance at 320 °F (160 °C) as a function of stress level for T300/934 graphite/epoxy laminate.

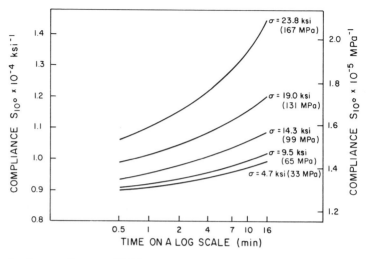

Fig. 5. $S_{10°}$ compliance at 350 °F (177 °C) as a function of stress level for T300/934 graphite/epoxy laminate.

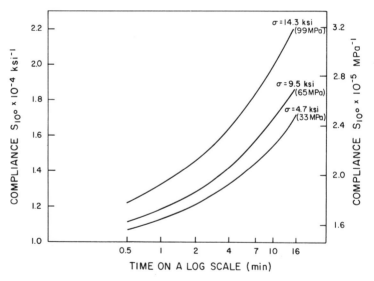

Fig. 6. $S_{10°}$ compliance at 380 °F (193 °C) as a function of stress level for T300/934 graphite/epoxy laminate.

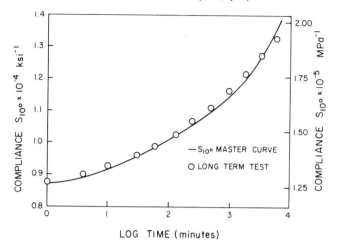

FIG. 7.   Comparison of $S_{10°}$ master curve for T300/934 graphite/epoxy laminate with a long-term test at 320°F (160°C) and $\sigma = 19\,500$ psi.

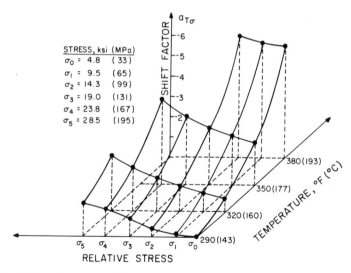

FIG. 8.   Shift surface for combined shift factor, $a_{T\sigma}$, for $S_{10°}$ for T300/934 graphite/epoxy laminate.

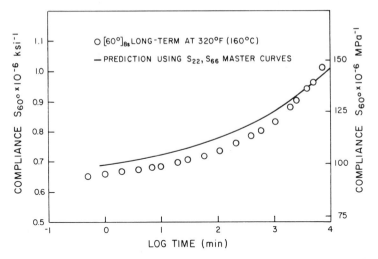

FIG. 9. Comparison of predicted and measured S_{60° compliance at 320 °F (160 °C).

DELAYED FAILURE MODEL

Most creep rupture criteria for homogeneous isotropic materials are based on a linearly decreasing logarithm of the time to rupture with increasing stress. This form, as exemplified by the Zhurkov, Larson and Miller, and Dorn methods, is given by

$$\log t_r = A - B\sigma \tag{1}$$

where t_r is the time to rupture for a constant creep load of σ and A and B are material constants for a given temperature.[11] Landel and Fedors[12] have noted that in some circles, the form

$$\log t_r = A - B \log \sigma \tag{2}$$

is viewed more favorably. Because of data scatter and the small range of stresses involved with our creep rupture data, however, the preference for one form over another becomes academic and eqn. (1) provides an adequate representation for the present analysis.

Experimentally, the creep stress level is the independent variable and the time to rupture at that stress level is the dependent variable. For the analysis, however, it is convenient to rearrange eqn. (1) to express the creep rupture strength, σ_r, as a function of the time to rupture

$$\sigma_r = (A - \log t_r)/B \tag{3}$$

Of the numerous orthotropic static failure theories available,[4] the Tsai–Hill criteria was chosen for the current analysis. If the Tsai–Hill criteria is modified to account for time dependent creep rupture strengths, the following form results[13]

$$\frac{\sigma_1^2}{[X(t_r)]^2} - \frac{\sigma_1 \sigma_2}{[X(t_r)]^2} + \frac{\sigma_2^2}{[Y(t_r)]^2} + \frac{\tau_{12}^2}{[S(t_r)]^2} = 1 \qquad (4)$$

Here, time independent strengths have been replaced by creep rupture strengths which result in failure at $t = t_r$. $X(t_r)$ represents the creep rupture strength for a uniaxial creep load parallel to the fiber direction. For the current material the assumption was made that delayed failures do not occur for $0°$ specimens and that $X(t_r) = X$. $Y(t_r)$ represents the functional relation with time of the creep rupture strength for a uniaxial creep load perpendicular to the fiber direction. $S(t_r)$ is a similar shear creep rupture strength. Theoretically, $S(t_r)$ can be determined from uniaxial creep rupture of off-axis specimens and prior knowledge of X and $Y(t_r)$. Such a procedure, though straight forward, proved unsatisfactory for the available data. As a result, the shear creep rupture strength was assumed to be of the form

$$S(t_r) = \alpha Y(t_r) \qquad (5)$$

Thus, all that must be determined from the data is the value of the proportionality constant, α, as the functional form of $S(t_r)$ has been established *a priori*. There is no rigorous justification to assume that the shear strength is proportional to the $90°$ strength, but this appears to be quite reasonable from an intuitive standpoint. Primarily, this procedure reduces the degrees of freedom to a more manageable level.

EXPERIMENTAL CREEP RUPTURE RESULTS

Griffith[14] obtained creep rupture data for $90°$, $60°$, and $45°$ off-axis specimens at several temperatures. His data for $320°F$ ($160°C$) has been replotted with the results for the three orientations and best fit lines have been drawn through the data points. The results are given in Fig. 10 and generally conform to the analytical model given by eqn. (3). The points denoted as 'Postcured 60 off-axis' were obtained during the present work to determine the effect of postcuring on creep rupture. While the magnitude remained about the same, there did seem to be a smaller decrease in the creep rupture strength with increasing rupture time. Because of the considerable data scatter, however, it is not known if this observation is

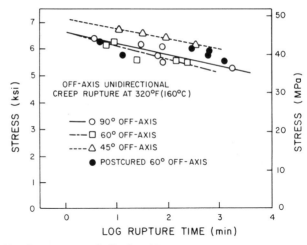

Fig. 10.   Creep rupture of off-axis unidirectional specimens at 320 °F (160 °C).

justified. These postcured data points were not used for the best fit lines.

Assuming that the creep rupture strengths may be represented by eqn. (3), determination of the slope and intercept for each best fit line in Fig. 10 allows the determination of the constants A and B. Thus the 90 ° data of Fig. 10 yields an appropriate expression for $Y(t_r)$ in eqn. (4).

To determine the appropriate value of α in eqn. (5), the unidirectional creep rupture data of Griffith[14] was again employed. A specific rupture time, $\hat{t}$, was selected within the range of the available data. Values of the creep rupture strengths for this particular rupture time were taken as the intercept values of the $t_r = \hat{t}$ line and the best fit lines. These represent the values of creep stress, for the 90 °, 60 °, and 45 ° specimens, which would result in rupture at time $\hat{t}$. The 60 ° and 45 ° creep rupture strengths may be normalized with respect to the 90 ° value at that particular rupture time and temperature.

These normalized creep rupture strengths have been plotted in Fig. 11 for several times and temperatures. Superimposed upon this data are normalized parametric curves representing the modified Tsai–Hill–Zhurkov predictions for various values of α. It should be noted that these curves will shift slightly depending on the ratio of the 90 ° strength to the 0 ° strength. For our material, this ratio is always very small, and this effect is completely negligible.

The data points shown in Fig. 11 indicate the change in the normalized strengths at a particular orientation with respect to temperature. The

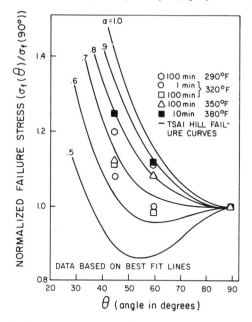

Fig. 11.   Normalized creep rupture versus fiber angle with parametric Tsai–Hill curves.

tendency of the strengths at a particular time and temperature to fall along a line of constant α indicates the appropriateness of the Tsai–Hill–Zhurkov criteria. A tendency for the points at different rupture times to fall along the same curve indicates the accuracy of the assumption that the time dependent shear strength is a constant proportion of the 90° strength. Presentation of information in this form provides a concise yet complete interpretation of the data. The results for the case of interest, 320 °F, indicate that the 45° and 60° creep rupture strengths for $t_r = 1$ and $t_r = 100$ minutes are closely clustered around an $\alpha = 0.65$ curve. For this reason, the formulation $S(t_r) = 0.65 Y(t_r)$ has been used for the failure model of eqn. (4) in the preceding section.

PREDICTION OF LAMINATE CREEP RUPTURES

The results of the compliance master curve formulations and the Tsai–Hill–Zhurkov failure criteria have been incorporated into a nonlinearly viscoelastic incremental lamination theory as specified by item F of Fig. 1.

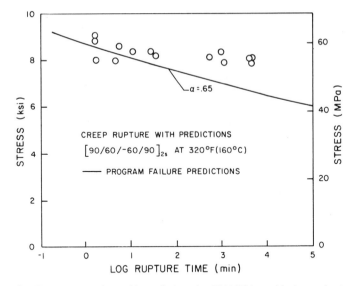

F~ig~. 12.   Creep rupture data with predictions for T300/934 graphite/epoxy laminate.

Details of this analysis may be found in reference 13 and are being reported elsewhere.[15] Comparison between predictions of the analysis and creep rupture test results for a $[90/60/-60/90]_{2s}$ laminate is shown in Fig. 12. As may be observed, reasonable correlation between prediction and experiment was obtained for short times but the model tends to be quite conservative for long times. Correlation for other laminates showed similar trends.[13]

SUMMARY AND CONCLUSIONS

A description of an accelerated characterization procedure for time dependent compliances and strengths of a viscoelastic composite laminate has been presented. A time–temperature–stress superposition principle has been described and applied to a T300/934 graphite/epoxy laminate. A modified time dependent Tsai–Hill–Zhurkov type failure theory has been described and experimental verification has been given. Compliance master curves and the time dependent failure criteria have been incorporated into an incremental nonlinearly viscoelastic lamination theory. Laminate creep rupture predictions have been compared to experimental test results.

Obviously, additional refinements to our accelerated characterization plan are needed. These might take the form of a better nonlinear viscoelastic analytical model, a better time dependent failure criteria, use of a finite element approach, etc. We do feel that viscoelastic failures in polymer based composite materials are quite important and that eventually a simple plan such as ours to predict long term failures will be needed.

ACKNOWLEDGEMENTS

The financial support provided for this work by NASA Grant NSG 2038 from the Materials and Physical Sciences Branch of NASA-Ames is gratefully acknowledged. Further, sincere appreciation is extended to Dr H. G. Nelson of NASA-Ames for his encouragement and helpful discussions.

REFERENCES

1. BRINSON, H. F., MORRIS, D. H. and YEOW, T. Y., *A new experimental method for the accelerated characterization and prediction of the failure of polymer-based composite laminates*, 6th International Conference for Experimental Stress Analysis, Munich, West Germany, Sept. 1978. Also, VPI-E-78-3, Feb. 1978.
2. YEOW, Y. T. and BRINSON, H. F., A comparison of simple shear characterization method for composite laminates, *Composites* (Jan. 1978), 49–55.
3. YEOW, Y. T., MORRIS, D. H. and BRINSON, H. F., The time–temperature behavior of a unidirectional graphite/epoxy laminate, *Composite materials: Testing and design* (5th Conference), STP 674, ASTM, Phil., 1979, 263–81. Also, VPI-E-78-4, Feb. 1978.
4. SANDHU, R. S., A survey of failure theories of isotropic and anisotropic materials, *Tech. Rep. AFFDL-TR-72-71*, Air Force Flight Dynamics Lab., Wright-Patterson Air Force Base, Ohio, USA.
5. MORRIS, D. H., BRINSON, H. F. and YEOW, Y. T., The viscoelastic behavior of the principal compliance matrix of a unidirectional graphite/epoxy composite, *Polymer Composites*, **1** No. 1 (Sept. 1980), 32–6. Also, VPI-E-79-9, Feb. 1979.
6. MORRIS, D. H., BRINSON, H. F., GRIFFITH, W. I. and YEOW, Y. T., The viscoelastic behavior of a composite in a thermal environment, In: *Severe environments* (Hasselman, D. P. H. and Heller, R. A. (eds), NY, Plenum Press, 1980, pp. 693–707. Also, VPI-E-79-40, Sept. 1979.
7. DAUGSTE, C. L., Joint application of time–temperature and time–stress analogies to constructing unified curves, *Polymer Mechanics*, **10** No. 3 (1974), 359–62.

8. CROSSMAN, F. W. and FLAGGS, D. L., LMSC-D33086, Lockheed Palo Alto Research Laboratory, November 1978.
9. GRIFFITH, W. I., MORRIS, D. H. and BRINSON, H. F., The accelerated characterization of viscoelastic composite materials, VPI-E-80-15, April 1980.
10. GRIFFITH, W. I., MORRIS, D. H. and BRINSON, H. F., *Accelerated characterization of graphite/epoxy composites*, Proceedings of the Third International Conference on Composite Materials, Palais des Congrès, Paris, France, Aug. 25–30, 1980 (in press). Also, VPI-E-80-27, Sept. 1980.
11. CONWAY, J. B., *Stress–rupture parameters: Origin calculation and use*, NY, Gordon and Breach, 1969.
12. LANDEL, R. F. and FEDORS, R. F., Rupture of amorphous unfilled polymers, *Fracture processes in polymeric solids*, Rosen, B. (ed.), NY, Interscience Publishers, 1964.
13. DILLARD, D. A., Creep and creep rupture of laminated graphite/epoxy composites, Ph.D. Thesis, March 1981. Also, VPI-E-81-3.
14. GRIFFITH, W. I., The accelerated characterization of viscoelastic composite materials, Ph.D. Thesis, May 1979. Also VPI-E-80-15, April 1980.
15. DILLARD, D. A., MORRIS, D. H. and BRINSON, H. F., *Predicting viscoelastic response and delayed failures in general laminated composites*, 6th Conference on Composite Materials Testing and Design, Phoenix, Arizona, USA, May 12–13, 1981.

20

Viscoelastic Properties of Composite Materials

A. Cardon and Cl. Hiel

*Vrije Universiteit Brussel, Faculty of Applied Sciences,
Department of Continuum Mechanics, Pleinlaan, 2, B-1050,
Brussels, Belgium*

ABSTRACT

*Complex modulus components of carbon–epoxy matrix laminated laminae,
are frequency dependent. This effect becomes more pronounced at higher
temperatures.*

*The dynamic elastic moduli and damping coefficients of laminates are
measured over a frequency of 0·1 to 10 Hz and a temperature range of 25°C
to 200°C by means of a forced non-resonance method.*

*Corrections were introduced in order to eliminate the influence of the
stiffness effects of elements of the test column.*

*A combination of micro- and macro-mechanics methods, suggested by
Sims and Halpin, which requires the measurement of only one viscoelastic
property, was used.*

The predicted results were compared with the experimental results.

1. Experimental results on carbon–epoxy laminates indicate that matrix
dominated laminae, such as angle plies, have a complex moduli frequency
function. At higher temperatures this frequency dependence becomes more
pronounced. The matrix in a composite material must not only organise the
stress transfer between fibers, but must present a good environmental
stability under normal service conditions.

Temperature is one of the most important environmental parameters
influencing the viscoelastic characterization of a composite matrix.

In this research, the viscoelastic behaviour of a composite matrix is studied over a frequency range from 0·1 to 10 Hz* and a temperature range from 25 °C to 200 °C.

2. In order to apply classical continuum mechanics methods to structural problems with composites, it is necessary to have some general method of calculating the moduli of the constitutive equations of the composite, starting from the mechanical properties of fibers and matrix, and some information on the fiber density, the fiber distribution and the fiber form.

A general method for predicting such moduli exactly would require a very considerable amount of information about the details of the fiber geometry, the fiber distribution, the fiber density, matrix homogeneity, adhesion, etc.

Two simplified methods—boundary methods and modelling methods—are generally proposed in order to develop some combining rules based on a partial schematic description of the microstructure. Bounding methods have the advantage that the calculated bounds are always valid, no matter what the unknown part of the structural information may be, but also the important disadvantage that this unknown (or inaccessible) information has generally so much influence that the difference between the obtained bounds may be very important.

Modelling methods start from a schematic construction of the composite, with the disadvantage that an important gap can exist between the model and the real composite, but with the advantage of a rapid computational procedure for the estimation of the ply properties, like the interpolation procedure proposed by Halpin and Tsai.[1]

3. We start from the general relations:

$$E_{11} = E_f V_f + E_m V_m \tag{1}$$

$$v_{12} = v_f V_f + v_m V_m \tag{2}$$

$$E_{22} = E_m \frac{E_f(1 + \xi_1 V_f) + \xi_1 E_m(1 - V_f)}{E_f(1 - V_f) + \xi_1 E_m(1 + V_f/\xi_1)} \tag{3}$$

$$G_{12} = G_m \frac{G_f(1 + \xi_2 V_f) + \xi_2 G_m(1 - V_f)}{G_f(1 - V_f) + \xi_2 G_m(1 + V_f/\xi_2)} \tag{4}$$

* The available laboratory equipment could extend the frequency range to 100 Hz but with different measuring devices. In a first stage we actually limit ourselves to 10 Hz.

where E_{11} = elastic modulus in the fiber direction (1), E_{22} = longitudinal modulus in the transverse direction (2), G_{12} = shear modulus in the 1–2 plane, v_{12} = Poisson ratio, E_f, E_m = elastic moduli of fiber and matrix, G_f, G_m = shear moduli of fiber and matrix, v_f, v_m = Poisson's ratios of fiber and matrix, V_f = fiber volume fraction, ξ = is an empirical factor which indicates the improvement of the material from its lower performance limit (Reuss bound) to its upper performance limit (Voigt bound) and ξ_1, ξ_2 = factors depending on fiber geometry, fiber distribution and loading conditions.

If, in eqn (4), $\xi_2 = 1$, this relation reduces to an equation given by Hashin and Rosen,[2] based on the analysis of the composite cylinder assemblage model. A more detailed analysis of complex moduli of viscoelastic composites was given by Hashin.[3,4] $\xi_2 \neq 1$ in eqn (4) does not correspond to a known model (Hashin, private communication, March, 1981), but this is not a restriction on the applicability of our starting equations.

4. We use a combination of macro- and micromechanics, an alternative to the method suggested by Sims and Halpin.[5] This method requires the measurement of only one viscoelastic characteristic function, $E_{22}(f, \theta)$, where f = frequency and θ = temperature. The data obtained on a transverse specimen are used in order to calculate $F_m(f, \theta)$ and $G_m(f, \theta)$ by means of the Halpin–Tsai equations.

This procedure is useful because of the following practical considerations.

(a) The simple unidirectional composite is generally supplied by the manufacturer in tape form.

(b) The effect of stress concentrations around the fibers is implicitly taken into account.

(c) The connection between the mechanical properties of the resin in the matrix and the resin in bulk form is not easy because of the influence of chemical composition, curing procedure and dimensions.

5. EXPERIMENTAL PROBLEMS

5.1. The samples we studied contain Hyfil-Torayca high strength, low loss carbon fibers. They were produced in a compression moulded plate by Hyfil Limited.[6] The fibers are continuous and uniaxially aligned in Shell R7B-resin. Cutting of the samples from the plates was done with a diamond saw.

5.2. The measuring method was the forced non-resonance method. This classical method in experimental viscoelasticity (see, for example, Murayama[7]) consists in the application of time harmonic stresses:

$$\sigma = \sigma_0 \cos \omega t = \sigma_0 \cos(2\pi f t) \qquad (5)$$

Under steady state conditions the strain will alternate sinusoidally but out of phase with the stress:

$$\varepsilon = \varepsilon_0 \cos(\omega t - \delta) \qquad (6)$$

The complex modulus, E, may be expressed as:

$$E = \frac{\sigma_0}{\varepsilon_0} e^{i\delta} = \left(\frac{\sigma_0}{\varepsilon_0} \cos \delta + i \frac{\sigma_0}{\varepsilon_0} \sin \delta \right) = E_1 + iE_2 \qquad (7)$$

E_1 is the storage modulus and E_2 is the loss modulus.

Generally the measured data are:

$$E_1 \qquad \text{and} \qquad tg\, \delta = \frac{E_2}{E_1} \qquad (8)$$

The instrumentation we used is commercially available from IMASS (see Fig. 1).

5.3. A transformation of the normal equipment gives us a direct output connection to an HP-1000 computer and a direct utilization of our data analysis programs. This transformation permits us the possibility of obtaining very rapidly a complete analysis of the different characteristic functions of the sample.

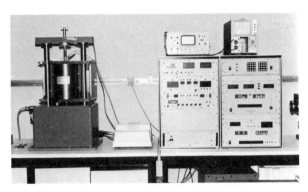

FIG. 1.

5.4. The equipment is generally used for the characterization of rubber-like materials and for that purpose the compliance of the test column is relatively low

$$\left(K = \frac{1}{J} \sim 5 \cdot 10^6 \, \text{N/m}\right) \qquad (F = Ku)$$

Figure 2 shows a schematic representation of the various components along the test column. The test specimen is only a part of the total chain and the total measured stiffness, K_m, is a combination of the stiffness of the different elements and the sample.

It is easy to show that we have the following relationship:

$$\frac{1}{K_m^*} = \frac{1}{K_{ft}} + \frac{1}{K_{c_1}} + \frac{1}{K_{sample}^*} + \frac{1}{K_{c_2}} \qquad (9)$$

Simple measurements give us the amount of:

$$\frac{1}{K_{c_1}} + \frac{1}{K_{c_2}} + \frac{1}{K_{ft}} = A \qquad (10)$$

and eqn 9 gives us the correction formula.

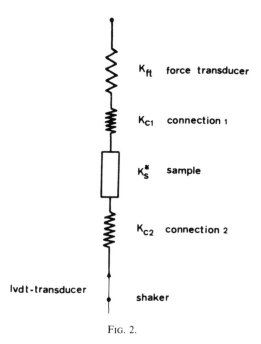

K_{ft} force transducer

K_{c1} connection 1

K_s^* sample

K_{c2} connection 2

lvdt-transducer   shaker

Fig. 2.

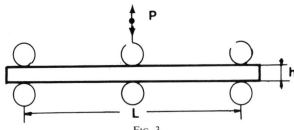

FIG. 3.

5.5. A very current deformation mode is three-point bending (see Fig. 3). Generally it is assumed that measurements in bending are inherently accurate and easy to carry out on a simple standard apparatus. This is not so and Bonnin *et al.*,[8] whose work was the basis of our experimental device, give results accurate to 1 % but only with very sophisticated experimental precautions.

The basic idea of the Bonnin apparatus is that the deflections, Δ, should be small enough so that none of the conditions of eqn (8) is violated, i.e. so that geometrical and physical linearity are preserved.

$$E = \frac{1}{48}\frac{PL^3}{I\Delta} \tag{11}$$

5.6. In three-point bending there is an additional shear deformation, as pointed out by Timoshenko and Goodier.[9] As a consequence, the modulus measured is only an apparent one and some shear correction is necessary.

Figure 4 shows the ratio of the normalized apparent modulus as a function of span to thickness ratio for different values of E/G.

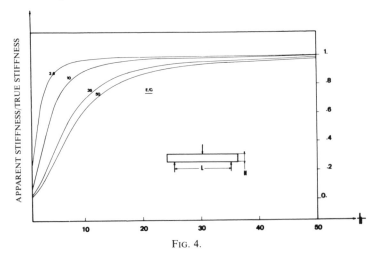

FIG. 4.

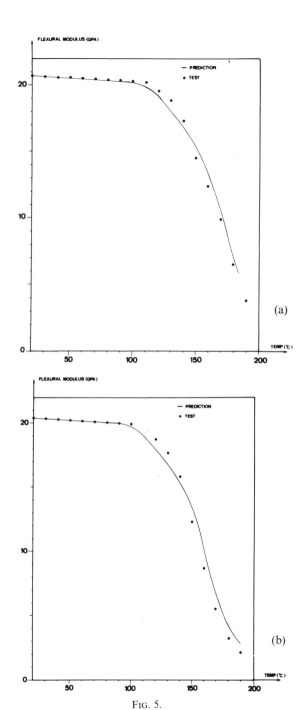

FIG. 5.

6.   EXPERIMENTAL RESULTS

6.1. A first series of experiments was made for $E_{11}(f, \theta)$ with a $[0]_8$ laminate, and $E_{22}(f, \theta)$ with a $[90]_8$ laminate.[10]

Using a spline function program, the surface $E_{22}(f, \theta)$ was smoothed out and the surface, $E_m(f, \theta)$, was calculated by means of eqn (3).

A laminated plate analysis program was connected to predict $E_{xx}(f, \theta)$, flexural modulus for any general laminate, and Figs. 5(a) and (b) present the results for $f = 1$ Hz and $f = 10$ Hz on a matrix dominated lamina $[+45 \ -45 \ +45 \ -45]_S$.

A very good agreement is observed between predicted and experimental results.

6.2. Another series of experiments was made on $[90]_8$ laminate in tension–compression for seven temperatures and twenty-eight frequencies

The results obtained, after correction (see paragraph 5.4) give us the values of E_1 and E_2 and, finally, a complete expression of the damping characteristic, $tg\ \delta$, as a function of f and θ (Fig. 6).

Figure 7 shows the sections $f = cte$ in the temperature function.

6.3. For the same specimen, measurements of the damping characteristic, $tg\ \delta$, obtained by the three-point bending method, are given in Fig. 8

6.4. On comparing the results from the tension test (Fig. 7) with those from the three-point bending test (Fig. 8), some discrepancies appear. I

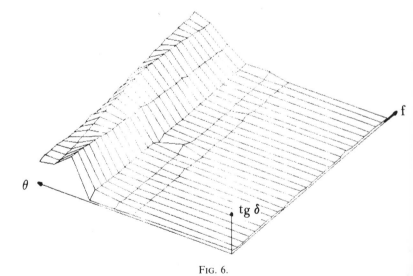

FIG. 6.

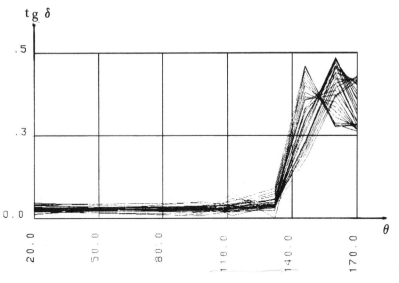

FIG. 7.

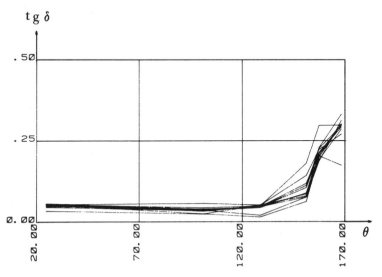

FIG. 8.

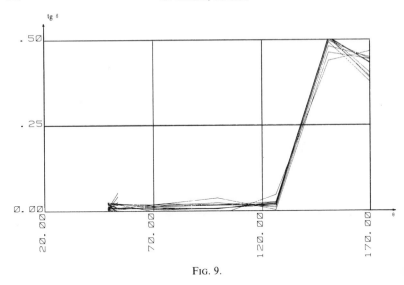

FIG. 9.

must be observed that the tension test was entirely in tension oscillation between two levels, and in the three-point bending test there is a change from tension to compression. The composites do not behave in the same way in tension and compression. In order to investigate this, we also carried out a full compression test, the results being given in Fig. 9.

Actually we try to explain the combined interpretation of the results of Figs. 7 and 9 in order to explain the results given in Fig. 8.

7. CONCLUSIONS

A prediction method for the viscoelastic behaviour of composite materials has been presented. It is based on a forced non-resonance micromechanics method. The experimental results and comparison with predictions show good agreement and we can conclude that this method is very useful.

Extension to more complex laminates has also been studied.

ACKNOWLEDGEMENT

This research is partially sponsored by the Belgian National Science Foundation (N.F.W.O.–F.K.F.O.).

The collaboration of F. Boulpaep (technical work), A. Vrijdag (Figures and photographs) and M. Bourlau (typing) is greatly appreciated.

REFERENCES

1. ASHTON, J. E., HALPIN, J. C. and PETIT, P. H., *Primer on composite material*, Technomic, 1969. (Chapter 5, p. 77).
2. HASHIN, Z. and ROSEN, B. W., The elastic moduli of fiber reinforced materials, *Journal of Appl. Mechs.*, **31** (1964) 223.
3. HASHIN, Z., Complex moduli of viscoelastic composites—I, *Int. J. Solids & Structures*, **6** (1970) 539–52.
4. HASHIN, Z., Complex moduli of viscoelastic composites—II, *Int. J. Solids & Structures*, **6** (1970) 797–807.
5. SIMS, D. F. and HALPIN, J. C., *Methods for Determining the Elastic and Viscoelastic Response of Composite Materials: Testing and Design (3rd Conference)*, ASTM-STP 546.
6. HYFIL LIMITED, Technical Data Sheets on Hyfil-Torayca Carbon Fiber.
7. MURAYAMA, T., *Dynamic mechanical analysis of polymeric materials*, Elsevier, 1978.
8. BONNIN *et al.*, A comparison of torsional and flexural deformations in plastics, *Plast. Polym.*, **37** (1969) 517.
9. TIMOSHENKO, S. and GOODIER, J., *Theory of elasticity* (2nd edn.), New York, McGraw-Hill, 1959.
10. CARDON, A. and HIEL, CL., *Forced non-resonance method for the determination of viscoelastic behaviour of composite materials.* Paper presented to International Symposium on the Mechanical Behaviour of Structural Media, Ottawa, Canada, 18–21 May, 1981.

21

Advances in Vibration, Buckling and Postbuckling Studies on Composite Plates

ARTHUR W. LEISSA

Department of Engineering Mechanics, The Ohio State University, Columbus, Ohio 43210, USA

ABSTRACT

Advances in the understanding of vibration and buckling behavior of laminated plates made of filamentary composite material are summarized in this survey paper. Depending upon the number of laminae and their orientation, vibration and buckling analyses of composite plates may be treated with: (1) orthotropic theory, (2) anisotropic theory, or (3) more complicated, general theory involving coupling between bending and stretching of the plate. The emphasis of the present overview is upon the last. Special consideration is given to the complicating effects of: inplane initial stresses, large amplitude (nonlinear) transverse displacements, shear deformation, rotary inertia, effects of surrounding media, inplane nonhomogeneity and variable thickness. Nonclassical buckling considerations such as initial imperfections are included, as well as postbuckling behavior.

1. INTRODUCTION

The term 'composite plate' has various meanings in the literature. It is occasionally used in connection with plates having step-wise thickness variation. It is often used also to denote layered plates, where each layer is made of an isotropic material. 'Sandwich plate' typically is used to describe a plate having a core material which separates two relatively thin face sheets of higher modulus material.

312

In the present paper a 'composite plate' will be made up of layers (or laminae), each lamina being composed of straight, parallel fibers (e.g., glass, boron, graphite) embedded in a matrix material (e.g., epoxy resin). Each lamina can be considered as a homogeneous, orthotropic material having a value of Young's modulus (E) considerably greater in the longitudinal direction (E_L) than in the transverse direction (E_T). Adjacent laminae will have longitudinal axes generally not parallel. Cross-ply laminated plates arise in the special case when the longitudinal axes of adjacent laminae are perpendicular, whereas angle-ply laminates occur when adjacent layers are alternately oriented at angles of $+\theta$ and $-\theta$ with respect to the coordinates of the plate.

The thicknesses and orientation of laminae will generally result in an unsymmetric geometry with respect to the midplane of the plate, causing coupling between bending and stretching. For the special case of symmetric laminates the coupling vanishes, and the composite plate can be represented by homogeneous, orthotropic or anisotropic plate theory. Finally, in this short description it should also be noted that a plate is considered perfectly flat. 'Curved plates', which are actually shells, will be beyond the scope of the present paper.

Approximately 14 years ago the writer completed a monograph which summarized the world literature dealing with free vibrations of plates.[1] The literature search was quite thorough, resulting in approximately 500 references that appeared in 1965 being found and summarized; however, none of them dealt specifically with composite plates. Although composite plates were clearly in use at the time, it is clear that the best analyses available treated them as orthotropic, homogeneous plates for purposes of large-scale phenomena, such as vibrations and buckling.

The coupling between bending and stretching in general composite plates had been observed experimentally and in application, and in 1961 Reissner and Stavsky[2] published a paper based upon the latter's dissertation[3] which provided the theoretical groundwork for the study of vibrations and buckling of composite plates. But it was not until the end of that decade that considerable further progress began. Since 1969 literally scores of published papers have appeared dealing with vibrations and buckling of composite plates, as well as excellent books by Ashton and Whitney[4] and by Jones.[5] Many of these references have been described in the outstanding literature surveys of Bert,[6-9] by himself, and together with Francis.[10]

The purpose of the present paper is to provide an overview of advances made in the understanding of vibration and buckling of composite plates. Because of the ability of plates to carry significant additional load beyond their buckling loads, special attention will also be devoted to postbuckling behavior.

This overview is by no means intended to be complete. The writer is currently in the early stages of a three-year research effort which will culminate in a monograph providing a thorough summarization of current knowledge on the subjects of buckling and postbuckling of flat plates and curved panels made of composite materials. That research will also result in a comprehensive survey of the free vibrations of flat and curved composite panels.

2. VIBRATIONS

As an aid in organizing the discussion of vibration problems arising with composite plates, an outline similar to that of Ref. 1 will be followed. That is, the discussion will begin with 'classical theory', and will subsequently be supplemented by 'complicating effects' which may also arise. Only free, undamped vibrations will be considered, and the principal concern is to determine the frequencies (eigenvalues) and mode shapes (eigenfunctions) of free vibration.

2.1 Classical Theory

For a composite plate the most simple theory is for the cross-ply, symmetrically laminated, thin plate. In this case the differential equation of transverse motion can be written as

$$D_{11}\frac{\partial^4 w}{\partial x^4} + 2(D_{12} + D_{66})\frac{\partial^4 w}{\partial x^2 \partial y^2} + D_{22}\frac{\partial^4 w}{\partial y^4} + \rho\frac{\partial^2 w}{\partial t^2} = 0 \qquad (1)$$

where the D_{ij} are the customary bending stiffness coefficients involving the elastic constants and thicknesses of the laminae (cf. Refs. 4–6), w is the transverse displacement, a function of the x, y coordinates and of time (t), and ρ is the mass density per unit of surface area of the entire plate. It is observed that eqn. (1) has exactly the same form as for homogeneous, orthotropic plates (cf. Ref. 1, p. 250), and therefore all the numerous results for the free vibration of classical orthotropic plates are applicable to this type of composite plate. Reference 1, for example, identified 21 publications dealing with the vibrations of orthotropic plates having rectangular orthotropy which appeared up to the year 1965. But research in this topic area increases at an increasing rate. For example, a more recent survey[11] identified 17 references published between 1973 and 1976.

A particularly useful piece of work in this area was presented by Bert,[12] who derived a formula for the approximate values of fundamental frequency of orthotropic plates of arbitrary shape and boundary conditions, if the frequencies of the corresponding isotropic plates are known.

Symmetrically laminated angle-ply plates yield an equation of motion more complicated than eqn. (1) by the addition of terms $4D_{16} \partial^4 w/\partial x^3 \partial y$ and $4D_{26} \partial^4 w/\partial x \partial y^3$ to its left-hand side. This is due to the coupling between inplane stretching and shear that exists in each lamina when the principal stresses (due to vibration, in this case) are not coincident with the axes of orthotropy of the lamina. The resulting equation is now identical to that of the classical, anisotropic plate. But the equation has no exact solutions; its variables cannot be separated. Solutions can be found by approximate methods such as Ritz–Galerkin, but they will be algebraically quite complicated, generally requiring the use of digital computers. Indeed, no solutions for generally anisotropic plate vibration problems were available to be reported in Ref. 1.

However, the importance of composite materials has served to spur anisotropic plate analysis. A particularly early and notable effort in this direction was made by Ashton and Waddoups[13] who demonstrated how the Ritz method and vibrating beam eigenfunctions can be used to analyze the vibrations of rectangular, anisotropic plates. Whitney[14,15] subsequently showed how a generalized Fourier series solution, which leads to an infinite frequency determinant different than that of the Ritz approach, can yield more rapidly convergent numerical results. This was demonstrated for square plates having clamped[14] and simply supported[15] boundaries.

Bert[16] derived a formula for the frequencies of a generally anisotropic rectangular plate having arbitrary boundary conditions, provided the frequencies of the corresponding rectangular isotropic plate are known. He then used this result to derive an optimal design procedure to maximize the fundamental frequencies of such plates.[16,17]

Rajamani and Prabhakaran[18,19] investigated the effect of square cutouts on the frequencies of square, anisotropic plates and presented extensive numerical results for plates simply supported[18] and clamped[19] all around.

But unsymmetrical laminated composite plates introduce a more serious complexity into the problem; namely, coupling between bending and stretching. The resulting set of equations for a thin plate is of eighth order (cf. Refs. 20 and 21), involving the tangential displacements u and v of the plate, as well as w, and requiring four boundary conditions to be stated per edge, instead of merely two. In this way the plate behaves like a shell, having coupling between bending and stretching.

Closed form, exact solutions for the free vibration frequencies and mode shapes were presented in some relatively early works by Whitney and Leissa[20,21] for certain cases of rectangular composite plates having antisymmetric layering and simply supported boundary conditions. Two types of problems were found to yield closed form, exact solutions. One was the antisymmetric, cross-ply plate having the boundary conditions

$$w = M_n = u_t = N_n = 0 \tag{2}$$

along all edges, where M_n is the normal bending moment, u_t is the inplane displacement tangent to the edge, and N_n is the normal inplane force (see Fig. 1). The simple displacement functions

$$u(x, y, t) = A_{mn} \cos \alpha_m x \sin \beta_n y \sin \omega_{mn} t$$

$$v(x, y, t) = B_{mn} \sin \alpha_m x \cos \beta_n y \sin \omega_{mn} t$$

$$w(x, y, t) = C_{mn} \sin \alpha_m x \sin \beta_n y \sin \omega_{mn} t \tag{3}$$

were chosen, where A_{mn}, B_{mn} and C_{mn} are undetermined amplitude coefficients, and ω_{mn} is the circular natural frequency. Letting $\alpha_m = m\pi/a$ and $\beta_n = n\pi/b$, where m and n are integers, satisfies the boundary condition equations, eqns. (2), exactly. Substituting eqns. (3) into the coupled equations of motion yields a closed form expression for the frequencies.

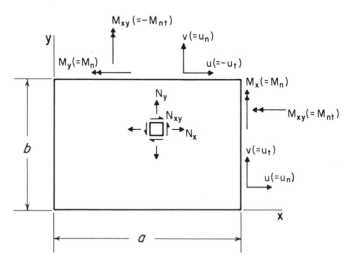

FIG. 1. Rectangular composite plate, showing positive directions of displacements, moments and inplane forces.

The analogy with vibrations of a shell is further reinforced by eqns. (2) and (3). The former are the classical shear diaphragm boundary conditions for a shell (cf. Ref. 22, p. 43), whereas the latter are the displacement forms yielding exact solutions for the frequencies of closed, circular cylindrical shells (cf. Ref. 22, p. 44) or of shallow shells having arbitrary curvature and shear diaphragm supports along rectangular edges.[23]

Other closed form, exact solutions for frequencies were presented in Refs. 20 and 21 for unsymmetrically laminated angle-ply plates having the boundary conditions.

$$w = M_n = u_n = N_{nt} = 0 \tag{4}$$

which also represents a simply supported edge, but having different inplane constraints than those of eqns. (2).

But the most important contribution of Refs. 20 and 21 is that they demonstrated clearly the decrease in natural frequencies, compared with uncoupled orthotropic plate theory, due to the coupling between bending and stretching. Frequencies of the order of 100 % too large can be predicted by the orthotropic theory. Because the solutions presented were exact, no questions of their accuracy or convergence can be raised. The results for unsymmetrically laminated, simply supported, cross-ply plates were subsequently further extended by Jones.[24]

Still another extension was made by Lin and King[25] to show how exact solutions can be obtained for classes of unsymmetrically laminated plates having two opposite sides simply supported, with arbitrary edge conditions on the other two sides. One assumes the following generalizations of eqns. (3)

$$u(x, y, t) = F_{mn}(y) \cos \alpha_m x \sin \omega_{mn} t$$

$$v(x, y, t) = G_{mn}(y) \sin \alpha_m x \sin \omega_{mn} t$$

$$w(x, y, t) = H_{mn}(y) \sin \alpha_m x \sin \omega_{mn} t \tag{5}$$

where F, G and H are functions of y arising from the solution of an eighth order set of coupled ordinary differential equations. Again, the analogy to solutions for circular cylindrical shells having two opposite sides supported by shear diaphragms (cf. Ref. 21, p. 83) is evident. This generalization is also analogous to Voigt's[26] extension of the earlier Navier solution for simply supported isotropic plates and is applicable to cross-ply plates. A similar generalization is possible for angle-ply plates. Numerical results were presented in Ref. 25 for plates having two opposite sides clamped or free.

For symmetrically laminated plates which, as described previously, can be represented as orthotropic or anisotropic homogeneous plates, thereby permitting uncoupling of the flexural equations from the inplane stretching equations, the classical types of boundary conditions (clamped, simply supported or free) which can exist on each edge can yield a total of 21 independent problems for rectangular shapes. For unsymmetrically laminated plates the number is far greater. Indeed, there exist 16 different possibilities for each edge, of which eqns. (2) and (4) identify only two.

Approximate methods have been used to treat a few of these cases. The important case of the rectangular plate having all sides clamped was examined by a number of people using various theoretical and experimental methods.[25,27–29] The case of a plate resting freely upon simple supports ($N_{nt} = 0$, rather than $u_t = 0$, in eqns. 2) was analyzed by Whitney and Leissa[30] using generalized Fourier series. Minich and Chamis[31] used a finite element method to obtain frequencies and nodal patterns for cantilevered plates.

2.2 Complicating Effects

It has been seen previously that the 'classical theory' of composite plates is already considerably more complicated than that of isotropic plates, partly because the laminae are themselves orthotropic, and partly because of additional elastic coupling that exists in each lamina and across the laminae in typical situations. However, just as in the case of isotropic plates (cf. Ref. 1), there exist several other effects which serve to complicate further the analysis and understanding of plate vibrational behavior. These include:

(a) inplane initial stresses
(b) large amplitude (nonlinear) transverse displacements
(c) shear deformation
(d) rotary inertia
(e) effects of surrounding media
(f) inplane nonhomogeneity
(g) variable thickness

From the lack of information available, it would appear that there is so far no interest in variable thickness composites.

Initial inplane stresses are caused by static forces applied in the plane of the plate. The forces may act along the edges of the plate or as body forces arising internally due to gravitational, acceleration or thermal fields. In general they cause internal stresses which vary in the plane of the plate (i.e. as functions of x and y). These stresses are superimposed. Tensile stresses tend to increase the free vibration frequencies, compressive stresses to decrease them, and shear stresses to decrease them to a lesser extent.

Generalization of the equations of motion to include the effects of initial stresses is relatively easy, resulting in terms

$$N_x \frac{\partial^2 w}{\partial x^2} + 2N_{xy} \frac{\partial^2 w}{\partial x \, \partial y} + N_y \frac{\partial^2 w}{\partial y^2} \tag{6}$$

being added to the right-hand side of the equation of motion for the transverse displacement, w. Here N_x, N_y and N_{xy} are the initial stress resultants (forces per unit length) and are shown in Fig. 1. Closed form, exact solutions were obtained by Whitney[15] and Jones[24] for symmetrically and unsymmetrically laminated plates, respectively.

It is generally known for isotropic plates that the effect of significant displacements (on the order of the plate thickness) serves to increase the effective stiffness of the plate, thereby increasing the free vibration frequencies as well. The resulting equations of motion are the dynamic generalizations of those of von Kármán,[32] they exhibit coupling between bending and stretching and, moreover, are nonlinear. The dominant nonlinear terms are of the 'hard spring' type, thereby causing the stiffness increase.

Generalizations of the dynamic von Kármán equations to composite plates of the most general (i.e. unsymmetrically laminated) type were presented in the previously mentioned works by Whitney and Leissa.[20,21] The first known solutions to the nonlinear, coupled equations were obtained by Bennett[33] for angle-ply, rectangular, composite plates having simply supported edges restrained against inplane displacement. He showed that the bending–stretching coupling does not directly enter the nonlinear term in the frequency equation; rather, the coupling appears only in the linear terms, thereby affecting the nonlinearity only indirectly. A more simple, approximate solution was subsequently obtained.[34]

Other results for the nonlinear vibrations of composite plates have been obtained by various researchers. Mayberry and Bert[35] conducted an experimental investigation on unsymmetrically laminated cross-ply and angle-ply plates having all edges clamped. Attempts were also made[35,36] to obtain satisfactory theoretical solutions to correlate with the experimental data. Chandra and Basava Raju[37] used the Galerkin method to make a study of angle-ply laminated plates. Numerical results were given for plates simply supported or clamped all around, having movable or immovable edges. Their work was extended further to cross-ply plates having the same boundary condition[38] and to ones having two opposite sides simply supported and the others clamped.[39] Chia and Prabhakara[40] showed that modal coupling in the solutions for large amplitude vibrations is more important for composite plates than for isotropic ones. Numerical results were presented for simply supported and clamped plates having movable edges.

The effects of shear deformation and rotary inertia both become significant for plates which are relatively thick compared with their inplane dimensions. For example, for isotropic plates classical theory is typically employed for ratios of thickness to minimum plate dimension of 1/20 or less. The addition of rotary inertia to the translational inertia, and the shear flexibility to the bending flexibility, both serve to decrease the frequencies of the plate.

Considerable research has taken place to develop theories suitable for moderately thick composite plates, including the effects of shear deformation and rotary inertia. Because of their complexity, discussion of these theories and the assumption made will not be considered in the present paper.

An early application of thick plate theory to the vibrations of composite plates was made by Whitney and Pagano[41] for the case of simply supported boundaries. Although the inclusion of shear deformation raises the differential order of the set of governing equations of motion to ten, an exact solution for simply supported edge conditions is still possible for antisymmetrical cross-ply and angle-ply plates by generalizations of eqns. (3) for the assumed displacements. Numerical results were obtained for the fundamental frequencies of antisymmetrical angle-ply ($\pm 45°$) plates made of four layers, each having the modulus ratio $E_L/E_T = 40$. It was seen that the effect of shear deformation can be considerably greater in a laminated plate than in homogeneous, isotropic plates.

Noor[42] also considered simply supported cross-ply plates having symmetric or antisymmetric lamination. He found the effect of rotary inertia to be relatively insignificant, compared with the shear flexibility. Siu and Bert[43] made an interesting study of the forced, damped response of completely free rectangular plates and, in the process, used the Ritz method to obtain natural frequencies and mode shapes for parallel-ply plates having various angles of fiber orientation. Numerical results were compared with the experimental ones obtained earlier by Clary,[44] and good agreement was found. Other early work on vibrations of thick laminated plates was reported by Srinivas and Rao.[45,46] Bert and Chen[47] presented results for simply supported plates having antisymmetric, angle-ply layups.

Results from a study using various thick plate theories were recently given by Hirashima.[48]

A finite element analysis of the vibration of thick, composite plates was undertaken by Mau *et al.*[49] using a hybrid stress element. Numerical comparison was made for the thin, two-layered, clamped, cross-ply plate studied earlier by Whitney.[29] A large number of degrees of freedom was used (405), but the finite element results did not converge monotonically, nor in good agreement.

Applications of finite elements to thick, composite plates were recently made by Reddy.[50-52] In Ref. 50 a rectangular finite element was developed having 40 degrees of freedom. Extensive numerical results were obtained for simply supported, antisymmetrically layered, angle-ply plates and were compared with the continuous function solutions of Ref. 47. In Ref. 51 a finite element based upon penalty functions was developed and also applied to the problems of Ref. 47, as well as those of Ref. 41. Further results using this element were reported in Ref. 52.

The inclusion of bimodulus effects (different elastic properties, depending upon whether the strain in the fiber direction is tensile or compressive) into a thick, laminated plate theory was made by Reddy[53] for application to vibration problems. Subsequent generalization of the finite element method to include these effects was then carried out.[54]

The combined effects of large amplitude (nonlinear) displacements and shear deformation and rotary inertia were studied in an early paper by Wu and Vinson.[55] Reddy and Chao[56] also developed a finite element method including all these effects, and applied it to a series of vibration problems for symmetrically laminated, cross-ply and angle-ply plates having simply supported and clamped edges.

For the free vibration problem, coupling with the surrounding media can enter in at least two ways: (1) an elastic foundation and/or (2) a surrounding fluid. Traditionally, the former is treated as linear, elastic coupling, while the latter coupling enters because of the mass of the fluid being moved during vibratory motion. For thin, homogeneous, isotropic plates the accommodation necessary in the analysis to include the elastic foundation stiffness is a minor one (cf. Ref. 1, p. 1). However, for composite plates the change is more significant, because of the presence of additional cross-inertia terms. Theoretical analyses are based upon vibration in a vacuum, and applications and experiments generally take place in air (or, worse yet, a liquid such as water). The effects of the surrounding fluid are to decrease the frequencies, and these effects can be significant, especially for thin plates of less dense material (cf. Ref. 1, pp. 299–306). It appears that neither the effects of elastic foundations nor of surrounding fluids have yet been considered for composite plates.

Finally, the author cannot resist pointing out the optimization of vibration frequencies (and of buckling loads) which may be possible by arranging fibers in a nonuniform manner in the laminae, yielding plates which are nonhomogeneous in their planar (x and y) directions, as well as in the z direction. The resulting problem is one of designing the material to provide the desired optimization. Such a procedure has been shown by Leissa and Vagins[57,58] to be possible for optimization of stresses by suitable spacewise variation in material properties.

3. BUCKLING

The subject of buckling is a complicated one. Indeed, even for the relatively simple isotropic plate a definition of what is meant by the word is not an easy one to make. Other words and phrases such as 'instability', 'collapse' and 'limit load' may or may not have the same meaning, depending on the particular problem being studied. The writer is currently preparing a general monograph on the buckling of plates, and it seems that the subject can be divided into three parts with reasonable clarity:

(a)   classical buckling analysis
(b)   classical complicating effects
(c)   nonclassical phenomena

Classical buckling is described by the path I–III shown in Fig. 2. There the inplane load (indicated as P) is plotted versus the transverse (w) displacement of a typical point on the plate. Assuming that the load acts perfectly in the midplane of the plate, its application causes no transverse displacement and, under conditions of perfect symmetry, the load could be increased to the compressive failure point of the material as shown by the path I–II in Fig. 2. But above a certain critical load (P_{cr}) this path is unstable. At $P = P_{cr}$ a bifurcation point exists. That is, the load–displacement curve IV can also be followed, which is a stable one. At or slightly above the load P_{cr} the slightest disturbance will result in a transverse displacement. The classical linear analysis, which is a generalization of the Euler buckling analysis for columns, would indicate that, at $P = P_{cr}$, w increases without bound (curve III). In actuality, nonlinear effects enter the problem and, after a finite initial displacement, P increases with further w,

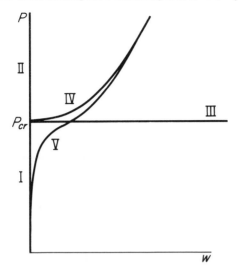

FIG. 2. Load–displacement curves for classical buckling and postbuckling, with (V) and without (IV) initial imperfection.

and the plate is able to carry loads greater than P_{cr} in this postbuckling condition (curve IV).

The complicating effects in classical plate buckling analysis are some of those already mentioned earlier in connection with vibration analysis; viz., shear deformation, elastic foundations, nonhomogeneity and variable thickness. Nonclassical buckling involves considerations such as imperfections, inelastic material behavior, dynamic (including parametric) load excitation and follower forces (P not remaining directed in the original plane of the plate).

3.1 Classical Buckling Analysis

The classical buckling analysis of laminated composite plates has an order of difficulty the same as discussed previously for vibration problems, depending upon the arrangement of laminae. That is, depending upon the relative angular orientations of the laminae and their stacking sequence, theoretical solutions are required for differential equations of equilibrium which, in increasing order of complexity, can be represented by

(a) uncoupled orthotropic plate theory
(b) uncoupled anisotropic plate theory
(c) coupled theory, involving both bending and stretching.

Thus, for example, eqn. (1) for the vibrations of cross-ply, symmetrically laminated, thin plates would be replaced by

$$
D_{11} \frac{\partial^4 w}{\partial x^4} + 2(D_{12} + D_{66}) \frac{\partial^4 w}{\partial x^2 \, \partial y^2} + D_{22} \frac{\partial^4 w}{\partial y^4}
$$

$$
= N_x \frac{\partial^2 w}{\partial x^2} + 2N_{xy} \frac{\partial^2 w}{\partial x \, \partial y} + N_y \frac{\partial^2 w}{\partial y^2} \qquad (7)
$$

for the classical buckling analysis. A mathematical eigenvalue problem then results from requiring the solution for w to satisfy eqn. (7) and the boundary conditions, where the eigenvalues are the nondimensional critical load parameters. Or the solution can be found from the somewhat more complicated free vibration problem including inplane forces, by determining at what values of force the frequencies become zero.

Some progress has been made in determining critical loads for composite plates, although the availability of results is still quite small when compared with those for isotropic, homogeneous plates. The year 1969 marks the beginning of a large rate of progress in composite plate buckling analysis. In that year the closed form, exact solutions for antisymmetric cross-ply and angle-ply plates having simply supported rectangular boundaries were presented by Whitney and Leissa,[21] based upon the former's dissertation.[20] Whitney[59] also solved the problem for unsymmetrically laminated, cross-ply plates loaded in shear. Simultaneously, Ashton and Waddoups[13] showed the results of applying the Ritz method for rectangular plates loaded in biaxial compression and shear and compared the theoretical results with experimental ones obtained by Ashton and Love.[60,61] Another important contribution in 1969 was made by Chamis[62] who used the Galerkin method to analyze orthotropic laminates of rectangular shape having their axes of orthotropy not parallel to the plate edges. The equilibrium equation then takes the more general anisotropic form. Numerical results were obtained for simply supported edges subjected to various combinations of constant inplane loads, and compared with experiment.

Prior to this time Sarkisyan and Movsisyan[63] had proposed a perturbation method for the buckling analysis of generally anisotropic plates having simply supported edges. Theoretical and experimental results for buckling were also obtained as a small part of a major research effort[64] devoted to carbon composite materials.

Immediately after 1969 considerable further results from buckling studies on composite plates were presented by Whitney, together with Leissa,[30] and by himself.[14,15,65,66] These additional results were both symmetrically[14,15,66] and unsymmetrically[30,65] laminated plates having various combinations of edge conditions and loadings. In these and the earlier investigations[20,21] the decrease in effective stiffness in the problems involving coupling between bending and stretching was clearly seen. The buckling load for a small number of antisymmetrically oriented laminae is significantly decreased due to strong coupling.

Extensive experimental results were obtained by Kicher and Mandel[67,68] for simply supported plates subjected to uniformly distributed compressive load in one direction, and compared with existing theory. Experimental results were also presented by Viswanathan *et al.*[69,70] for symmetrically and unsymmetrically laminated plates having their loaded opposite edges clamped and the others either free or simply supported. Jones[24] showed that the effects of unsymmetric laminating can be more severe in reducing the buckling load than exist for the antisymmetric case. Housner and Stein[71] made parametric studies and gave extensive numerical results for orthotropic representations of simply supported and clamped composite plates loaded axially and in shear. Their work showed the relative buckling efficiency of composite plates over those made of aluminium. Rotational elastic edge constraints were also treated.[71] Chailleux *et al.*[72] used the Southwell plot static method, and a procedure based upon change in free vibration frequencies, to determine experimental critical loads of composite plates made of boron or glass fibers in an epoxy or aluminium matrix. Results were given for two opposite sides simply supported and the others clamped or simply supported. Additional theoretical results for inplane normal and shear critical loads for simply supported plates were presented by Sawyer.[73]

Optimization studies based upon buckling criteria for composite panels were undertaken by several researchers. Chao *et al.*[74] developed a procedure for optimizing the orientation angle of simply supported, angle-ply, symmetrically laminated plates so as to maximize the critical load. Hayashi[75] and Bert[17] performed similar analyses. The problem was also investigated by Schmit and Farshi.[76] Tsai[77] proved the following theorem: 'A laminated plate has a stable natural state even though its constituent laminae do not have any stable natural state.'

A number of publications treat the buckling of stiffened plates.[69,70,78–82] Both analytical and test results for critical loads were determined. Because of the complexity of the geometrical description required, no further comments will be made here about this work.

3.2 Complicating Effects and Nonclassical Phenomena

The effect of transverse shear deformation on buckling loads was investigated relatively early by Whitney[83] for symmetrically laminated plates tractable by orthotropic plate theory, for the case of simply supported edges. This case, along with three others having two opposite sides simply supported, was also studied by Vinson and Smith.[84] They also showed that shear deformation can be significant, not only in altering the magnitudes of the buckling loads, but in changing their mode shapes. Turvey[85] demonstrated how the reduced stiffness concept may be used with shear deformation theory to deal with unsymmetrically laminated plates. Extensive numerical results were presented for simply supported plates.

One buckling analysis for variable thickness composite plates is known. Ashton[86] obtained results for linearly tapered plates subjected to uniaxial loading.

Nonlinear stress–strain relationships were recently incorporated into buckling analysis by Morgan and Jones.[87] Numerical results were obtained for simply supported plates of symmetric or unsymmetric cross-ply layering. It was found that the buckling loads so determined were significantly less than the elastic buckling loads for boron–aluminium composites, although the effect for boron–epoxy and graphite–epoxy composites was nearly insignificant. Furthermore, the nonlinear stress–strain behavior was seen to accentuate the coupling effects.

Hygrothermal effects due to moisture diffusion and heating on one or both surfaces of composite plates were analyzed in a recent study by Flaggs and Vinson.[88] An elevated hygrothermal environment serves to decrease the critical loads. A theory was formulated which accounts for hygrothermal effects, transverse shear deformation, normal deformation and bending–stretching coupling. Parametric studies were performed for a symmetrically laminated graphite–epoxy plate having either simply supported or clamped boundary conditions.

Initial imperfections can be caused by, for example, a very slight variation from flatness, material imperfections or load eccentricity. In such cases buckling cannot be established by a distinct bifurcation point at the critical load. Rather, a representative load–displacement curve followed may be like that designated as V in Fig. 2. A recent analysis by Bauld and Satyamurthy[89] can be applied to such problems. The effect of initial imperfections was also considered by Meffert *et al.*,[90] for both short duration and long duration (creep buckling) situations. Stroud *et al.*[91] made studies of the effects of initial curvature as an imperfection upon the buckling characteristics of stiffened composite panels.

4. POSTBUCKLING

The postbuckling behavior of plates has already been partially demonstrated by the load–displacement curves shown in Fig. 2. There curve IV is seen to characterize the behavior according to classical theory after the bifurcation point at the critical load (P_{cr}) is reached, whereas curve V shows how the presence of imperfections eliminates the clearly defined bifurcation point. In either case it is well known for a classical, isotropic plate that loads considerably greater than P_{cr} can typically be applied before the plate collapses. The additional loading is generally assumed to occur in the elastic stress–strain range of the material, and is due to stretching of the middle surface of the plate, a nonlinear phenomenon. In addition to the load versus transverse displacement plot illustrated by Fig. 2, another very useful plot is that of 'load–shortening' where, for example, the axial load (or stress) is plotted versus axial displacement (or strain). These curves are typically linear (i.e. constant slope) to the birfucation point (P_{cr}), where a marked decrease in the slope occurs.

Some rather extensive, early studies[64,92–96] of postbuckling behavior of laminated composite plates took place a decade ago. A finite element analysis capable of dealing with the geometrical nonlinearities present was developed by Schmit and Monforton.[92,93] Load–displacement curves were plotted for uniaxially loaded, simply supported plates having cross-ply, symmetric layering (graphite–epoxy) and angle-ply, antisymmetric layering (boron–epoxy) and compared with experiment,[92] showing good agreement. Islam[95] and Nara[96] developed an analysis using the Ritz method and compared theoretical results for simply supported plates with experiment. Symmetric cross-ply and angle-ply specimens of glass–epoxy were used.

Turvey and Wittrick[97,98] made a postbuckling study of both symmetrically and unsymmetrically laminated plates. For the unsymmetric case it was found that, although coupling between bending and stretching caused the (now) well-known decrease in buckling load, the slope of the load–shortening curves in the postbuckling range was the same as for a symmetric layup, indicating no difference in postbuckling stiffness. A subsequent analysis by Harris[99] for different boundary conditions showed that the stiffness change after buckling was mainly due to a change in the mode shape of deformation. Chandra[100] found that the postbuckling load–displacement curves of simply supported, cross-ply *square* plates were only slightly affected by bending–stretching coupling, but that for rectangular plates of aspect ratio (a/b) of 2, considerable difference could be seen.

Chia and Prabhakara[101] presented extensive results for the postbuckling behavior of clamped and simply supported, unsymmetrical, angle-ply plates and clamped, unsymmetrical, cross-ply plates using a series approach. Another solution procedure for cross-ply plates was also proposed,[102] utilizing beam functions to represent the inplane effects.

Noor *et al.*[103] examined the symmetries of deformation possible in postbuckling analysis, with the aid of reducing the computational size of the problem where possible. They also developed a finite element method of analysis based upon nonlinear, von Kármán plate theory, including the effects of shear deformation. The method was used to produce load–displacement and load–shortening curves for various biaxially loaded square plates. The load–shortening curves showed very large decrease in axial stiffness after buckling.

The postbuckling behavior of a square, symmetrically laminated composite plate having an internal square cutout was studied by Ter-Emmanuil'yan.[104] Uniaxial load was distributed along two opposite, hinged edges, the other two edges being free.

5. CONCLUDING REMARKS

The vibration, buckling and postbuckling behavior of composite plates is a very complicated subject. All the complexities for classical, isotropic, homogeneous plates are present, but further complicated by the complexities of laminate description and bending–stretching coupling phenomena. Yet, the efficiency of composites for high performance design applications is being more widely demonstrated, and thus the need for accurate theoretical and experimental information becomes increasingly important.

The rational analysis of laminated composite plates is still a relatively new subject, having only begun in earnest little over a decade ago. Although some noteworthy progress has been made, it is still very small in comparison with what has been done with ordinary plates, where on the order of 2000 references dealing with vibration, buckling and postbuckling are available.

ACKNOWLEDGEMENTS

This study was performed with the support of the Aeronautical Systems Division of the US Air Force, Wright Patterson Air Force Base under contract F33615-81-K-3203.

The author wishes to thank Mr Chandru Kalro, Graduate Research Associate, for his help in searching the literature and procuring publications.

REFERENCES

1. LEISSA, A. W., *Vibration of plates*, NASA SP-160, Washington, DC, US Govt. Printing Off., 1969.
2. REISSNER, E. and STAVSKY, Y., Bending and stretching of certain types of heterogeneous aeolotropic elastic plates, *Trans. ASME, J. Appl. Mech.*, **28** (1961) 402–8.
3. STAVSKY, Y., *On the theory of heterogeneous anisotropic plates*, D.Sc. Thesis, Mass. Inst. of Tech., 1959.
4. ASHTON, J. E. and WHITNEY, J. M., *Theory of laminated plates*, Stamford, Conn., Technomic Publishing, 1970.
5. JONES, R. M., *Mechanics of composite materials*, New York, McGraw-Hill, 1975.
6. BERT, C. W., Dynamics of composite and sandwich panels. Part I, *Shock Vib. Dig.*, **8** (1976) 37–48.
7. BERT, C. W., Dynamics of composite and sandwich panels. Part II, *Shock Vib. Dig.*, **8** (1976) 15–24.
8. BERT, C. W., Recent research in composite and sandwich plate dynamics, *Shock Vib. Dig.*, **11** (1979) 13–23.
9. BERT, C. W., *Vibration of composite structures*, Proc. of the Inter. Conf. on Recent Advances in Structural Dynamics, University of Southampton, Southampton, England, July 7–11, 1980.
10. BERT, C. W. and FRANCIS, P. H., Composite material mechanics: structural mechanics, *AIAA J.*, **12** (1974) 1173–86.
11. LEISSA, A. W., Recent research in plate vibrations. 1973–1976: complicating effects, *Shock Vib. Dig.*, **10** (1978) 21–35.
12. BERT, C. W., Fundamental frequencies of orthotropic plates with various planforms and edge conditions, *Shock Vib. Bull.*, **47** (1977) 89–94.
13. ASHTON, J. E. and WADDOUPS, M. E., Analysis of anisotropic plates, *J. Composite Mat.*, **3** (1969) 148–65.
14. WHITNEY, J. M., Fourier analysis of clamped anisotropic plates, *Trans. ASME, J. Appl. Mech.*, **38** (1971) 530–2.
15. WHITNEY, J. M., *On the analysis of anisotropic rectangular plates*, Air Force Materials Lab. Report TR-72-76, 1972, 24 pp.
16. BERT, C. W., Optimal design of a composite-material plate to maximize its fundamental frequency, *J. Sound Vib.*, **50** (1977) 229–37.
17. BERT, C. W., *Optimal design of composite-material panels for business aircraft*, Presented at Society of Automotive Engineers Business Aircraft Meeting, Wichita, Kansas, Mar. 29–Apr. 1, 1977.
18. RAJAMANI, A. and PRABHAKARAN, R., Dynamic response of composite plates with cut-outs: Part I, Simply-supported plates, *J. Sound Vib.*, **54** (1977) 549–64.

19. RAJAMANI, A. and PRABHAKARAN, R., Dynamic response of composite plates with cut-outs: Part II, Clamped–clamped plates, *J. Sound Vib.*, **54** (1977) 565–76.
20. WHITNEY, J. M., *A study of the effects of coupling between bending and stretching on the mechanical behavior of layered anisotropic composite materials*, Ph.D. Dissertation, The Ohio State University, Dept. of Engrg. Mech., 1968.
21. WHITNEY, J. M. and LEISSA, A. W., Analysis of heterogeneous anisotropic plates, *Trans. ASME, J. Appl. Mech.*, **36** (1969) 261–6.
22. LEISSA, A. W., *Vibration of shells*, NASA SP-288, Washington, DC, US Govt. Printing Off., 1973.
23. LEISSA, A. W. and KADI, A. S., Curvature effects on shallow shell vibrations, *J. Sound Vib.*, **16** (1971) 173–87.
24. JONES, R. M., Buckling and vibration of unsymmetrically laminated cross-ply rectangular plates, *AIAA J.*, **11** (1973) 1626–32.
25. LIN, C.-C. and KING, W. W., Free transverse vibrations of rectangular unsymmetrically laminated plates, *J. Sound Vib.*, **36** (1974) 91–103.
26. VOIGT, W., Observations on the problem of the transverse vibrations of rectangular plates (in German), *Nachr. Ges. Wiss (Göttingen)*, **6** (1893) 225–30.
27. ASHTON, J. E. and ANDERSON, J. D., The natural modes of vibration of boron-epoxy plates, *Shock Vib. Bull.*, April (1969) 81–91.
28. BERT, C. W. and MAYBERRY, B. L., Free vibrations of unsymmetrically laminated composite plates with clamped edges, *J. Composite Mat.*, **3** (1969) 282–93.
29. WHITNEY, J. M., The effect of boundary conditions on the response of laminated composites, *J. Composite Mat.*, **4** (1970) 192–203.
30. WHITNEY, J. M. and LEISSA, A. W., Analysis of a simply supported laminated anisotropic rectangular plate, *AIAA J.*, **8** (1970) 28–33.
31. MINICH, M. D. and CHAMIS, C. C., *Analytical displacements and vibrations of cantilevered unsymmetric fiber composite laminates*, NASA Tech. Memo., NASA TM X-71699, 1975, 13 pp.
32. VON KÁRMÁN, T., Strength of materials problems in mechanical engineering (in German), *Encyklopädie der Math. Wissenschaften*, Bd. 4 (1910) 311–85.
33. BENNETT, J. A., Nonlinear vibration of simply supported angle ply laminated plates, *AIAA J.*, **9** (1971) 1997–2003.
34. BENNETT, J. A., Some approximations in the nonlinear vibrations of unsymmetrically laminated plates, *AIAA J.*, **10** (1972) 1145–6.
35. MAYBERRY, B. L. and BERT, C. W., Experimental investigation of nonlinear vibrations of laminated anisotropic panels, *Shock Vib. Bull.*, (1969) 191–9.
36. BERT, C. W., Nonlinear vibration of a rectangular plate arbitrarily laminated of anisotropic material, *Trans. ASME, J. Appl. Mech.*, **40** (1973) 452–8.
37. CHANDRA, R. and BASAVA RAJU, B., Large deflection vibration of angle ply laminated plates, *J. Sound Vib.*, **40** (1975) 393–408.
38. CHANDRA, R. and BASAVA RAJU, B., Large amplitude flexural vibration of cross ply laminated composite plates, *Fibre Sci. Tech.*, **8** (1975) 243–63.
39. CHANDRA, R., Large deflection vibration of cross-ply laminated plates with certain edge conditions, *J. Sound Vib.*, **47** (1976) 509–14.

40. CHIA, C. Y. and PRABHAKARA, M. K., A general mode approach to nonlinear flexural vibrations of laminated rectangular plates, *Trans. ASME, J. Appl. Mech.*, **45** (1978) 623–8.
41. WHITNEY, J. M. and PAGANO, N. J., Shear deformation in heterogeneous anisotropic plates, *Trans. ASME, J. Appl. Mech.*, **37** (1970) 1031–6.
42. NOOR, A. K., Free vibrations of multilayered composite plates, *AIAA J.*, **11** (1973) 1038–9.
43. SIU, C. C. and BERT, C. W., *Sinusoidal response of composite-material plates with material damping*, ASME Paper No. 73-DET-120, 1973, 8 pp.
44. CLARY, R. R., *Vibration characteristics of unidirectional filamentary composite material panels*, Composite Materials: Testing and Design (Second Conf.), ASTM STP 497, 1972, pp. 415–38.
45. SRINIVAS, S., *A refined analysis of thick anisotropic laminates*, Indian Inst. Sci., Bangalore Aero. Engrng. Dept. Report, 1971, 68 pp.
46. SRINIVAS, S. and RAO, A. K., Bending, vibration and buckling of simply supported thick orthotropic rectangular plates and laminates, *Int. J. Solids Struc.*, **6** (1970) 1463–81.
47. BERT, C. W. and CHEN, T. L. C., Effect of shear deformation on vibration of antisymmetric angle-ply laminated rectangular plates, *Int. J. Solids Struc.*, **14** (1978) 465–73.
48. HIRASHIMA, K., *General higher-order equations of two-dimensional static and dynamic theories for homogeneous and laminated plates*, Report, Princeton Univ., Dept. of Civil Engrng., July, 1980, 183 pp.
49. MAU, S.-T., PIAN, T. H. H. and TONG, P., Vibration analysis of laminated plates and shells by a hybrid stress element, *AIAA J.*, **11** (1973) 1450–2.
50. REDDY, J. N., Free vibration of antisymmetric, angle-ply laminated plates including transverse shear deformation by the finite element method, *J. Sound Vib.*, **66** (1979) 565–76.
51. REDDY, J. N., A penalty plate-bending element for the analysis of laminated anisotropic composite plates, *Int. J. Numer. Methods Engr.*, (1980).
52. REDDY, J. N., *A comparison of closed-form and finite-element solutions of thick, laminated, anisotropic rectangular plates*, School of Aerospace, Mechanical and Nuclear Engr., Univ. of Oklahoma (Norman), Report No. OU-AMNE-79-19, 1979, 39 pp.
53. REDDY, V. S., *Analyses of cross-ply rectangular plates of bimodulus composite material*, M.Sc. Thesis, Univ. of Oklahoma, 1980, 101 pp.
54. BERT, C. W., REDDY, J. N., CHAO, W. C. and REDDY, V. S., *Vibration of thick rectangular plates of bimodulus composite material*, Off. Naval Res., Tech. Report No. 15, Contract N00014-78-C-0647, 1980, 26 pp.
55. WU, C. and VINSON, J. R., Nonlinear oscillations of laminated specially orthotropic plates with clamped and simply supported edges, *J. Acoust. Soc. Amer.*, **49** (1971) 1561–7.
56. REDDY, J. N. and CHAO, W. C., *Large deflection and large-amplitude free vibrations of laminated composite-material plates*, School of Aerospace, Mechanical and Nuclear Engr., Univ. of Oklahoma, Report No. OU-AMNE-80-7, 1980, 25 pp.
57. LEISSA, A. and VAGINS, M., Stress optimization in nonhomogeneous materials, *Developments in theoretical and applied mechanics*, Proceedings of SECTAM 8, vol. 8, 1976, pp. 13–22.

58. LEISSA, A. W. and VAGINS, M., The design of orthotropic materials for stress optimization, *Int. J. Solids Struc.*, **14** (1978) 517–26.

59. WHITNEY, J. M., Shear buckling of unsymmetrical cross-ply plates, *J. Composite Mat.*, **3** (1969) 359–63.

60. ASHTON, J. E. and LOVE, T. S., Shear stability of laminated anisotropic plates, *Composite materials: testing and design*, ASTM STP 460, 1969, pp. 352–61.

61. ASHTON, J. E. and LOVE, T. S., Experimental study of the stability of composite plates, *J. Composite Mat.*, **3** (1969) 230–42.

62. CHAMIS, C. C., Buckling of anisotropic composite plates, *Proc. ASCE, J. Struc. Div.*, **95** (1969) 2119–39.

63. SARKISYAN, V. S. and MOVSISYAN, L. A., A method for determining the critical loads on anisotropic plates, *Soviet Engr. J.* (*Inzhenernyi Zhurnal*), **5** (1965) 600–2.

64. Union Carbide Corporation, Case Western Reserve University and Bell Aerosystems Company, *Integrated research on carbon composite materials*, Air Force Materials Lab., Report AFML-TR-66-310, Part I (Oct. 1966), Part II (Dec. 1967), Part III (Jan. 1969), Part IV (Oct. 1969), Part V (1970).

65. WHITNEY, J. M., *Bending, vibrations, and buckling of laminated anisotropic rectangular plates*, Air Force Systems Command, Wright Patterson AFB, Ohio, Materials Lab. Report No. AFML-TR-70-75, Aug. 1970, 35 pp.

66. WHITNEY, J. M., Analysis of anisotropic rectangular plates, *AIAA J.*, **10** (1972) 1344–5.

67. KICHER, T. P. and MANDELL, J. F., A study of the buckling of laminated composite plates, *AIAA J.*, **9** (1971) 605–13.

68. MANDEL, J. F., *An experimental study of the buckling of anisotropic plates*, M.S. Thesis, Case Western Reserve University, 1968.

69. VISWANATHAN, A. V., SOONG, T.-C. and MILLER, R. E., JR., *Buckling analysis for axially compressed flat plates, structural sections, and stiffened plates reinforced with laminated composites*, NASA Contractor Report CR-1887, Nov. 1971, 75 pp.

70. VISWANATHAN, A. V., SOONG, T.-C. and MILLER, R. E., JR., Buckling analysis for structural sections and stiffened plates reinforced with laminated composites, *Int. J. Solids Struc.*, **8** (1972) 347–67.

71. HOUSNER, J. M. and STEIN, M., *Numerical analysis and parametric studies of the buckling of composite orthotropic compression and shear panels*, NASA TN D-7996, Oct. 1975, 102 pp.

72. CHAILLEUX, A., HANS, Y. and VERCHERY, G., Experimental study of the buckling of laminated composite columns and plates, *Int. J. Mech. Sci.*, **17** (1975) 489–98.

73. SAWYER, J. W., Flutter and buckling of general laminated plates, *J. Aircraft*, **14** (1977) 387–93.

74. CHAO, C. C., KOH, S. L. and SUN, C. T., Optimization of buckling and yield strengths of laminated composites (synoptic), *AIAA J.*, **13** (1975) 1131–2. Full paper available from NTIS as N75-19370, May 1974, 33 pp.

75. HAYASHI, T., Optimum design of cross- and angle-plied laminated composite plates under compression, *Composite Mat. Struc.* (*Japan*), **3** (1974) 18–20.

76. SCHMIT, L. A., JR. and FARSHI, B., Optimum design of laminated fibre composite plates, *Int. J. Numer. Methods Engr.*, **11** (1977) 623–40.

77. TSAI, W. T., A theorem of stability of laminated plates, *Trans. ASME, J. Appl. Mech.*, **42** (1975) 237–9.
78. AGARWAL, B. and DAVIS, R. C., *Minimum-weight designs for hat-stiffened composite panels under uniaxial compression*, NASA TN D-7779, Nov. 1974, 44 pp.
79. STROUD, W. J. and AGRANOFF, N., *Minimum-mass design of filamentary composite panels under combined loads: design procedure based on simplified buckling equations*, NASA TN D-8257, Oct. 1976, 51 pp.
80. STROUD, W. J., AGRANOFF, N. and ANDERSON, M. S., *Minimum-mass design of filamentary composite panels under combined loads: design procedure based on a rigorous buckling analysis*, NASA TN D-8417, July 1977, 40 pp.
81. WILLIAMS, J. G. and MIKULAS, M. M., JR., *Analytical and experimental study of structurally efficient composite hat-stiffened panels loaded in axial compression*, NASA TM X-72813, Jan. 1976, 21 pp.
82. WILLIAMS, J. G. and STEIN, M., Buckling behavior and structural efficiency of open-section stiffened composite compression panels, *AIAA J.*, **14** (1976) 1618–26.
83. WHITNEY, J. M., The effect of transverse shear deformation on the bending of laminated plates, *J. Composite Mat.*, **3** (1969) 534–47.
84. VINSON, J. R. and SMITH, A. P., JR., *The effect of transverse shear deformation on the elastic stability of plates of composite materials*, AFOSR TR-75-1628, March 1975, 77 pp.
85. TURVEY, G. J., Biaxial buckling of moderately thick laminated plates, *J. Strain Analysis*, **12** (1977) 89–96.
86. ASHTON, J. E., Analysis of anisotropic plates II, *J. Composite Mat.*, **3** (1969) 470–9.
87. MORGAN, H. S. and JONES, R. M., Buckling of rectangular cross-ply laminated plates with nonlinear stress–strain behavior, *Trans. ASME, J. Appl. Mech.*, **46** (1979) 637–43.
88. FLAGGS, D. L. and VINSON, J. R., *Elastic stability of generally laminated composite plates including hygrothermal effects*, AFOSR TR 78-1349, July 1977, 68 pp.
89. BAULD, N. R., JR. and SATYAMURTHY, K., *Collapse load analysis for plates and panels*, Air Force Flight Dynamics Lab., Wright Patterson AFB, Ohio, AFFDL-TR-79-3038, May 1979, 209 pp.
90. MEFFERT, B., DEREK, H. and MENGES, G., Stress deformation behavior of orthotropic plates with initial curvature made of glass fiber-reinforced unsaturated polyester plastics (GFUP) under uniaxial load in the plane of the plate (in German), *Bauingenieur*, **52** (1977) 211–16.
91. STROUD, W. J., ANDERSON, M. S. and HENNESSY, K. W., *Effect of bow-type initial imperfection on the buckling load and mass of graphite–epoxy blade-stiffened panels*, NASA Tech. Memo., NASA TM-74063, 1977, 25 pp.
92. MONTFORTON, G. R., *Discrete element finite displacement analysis of anisotropic sandwich shells*, Ph.D. Thesis, Case Western Reserve Univ., Cleveland, Ohio, 1970.
93. SCHMIT, L. A., JR. and MONTFORTON, G. R., Finite deflection discrete element analysis of sandwich plates and cylindrical shells with laminated faces, *AIAA J.*, **8** (1970) 1454–61.

94. CHAN, D. P., *An analytical study of the postbuckling of laminated, anisotropic plates*, Ph.D. Thesis, Case Western Reserve Univ., Cleveland, Ohio, 1971.

95. ISLAM, M. R., *Buckling and postbuckling strength of anisotropic plates*, Ph.D. Thesis, Case Western Reserve Univ., Cleveland, Ohio, 1971.

96. NARA, H. R. (Ed.), *Interface and mechanics research in fiber reinforced composites*, Air Force Materials Lab., Wright Patterson AFB, Ohio, Tech. Report. AFML-TR-71-260, March 1972, 262 pp.

97. TURVEY, G. J., *A contribution to the elastic stability of thin-walled structures fabricated from isotropic and orthotropic materials*, Ph.D. Thesis, Dept. of Civil Engr., Univ. of Birmingham, 1971.

98. TURVEY, G. J. and WITTRICK, W. H., The large deflection and postbuckling behavior of some laminated plates, *Aero. Quart.*, **24** (1973) 77–86.

99. HARRIS, G. Z., The buckling and post-buckling behavior of composite plates under biaxial loading, *Int. J. Mech. Sci.*, **17** (1975) 187–202.

100. CHANDRA, R., Postbuckling analysis of crossply laminated plates, *AIAA J.*, **13** (1975) 1388–9.

101. CHIA, C. Y. and PRABHAKARA, M. K., Postbuckling behavior of unsymmetrically layered anisotropic rectangular plates, *J. Appl. Mech.*, **41** (1974) 155–62.

102. PRABHAKARA, M. K., Post-buckling behaviour of simply-supported cross-ply rectangular plates, *Aero. Quart.*, **27** (1976) 309–16.

103. NOOR, A. K., MATHER, M. D. and ANDERSON, M. S., Exploiting symmetries for efficient postbuckling analysis of composite plates, *AIAA J.*, **15** (1977) 24–32.

104. TER-EMMANUIL'YAN, N. YA., Stability of an orthotropic flexible square plate weakened by a square opening, *Polymer Mech.*, **7** (1971) 425–9.

22

On the Use of the Effective Width Concept for Composite Plates

J. Rhodes

Senior Lecturer, University of Strathclyde,
Glasgow G1 1XJ, Scotland

AND

I. H. Marshall

Lecturer, Paisley College of Technology,
Paisley PA1 2BE, Scotland

ABSTRACT

An examination of the effective width of orthotropic plates in the post-local-buckling range is presented. Three types of boundary conditions are considered as those most suitable for use in design, and results are given for each type.

It is found that for any specified buckle half-wavelength there can be a substantial degree of difference between the behaviour of plates with different properties. However, if infinite plates are examined then all the plates have similar post-buckling behaviour. It is postulated, on this basis, that effective width evaluated for isotropic plates can be used for the design analysis of orthotropic plates.

NOTATION

A	magnitude of local buckling deflection
a	buckle half-wavelength
b	plate width
D_{11}, D_{22}	plate flexural rigidity factor: $D_{ii} = E_{ii}t^3/12(1 - v_{12}v_{21})$
D_{33}	twisting rigidity of plate about x and y axes given by $D_{33} = G_{12}t^3/12$

D_3	defined as $D_3 = G_{12}t^3/6 + v_{12}/D_{22}$
E_{11}, E_{22}	plate moduli of elasticity in x and y directions respectively
E^*	ratio of plate compressional stiffness after buckling to that
$\overline{E}$	before buckling
F	stress function
G_{12}	elastic shear modulus in x–y plane
H	defined as $(1/G_{12} - 2v_{12}/E_{11})$
K	buckling coefficient, $K = \sigma_{CR} \left/ \dfrac{\pi^2 D_{11}}{b^2 t} \right.$
P	end load on plate
t	plate thickness
x, y	co-ordinates
ε	average strain in direction of loading
$\sigma_x, \sigma_y, \tau_{xy}$	membrane stresses in x and y directions and shear stress in x–y plane respectively
v_{12}, v_{21}	Poisson's ratio in the x and y directions respectively
η	ratio of E_{22} to E_{11}
$P_{CR}, \varepsilon_{CR}, \sigma_{CR}$	values of P, σ_x and ε at local buckling
$\bar{P}, \bar{\varepsilon}$	non-dimensional forms of P and ε defined as

$$\bar{P} = \frac{Pb}{\pi^2 D_{11}} \qquad \bar{\varepsilon} = \frac{\varepsilon b^2 t E_{11}}{\pi^2 D_{11}}$$

INTRODUCTION

The behaviour of thin-walled structural members is well-known to be highly influenced by local buckling and its effect on member strength and stiffness. Because of the high strength to stiffness ratios obtainable from the use of composite materials thin-walled members of such materials are particularly prone to local buckling, and a great deal of research has been carried out into the buckling and post-buckling behaviour of composite plates and shells.

The mechanics of local buckling of orthotropic plates is well-known,[1,2] and a number of analyses of particular post-buckling problems have been obtained.[3–6]

While detailed and rigorous analysis of particular problems is invaluable and necessary to the understanding of the subject, it is also very important that simplified analytical methods be available to the designer enabling him to design plates in the post-buckling range with confidence. In the case of metal structures there are various design codes in use,[7,8] which permit evaluation of plate behaviour at loads far beyond buckling using the concept of 'effective width' or similar such devices.

A number of researchers in the past have obtained effective width curves for particular orthotropic plates. It is a worthwhile goal, however, in the authors' opinion, to attempt to obtain generally applicable formulae suitable, perhaps with modifications, for all plates. This is the aim of this paper.

There are two quite distinct aspects of plate behaviour which require investigation in the post-buckling range; (a) strength, and (b) stiffness. The necessity for examining the strength aspect is obvious. The stiffness aspect is also of great importance because (1) deflection limitations often govern design, (2) in multiredundant structures strength is often greatly affected by the stiffness of individual members, and (3) a wide variety of column and beam column type structures can fail elastically, often dynamically due to the sudden loss of stiffness caused by local buckling.

The investigation presented here is confined to the examination of the stiffness aspect of post-buckling behaviour. While for metal plate elements the same effective widths for strength and stiffness are often used in design, it is recognised that the strength aspect is more complex in the case of composite materials, and that such approximations can not be justified without extreme caution.

The most important requirements for generally applicable effective width formulae are that they should be simple to use, have a reasonable degree of accuracy and be conservative and applicable even under the most adverse conditions which they could reasonably be expected to deal with. It is also important that the range of applicability be large if possible and any restrictions be specified.

There are three main types of plate boundary conditions examined in design, and these will be investigated in this paper. These are, with reference to Fig. 1:

(a) Simply supported plates, uniformly compressed in one direction, with the unloaded edges kept straight due to the presence of a self-equilibrating normal stress system, Fig. 1(a).

(b) Simply supported plates, uniformly compressed as in (a), but with the unloaded edges free from normal and shear stresses, Fig. 1(b).

(c) Uniformly compressed plates which have one unloaded edge simply supported and the other edge free to deflect, Fig. 1(c).

In all cases the loaded edges are assumed to be simply supported.

J. Rhodes, I. H. Marshall

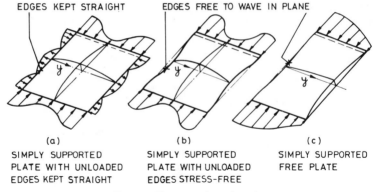

EDGES KEPT STRAIGHT   EDGES FREE TO WAVE IN PLANE

(a)   (b)   (c)
SIMPLY SUPPORTED   SIMPLY SUPPORTED   SIMPLY SUPPORTED
PLATE WITH UNLOADED   PLATE WITH UNLOADED   FREE PLATE
EDGES KEPT STRAIGHT   EDGES STRESS-FREE

FIG. 1.   Plate types investigated.

In this paper the investigation will be limited to the examination of specially orthotropic plates which have flexural and membrane stiffnesses defined as in the notation.

GOVERNING PLATE EQUATIONS

For a specially orthotropic plate the strain energy of bending is given by the expression[3]

$$V_B = \frac{1}{2} \iint \left[D_{11} \left(\frac{\partial^2 w}{\partial x^2} \right)^2 + 2v_{21}D_{11} \frac{\partial^2 w}{\partial x^2} \frac{\partial^2 w}{\partial y^2} \right.$$
$$\left. + D_{22} \left(\frac{\partial^2 w}{\partial y^2} \right)^2 + 4D_{33} \left(\frac{\partial^2 w}{\partial x \partial y} \right)^2 \right] dx\,dy \qquad (1)$$

The strain energy of membrane actions is

$$V_D = \frac{t}{2} \iint \left[\frac{1}{E_{11}} \left(\frac{\partial^2 F}{\partial y^2} \right)^2 - \frac{2v_{12}}{E_{11}} \frac{\partial^2 F}{\partial x^2} \frac{\partial^2 F}{\partial y^2} \right.$$
$$\left. + \frac{1}{G_{12}} \left(\frac{\partial^2 F}{\partial x \partial y} \right)^2 + \frac{1}{E_{22}} \left(\frac{\partial^2 F}{\partial x^2} \right)^2 \right] dx\,dy \qquad (2)$$

where F is a stress function such that

$$\frac{\partial^2 F}{\partial x^2} = \sigma_y \qquad \frac{\partial^2 F}{\partial y^2} = \sigma_x \qquad \frac{\partial^2 F}{\partial x \partial y} = -\tau_{xy}$$

The membrane actions are related to the out-of plane deflections by the compatibility equation

$$\frac{1}{E_{22}} \frac{\partial^4 F}{\partial x^4} + H \frac{\partial^4 F}{\partial x^2 \partial y^2} + \frac{1}{E_{11}} \frac{\partial^4 F}{\partial y^4} = \left(\frac{\partial^2 w}{\partial x \partial y}\right)^2 - \frac{\partial^2 w}{\partial x^2} \frac{\partial^2 w}{\partial y^2} \tag{3}$$

The general method of analysis is to postulate a deflected form w, obtain a suitable stress function F, using eqn. (3), and henceforth evaluate the total strain energy of the system and apply the Principle of Minimum Potential Energy to furnish the required solution.

SIMPLY SUPPORTED PLATE WITH UNLOADED EDGES RESTRAINED IN-PLANE

Consider the half-wavelength of the plate shown in Fig. 1(a) loaded in compression with uniform end displacement u causing an average strain ε in the direction of loading.

It is well-known that the deflected form at buckling is sinusoidal in the x and y directions. Therefore if the buckled form is taken as

$$w = A \sin \frac{\pi y}{b} \cdot \sin \frac{\pi x}{a} \tag{4}$$

substitution into the compatibility equation, eqn. (3), and satisfaction of the boundary conditions results in the following expressions for the stresses:

$$\sigma_x = E_{11} \left[\frac{u}{a} - \frac{1}{4} \left(\frac{\pi}{a}\right)^2 A^2 \sin^2 \frac{\pi y}{b} \right] \tag{5}$$

$$\sigma_y = E_{22} \left[\frac{1}{8} \left(\frac{\pi}{b}\right)^2 A^2 \cos \frac{2\pi x}{a} \right] \tag{6}$$

$$\tau_{xy} = 0 \tag{7}$$

Using these expressions for deflections and stresses in the energy equations, applying the Principle of Minimum Potential Energy and non-dimensionalising gives the following results:

$$\left(\frac{A}{t}\right)^2 = \frac{\bar{\varepsilon} - K}{(1 - v_{12} v_{21}) \left(\frac{b}{a}\right)^2 \left[\frac{9}{4} + \frac{3}{4} \eta \left(\frac{a}{b}\right)^4\right]} \tag{8}$$

where K is the non-dimensional buckling coefficient, given by

$$K = \left(\frac{b}{a}\right)^2 + 2\frac{D_3}{D_{11}} + \left(\frac{a}{b}\right)^2 \eta \tag{9}$$

and $\bar{\varepsilon}$ is the non-dimensional average strain defined in the notation.

$$\bar{P} = K + \frac{E^*}{E}(\bar{\varepsilon} - K) \tag{10}$$

where E^*/E is the reduced stiffness in the post-buckling range and is given by

$$\frac{E^*}{E} = 1 - \frac{2}{3 + \eta\left(\dfrac{a}{b}\right)^4} \tag{11}$$

The expression for K, eqn. (9), is known to be exact. The expression for E^*/E was derived by Benthem[9] for isotropic plates, in which case $\eta = 1$, and by Harris[4] for the more general case. From this equation it can be seen that the maximum possible value for E^*/E is unity and the minimum possible value is $\frac{1}{3}$ which occurs when (a/b) tends to zero regardless of the value of η

For an unbuckled plate of width b_e (rather than b) the load required to cause an average longitudinal strain ε is

$$P = \varepsilon t E b_e \tag{12}$$

or, in non-dimensional form

$$\frac{Pb}{\pi^2 D_{11}} = \frac{\varepsilon b^2 t E_{11}}{\pi^2 D_{11}} \cdot \frac{b_e}{b} \tag{13}$$

i.e.

$$\bar{P} = \bar{\varepsilon} \cdot \frac{b_e}{b} \tag{14}$$

Equating $\bar{P}$ from eqn. (10) with that from eqn. (14) and rearranging gives the relevant expression for b_e

$$\frac{b_e}{b} = \frac{E^*}{E} + \left(1 - \frac{E^*}{E}\right)\frac{K}{\bar{\varepsilon}} \tag{15}$$

As $\bar{\varepsilon}$ tends to infinity the ratio of effective width to full width tends to E^*/E with a minimum value of $\frac{1}{3}$. While the results obtained from this analysis are exact at the point of buckling and very accurate at loads no greatly in excess of buckling (i.e. $P - 2P_{CR}$) it is known that far into the post-buckling range the effective width can be very much less than one-third of the full width. To obtain accuracy at strains far beyond buckling the effective width should tend to zero at very high strains. This arises because of changes in the buckled form after buckling.

To perform a rigorous analysis at loads very far beyond buckling becomes extremely laborious and to avoid the labour and complexity a relatively simple analysis is used, as follows. It has been shown[10] that the major factors affecting the behaviour far into the post-buckling range are (1) a reduction in the natural half-wavelength of the buckle (if this is possible), and (2) a flattening of the buckled form across the plate.

Reduction in buckle half-wavelength can be taken into account by allowing the ratio (a/b) in eqn. (11) to vary. However, this equation cannot describe the flattening across the plate. In order to describe this flattening the deflected form indicated in Fig. 2 may be used. This was first postulated by Cox[11] in an early analysis of isotropic plates and may be specified thus:

$$w = A \sin \frac{\pi y}{\alpha b} \sin \frac{\pi x}{a} \qquad \begin{array}{l} \text{for } 0 < y < \dfrac{\alpha b}{2} \\[2mm] \text{and } \left(1 - \dfrac{\alpha b}{2}\right) < y < b \end{array} \qquad (16)$$

$$w = A \sin \frac{\pi x}{l} \qquad \text{for } \frac{\alpha b}{2} < y < \left(1 - \frac{\alpha b}{2}\right) \qquad (17)$$

Suitable stresses which satisfy the membrane boundary conditions and overall equilibrium are

$$\sigma_x = E_{11} \left[\frac{u}{a} - \frac{1}{4} \left(\frac{\pi}{a}\right)^2 A^2 \sin^2 \frac{\pi y}{\alpha b} \right] \qquad \begin{array}{l} \text{for } 0 < y < \dfrac{\alpha b}{2} \\[2mm] \text{and } \left(1 - \dfrac{\alpha b}{2}\right) < y < b \end{array} \qquad (18)$$

$$\sigma_x = E_{11} \left[\frac{u}{a} - \frac{1}{4} \left(\frac{\pi}{a}\right)^2 A^2 \right] \qquad \text{for } \frac{\alpha b}{2} < y < \left(1 - \frac{\alpha b}{2}\right) \qquad (19)$$

$$\sigma_y = E_{22} \left[\frac{1}{8\alpha} \left(\frac{\pi}{b}\right)^2 A^2 \cos \frac{2\pi x}{a} \right] \qquad \tau_{xy} = 0 \qquad (20)$$

It is of note that the functions for w introduce moment discontinuity at $y = \alpha(b/2)$. This was investigated by Koiter[10] who found that it had little effect on the solution.

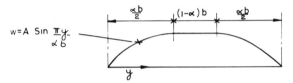

FIG. 2.   Assumed deflected form.

The stresses postulated do not satisfy the compatibility equation exactly except in the case $\alpha = 1$ and therefore the solution resulting cannot be termed a strict upper bound, although it is unlikely that the slight violation of compatibility involved should lead to significant discrepancies.

Performing the analysis as before yields the modified equations

$$\left(\frac{A}{t}\right)^2 = \frac{\bar{\varepsilon}(2 - \alpha) - \bar{K}}{(1 - v_{12}v_{21})\left(\frac{b}{a}\right)^2\left[6 - \frac{15}{4}\alpha + \frac{3\eta}{4\alpha^2}\left(\frac{a}{b}\right)^4\right]} \tag{21}$$

$$\bar{K} = \left[(2 - \alpha)\left(\frac{b}{a}\right)^2 + \frac{2}{\alpha}\frac{D_3}{D_{11}} + \left(\frac{a}{b}\right)^2\frac{D_{22}}{\alpha^3 D_{11}}\right]\Big/(2 - \alpha) \tag{22}$$

$$\frac{E^*}{E} = 1 - \frac{(2 - \alpha)^2}{4 - \frac{5}{2}\alpha + \frac{\eta}{2\alpha^2}\left(\frac{a}{b}\right)^4} \tag{23}$$

Equations (10) and (15) should also be generalised by replacing K by $\bar{K}$.

For $\alpha = 1$ all equations reduce to those obtained previously. Examination of eqn. (22) shows that for a given value of (a/b) a reduction in α increases $\bar{K}$. Also from eqn. (23) it can be seen that as α and (a/b) tend towards zero E^*/E also tends toward zero.

An indication of the predictions of the approach is given in Fig. 3 which compares the load–end shortening curves obtained using this method with those of Chandra and Raju.[5] Results for three plates are shown: one isotropic, one very stiff in the direction of loading and the third very stiff in the direction perpendicular to loading. The properties used for the two orthotropic plates were those of Ref. 7 which had greatest directionality. The results are in good agreement. The buckle half-wavelength used here was equal to the plate width, and it can be seen that for this set value of half-wavelength the post-buckling stiffness is very highly dependent on the degree of orthotropy, with stiffness increasing as E_{22}/E_{11} increases. This of course can be seen from eqn. (23). However, the use of an arbitrary fixed value of (a/b) can result in misleading information. For $a/b = 1$ the

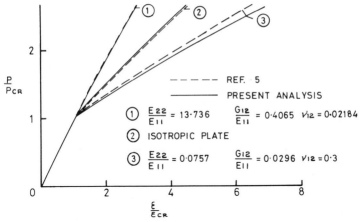

FIG. 3. Load–end displacement curves for plates with $a/b = 1$.

isotropic plate in Fig. 3 has its natural wavelength for minimum buckling load, whereas plate 3 has a natural wavelength almost twice this value. Furthermore plate 1 would naturally buckle into two waves in a square plate, so that this plate would require physical constraint to induce buckling into this wavelength.

For purposes of design it is desirable that the need to take wavelength into account be eliminated and that any overestimation of stiffness due to artifically fixed wavelength be avoided. Therefore it is advisable to be able to evaluate the minimum load for any value of end displacement, i.e. the minimum effective width. This can be done by allowing the buckle half-wavelength to take its natural half-wavelength at buckling and thereafter to change in the post-buckling range. This always results in a shortening of the half-wavelength after buckling, and assumes that the plate is infinitely long so that continuous change in the wavelength can occur.

At the point of initial buckling the natural buckle half-wavelength can be obtained by differentiating eqn. (9) for K with respect to a/b and equating to zero, to obtain

$$\frac{a}{b} = \frac{1}{\eta^{0.25}} \tag{24}$$

Substitution into eqns. (9) and (11) gives

$$K = 2\left[\eta + \frac{D_3}{D_{11}}\right] \tag{25}$$

$$\frac{E^*}{E} = \frac{1}{2} \tag{26}$$

It may be noted that these results are exact at buckling. The result for E^*/E indicates the important fact that when an orthotropic plate of this type buckles into its natural half-wavelength then the compressional stiffness reduces by half irrespective of the degree of orthotropy. Furthermore, if this wavelength is maintained after buckling then the degree of orthotropy does not affect the stiffness as the deflected form across the plate changes.

To permit evaluation of the minimum stiffness in the post-buckling range the half-wavelength must be allowed to shorten after buckling to give the minimum load for a given displacement. This can be done either by numerical minimisation of eqn. (10) using eqns. (22) and (23), or graphically by drawing curves for a variety of values of α and a/b and obtaining the curve of minimum load from these. In this investigation a numerical minimisation was used with the aid of a microcomputer.

Figure 4 shows the minimum load versus end displacement curves obtained for a variety of orthotropic plates, using properties given in Ref. 5, together with the curve obtained, applicable to all plates, for the half-wavelength fixed at $a/b = \eta^{0.25}$.

Two main points are worthy of note: (1) shortening of the wavelength after buckling has a substantial effect on stiffness, and (2) the degree of orthotropy does not have a substantial effect even when the wavelength and

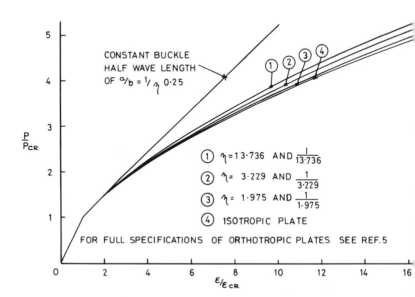

FIG. 4.   Load–end displacement curves for infinite plates.

cross-sectional buckle shape are allowed to vary far into the post-buckling range.

It should be pointed out here that although the results of this investigation are in agreement with those of Chandra and Raju[5] for plates with fixed half-wavelengths, they are in strong disagreement with the results of those authors for infinite plates. In Ref. 5 it is concluded that the degree of orthotropy has a great deal of effect on the post-buckling behaviour of an infinitely long plate. The equations given in that paper are lengthy, complex and difficult to check. However, there are various points of considerable uncertainty regarding the load–shortening graph presented; for example, the plate which is consistently stiffest for any stated constant half-wavelength is by far the most flexible if the wavelength is allowed to change. This and other points of doubtful veracity cause the writers to disregard the results of Ref. 5 pertaining to infinite plates.

A further point of note from Fig. 4 is that the curve obtained for the isotropic plate forms a lower estimate of post-buckling stiffness for all orthotropic plates shown. This is to some extent due to the fact that there is slightly more resistance to change in wavelength of the orthotropic plates after buckling. It can be shown that if $D_3/D_{11} < \sqrt{\eta}$ there will be a slight reduction in the relative stiffness after buckling from the isotropic case, as

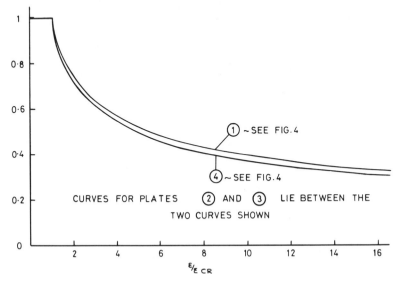

FIG. 5.   Effective widths for infinite plates.

occurs in Fig. 4. Indeed the quantity $(D_3/D_{11})\sqrt{\eta}$ completely specifies the variation in orthotropic curves from the isotropic case. Note that this value is the same whether the plate is loaded in its direction of maximum or minimum stiffness. This results in identical curves being obtained for the same orthotropic plate loaded in either direction, as shown in Fig. 4.

Figure 5 shows the variation in effective width with variation in $\varepsilon/\varepsilon_{CR}$, indicating that the isotropic case gives a close lower estimate to the orthotropic case.

SIMPLY SUPPORTED PLATE WITH UNLOADED EDGES FREE TO WAVE IN-PLANE

In this case the buckling situation is identical to that for the previous case. The post-buckling behaviour is slightly more difficult to analyse rigorously and using rigorous analysis does not result in a simple expression for E^*/E as in the previous case. It is of note, however, that a lower bound to the initial post-buckling stiffness can be obtained by omitting the term $\eta(a/b)^4$ from eqn. (11), which amounts to neglecting transverse and shearing stresses, to obtain $E^*/E = \frac{1}{3}$ regardless of the degree of orthotropy. This is a true lower bound at the instant of buckling.

A better approximation can be obtained by postulating a form for the stress function which satisfies all boundary conditions, using Galerkin's method to obtain approximate satisfaction of the compatibility equation and henceforth using the Energy method to complete the solution.

A stress function which satisfies all boundary conditions is

$$F = \frac{E_{11}}{2} y^2 \left(\frac{u}{a} - \frac{A^2}{8} \left(\frac{\pi}{a} \right)^2 \right) + \frac{E_{11}}{32} A^2 \left(\frac{b}{a} \right)^2 \cos \frac{2\pi y}{b} + C \cos \frac{2\pi x}{a} \sin^2 \frac{\pi y}{b} \tag{27}$$

Applying the analysis as outlined gives the following results:

$$C = \frac{A^2 E_{11}}{8 \left[\left(\frac{a}{b} \right)^2 + E_{11} H + 3 \frac{E_{11}}{E_{22}} \left(\frac{b}{a} \right)^2 \right]} \tag{28}$$

and

$$\frac{E^*}{E} = \frac{2 + \left(1 + E_{11} H \left(\frac{b}{a} \right)^2 + 3 \frac{E_{11}}{E_{22}} \left(\frac{b}{a} \right)^4 \right)}{2 + 3 \left(1 + E_{11} H \left(\frac{b}{a} \right)^2 + 3 \frac{E_{11}}{E_{22}} \left(\frac{b}{a} \right)^4 \right)} \tag{29}$$

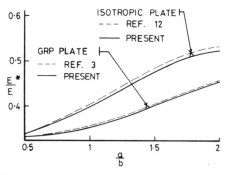

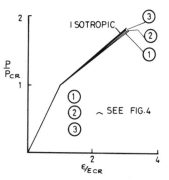

FIG. 6. Variation of E^*/E with buckle half-wave-length.

FIG. 7. Load–end displacement curves for plates with stress-free unloaded edges.

The accuracy of this equation in predicting the value of the initial post-buckling reduction in stiffness is examined in Fig. 6. For the isotropic case comparison is made with Rhodes' results of rigorous analysis[12] and a maximum error of about 2 % is observed in this case. Agreement with the rigorous analysis of Banks[3] for the case of an orthotropic GRP plate is even better as indicated in the figure. This inspires confidence in eqn. (29).

Setting $a/b = 1/\eta^{0.25}$ to obtain the minimum buckling load and substituting into eqn. (29) results in the curves of P/P_{CR} against $\varepsilon/\varepsilon_{CR}$ as shown in Fig. 7 in the initial post-buckling range. In this range the value of E^*/E is obtained from

$$\frac{E^*}{E} = \frac{6 + H\sqrt{E_{11}E_{22}}}{14 + H\sqrt{E_{11}E_{22}}} \tag{30}$$

As can be observed from this equation and Fig. 7 there is a slight variation in initial post-buckling behaviour for different plate properties. This is influenced by the quantity $H\sqrt{E_{11}E_{22}}$. If this quantity is equal to 2, the isotropic case results, regardless of the value of η. Variation in $H\sqrt{E_{11}E_{22}}$ does not have a great deal of effect; if this quantity is zero then $E^*/E = 6/14 = 0.428$, and if it is infinite then $E^*/E = 1/3$. For the plates examined a slight degree of relative loss in stiffness compared to the isotropic plate may be observed. Although the analysis was not continued into the far post-buckling range, it is evident that for the plates considered this slight initial increase in E^*/E will offset the slightly higher relative stiffness at loads well beyond buckling to make the isotropic and orthotropic curves very close.

SIMPLY SUPPORTED FREE PLATE

For these plates it is known that neglect of shear stresses and transverse stresses have very little effect on post-buckling stiffness. This was shown for isotropic plates in Ref. 12 and the results confirmed for orthotropic plates in Ref. 3.

Assuming the deflected form is approximated to by

$$w = A \frac{y}{b} \sin \frac{\pi x}{a} \tag{31}$$

for the plate of Fig. 1(c), and analysing as for the previous cases, in this instance considering only the effects of σ_x gives the results

$$K = \left(\frac{b}{a}\right)^2 + \frac{12}{\pi^2} \frac{D_{33}}{D_{11}} \tag{32}$$

$$\frac{E^*}{E} = \frac{4}{9} \tag{33}$$

Therefore in this case the ratio of post-buckling stiffness to pre-buckling stiffness does not depend on the degree of orthotropy. The buckling coefficient reduces with increase in half-wavelength, tending to a value of $(12/\pi^2)(D_{33}/D_{11})$. The deflections for such plates are very much larger than for the other plates examined and it is known that the expression for post-buckling stiffness extends, with accuracy, further than the other plates. Therefore there is no need to carry out an analysis taking into account change in the buckled form as in general the expression given will be adequate until gross deflections become unacceptable.

It may also be mentioned that eqn. (33) gives a lower estimate for the initial post-buckling stiffness of a plate with one free edge and the other supported and fixed against rotation, either fully or partially. In such a case the buckling coefficient corresponding to the condition under examination should be used.

EFFECTS OF IMPERFECTIONS

The effects of imperfections on plate post-buckling behaviour are well known in the case of isotropic plates. It is known that the maximum adverse effects on plate strength and stiffness are obtained for a given imperfection magnitude if the form of the imperfections is the same as the locally buckled form of the plate.

Analysis of this type of imperfection using the methods of Refs. 13 or 14 can be carried out without untoward difficulty. Because of space restrictions such an analysis is not included in this paper. However the results of an investigation of this type indicate that while there is some degree of dependency of imperfection effects on the orthotropy of a plate this is not very substantial. To incorporate a generalised imperfection magnitude into any design analysis requires, for accuracy, a knowledge of the typical magnitudes encountered in practice. In the case of cold-formed steel plates, Walker[14] derived a suitable magnitude of maximum imperfection which applied to commercial products. Whether this value also applies to composite plates is a matter for conjecture.

SUITABLE EFFECTIVE WIDTH CURVES FOR DESIGN USE

The effective width curves obtained for various types of plates are shown in Fig. 8. Those for plates of the second type investigated have only been drawn for $u/u_{CR} < 6$ since they were computed on the basis of unchanging buckled form and wavelength and have reached their limit of reasonable accuracy when u/u_{CR} has this value.

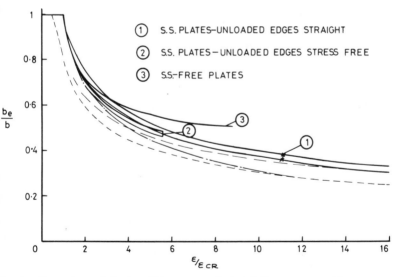

FIG. 8. Comparison of effective widths. — — — (Koiter) $b_e/b = 0.785 (\varepsilon_{CR}/\varepsilon)^{1/3}$; — · — · — (von Karman) $b_e/b = \sqrt{(\varepsilon_{CR}/\varepsilon)}$; — — — — (AISI) $b_e/b = \sqrt{(\varepsilon_{CR}/\varepsilon)}(1 - 0.22\sqrt{(\varepsilon_{CR}/\varepsilon)})$.

It may be observed that, as discussed previously, the differences between plates of different orthotropy are not substantial if the minimum effective width is plotted to a base of u/u_{CR} where u_{CR} is derived using orthotropic theory. On this basis it may be suggested that effective width curves derived for isotropic plates may also be applied within engineering accuracy to orthotropic plates.

Three well-known isotropic plate curves of long standing are also shown in Fig. 8. The curve due to Koiter was given as being accurate at strains far beyond buckling, and it is noticeable that this curve becomes almost coincident with that obtained for isotropic plates in this paper at high strains. The curve due to von Karman was possibly the first theoretically derived effective width curve. As the analysis involved completely neglected the effects of bending, etc. in the ineffective part of the plate this curve becomes somewhat conservative at high strains.

The AISI curve is an empirical modification of von Karman's curve, taking initial imperfections into account, and has been used in the USA and in a number of other countries for many years as a conservative estimate of effective width for use in the design of metal plates. The indications of the results of this paper are that the AISI effective width expression may also be used with confidence as a conservative estimate of the stiffness after buckling of orthotropic plates. There are other effective width expressions which may be employed with a similar degree of confidence, and these may well be worth investigation. However for the purposes of this paper the AISI approach provides the required simplicity and conservatism to be worthy of design use.

It should also be mentioned that Turvey and Wittrick[15] found that for particular types of laminated plates coupling behaviour, while affecting the local buckling stresses, has little effect on post-buckling behaviour. In view of this it may be that the results obtained here have a wider applicability than has been investigated in this paper.

CONCLUSIONS

The stiffness after buckling of orthotropic plates with three different boundary conditions commonly used in design has been investigated. It has been found that the stiffnesses are in most cases highly dependent on degree of orthotropy and buckle half-wavelength. However, if the post-buckling behaviour of infinite plates is considered, in which case the minimum load for any given strain is required in the post-buckling range, then the buckle half-wavelength which naturally occurs is such that post-buckling stiffness is virtually independent of the degree of orthotropy.

This is the condition which is of greatest importance to designers. In view of this it is suggested that the effective width formulae produced in the past for isotropic plates can be safely used to obtain suitable values, with engineering accuracy, for use with orthotropic plates. It is important, however, that the buckling coefficients used should be obtained from the relevant orthotropic plate formulae.

REFERENCES

1. LEKHNITSKII, S. G., *Anisotropic plates*, Gordon and Breach, 1968.
2. WITTRICK, W. H., Correlation between some stability problems for orthotropic and isotropic plates under bi-axial and uniaxial direct stress, *Aero Quarterly*, **IV** (August, 1952) 83–92.
3. BANKS, W. M., 'A contribution to the geometrical nonlinear behaviour of orthotropic plates', Ph.D. Thesis, University of Strathclyde, 1977.
4. HARRIS, G. Z., *Instability of laminated composite plates*, AGARD Conf Proc No. 112, Paper 14, 1973.
5. CHANDRA, R. and RAJU, B. B., Post buckling analysis of rectangular orthotropic plates, *Int. J. Mech. Sci.*, **15** (1973) 81–97.
6. MARSHALL, I. H., 'The non-linear behaviour of thin-initially curved orthotropic plates', Ph.D. Thesis, University of Strathclyde, 1976.
7. B.S. 449: Addendum No. 1 (1975). Specification for the use of cold-formed steel sections in building.
8. American Iron and Steel Institute: 'Specification for the design of cold-formed steel structural members', 1968 Edition.
9. BENTHEM, J. P., 'The reduction in stiffness of combinations of rectangular plates in compression after exceeding the buckling load', Nat Aero Research Inst., Amsterdam, NLL-TRS 539, 1959.
10. KOITER, W. T., *Introduction to the post-buckling behaviour of flat plates*, Colloquium on the Post-Buckling Behaviour of Plates used in Metal Structures, University of Liege, Belgium, 1963.
11. COX, H. L., 'Buckling of thin-plates in compression' Aero Research Council R & M No. 1554, 1934.
12. RHODES, J., 'The post-buckling behaviour of thin-walled beams', Ph.D. Thesis, University of Strathclyde, 1969.
13. RHODES, J. and HARVEY, J. M., Examination of plate post-buckling behaviour, *J. Eng. Mech. Div. Am. Soc. Civ. Engrs*, **103**, EM3 (June, 1977), 461–78.
14. WALKER, A. C., Post buckling behaviour of simply-supported square plates, *Aero. Quarterly*, **XX** (August 1969) 203–22.
15. TURVEY, G. J. and WITTRICK, W. H., The large deflection and post-buckling behaviour of some laminated plates, *Aero Quarterly*, **XXIV** (1973) 77–86.

23

Unsymmetrical Buckling of Laterally Loaded, Thin, Initially Imperfect, Orthotropic Plates

I. H. MARSHALL

Department of Mechanical and Production Engineering,
Paisley College of Technology, High Street, Paisley PA1 2BE, Scotland

ABSTRACT

A theoretical analysis is presented for the phenomenon of unsymmetrical or bifurcation buckling of thin, initially imperfect, orthotropic plates when loaded laterally on the convex face. In the cases analysed herein, unsymmetrical buckling is shown to be a function of the plate initial geometry and to greatly reduce the effective load-carrying capacity of such structural members when it occurs. For a range of plate geometries it is shown that design criteria based on symmetrical buckling considerations can be greatly in error.

NOTATION

a_{11} ω_{11}/h non-dimensional plate central deflection

$a_{0_{11}}$ $\omega_{0_{11}}/h$ non-dimensional plate initial symmetrical deflection component

$a_{0_{21}}$ $\omega_{0_{21}}/h$ non-dimensional plate initial unsymmetrical deflection component

a, b plate length/breadth

D_1 plate flexural rigidity in the major direction

D_2 plate flexural rigidity in the minor direction

D_3 plate twisting rigidity

E_x Young's modulus of elasticity for x direction

E_y Young's modulus of elasticity for y direction

F Airy stress function

G_{xy} modulus of rigidity

h plate thickness

$\bar{U}$ non-dimensional pressure loading— $\dfrac{q_z b^4}{E_y h^4}$

U_B bending energy

U_p load potential energy

U_S mid-plane energy

U_T total energy of the system

x, y Cartesian coordinates

α $\dfrac{E_y}{G} - 2v_y$

β $\dfrac{E_y}{E_x}$ stiffness ratio

λ plate aspect ratio $= a/b$

v_x Poisson's ratio with respect to the major direction

ω_{11} symmetrical plate deflection coefficient

ω_{21} unsymmetrical deflection coefficient

INTRODUCTION

When thin plates are loaded laterally, irrespective of the edge boundary conditions or the form of lateral loading, stable equilibrium conditions will prevail throughout the range of useful loading. If the maximum lateral deflection is limited to approximately half the plate thickness a linear relationship between load and lateral deflection can be reasonably assumed, i.e. 'small deflection theory'. At larger magnitudes of lateral deflection non-linear load/deflection tendencies become increasingly apparent. These are directly attributable to the increasing significance of membrane or mid-surface strains over the strains induced by flexural effects, i.e. 'large deflection theory' must be employed. Such structural members have been the subject of a vast number of previous theoretical and experimental studies from which has resulted a wealth of useful design information.

The presence of geometrical imperfections or 'imperfections in shape' in plates has also received considerably attention in the past.[1,2,3,4] However, by and large, previous investigations have been concerned with isotropic material characteristics with the equivalent problem for orthotropic or anisotropic material properties receiving comparatively little attention.

Certain general characteristics are apparent in imperfect plate analysis, irrespective of the particular material of construction. If lateral deflections, subsequent to applied loading, are of the same sense as the initially deflected shape, i.e. the magnitude of the initial mode shape increases, a loss in plate stiffness will generally result. In the case of in-plane compression the imperfection or 'lack of flatness' induces a stable equilibrium configuration thereby negating the possibility of buckling. A comprehensive study of the effects of such imperfections in isotropic plates is contained in references 5, 6 and 7.

Also, when such plates are loaded laterally on the concave surface a stable equilibrium state results. However, on loading on the convex face an unstable equilibrium state may result depending on the plate initial geometrical conditions, e.g. aspect ratio, thickness, form and magnitude of imperfection. This phenomenon of snap-buckling is well understood and has been extensively analysed for, amongst other structural members, circular arches,[8,9] imperfect plates[10,11] and initially curves plates.[12,13] These references are by no means complete and are cited purely as examples of past work. Again, relatively speaking, previous investigations have concentrated on isotropic material characteristics. Snap-buckling of initially curved orthotropic plates has been the subject of a number of theoretical and experimental investigations.[14,15,17]

Again previous studies on the phenomenon of unsymmetrical or bifurcation buckling have been largely confined to isotropic structural members and in particular arch-type structures. In a previous paper the parameters affecting the onset of unsymmetrical buckling of rectangular planform reinforced plastic shells were investigated by the author.[18] Consequently in view of the increasing significance of reinforced plastics as structural members, particularly as plate-type structural members, a study of the effects of geometrical imperfections would seem appropriate. Bearing in mind the inevitability of shape imperfections in a predominantly labour intensive fabricating operation, their effects are clearly of significance in the design of platework structures. Before going any further it may be advantageous to consider, in a general sense, the phenomenon of snap-buckling.

DESCRIPTION OF THE SNAP-BUCKLING EFFECT

The type of load–deflection relationship typical of curved plates, laterally loaded on the convex face, is shown in Fig. 1. In this figure it is assumed that the plate will deform symmetrically. The extent of the non-linearity of the above relationship depends on a number of factors, viz. type of lateral

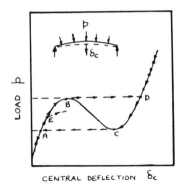

FIG. 1.   Typical load–deflection relationship.

loading, edge boundary conditions, plate initial curvature, plate thic!.ness, material properties, etc.

On incremental application of the load, the load–deflection curve (equilibrium path) will follow the stable portion of the curve from 0 to B. At point B (the critical point), any increase in load will cause the panel to snap-buckle to point D. If the load is further increased, the panel stiffness will increase as indicated.

On reduction of the load, the curve will again pass through point D but this time it will continue to point C. Further reduction of the load will cause a second snap from point C to point A.

Thus for any value of load in region $P_A \geq P \geq P_B$ there will be three possible values of equilibrium displacement. It should be noted that if unsymmetrical behaviour occurs during the 'snapping' process it will have the effect of causing bifurcation at (say) point E.

THEORETICAL ANALYSIS

Governing Equations

The von Karman-type large deflection compatibility equation for the case of an orthotropic plate with initial imperfections can be written as:

$$\frac{\partial^4 F}{\partial x^4} + \alpha \frac{\partial^4 F}{\partial x^2 \partial y^2} + \beta \frac{\partial^4 F}{\partial y^4}$$
$$= E_y \left[\left(\frac{\partial^2 \omega}{\partial x \partial y} \right)^2 - \frac{\partial^2 \omega}{\partial x^2} \cdot \frac{\partial^2 \omega}{\partial y^2} + 2 \frac{\partial^2 \omega_0}{\partial x \partial y} \cdot \frac{\partial^2 \omega}{\partial x \partial y} \right.$$
$$\left. + \frac{\partial^2 \omega_0}{\partial x^2} \cdot \frac{\partial^2 \omega}{\partial y^2} + \frac{\partial^2 \omega_0}{\partial y^2} \cdot \frac{\partial^2 \omega}{\partial x^2} \right] \quad (1)$$

An equation similar to eqn (1) can be formed from equilibrium considerations. However, as an alternative to solving the equilibrium equation, the total energy of the system will be minimised according to the Ritz technique. The total energy of the system comprises the energy of mid-plane straining U_S, the bending energy U_B and the load potential energy U_p.

Thus:

$$U_T = U_S + U_B + U_p \tag{2}$$

where

$$U_S = \frac{h}{2} \int_0^a \int_0^b \left[\frac{1}{E_x} \left(\frac{\partial^2 F}{\partial y^2} \right)^2 - 2 \frac{v_x}{E_x} \left(\frac{\partial^2 F}{\partial x^2} \right) \left(\frac{\partial^2 F}{\partial y^2} \right) \right.$$
$$\left. + \frac{1}{G} \left(\frac{\partial^2 F}{\partial x\, \partial y} \right)^2 + \frac{1}{E_y} \left(\frac{\partial^2 F}{\partial x^2} \right)^2 \right] dx\, dy \tag{3}$$

$$U_B = \frac{1}{2} \int_0^a \int_0^b \left[D_1 \left(\frac{\partial^2 \omega}{\partial x^2} \right)^2 + 2D_3 \left(\frac{\partial^2 \omega}{\partial x\, \partial y} \right) + D_2 \left(\frac{\partial^2 \omega}{\partial y^2} \right)^2 \right] dx\, dy \tag{4}$$

$$U_p = - \int_d^e \int_f^g \omega \cdot q_z\, dx\, dy \tag{5}$$

where d, e, f and g are the limits of integration and q_z is the transverse load on the plate per unit area.

The plate lateral deflection and membrane stresses can be stated as general Fourier series of the form:

$$\omega(x, y) = \sum_m^\infty \sum_n^\infty \omega_{mn} m(x) n(y) \tag{6}$$

$$F(x, y) = \sum_j^\infty \sum_k^\infty F_{jk} j(x) k(y) \tag{7}$$

The plate edge boundary conditions will be considered as those consistent with 'simple supports'. With reference to Fig. 2 these can be written as:

	along $x = 0, a$	along $y = 0, b$

Flexural bcs

$$\omega = 0 \qquad\qquad \omega = 0$$

$$\frac{\partial^2 \omega}{\partial x^2} + v_y \frac{\partial^2 \omega}{\partial y^2} = 0 \qquad\qquad \frac{\partial^2 \omega}{\partial y^2} + v_x \frac{\partial^2 \omega}{\partial x^2} = 0 \tag{8}$$

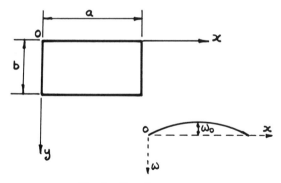

FIG. 2. Coordinate axes.

Membrane bcs

$$\frac{\partial^2 F}{\partial y^2} = 0 \qquad\qquad \frac{\partial^2 F}{\partial x^2} = 0$$

$$\frac{\partial^2 F}{\partial x\,\partial y} = 0 \qquad\qquad \frac{\partial^2 F}{\partial x\,\partial y} = 0 \tag{9}$$

It will be noted that the boundary conditions of stress-free plate edges as stated in eqn (9) will allow tangential edge displacements to occur.

Equation (8) can be satisfied by stating eqn (6) in the form:

$$\omega = \sum_{m}^{\infty}\sum_{n}^{\infty} \omega_{mn} \sin\frac{m\pi x}{a} \sin\frac{n\pi y}{b} \qquad m,n = 1,2,3 \tag{10}$$

Also, eqn (9) can be satisfied by stating eqn (7) as:

$$F = \sum_{j}^{\infty}\sum_{k}^{\infty} F_{jk} \left[\cos\frac{j\pi x}{a} - 1\right]\left[\cos\frac{k\pi y}{b} - 1\right] \qquad j,k = 1,2,3 \tag{11}$$

The plate initial deflections can be similarly considered as Fourier series of the type:

$$\omega_0 = \sum_{p}^{\infty}\sum_{q}^{\infty} \omega_{pq} \sin\frac{p\pi x}{a} \sin\frac{q\pi y}{b} \qquad p,q = 1,2,3 \tag{12}$$

It will be noted that eqn (12) allows symmetrical and unsymmetrical forms of initial plate deflections to be considered.

I. H. Marshall

In previous work[14] it has been shown that the use of relatively few terms in the aforementioned Fourier series yields acceptable accuracy in the final analysis. Thus, since the present analysis is not really concerned with loads appreciably greater than the snap-buckling or critical load the following series will be deemed suitable.

$$\omega = \sum_{m=1}^{2} \sum_{n=1}^{1} \omega_{mn} \sin \frac{m\pi x}{a} \sin \frac{n\pi y}{b} \tag{13}$$

$$\omega_0 = \sum_{p=1}^{2} \sum_{q=1}^{1} \omega_{0_{pq}} \sin \frac{p\pi x}{a} \sin \frac{q\pi y}{b} \tag{14}$$

$$F = \sum_{j=1}^{1} \sum_{k=1}^{1} F_{jk} \left[\cos \frac{j\pi x}{a} - 1 \right] \left[\cos \frac{k\pi y}{b} - 1 \right] \tag{15}$$

For this case, the boundary eqns (8) and (9) cannot be satisfied by any finite solution of eqn (1). Hence a solution to the compatibility eqn (1) will be sought using the Galerkin integration technique. A full explanation of this method is contained in reference 19.

The Galerkin method requires satisfaction of the following series of equations.

$$\int_0^a \int_0^b \left[\frac{\partial^4 F}{\partial x^4} + \alpha \frac{\partial^4 F}{\partial x^2 \partial y^2} + \beta \frac{\partial^4 F}{\partial y^4} \right] l(x) t(y) \, dx \, dy$$

$$= \int_0^a \int_0^b \left[\left(\frac{\partial^2 \omega}{\partial x \, \partial y} \right)^2 - \frac{\partial^2 \omega}{\partial x^2} \frac{\partial^2 \omega}{\partial y^2} - 2 \frac{\partial^2 \omega_0}{\partial x \, \partial y} \cdot \frac{\partial^2 \omega}{\partial x \, \partial y} \right.$$

$$\left. + \frac{\partial^2 \omega_0}{\partial x^2} \cdot \frac{\partial^2 \omega}{\partial y_2} + \frac{\partial^2 \omega_0}{\partial y^2} \cdot \frac{\partial^2 \omega}{\partial x^2} \right] l(x) t(y) \, dx \, dy \tag{16}$$

where $l(x) t(y)$ is the differential of the stress function with respect to each chosen coefficient in turn. Solution of eqn (16) yields a relationship between stress function coefficients F_{jk} and deflection function coefficients ω_{mn} and $\omega_{0_{pq}}$. Thus, by substituting eqns (13), (14) and (15) into energy eqns (3), (4) and (5) and noting the results of eqn (16) the total energy of the system can be written in terms of deflection function coefficients only. For the case of uniform pressure loading this can be written as:

$$
U_T = \frac{\pi^4 \lambda E_y h}{8(3 + \alpha\lambda^2 + 3\beta\lambda^4)_0^2} \left[\frac{\omega_{11}^4}{4} + \omega_{11}^2 \omega_{21}^2 - \omega_{11}^3 \omega_{0_{11}} \right.
$$

$$
- 2\omega_{11}^2 \omega_{21} \omega_{0_{21}} + \omega_{21}^4 - 2\omega_{11} \omega_{0_{11}} \omega_{21}^2 - 4\omega_{21}^3 \omega_{0_{21}}
$$

$$
\left. + \omega_{11}^2 \omega_{0_{11}}^2 + 4\omega_{11} \omega_{0_{11}} \omega_{21} \omega_{0_{21}} + 4\omega_{21}^2 \omega_{0_{21}}^2 \right]
$$

$$
+ \frac{\pi^4 E_y h^3}{96\lambda^3(1 - v_x v_y)b^2} \left[\omega_{11}^2 \left(\frac{1}{\beta} + \frac{2D_3}{D_2}\lambda^2 + \lambda^4 \right) \right]
$$

$$
+ \frac{\pi^4 E_y h^3}{12\lambda^3(1 - v_x v_y)b^2} \left[\omega_{21}^2 \left(\frac{2}{\beta} + \frac{D_3}{D_2}\lambda^2 + \frac{\lambda^4}{8} \right) \right] - 4\omega_{11} \frac{ab}{\pi^2} q_z
$$

$$
\tag{17}
$$

Minimising the resulting expression for the total energy (U_T) of the system with respect to each of the chosen deflection function coefficients yields a set of simultaneous non-linear algebraic equations equal in number to the chosen number of terms in the deflection function series, i.e. in the present analysis two such equations are formed. A solution to these equations was found by utilising iterative techniques involving successive approximations. In the final analysis all parameters were non-dimensionalised to preserve generality.

DISCUSSION OF RESULTS

Symmetrical Initial Deformations ($a_{0_{11}} \neq 0$, $a_{0_{21}} = 0$)

In order to illustrate the effects of increasing magnitudes of symmetrical modes of initial deflection on the stability characteristics of rectangular plates a typical aspect ratio, $\lambda = 1.5$, was considered (Fig. 3). From this figure it is apparent that stable equilibrium conditions will prevail at values of $a_{0_{11}} \leq 3$, i.e. a continuous load/displacement relationship is predicted.

As the imperfection magnitude increases, $a_{0_{11}} \geq 4$, snap-buckling conditions become increasingly possible, i.e. dynamic changes of shape are theoretically predicted. Also at $a_{0_{11}} \geq 6$ the plate would retain its buckled

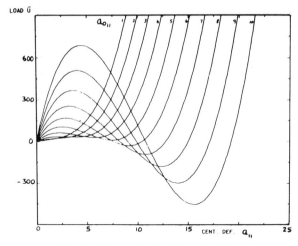

FIG. 3.   Symmetrical buckling ($\lambda = 1 \cdot 5$).

form after unloading, i.e. permanent deformation would result and could only be removed by lateral loading in the opposite direction. Although the foregoing results are only applicable to aspect ratio $\lambda = 1 \cdot 5$, certain general characteristics are apparent. Conceivably plates of this nature, subjected to lateral loading on the convex face, could be designed using the present analysis with appropriate factors of safety employed to negate the possibility of unstable behaviour. However, an initially improbable phenomenon, i.e. a symmetrical structure, loaded symmetrically, buckling unsymmetrically, can greatly undermine any such design criteria. Figure 4 shows the same plate with bifurcation of the equilibrium path, corresponding to unsymmetrical buckling, shown by the superpositioned linear relationships. Clearly as the magnitude of symmetrical initial deformation increases the possibility of unsymmetrical bifurcation similarly increases. Consequently if unacceptably high plate deflections are to be avoided any design criteria must take cognizance of the bifurcation load.

Unsymmetrical Initial Deformations ($a_{0_{11}} \neq 0$, $a_{0_{21}} \neq 0$)

The presence of an unsymmetrical component of initial deformation can substantially alter the results found in the previous section. Consider again plate aspect ratio $\lambda = 1 \cdot 5$ and a typical symmetrical initial deflection

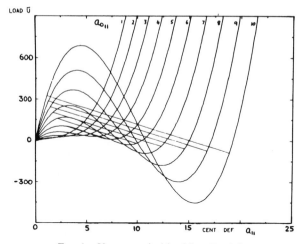

FIG. 4.   Unsymmetrical buckling ($\lambda = 1\cdot5$).

$a_{0_{11}} = 6$. Figure 5 illustrates the effect of superpositioning an unsymmetrical mode of initial deflection $a_{0_{21}}$ on this plate. In the case of $a_{0_{21}} = 0\cdot5$, i.e. approximately 8 % of the symmetrical coefficient, little change in the stability of characteristics is predicted. However, at $a_{0_{21}} \geq 3$ (50 % of the symmetrical coefficient) a stable equilibrium path results where previously snap-buckling characteristics were predicted. In the case of higher aspect ratio plates, e.g. $\lambda = 2\cdot0$, essentially similar characteristics prevail (Fig. 6).

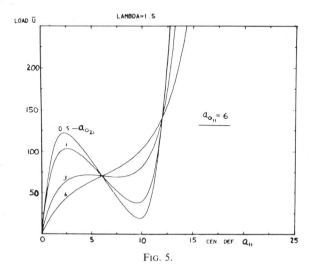

FIG. 5.

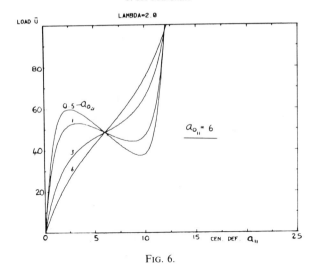

FIG. 6.

The results shown in Figs 5 and 6 are entirely consistent with earlier theoretical and experimental work in which initially curved orthotropic plates were considered.[18]

PLATEWORK DESIGN IMPLICATIONS

The inevitability of geometrical imperfections, i.e. deviations from flatness, in manually laminated platework structures is undeniable. Consequently, a technique which allows such imperfections to be accounted for in the design

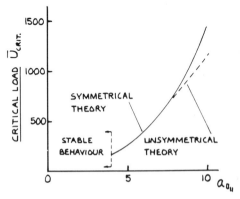

FIG. 7. Symmetrical Initial Deflection ($\lambda = 1 \cdot 0$).

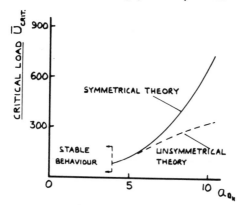

FIG. 8. Symmetrical Initial Deflection ($\lambda = 1\cdot5$).

or post-design of such structural members has considerable practical importance. Although the present analysis is capable of analysing such members subjected to lateral loading on the concave face the dynamic changes of shape associated with similar loading on the convex face clearly has more significance.

The parameters governing the onset of symmetrical or unsymmetrical buckling are outlined herein. However, in order to illustrate the importance of unsymmetrical buckling as a design criteria consider Figs 7, 8 and 9. In each case the critical load is defined as the load necessary to initiate unacceptably large lateral deflections. From these figures it is apparent that, in the case of a symmetrical initial imperfection, snap-buckling as a

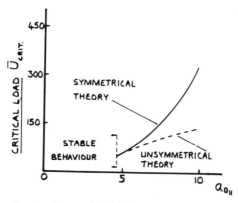

FIG. 9. Symmetrical Initial Deflection ($\lambda = 2\cdot0$).

I. H. Marshall

TABLE 1

Imperfection magnitude $a_{0_{11}}$	Plate aspect ratio (λ)		
	1·0	1·5	2·0
0	Stable behaviour		
1			
2			
3			
4			
5	Symmetrical snap-buckling		
6			
7			
8	Unsymmetrical buckling		
9			
10			

design criteria has limited usefulness particularly as the plate aspect ratio increases. The results of these figures can be usefully summarised in Table 1.

CONCLUSIONS

The present theoretical analysis is capable of analysing thin, orthotropic plates with symmetrical or unsymmetrical forms of initial imperfection when loaded laterally. The relative magnitude and form of imperfection has been shown to govern the stability characteristics of such structural members. In particular the presence of an unsymmetrical component in the plate initially deflected form has a marked effect on its load/deflection characteristics. Since a wide spectrum of initially deflected shapes can be reasonably modelled using a Fourier series containing one symmetrical and one unsymmetrical term the versatility of the present analysis is self evident.

REFERENCES

1. LEVY, S., Bending of rectangular plates with large deflections, NACA Tech. note No. 846, 1942.
2. COAN, J. M., Large-deflection theory for plates with small initial curvature loaded in edge compression, *Trans. ASME*, **73**, 1951, 143–51.
3. HU, P. C. *et al.*, Effect of small deviations from flatness on effective width and buckling of plates in compression, NACA Tech. note No. 1124, 1951.

 4. DAWSON, R. G. and WALKER, A. C., Post-buckling behavior of geometrically imperfect plates, *Proc. ASCE*, Jan. 1972, 75–94.
 5. YAMAKI, N., Post-buckling behavior of rectangular plates with small initial curvature loaded in edge compression, *J. Appl. Mech.*, **26**, 1959, 407–14.
 6. YAMAKI, N., ibid. (continued) *J. Appl. Mech.*, 1960, 335–42.
 7. YAMAKI, N., Experiments in the post-buckling behavior of square plates loaded in edge compression, *J. Appl. Mech.*, 1961, 238–44.
 8. FUNG, Y. C. and KAPLAN, A., Buckling of low arches or beams of small curvature, NACA Tech. note No. 2840, Nov. 1952.
 9. KERR, A. D. and EL-BAYOUMY, L., On the nonunique equilibrium states of a shallow arch subjected to a uniform lateral load, *Quarterly of Applied Mathematics*, Oct. 1970, 399–409.
10. RUSHTON, K. R., Large deflection of plates with initial curvature, *Int. J. Mech. Sci.*, **12**, 1970, 1037–51.
11. RUSHTON, K. R., Buckling of laterally loaded plates having initial curvature, *Int. J. Mech. Sci.*, **14**, 1972, 667–80.
12. MARSHALL, I. H. and RHODES, J., Snap-buckling of thin shells of rectangular planform, *Proc. Stability Problem in Eng. Structures and Component Conf.*, Cardiff, Sept. 1978, pp. 249–64.
13. RHODES, J., TOOTH, A. S. and MARSHALL, I. H., Snap-out buckling of imperfections in cylindrical panels used in storage vessels, ibid., pp. 199–216.
14. MARSHALL, I. H., RHODES, J. and BANKS, W. M., The nonlinear behaviour of thin, orthotropic, curved panels under lateral loading, *J. Mech. Eng. Sci.*, **19** No. 1, Feb. 1977, 30–7.
15. MARSHALL, I. H., RHODES, J. and BANKS, W. M., Experimental snap-buckling behaviour of thin GRP curves panels under lateral loading, *Composites*, April 1977, 81–6.
16. BURMISTROV, E. F., Calculation of flat orthotropic shells with allowance for final strains, *Inzh. Sb.*, **22**, 1955 (in Russian).
17. MARSHALL, I. H., RHODES, J. and BANKS, W. M., General investigation of snap-through buckling of shallow orthotropic shells, *Acta Technica Academiae Scientiarum Hungaricae*, **87**, 1978, 69–86.
18. RHODES, J. and MARSHALL, I. H., Unsymmetrical buckling of laterally loaded reinforced plastic shells, *Proceedings I.CC.M/2*, Toronto, 1978, 303–15.
19. DUNCAN, W. J., The principles of the Galerkin method, *ARC Tech. report*, 1938.

24

The Effect of Mode Interaction in Orthotropic Fibre Reinforced Composite Plain Channel Section Columns

A. R. UPADHYA

Structures Division, National Aeronautical Laboratory,
Bangalore-560017, India

AND

J. LOUGHLAN

Cranfield Institute of Technology, College of Aeronautics,
Department of Aircraft Design, Cranfield, Bedford MK43 0AL, England

ABSTRACT

A theoretical investigation into the interactive buckling behaviour of thin-walled orthotropic plain channel section columns is presented. A semi-energy method of analysis is used and the effects of local buckling in all elements of the section are taken into consideration. Changes in the locally buckled form after buckling are taken into account in an approximate manner and the effects of local and overall imperfections are also considered.

In the paper particular attention is paid to the analysis of columns with coincident local and Euler buckling loads, since these columns exhibit unstable post-buckling equilibrium behaviour and due to this are highly sensitive to geometrical imperfections.

Results are presented for orthotropic glass fibre reinforced composite columns in the form of non-dimensional graphs. The graphs depict such aspects as flexural stiffness variation at buckling with change in cross-sectional geometry, post-buckling equilibrium behaviour for concentrically and eccentrically loaded columns and the effect of imperfections on the maximum carrying capacity of coincident mode designs.

NOTATION

A, A_0	magnitude coefficients for local deflection and local imperfection, respectively
A_{11}, A_{12}, A_{22}, A_{66}	elements of in-plane stiffness matrix $[A]$ for orthotropic plate
a_{11}, a_{12}, a_{22}, a_{66}	elements of matrix $[a] = [A]^{-1}$
a_{33}	$(2a_{12} + a_{66})$
D_{11}, D_{12}, D_{22}, D_{66}	elements of bending stiffness matrix $[D]$ for orthotropic plate
D_{33}	$(D_{12} + 2D_{66})$
$\bar{e}$	non-dimensional load eccentricity $= e/d$
N_x, N_y, N_{xy}	in-plane stress resultants
P_M	maximum load
R, R_0	radius of curvature of overall deflected form, and overall imperfection, respectively
$\bar{u}$	web compression
w, w_0	local deflection and local imperfection, respectively
w_c, w_{0c}	amplitude of w and w_0 respectively at web mid-point
x	co-ordinate along length of column
α	compression eccentricity
$\bar{\varepsilon}$	web compressive strain $= 2\bar{u}/\lambda$
δ, δ_0	overall deflection and imperfection respectively
δ_c, δ_{0c}	amplitude of δ and δ_0 respectively at column centre
i	suffix relating to particular plate element of section ($i = 1$ for web, $= 2, 3$ for flanges)
CR	suffix relating to local buckling condition

INTRODUCTION

Thin-walled columns under compression can exhibit local buckling of the plate elements comprising the cross-section or flexural buckling of the whole column. Although these two types of buckling are in themselves essentially stable, coupling of the modes can result in either stable or unstable post-buckling behaviour.

If a column is designed such that the critical loads of the individual modes are coincident, or relatively close, non-linear coupling of the modes can result in failure at buckling due to the inherent unstable state of the equilibrium of such columns when buckling occurs. Further, it has been observed during experiment that the nature of the failure or collapse of near simultaneous mode designs is explosive, resulting in a sudden snap or bang of the structure followed by severe destruction. These designs are very imperfection-sensitive and their maximum loads are very quickly eroded by the effects of both local and overall imperfections.

The interaction behaviour of thin-walled isotropic structural elements in compression has received a great deal of attention in past years by many researchers. Neut[1] and Thompson and Lewis[2] studied the phenomena of interactive buckling using a simplified two flange column model. Fok *et al.*[3,4] examined the behaviour of integrally stiffened panels with plain flat outstands. Rhodes and Harvey[5] considered the interaction characteristics of plain channel section columns and the behaviour of lipped channel columns was analysed by Loughlan[6] and Loughlan and Rhodes.[7,8]

With the increasing use of composite materials in structural design, especially in the aircraft and aerospace industries, a study of the post-buckling and interactive buckling behaviour of structural elements made of fibre reinforced composites becomes relevant.

The initial compressive buckling stresses and associated buckling modes for a fairly wide range of structural elements manufactured from orthotropic carbon fibre reinforced plastic (CFRP) have been obtained by Turvey and Wittrick.[9] Some local instability results for thin-walled orthotropic sections have also been obtained by Lee[10] and Lee and Hewson.[11] The post-buckling behaviour of composite box sections is being considered at this conference by Banks and Rhodes.[12] To the authors' knowledge, however, little or no information exists in the literature on the post-buckling interactive behaviour of thin-walled composite structural elements and the aim of this paper is to make a contribution in this respect.

In this paper results are presented which describe the post-local-buckling interactive behaviour of pin-ended plain channel section columns manufactured from glass reinforced plastic (GRP). In the analysis it is assumed that the material of the column is orthotropic with its maximum stiffness along the length of the column. For the case of locally perfect columns, the effect of allowing the locally buckled form to change in the post-buckling range, is considered in an approximate manner.

LOCAL BUCKLING

Consider the plane channel column shown in Fig. 1, loaded eccentrically towards the flange free edges by an amount, *e*, and which has just developed

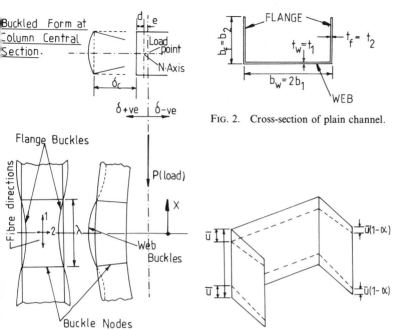

FIG. 2. Cross-section of plain channel.

FIG. 1. Central portion of column showing local buckles.

FIG. 3. Compression system at nodes of central buckle.

local buckles along its length. It is assumed, as is shown in Fig. 1, that a local buckle at the centre of the column has a sinusoidal form in the longitudinal direction with a half-wavelength equal to λ. The column length is L and its geometrical cross-sectional parameters are as indicated in Fig. 2. The ends of the column are pinned and it is assumed that the column contains an overall geometrical imperfection described by the expression

$$\delta_0 = \delta_{0c} \cos \frac{\pi x}{L} \qquad (1)$$

The analysis considers the central buckle as being a representative section of the column and treats this as a short strut of length λ which has its ends displaced according to the displacement system shown in Fig. 3. It will be noted that for positive column deflection δ as indicated in Fig. 1, the factor α in the compression system will be negative. The analysis of the local buckling and post-buckling behaviour of a short orthotropic plain channel section strut is given in detail by Upadhya[13] and only a brief outline of this work is given in the paper. The Rayleigh–Ritz method of analysis is used to

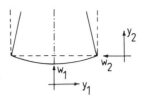

FIG. 4.   Co-ordinate system for local deflections and buckled form.

determine initial local buckling and in this the local deflections w_i of each
plate element of the section are assumed to be of the form

$$w_i = Y_i(y_i)\cos\frac{\pi x}{\lambda} \qquad (2$$

where

$$Y_i(y_i) = \sum_{n=1}^{N} B_n Y_{in}(y_i) \qquad (3$$

The locally deflected form $Y_i(y_i)$ is shown in Fig. 4 as well as the co-ordinate
system used for the local deflections. The functions $Y_{in}(y_i)$ are postulated in
terms of algebraic polynomials, each polynomial set satisfying the
compatibility and equilibrium conditions at the flange–web junction and a
the flange free edges. The total potential energy change at buckling is then
minimised with respect to the deflection coefficients B_n and using four term
in the solution the required critical web strain $\bar{\varepsilon}_{CR}$ to cause buckling and the
corresponding locally deflected form at buckling are obtained with very
good accuracy.

   This analysis of the column is carried out for different values of λ or in
other words, for different numbers of buckles along the column length, to
obtain that value of λ which induces the lowest buckling strain $\bar{\varepsilon}_{CR}$. In the
column solution an iterative procedure is used to obtain the values of α,
and hence λ at buckling such that the corresponding internal stress system
provides overall axial and moment equilibrium for the column.

POST-LOCAL-BUCKLING

In the post-local-buckling analysis the local buckling deflections are
approximated by the following expression

$$w_i = AY_i(y_i)\cos\frac{\pi x}{\lambda} \qquad (4$$

where the deflected form $Y_i(y_i)$ is assumed to be the same as that obtained a
local buckling as described by eqn. (3).

   After local buckling the changes of the stress system within the section are determined by satisfying von Karman's compatibility equation for a specially orthotropic plate, given by

$$(a_{22})_i \frac{\partial^4 F_i}{\partial x^4} + (a_{33})_i \frac{\partial^4 F_i}{\partial x^2 \partial y_i^2} + (a_{11})_i \frac{\partial^4 F_i}{\partial y_i^4} = \left(\frac{\partial^2 w_i}{\partial x \partial y_i}\right)^2 - \frac{\partial^2 w_i}{\partial x^2} \cdot \frac{\partial^2 w_i}{\partial y_i^2} \quad (5)$$

where F_i is a middle surface force function for the ith plate such that

$$(N_x)_i = \frac{\partial^2 F_i}{\partial y_i^2} \qquad (N_y)_i = \frac{\partial^2 F_i}{\partial x^2} \qquad (N_{xy})_i = -\frac{\partial^2 F_i}{\partial x \partial y_i} \quad (6)$$

Substituting for w_i from eqn. (4) into eqn. (5) and satisfying the relevant in-plane boundary conditions at the plate junctions and the flange free edges along with the conditions of in-plane movement at the nodes of the central buckle (i.e. at the ends of the short strut) yields the middle surface stresses in terms of $\bar{u}$, α and the magnitude coefficient A of the locally deflected form.

The principle of minimum strain energy is now used to obtain the final relationships between end displacements, out-of-plane deflections and middle surface forces. The strain energy stored in the short strut due to in-plane forces and plate bending is expressed in terms of the functions F_i and w_i and minimised with respect to the deflection coefficient A to obtain A as follows:

$$A^2 = \frac{1}{2C_3} [\bar{\varepsilon} C_1 - (\bar{\varepsilon}\alpha) C_2 - C_4] \quad (7)$$

The coefficients C_1 to C_4 in eqn. (7) are given by the following expressions:

$$C_1 = \frac{\pi^2}{4\lambda^2} \sum_{i=1}^{3} \frac{1}{(a_{11})_i} \int_0^{b_i} Y_i^2 \, dy_i$$

$$C_2 = \frac{\pi^2}{4\lambda^2} \sum_{i=2}^{3} \frac{1}{(a_{11})_i} \int_0^{b_i} Y_i^2 \frac{y_i}{b_i} \, dy_i$$

$$C_3 = \frac{\pi^4}{32\lambda^4} \sum_{i=1}^{3} \frac{1}{(a_{11})_i b_i^4} \int_0^{b_i} \left[b_i^4 Y_i^4 + 2(\phi_i'')^2 \right.$$
$$\left. + 2\frac{(a_{22})_i}{(a_{11})_i}\left(\frac{2\pi}{\lambda}\right)^4 \phi_i^2 + 2\frac{(a_{33})_i}{(a_{11})_i}\left(\frac{2\pi}{\lambda}\right)^2 (\phi_i')^2 \right] dy_i$$

$$C_4 = \frac{1}{4} \sum_{i=1}^{3} (D_{11})_i \left\{ \int_0^{b_i} \left[\left(\frac{\pi}{\lambda}\right)^4 Y_i^2 + 2\left(\frac{\pi}{\lambda}\right)^2 \frac{(D_{33})_i}{(D_{11})_i}(Y_i')^2 \right. \right.$$
$$\left. \left. + \frac{(D_{22})_i}{(D_{11})_i}(Y_i'')^2 \right] dy_i - 2\frac{(D_{12})_i}{(D_{11})_i}\left(\frac{\pi}{\lambda}\right)^2 [Y_i Y_i']_0^{b_i} \right\} \quad (8)$$

The function $\phi_i(y_i)$ in the expression for C_3 is obtained from the solution of eqn. (5). The complementary function part of this solution contains four unknown constants for each plate element of the section and these are determined from the in-plane boundary conditions prevailing at the edges of each plate.

Integration of the longitudinal membrane forces around the cross-section and of the moments of these forces about an axis through the web mid-surface gives the axial load P on the section and the moment M about the web mid-surface in terms of $\bar{\varepsilon}$, α and A as follows:

$$P = \bar{\varepsilon}X_c + b_2 \frac{(\bar{\varepsilon}\alpha)}{(a_{11})_2} - A^2 C_1 \tag{9}$$

$$M = \frac{b_2^2}{(a_{11})_2}\bar{\varepsilon} - \frac{2}{3}\frac{b_2^2}{(a_{11})_2}(\bar{\varepsilon}\alpha) - A^2 b_2 C_2 \tag{10}$$

where X_c is given by

$$X_c = \frac{2b_1}{(a_{11})_1} + \frac{2b_2}{(a_{11})_2} \tag{11}$$

The average axial stress distribution around the section after the occurrence of local buckling is shown in Fig. 5. The loss in load in the web and flanges due to local buckling, indicated in the figure by the departure from linearity of the stress system, is accounted for in eqn. (9) by the term $A^2 C_1$.

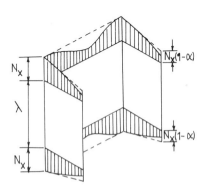

Fig. 5.   Average stress system after local buckling.

POST-LOCAL-BUCKLING INTERACTION

For a locally unbuckled orthotropic column the relationship between P and M can be obtained by setting $A = 0$ in eqns (9) and (10), evaluating $\bar{\varepsilon}$ in terms of P and $\bar{\varepsilon}\alpha$ from eqn. (9), and substituting in eqn. (10) to obtain

$$M = \frac{Pb_2^2}{(a_{11})_2 X_c} - \frac{\bar{\varepsilon}\alpha}{b_2}\left\{\frac{b_2^3}{(a_{11})_2}\left(\frac{2}{3} - \frac{b_2}{(a_{11})_2 X_c}\right)\right\} \tag{12}$$

It can easily be shown that the coefficient of P in eqn. (12) defines the distance d of the section neutral axis from the web and that the coefficient of $\bar{\varepsilon}\alpha/b_2$ is EI, the section stiffness against bending about its neutral axis. Also, from geometrical considerations, assuming that plane sections remain plane after bending, the change of curvature of an element in the longitudinal direction due to bending is given by

$$\frac{1}{R} - \frac{1}{R_0} = \frac{d^2(\delta - \delta_0)}{dx^2} = \frac{\bar{\varepsilon}\alpha}{b_2} \tag{13}$$

For a pin-ended column, the condition of equilibrium of internal and external moments is given by

$$M = P(d + e + \delta) \tag{14}$$

Substituting eqn. (12) in eqn. (14) and then utilising eqn. (13) results in the expression

$$EI\frac{d^2(\delta - \delta_0)}{dx^2} + P\delta = -Pe \tag{15}$$

This is the differential equation for the lateral equilibrium of an eccentrically loaded column with an initial overall geometrical imperfection.

In the post-local-buckling range, the validity of the geometrical relationship, eqn. (13), is only true at the buckle nodes. However, by assuming that the relationships evolved so far apply to an infinitesimally small length of the column, it is possible to express the equilibrium behaviour of a locally buckled column by means of an expression similar in form to eqn. (15).

Substituting for A from eqn. (7) into eqns (9) and (10) and manipulating these as for the nonlocally buckled column yields the following expression:

$$M = Pd^* - \frac{\bar{\varepsilon}\alpha}{b_2}(EI)^* + P_{\text{CRUC}}(d - d^*) \tag{16}$$

In this equation d^* defines the effective position of the neutral axis of the section in the post-buckling range and EI^* is a reduced value of the bending stiffness of the section which takes local buckling into account. The final term in eqn. (16) is an internal moment caused by the effects of local buckling and in this, the parameter P_{CRUC} is the critical load to cause local buckling of a uniformly compressed column according to the current locally deflected form.

Expressions for d^*, $(EI)^*$ and P_{CRUC} for an orthotropic column are as follows:

$$d^* = \frac{b_2 \left\{ \frac{b_2}{(a_{11})_2} - \frac{C_1 C_2}{2C_3} \right\}}{\left(X_c - \frac{C_1^2}{2C_3} \right)} \tag{17}$$

$$(EI)^* = \frac{\left\{ \frac{b_2^3}{(a_{11})_2} \left[\frac{2}{3} - \frac{1}{X_c} \left[\frac{b_2}{(a_{11})_2} + \frac{1}{3} \frac{C_1^2}{C_3} - \frac{C_1 C_2}{C_3} \right] \right] - \frac{b_2^2 C_2^2}{2C_3} \right\}}{\left(1 - \frac{C_1^2}{2C_3 X_c} \right)} \tag{18}$$

$$P_{CRUC} = \frac{C_4}{C_1} X_c \tag{19}$$

Substitution of $\bar{\varepsilon}\alpha/b_2$ and M from eqns (13) and (14) respectively in eqn (16) gives the differential equation for the lateral equilibrium of a locally buckled column as follows:

$$(EI)^* \frac{d^2(\delta - \delta_0)}{dx^2} + P\delta = -P[e + (d - d^*)] + P_{CRUC}(d - d^*) \tag{20}$$

With the initial overall geometrical imperfection as specified by eqn. (1), the column central deflection δ_c is obtained from the solution of eqn. (20) in the form

$$\delta_c = \left[e + (d - d^*)\left(1 - \frac{P_{CRUC}}{P} \right) \right] \left[\sec \frac{\pi}{2} \left(\frac{P}{P_E^*} \right)^{1/2} - 1 \right] + \frac{\delta_{0c}}{\left(1 - \frac{P}{P_E^*} \right)} \tag{21}$$

where P_E^* is the reduced Euler load for the column given by

$$P_E^* = \frac{\pi^2 (EI)^*}{L^2} \tag{22}$$

In the pre-local-buckling range, the corresponding central deflection is given by the solution of eqn. (15) as

$$\delta_c = \left[e \sec \frac{\pi}{2} \left(\frac{P}{P_E} \right)^{1/2} - 1 \right] + \frac{\delta_{0c}}{\left(1 - \dfrac{P}{P_E} \right)} \tag{23}$$

where P_E is the true Euler load of the column. Equilibrium of the internal stress system, caused by a given compression system represented by $\bar{u}$ and α is obtained by equating the internal moment due to these stresses to the external moment caused by the applied load P on the column. This condition is written as

$$M - P[e + d + \delta_c] = 0 \tag{24}$$

Using eqns (7), (9), (10) and (21), equilibrium eqn. (24) can be expressed as a function of $\bar{\varepsilon}$ and α in the form

$$f(\bar{\varepsilon}, \alpha) = 0 \tag{25}$$

Equation (25) is then solved numerically to obtain the equilibrium values of α for different values of $\bar{\varepsilon} > \bar{\varepsilon}_{CR}$. Having solved eqn. (25) for any given $\bar{\varepsilon}$, eqns (7), (9), (10) and (21) then give the corresponding values of local deflection coefficient A, load P, moment M and central deflection δ_c. Following this procedure the complete post-buckling load–deflection equilibrium path for the column can be established.

VARIATION IN LOCALLY BUCKLED FORM AFTER BUCKLING

So far in the analysis the buckled form has been assumed to remain constant in the post-local-buckling range and consequently the values of d^*, $(EI)^*$, P_E^* and P_{CRUC} remain unaltered from those obtained at local buckling. In reality, however, changes in load and in load eccentricity due to overall deflections, cause the buckled form to change and hence the values of $|P_E^*$, d^* and P_{CRUC} are variable after local buckling. To take this into account in an approximate manner the buckling modes and initial post-buckling paths are first obtained for eccentrically compressed struts corresponding to a range of compression eccentricities. It is now possible to obtain several solutions to a column's complete load–deflection equilibrium path, each based on a different buckled form and hence on different values of P_E^*, d^*

and P_{CRUC}. The lowest envelope of all such curves is considered to be the most accurate solution, the justification being that for any given column deflection δ_c, the solution which gives the lowest axial load P is the most accurate since it prescribes approximately the least value of strain energy of the system.

INITIAL LOCAL IMPERFECTIONS

It is known that the effects of local imperfections are most marked at or around the local buckling load and are most detrimental to the load-carrying capacity if they are similar in form to the local buckling mode. The local imperfections in the present work are therefore assumed to be of the same form as the locally buckled shape of the perfect column and are given by

$$w_{0i} = A_0 Y_i(y_i) \cos \frac{\pi x}{\lambda} \tag{26}$$

where A_0 specifies the magnitude of the initial imperfection and $Y_i(y_i)$ is defined by eqn. (3).

The compatibility equation of von Karman, eqn. (5), now takes the form

$$(a_{22})_i \frac{\partial^4 F_i}{\partial x^4} + (a_{33})_i \frac{\partial^4 F_i}{\partial x^2 \partial y_i^2} + (a_{11})_i \frac{\partial^4 F_i}{\partial y_i^4}$$

$$= \left[\left(\frac{\partial^2 w_i}{\partial x \partial y_i} \right)^2 - \frac{\partial^2 w_i}{\partial x^2} \cdot \frac{\partial^2 w_i}{\partial y_i^2} \right] - \left[\left(\frac{\partial^2 w_{0i}}{\partial x \partial y_i} \right)^2 - \frac{\partial^2 w_{0i}}{\partial x^2} \cdot \frac{\partial^2 w_{0i}}{\partial y_i^2} \right] \tag{27}$$

and a semi-energy analysis similar to that for a locally perfect column gives the final relationships. For reasons of space the local imperfection analysis has only been outlined here. Results are presented, however, illustrating the effects of these imperfections, in particular, on column designs with simultaneous buckling modes.

TYPICAL RESULTS

The results presented here are for orthotropic columns made from glass reinforced plastic with typical material properties of

$$E_{11} = 30 \text{ GN/m}^2 \qquad E_{22} = 6 \text{ GN/m}^2 \qquad G_{12} = 5 \text{ GN/m}^2 \qquad v_{12} = 0\cdot33$$

and tensile strengths in the 1 and 2 directions (see Fig. 1) equal to $1000\ \text{MN/m}^2$ and $20\ \text{MN/m}^2$ respectively. The web and flanges are assumed to have the same thickness and stiffness properties and the material is assumed to be homogeneous across the thickness.

Figure 6 shows, for centroidally loaded columns, the change in flexural stiffness (P_E^*/P_E) and the new position of the neutral axis (d^*/b_f) at the instant of local buckling, as well as the neutral axis position (d/b_f) for the locally unbuckled column for different values of flange–web ratio (b_f/b_w). The variations shown are similar in form to those obtained by Rhodes and Harvey[5] for isotropic columns. For columns with narrow flanges local buckling of the section is initiated by the web and the neutral axis moves further away from the web resulting in overall column deflections in the negative direction (see Fig. 1) at local buckling. For columns with wide flanges local buckling of the section is initiated by the flanges and consequently the neutral axis approaches the web at local buckling resulting in overall deflections in the positive sense. The reduction in flexural stiffness for flange-initiated buckling is seen to be much greater than that for web-initiated buckling. Of particular interest is the cross-sectional shape $(b_f/b_w) \simeq 0.33$, the neutral axis position and flexural stiffness of which remains unaltered at buckling. This suggests that for this section the buckling behaviour of a simultaneous mode design would in effect be essentially stable. In reality, however, change in the locally deflected form due to overall bending in the immediate post-buckling range would result in a shift of the neutral axis and unstable buckling behaviour. Also for this section it would appear that for a design whose local buckling load is considerably less than its Euler buckling load the behaviour of the column would be such as to remain straight after local buckling, as in the

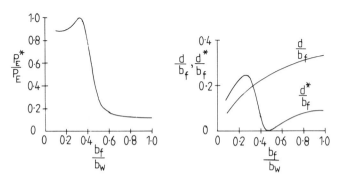

Fig. 6. Change of flexural stiffness and position of neutral axis for orthotropic GRP column.

case of a centroidally loaded box section of similar design. Again in reality, however, change in the locally deflected form in the immediate post-local-buckling range would result in a shift of the neutral axis and in overall deflections of the column after local buckling. A solution therefore which does not consider change in the locally deflected form after buckling would not be sufficient to account for the actual behaviour of the column whose cross-sectional shape factor is $(b_f/b_w) \simeq 0.33$. A direct method of allowing for change in the locally deflected form after buckling has been outlined by Loughlan[6] and Loughlan and Rhodes[7,8] for isotropic lipped channel columns.

Figure 7 presents load–deflection equilibrium paths for centroidally loaded columns with a flange–web ratio of 0·4 and various values of L/b_w. The results shown take into consideration in the approximate manner described earlier the change in locally deflected shape after buckling. From Fig. 6 it can be seen that the immediate change of flexural stiffness for this section at local buckling is given by $(P_E^*/P_E) \simeq 0.78$. If post-local-buckling behaviour is now based on the deflected shape obtained at buckling, i.e. if the deflected shape is not allowed to change after buckling, then the curves of Fig. 7 would tend asymptotically towards the P/P_E value of 0·78 for

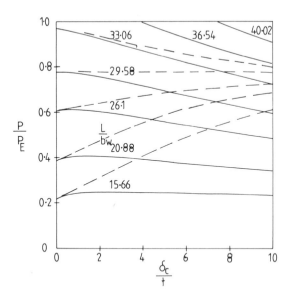

FIG. 7.   Load–deflection paths for centroidally loaded GRP channel columns ($b_f/b_w = 0.4$, $b_w/t = 100$, $\bar{e} = 0$, $\delta_{0c} = 0$, $w_{0c} = 0$). ----Constrained locally deflected form after buckling; ——— deflected form allowed to change after buckling.

critical local buckling loads $P_{CR} \leq P_E$ as indicated schematically by the dotted curves in the figure. The actual behaviour is seen, however, to be quite different from this due solely to the high sensitivity of this particular cross-sectional geometry to changes in the locally deflected shape after buckling. It will be noted from Fig. 7 that the curves signifying an unchanging locally deflected form after buckling, i.e. the dotted curves, suggest that for column designs with $(P_{CR}/P_E) < 0.78$, stable post-local-buckling interaction behaviour will ensue. The curves which allow for local form change, however, i.e. the full curves, show that the behaviour is initially stable after local buckling and for loads not much in excess of the local buckling load the behaviour changes from being stable to unstable. For designs with $0.78 < (P_{CR}/P_E) < 1$ unstable behaviour follows after local buckling, whereas for designs with $(P_{CR}/P_E) > 1$ essentially stable behaviour exists until overall deflections are such as to cause local buckling after which unstable behaviour occurs.

Figure 8 shows the load–deflection equilibrium paths of some of the columns of Fig. 7 when loaded eccentrically towards the flange free edges, thus causing positive overall deflections from the onset of loading. In this case overall bending causes higher compression towards the flange free

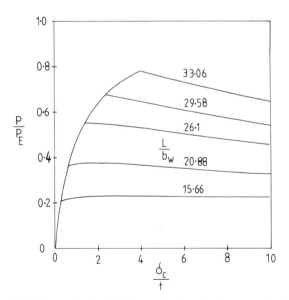

FIG. 8.   Load–deflection paths for GRP channel columns loaded eccentrically towards the flanges ($b_f/b_w = 0.4$, $b_w/t = 100$, $\bar{e} = +0.1$, $\delta_{0c} = 0$, $w_{0c} = 0$).

edges and due to this the local buckling loads are lower, to a varying degree, depending on column length, than those due to centroidal loading. The curves shown take change in locally deflected form into account. The columns with L/b_w ratios of 26·1 and 29·58, which exhibited initially stable and essentially stable behaviour after local buckling respectively under centroidal loading conditions, are seen to experience unstable behaviour at buckling for the eccentrically loaded case.

Figure 9 shows some imperfection sensitivity curves for a coincident mode design with $(b_f/b_w) = 0.75$. From Fig. 6 it is seen that the immediate change of flexural stiffness for this section at local buckling is given by $(P_E^*/P_E) \simeq 0.12$ and hence the sensitivity to changes in the locally deflected shape after buckling for this section, will be quite small, i.e. a post-buckling analysis based on the locally deflected form at buckling would probably be adequate for column designs using this section. The sensitivity to geometrical imperfections of a simultaneous mode design is seen from Fig. 9 to be quite high. For overall and local imperfection amplitudes of $(\delta_{0c}/t) = 1$ and $(w_{0c}/t) = 0.1$ the maximum load is seen to be reduced by as much as 60 %. This high imperfection sensitivity is due of course to the unstable nature of the equilibrium of these designs at buckling.

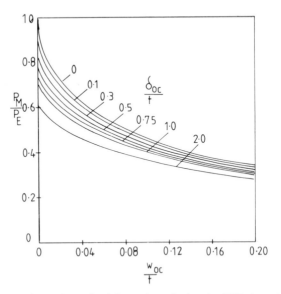

FIG. 9.   Reduction in maximum load due to imperfections in GRP channel column with coincident modes ($b_f/b_w = 0.75$, $b_w/t = 40$, $L/b_w = 41.19$, $\bar{e} = 0$, $P_{CR}/P_E = 1$).

CONCLUDING REMARKS

A theoretical analysis of the interactive buckling behaviour of orthotropic fibre reinforced composite, plain channel section columns has been presented.

For certain cross-sectional geometries it has been shown that the asymptotic approach to a reduced Euler load, based on an unchanging locally deflected form after buckling, is a severe approximation. For these geometries the behaviour after local buckling is accounted for in the analysis by allowing the locally deflected form to change in an approximate manner after local buckling.

The analysis includes the effects of initial local and overall imperfections and columns with coincident local and Euler buckling modes are shown to possess a high degree of imperfection sensitivity.

REFERENCES

1. NEUT, A. VAN DER. The interaction of local buckling and column failure of thin walled compression members. *Proceedings of the Twelfth International Congress of Applied Mechanics* (Stanford University, 26–31 August, 1968), Berlin, Heidelberg, New York, Springer-Verlag, 1969.
2. THOMPSON, J. M. T. and LEWIS, G. M. On the optimum design of thin walled compression members, *J. Mech. Phys. Solids*, **20** (May 1972) 101–9.
3. FOK, W. C., RHODES, J. and WALKER, A. C. Local buckling of outstands in stiffened plates, *Aeronautical Quarterly*, **XXVII** (May 1976).
4. FOK, W. C., WALKER, A. C. and RHODES, J. Buckling of locally imperfect stiffeners in plates, *Proceedings of the ASCE*, **103**(EM5) (October 1977) 895–911.
5. RHODES, J. and HARVEY, J. M. Interaction behaviour of plain channel columns under concentric or eccentric loading, *Second International Colloquium on the Stability of Steel Structures* (Liège, 13–15 April, 1977) pp. 439–44.
6. LOUGHLAN, J. Mode interaction in lipped channel columns under concentric or eccentric loading, Ph.D. Thesis, University of Strathclyde, December 1979.
7. LOUGHLAN, J. and RHODES, J. Interaction buckling of lipped channel columns, In *Stability Problems in Engineering Structures and Components*, Richards, T. H. and Stanley, P. (Eds), London, Applied Science Publishers Ltd, 179–98, 1979.
8. LOUGHLAN, J. and RHODES, J. *The interactive buckling of lipped channel columns under concentric or eccentric loading*, International Conference on Thin Walled Structures, University of Strathclyde, Glasgow, Scotland, 3–6 April 1979.
9. TURVEY, G. J. and WITTRICK, W. H. The influence of orthotropy on the stability of some multi-plate structures in compression, *Aeronautical Quarterly*, **XXIV** (Feb. 1973).

10. LEE, D. J. The local buckling coefficient for orthotropic structural sections, *Aeronautical Journal*, **82**(811) (July 1978) 313–20.

11. LEE, D. J. and HEWSON, P. J. The use of fibre reinforced plastics in thin walled structures, In *Stability Problems in Engineering Structures and Components*, Richards, T. H. and Stanley, P. (Eds), London, Applied Science Publishers Ltd, 23–55, 1979.

12. BANKS, W. M. and RHODES, J. *The post-buckling behaviour of composite box sections*, International Conference on Composite Structures, Paisley College of Technology, Paisley, Scotland, 16–18 September 1981. (Chapter 26 of this volume.)

13. UPADHYA, A. R. A study of the buckling behaviour of composite reinforced metal panels and composite plain channel section columns in compression, Ph.D. Thesis, Cranfield Institute of Technology, Cranfield, Bedford, England, Sept. 1980.

25

The Stability Analysis of a Continuum/Skeletal Fibre Matrix System

V. G. Ishakian and L. Hollaway

Department of Civil Engineering, University of Surrey,
Guildford, Surrey GU2 5XH, England

ABSTRACT

A single unit of a composite Vee section construction was manufactured from glass reinforced polyester and has been analysed analytically and experimentally to first buckling. The composite construction was made from two components, one of which was a skeletal system and the other a continuum one. The pultruded skeletal component had a glass fibre/polyester resin matrix ratio of 65–35% weight and the hand lay-up continuum component had a glass fibre/polyester resin matrix of 30–70% by weight.

The analytical analysis was undertaken by the finite element method using rectangular plate elements in combination with line elements, as proposed by Scordelis, and included buckling of the continuum and its effect on the post-buckling behaviour of the structure. The buckling mode is expressed by determining the corresponding eigen-vector.

It was found that provided the analytical modelling of the practical structure is carefully performed, good correlation is achieved between the experimental and analytical structures. In the stability analysis small displacements were assumed and it was found that the bifurcation behaviour was totally dependent upon the level of the axial stresses on the structure.

INTRODUCTION

Glass reinforced plastic (GRP) structures are generally manufactured as continuum systems from the hand lay-up methods using chopped strand mat and polyester resin; the composite has a fibre/matrix % ratio by weight of approximately 30/70. Stiffness of the overall system is developed by folding the composites. Another conventional method is provided by grid systems and as their members require unidirectional strength and stiffness a suitable manufacturing technique for their production is the pultrusion one using continuous unidirectional rovings and polyester resin; the fibre volume fraction in this case is about 65/70 wt %. By combining the above two systems into a structural form, which in this paper will be called a composite skeletal/continuum space structure, a method of providing very much greater stiffness than that from the two individual components is achieved.

The paper will demonstrate a theoretical approach to analysing the stability behaviour of a composite GRP structure and the results will be compared with an experimental investigation of a large model.

THEORETICAL ANALYSIS

Basic Concepts and Types of Elements

For the analysis of continuum/skeletal systems it is desirable to use the finite element discretisation in combination with the direct stiffness method. This procedure provides a systematic approach to the solution which is applicable to any shape, configuration and boundary condition. In addition it enables optimisation to be made of the configuration, the proportion and the cross-sectional properties of both the components which form the composite.

To analyse the composite structure, it is necessary to combine the line and plate types of elements. The plate element used in this paper is a four noded rectangular one and the skeletal members are represented by a two noded line element.

No discussion will be made of the necessary steps for the stiffness matrix derivation of the elements with an assumed displacement function across the element; these steps have been described elsewhere.[1-6]

However, an important consideration arises regarding the degrees of freedom per node for the two types of elements whcn combined in one overall stiffness matrix. For the system of equations to have a solution, compatibility should exist for each degree of freedom per node. The line element used in this paper is that of a two ended member in space, having six degrees of freedom per node (three translational and three rotational, as shown in Fig. 1a). It is derived from an assumed linear polynomial for the

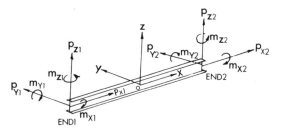

FIG. 1a. Degrees of freedom for a skeletal member in space in local coordinates.

axial and torsional displacements and a cubic polynomial for the transverse bending displacement, consistent with skeletal structural theory.[2,7,8] Consequently, the plate element in space should have six degrees of freedom per node, and this is achieved by the uncoupled combination of inplane degrees of freedom (u, v, θ_z) and the out-of-plane bending action (w, θ_x, θ_y) shown in Fig. 1b.

However, most thin plate finite element formulations include only two translational degrees of freedom for the inplane stress analysis (u and v) and

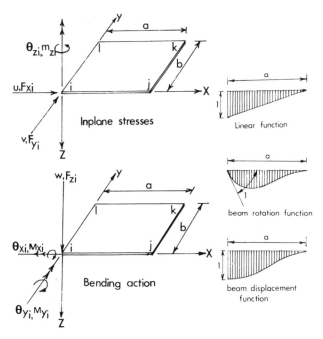

FIG. 1b. Degrees of freedom for plate elements.

neglect the inplane rotation θ_z (Fig. 1b), since it is necessarily arbitrary, because there is no unique value of such rotation (apart from a rigid body movement) at a point in a two dimensional continuum. The moment which corresponds to it is not fully tractable to physical explanation and hence combining such formulations (u and v) with the bending action (w, θ_x, θ_y) results in an element with only five degrees of freedom per node.

A method is proposed by Zienkiewicz[1] which overcomes the singularity of the stiffness matrix because of the omission of the inplane rotation in plate elements. This method inserts a fictitious set of inplane rotational stiffnesses, based on the fact that the real rotational stiffness is very high in comparison with the plate bending stiffness. This method has been discussed in reference 9. However, this approach lacks one important aspect, i.e. the inplane displacement field has two degrees of freedom whereas the bending displacement field has three. This makes it difficult to use the same shape functions for both the inplane stress and plate bending analysis.

The difference between the functional variations of the inplane displacement field and the transverse displacement field leads to gross violation of conformity between adjacent elements which do not lie in the same plane.[10]

The utilisation of the inplane rotation θ_z as an additional degree of freedom enables the same shape functions to be employed for both inplane stress and plate bending analysis. Such a formulation which achieves six degrees of freedom, is ideally suited to the analysis of three dimensional plate assemblies forming folded plate structures combined with skeletal systems.

Some formulations which include the inplane rotation θ_z as a degree of freedom have been developed by Macleod,[11] Tocher and Hartz,[12] Pole and Felippa[13] and Scordelis.[14]

The formulation which is used in this investigation is that of Scordelis[14] which was initially developed for the analysis of box-girder bridges. The element used is a rectangular one of four nodes each having six degrees of freedom formed by the uncoupled combination of inplane degrees of freedom (u, v, θ_z) and out of plane bending degrees of freedom (w, θ_x, θ_y). It is important to mention here that the same shape functions are used in setting the displacement functions for both inplane and out of plane deformations. These functions combine a linear function, a beam rotation (i.e. a beam clamped at one end and a unit rotation applied at the other, as shown in Fig. 1b) and a beam displacement function (i.e. a beam clamped at both ends but with a relative displacement of unity between them). At each of the four nodes the inplane rotation θ_z is defined as the average of the rotations of the two adjacent sides of the element at any particular node, i.e.

$$\theta_z = \frac{1}{2}\left[\frac{\partial_v}{\partial_x} - \frac{\partial_u}{\partial_y}\right]$$

Finite Element Formulations Considering Non-linear Behaviour

A general method for non-linear analysis is given by Zienkiewicz[1] in which he introduces a 'tangent stiffness matrix' for the element which includes all the sources of non-linearities as:

$$K_T = K_0 + K_\sigma + K_L$$

where $K_0 =$ linear elastic stiffness matrix; $K_\sigma =$ symmetric matrix dependent on the axial stress level and is called the 'initial stress matrix' or 'geometric matrix' (defined in reference 15) and $K_L =$ 'initial displacement matrix' or 'large deformation matrix' and contains only terms which are linear and quadratic in the displacement (defined in reference 15).

In situations such as perfectly straight struts, plates, shells, 'V' and 'box' sectioned members, etc., under inplane stresses only, the 'large deformation matrix', K_L, is identically zero because of the absence of the coupling effect of bending stresses with the axial stresses. In these situations the non-linear behaviour initiates as a distinct bifurcation point followed by a non-linear post-buckling path in the deflections and the stresses. The distinct bifurcation load can be found by solving the typical eigen-value problem for the equation:

$$\mathrm{d}F = (K_0 + \lambda K_\sigma)\,\mathrm{d}\delta \equiv 0 \tag{1}$$

where $\lambda =$ the load factor. The buckling mode is expressed by determining the corresponding eigen-vector.

Theoretically, folded plate structures interconnected to skeletal systems have very small bending stresses and these arise from the rigid connections under a loading system applied at the nodal joints. The start of non-linear behaviour, therefore, can be reasonably predicted by the assumption of a linear load–deflection relation up to a distinct bifurcation point which is a function of the axial stresses only.

A computer program has been written in Fortran 4 for a linear elastic analysis and the prediction of the first critical load by the reformulation of the global stiffness matrix at every load increment, according to eqn. (1), followed by the triangular decomposition; the critical load is reached as the number of negative pivots of the decomposed stiffness matrix changes from 0 to 1. A flow chart of the program is given in Fig. 2.

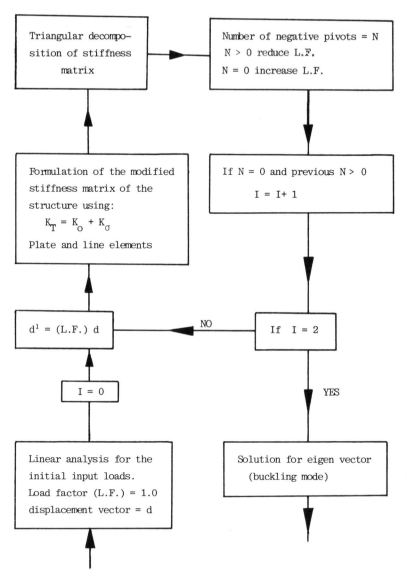

Fig. 2.   Flow chart of the computer program for determining the buckling load and the
corresponding buckling mode of the composite space structure.

THE COMPOSITE STRUCTURAL MODEL

The low modulus of elasticity of GRP manufactured from the hand lay-up or semi-mechanical process to form a quasi-isotropic material is well known. Consequently, the buckling of compression members which are made from this material is a critical factor in the design of a GRP structural system. The main reason for using a folded plate system manufactured from materials of low modulus of elasticity is to overcome the buckling aspect. If GRP, carbon fibre reinforced plastics (CFRP) or a hybrid of carbon and glass reinforced plastic composites are manufactured by the pultrusion technique and used in the skeletal system, the buckling criterion is still likely to be the critical factor in design but it does show an improvement over the former system because of the preferred fibre orientation and the increased glass:resin ratio. Further, in a composite construction consisting of a continuum/skeletal configuration the stiffness of the overall unit or structure is again increased above the former two cases, although buckling of the structural elements is still a critical factor.

The composite structure in question consisted of a continuum made from Vee sectioned members together with bars jointed to one another and to the folded plates at the nodes. The width of the structure is a function of the number of Vee sections; a photograph of a perspex model showing the system is given in Fig. 3.

The single GRP Vee unit had a width of 1076 mm and a vertical depth of 702 mm, tested over a span of 4580 mm. Figure 4 shows the dimensions and the support conditions of the structure and Fig. 5 shows a photograph of the structure.

The 4 mm continuum was manufactured from a chopped strand mat, polyester resin composite with a fibre/matrix % by weight of 30/70 and the skeletal system was fabricated from pultruded tubes of 25 mm external diameter and 2 mm thickness. The nodal points were made by forming parts of hollow spheres in the continuum at the time of fabrication, and when the various parts of the latter were assembled, these part spheres were then developed into full ones. The pultruded tubes passed through holes drilled in the spheres and the whole was filled with epoxy resin; thus the skeletal members were bonded to the continuum and the node was complete. Although this method of manufacturing nodal points is not ideal for prototype structures from an economics viewpoint, it proved extremely efficient for the experimental test model, with buckling and/or failure of the members of the composite structure occurring before fracture of the nodal point or pull out of the skeletal member from the joint.

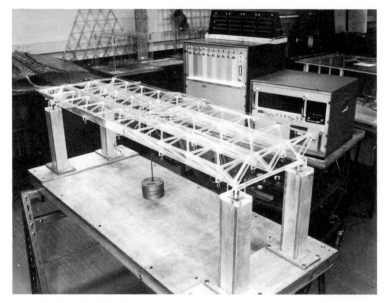

FIG. 3.   Photograph of a Perspex model showing Vee system.

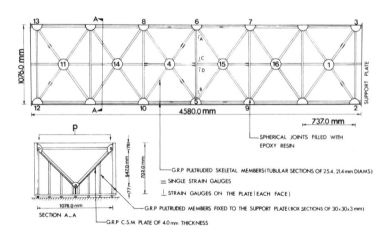

FIG. 4.   Plan and section of the GRP prototype structure showing the dimensions and the positions of strain gauges.

Fig. 5. A top view of the prototype GRP composite structure composed of a Vee shaped CSM laminate continuum connected to a skeletal system fabricated from pultruded glass fibre/polyester resin.

Diaphragm plates manufactured from chopped strand mat laminates of 4 mm thickness were attached to the Vee component at both sides. The joint was made by bolting to both parts GRP pultruded angles which ran the full length of the joint and were stiffened at intervals by steel angle pieces.

Steel members were bonded to the base of the diaphragms and these rested on reaction beams through 18 mm ball bearings; these are shown diagrammatically in Fig. 4. Both the steel members at the base of the diaphragms had Vee grooves in which the ball bearings were seated. At one end of the structure the reactions for the ball bearings were provided by similar Vee grooves and at the other end by flat plates. This allowed the system to have both translation and rotation at one end and rotation only at the other end.

EXPERIMENTAL PROCEDURE AND RESULTS

The load was applied to the structure by hydraulic jacks; only the top nodes of the system were loaded. Electrical resistance strain gauges were employed to measure strain and hence stresses at selected positions on the structure. Linear strain gauges were bonded diametrically opposite each other on the skeletal members and rosette strain gauges were bonded on either side of the continuum plate to predict bending moments. Deflections of the structure were measured by electrical transducers. The positions of the strain gauges and displacement transducers are shown in Fig. 4.

During preliminary testing of the single Vee unit structure in the linear region it was found necessary to erect a set of lateral supports along its length at joints 8, 6, 7, 10, 5, 9 as shown in Fig. 4 to avoid the possibility of torsional buckling. This was achieved by placing, on both sides of the structure, horizontal steel frames on which were bolted vertical steel members so that vertical movement only could take place.

The structure was loaded in increments of 50 kg and at a load of 218·8 kg per bay (corresponding to a total load of 1094·0 kg) buckling occurred in the panels of the continuum adjacent to the diaphragms. The deformation of the buckled zone increased as the external load increased with further buckling in the continuum at different positions; the final form was a wavy contour throughout the length of the structure. The test was stopped at a total load of 2780 kg as the capacity of the reaction-beam to which the hydraulic jacks were connected had been reached. At this external load no buckling or crushing of the pultruded skeletal members had occurred.

THEORETICAL PROCEDURE AND RESULTS

The computer program of the composite structural model, described earlier, was used to predict the buckling as a distinct bifurcation point in the GRP structure. The results from a previous investigation[9] using similar configuration perspex models to the current GRP structure showed that, by considering two mesh divisions representing a coarse mesh and a fine mesh size, only a small difference in the general behaviour of the structure resulted. It was therefore decided to choose a mesh division between the above two which would be adequate for the analysis. Figures 6a and 6b show the mesh division of 102 plate elements and the joint numbering system for one quarter of the structure.

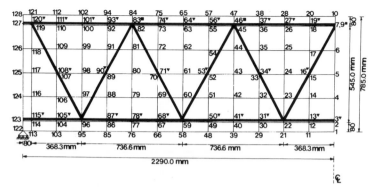

FIG. 6a.   Mesh division and joint numbering system for one quarter of the GRP prototype structure. ▼: Joints introduced in the skeletal members; members 7–9(■), 45–46(■) and 82–83(■) are perpendicular to the plane of the page.

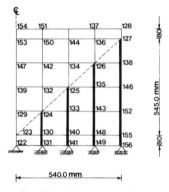

FIG. 6b.   The support plate.

The analytical predicted load was 194·4 kg on each bay. The eigen-vector, which represented the relative values of displacements at buckling, is given in Fig. 7 for one quarter of the structure. The wavy mode of buckling is basically similar to that obtained in the experimental solution resulting from gradual increase of load increments.

Comparison Between the Experimental and the Theoretical Results

Axial stresses in the continuum component

From Fig. 8 and Table 1 it may be seen that the values of the longitudinal axial stresses expressed as force per unit width at positions C and D (Fig. 4) are within 10 and 15% of the theoretical solution respectively; the

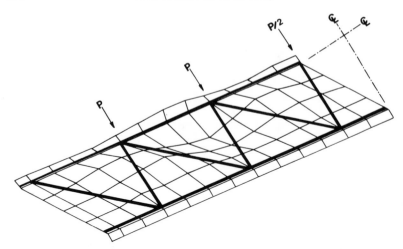

FIG. 7.   Finite element idealisation of the buckling mode (eigen-vector) in the continuum component of the GRP prototype structure (Fig. 4) at the buckling load of 194·4 kg on each bay.

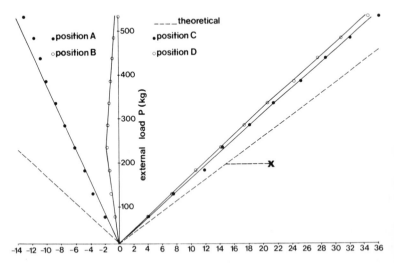

FIG. 8.   Experimental and theoretical axial stresses (given by the abscissa) in the continuum at the midspan of the GRP prototype structure expressed as force per unit width (N/mm). 'X' marks the point of theoretical buckling.

TABLE 1
Axial stresses at positions in the continuum (expressed as force per unit width)

Position	Experimental (N/mm)	Theoretical (N/mm)	Theoretical > < Experimental
C	+7·63	+8·37	>9·7%
D	+7·29	+8·37	>14·8%
A	−3·65	−6·75	>84·9%
B	−1·0	−6·75	>575%

percentage difference between the two experimental values is within 5%. The experimental value of the force per unit width at positions A and B varies greatly from that of the theoretical but this is not surprising as the modelling of the theoretical solution at this point is not accurate. The spherical nodal joint is represented in the theoretical analyses as a unique rigid joint where the line and plate elements meet the centroidal axis of the skeletal members coinciding with the neutral axis of the continuum. In reality, however, the skeletal members are eccentric to the continuum and when the structure is under load a different stress field is created around the hemispherical area of the joint. Consequently strain gauges which are situated under the spherical joints (as is the case at positions A and B) will give different results to those obtained from the theoretical analysis.

TABLE 2
Bending moments at mid length of the skeletal members at a load of 113·2 kg

Member	Experimental (N/mm)	Theoretical (N/mm)
1–2	+292·5	−26·1
1–3	+326·8	−26·1
11–12	+1 738·0	−26·1
11–13	−43·3	−26·1
4–6	+35·11	+39·4
4–5	−51·0	+39·4
6–8	+258·6	+269·2
6–7	+79·2	+269·2
5–10	+250·6	+269·2
5–9	+275·3	+269·2
4–14	+94·0	0·0
15–16	+882·2	0·0
4–15	+189·0	0·0
5–6	−189·5	0·0

TABLE 3
Bending stresses at positions in the continuum (expressed as BM per unit width)

Position	Experimental (N/mm)	Theoretical (N/mm)
C	−0·67	−0·024
D	+0·1	−0·024
A	−0·33	−0·01
B	−1·45	−0·01

Bending moments in the skeletal members and in the continuum

Tables 2 and 3 show the theoretical and experimental bending moments at mid length of the skeletal members and the bending stresses in the continuum expressed as bending moment per unit width. It can be seen from the tables that there is poor agreement between the two methods of measurements. The non-symmetric values of bending moments can be clearly seen in the members. Members 4–14, 15–16, 4–5 and 5–6 should have zero values as they lie in the plane of symmetry. Discussion of these results is made in the next section.

DISCUSSION

The Buckling Behaviour of the Structure

The theoretical buckling load of the continuum (as a bifurcation point) was 194·4 kg compared with the experimental value of 218·8 kg. As the applied load increased beyond this value the buckling deformations became excessive near the supports.

It is seen that the experimental buckling is 12·5 % greater than that predicted by the analytical analysis. This is an acceptable variation when the imperfections in the structure and the difficulty of accurately idealising the experimental model are realised.

More specifically the discrepancy between the theoretical and the experimental buckling load is due to two factors:

(i)   the differences in the axial forces in the skeletal members and the axial stress distribution in the continuum; the buckling is a direct function of the axial stresses;

(ii) the existence of secondary bending moments in the real model, coupled with the axial stresses, affects the value of the buckling load; in the program used no account is taken of the non-linear coupling effect of the bending moments with the axial stresses and hence the resulting geometric and stress matrices of the line and plate elements are functions of axial stresses only.

It can be seen from Figs 8, 9 and 10 that the tensile stresses in the continuum and the tensile axial forces in the skeletal members 4–15, 4–14 and 15–16 retain the same linear relationship with the external load before and after buckling of the continuum. This can be explained by the fact that when the continuum buckled, redistribution of stresses took place between the continuum and the compressive skeletal members at the top of the structure.

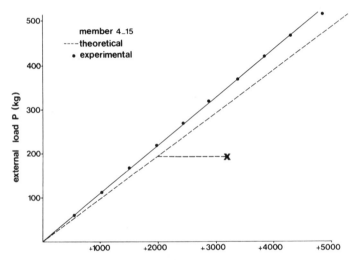

FIG. 9.   Experimental and theoretical axial force (given by the abscissa in N) in member 4–15 of the GRP prototype structure. 'X' marks the point of theoretical buckling.

To improve the resistance to buckling a rigid joint was introduced at the mid length of all the skeletal members connecting them to the continuum at this point. This condition resulted in a higher analytical buckling load (of value 280·8 kg), compared with the previous case (of value 194·4 kg) before the rigid joint was introduced, and a less severe buckling condition. The equivalent value for the experimental results were 310·0 and 218·8 kg respectively. The buckling mode for this case is represented diagrammatically in Fig. 11.

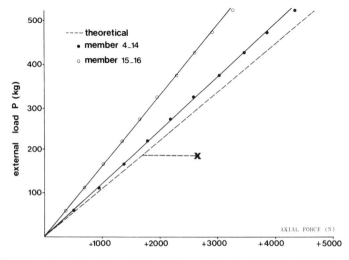

FIG. 10. Experimental and theoretical axial forces (given by the abscissa in N) in skeletal members in the GRP prototype structure. 'X' marks the point of theoretical buckling.

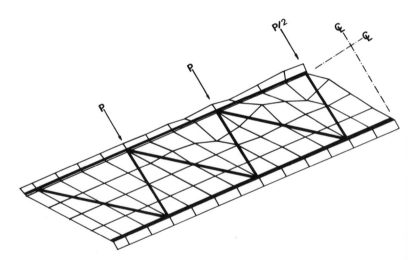

FIG. 11. Finite element idealisation of the buckling mode (eigen-vector) in the continuum component of the GRP structure (Fig. 4) in which additional rigid joints connect the skeletal members to the continuum at the midlength of all the skeletal members.

The Behaviour of the Structure at Low Load Levels

The experimental and theoretical solutions to the tensile and compressive forces in the bottom and top members of the skeletal component of the models agree to within 10 %; the forces in the diagonal members however, show poor agreement, with variations as great as 50 % between the two solutions.

Certain criticisms may be levelled at the experimental model and because of these errors may have crept into the experimental procedure.

The beam was a single Vee section virtually unsupported laterally, and as the system was symmetric it was possible in the analytical model to consider one quarter of the structure only. However, in the experimental model it was possible for the structure to rotate, thus causing a torsional couple in the cross-section and hence producing unequal forces in symmetrical members.

Both the pultruded tubes (the skeletal members) and the hand lay-up composite (the continuum plate) were manufactured as normal factory made units. During fabrication of the composite structure it was necessary, on occasions, to force members into position, and this undoubtedly caused some stresses to be built into the system.

In the experimental model the external loads were applied at the nodal points of the skeletal component which were eccentric to the continuum component. In the theoretical analysis, however, the external loads were applied at a common nodal point of the continuum/skeletal components; the centroid of the skeletal nodal points and the centre line of the continuum were coincident. Consequently the theoretical structure did not model the experimental one exactly and because of this, the comparison of the solutions for both structures and the symmetry of the results for the experimental one were affected to some degree.

OBSERVATIONS

In the present analysis the sixth degree of freedom (viz. θ_z) at each nodal point of a rectangular finite element has been considered. Both the linear and the stability analysis have produced good agreement between the analytical and experimental models. The high order formulation of the element and the use of the same shape function for both inplane stress and plate bending analysis assured the continuity at the folds and consequently relatively coarse mesh divisions were able to be used.

From previous investigations it has been found that the relative stiffness of the two component parts of the composite structure did influence the degree of correlation between the analytical and experimental solutions. The present composite structure had thin continuum components and this gave greater diversity between the two techniques than a similar perspex structure gave when the continuum components were thick in relation to the skeletal ones.

It has also been established that the analytical modelling of the practical structure should be undertaken as carefully as possible. For instance, it is unusual for the centre lines of skeletal members to meet at the centre of the nodal points and if this eccentricity is not considered in the analytical model discrepancies can be expected. Also, it is most unlikely that the centre lines of the skeletal and the continuum members in double layer grid systems will coincide and the present structure clearly shows this. Unless this eccentricity is considered, inaccuracies will occur in the analytical solution and particularly in the buckling analysis. In the present work the theoretical combination of the line and plate elements was effected by introducing a short stiff member connecting the continuum to the skeletal members at node points.

In the stability analysis, small displacements were assumed, i.e. linear behaviour up to the bifurcation point, and the bifurcation behaviour was totally dependent upon the level of axial stresses in the structure. This analysis has been shown to be reliable and adequate for design and optimisation purposes. The reason for this is that the stresses in the prototype model are mainly axial as is typical for most continuum/skeletal systems loaded at nodal joints.

ACKNOWLEDGEMENTS

The authors wish to acknowledge the support of the Science Research Council, UK who are sponsoring the work described.

REFERENCES

1. ZIENKIEWICZ, O. C. *The Finite Element Method in Engineering Science*, McGraw-Hill, New York, 1971.
2. NATH, B. *Fundamentals of the Finite Element Method*, Athlone, London, 1974.
3. ROĆKY, K. C., EVANS, H. R., GRIFFITHS, D. V. and NETHERCOT, D. A. *The Finite Element Method*, Crosby Lockwood, London, 1975.

4. BREBBIA, C. A. and CONNER, J. J. *Fundamentals of Finite Element Techniques for Structural Engineers*, Butterworths, London, 1973.
5. FENVES, S. J. *et al. Numerical and Computer Methods in Structural Mechanics*, Academic Press, London, 1973.
6. GALLAGHER, R. H. *Finite Element Analysis Fundamentals*, Prentice-Hall, Englewood Cliffs, New Jersey, 1975.
7. LIVESLEY, R. K. *Matrix Methods of Structural Analysis*, 2nd Edition, Pergamon Press, Oxford, 1975.
8. GERE, J. M. and WEAVER, W. JR. *Analysis of Framed Structures*, Van Nostrand, London, 1965.
9. ISHAKIAN, V. G. and HOLLAWAY, L. Application of the finite element method to the analysis of a skeletal/continuum GRP space structure, *Composites*, **10**, 1979, April, 2, 81–8.
10. WILLIAM, K. T. Finite element analysis of cellular structures, Ph.D. Thesis, University of California, Berkeley, 1969.
11. MACLEOD, I. A. New rectangular finite element for shear wall analysis, *Journal of the Structural Division, Proceedings of the American Society of Civil Engineers (ASCE)*, **95**, 1969, March.
12. TOCHER, H. L. and HARTZ, B. J. Higher-order finite element for plane stress, *Journal of the Engineering Mechanics, Proceedings of the American Society of Civil Engineers (ASCE)*, **93**, 1967, August.
13. POLE, G. M. and FELIPPA, C. A. Discussions on new rectangular finite element for shear wall analysis, *Journal of the Structural Division, Proceedings of the American Society of Civil Engineers (ASCE)*, **96**, 1970, January.
14. SCORDELIS, A. C. Analysis of continuous box girder bridges, Struct. Engng and Struct. Mech. Report No. SESM 67-25, University of California, Berkeley, USA, November 1967.
15. ISHAKIAN, V. G. Stability analysis of continuum/skeletal fibre matrix systems, University of Surrey, UK, 1980.

26

The Postbuckling Behaviour of Composite Box Sections

W. M. BANKS AND J. RHODES

Department of Mechanics of Materials,
University of Strathclyde, 75, Montrose Street,
Glasgow G1 1XJ, Scotland

ABSTRACT

The authors have already examined the buckling and postbuckling behaviour of reinforced plastic panels fabricated as orthotropic plates and subject to unidirectional in-plane loading[1] as well as an extension of that work to examine theoretically the buckling behaviour of orthotropic sections.[2] The present contribution extends this work further to examine the postbuckling behaviour. The sections are considered as a series of linked plates with rotationally restrained unloaded edges. These conditions have been considered for the plates alone. The linking procedure enables the instability of the sections to be evaluated.

After buckling the section is given a common end displacement. The moments and slopes at each edge are related to this and combined in such a way as to ensure that equilibrium and compatibility are satisfied at the plate edges using an iterative procedure. Thereafter the relevant postbuckling behaviour is evaluated.

The results are applied to a typical reinforced plastic box section fabricated from unidirectional composites. Sections of this nature are already being considered for structural applications and further markets should be found as the material potential is appreciated.

NOTATION

a, b plate dimensions in x and y directions respectively

D_{11}, D_{22} flexural rigidity of plate per unit width for bending about the y and x axes respectively, given by

$$D_{11} = E_{11}t^3/12(1 - v_{12}v_{21}), \quad D_{22} = D_{11}E_{22}/E_{11}$$

E_{11}, E_{22}	modulus of elasticity in the x and y directions respectively
e	ratio of buckle half wavelength to plate width
G_{12}	elastic shear modulus in x–y plane
K	elastic buckling coefficient for orthotropic plates
t	plate thickness
$Y(y)$	deflections across buckled plate
$v_{12}v_{21}$	Poisson's ratio in the x and y directions respectively
σ_{cr}	critical buckling stress
w	out-of-plane deflections of the plate

Other symbols used in the text are defined when they appear.

INTRODUCTION

The increasing use of reinforced plastics in structural applications has led to the need to examine their behaviour when subjected to compressive loading. Their low elastic modulus coupled with a high strength makes instability a major problem area. To increase and exploit applications to composite structures, this problem will need to be understood and appropriate steps taken at the design stage.

The purpose of the present paper is to make a contribution in this direction and to permit an understanding of the initial postbuckling behaviour of composite sections. The sections considered are fabricated from glass reinforced plastic (GRP) with unidirectional orientation. However, provided the properties of the composite are known, the method could be applied to other materials.

PLATE ANALYSIS

The method of analysis is firstly to examine the behaviour of single composite plates and obtain simple expressions governing their buckling and postbuckling behaviour. Thereafter the plates are linked together to form a section and using an iterative procedure the buckling and postbuckling behaviour of the section is obtained.

The Buckling of Reinforced Plastic Plates

The fundamental problem being addressed is shown in Fig. 1. The plate is uniformly compressed on the loaded ends which are considered to be simply

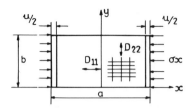

FIG. 1. Coordinate axes and sign convention.

supported. The unloaded edges are elastically restrained against rotation to an equal degree, although these can be altered to give unequal rotational restraints if required. Also, the unloaded edges are considered to be stress free while on the loaded ends the shear stress is considered to be zero.

The detailed analysis of this plate using a semi-energy approach is given in reference 1. The results for the buckling of plates with various rotational restraints varying from the simply supported case to the fixed case were obtained and comparison with existing solutions for particular values showed excellent agreement and gave confidence in the results.

At buckling, the deflection for the plate shown in Fig. 1 can be taken in the form

$$w = Y(y)\cos\frac{\pi x}{eb} \tag{1}$$

where e is a measure of the buckle half wavelength in the x direction and is introduced to enable the effect of changing the buckle wavelength to be studied. The function $Y(y)$ are polynomials satisfying the boundary conditions on the unloaded edges.

The critical buckling stress for the plate can be written in the form

$$\sigma_{cr} = \frac{K\pi^2\sqrt{D_{11}D_{22}}}{b^2 t} \tag{2}$$

where K is defined as the elastic buckling coefficient and is a function of the rotational restraint on the plate unloaded edges and the buckle half wavelength.

It has been shown[1] that the coefficient of restraint at the plate edge can be written as

$$R = \frac{\alpha b}{D_{22}} \tag{3}$$

where α is an elastic constant and for the general case may be different at each edge. For positive restraint on rotation R_0 on the edge $y = 0$ is positive and R_b on the edge $y = b$ is negative.

Using the above approach, it is possible to obtain the buckling coefficients for a large number of different plates, i.e. with different aspect ratios and different restraints on the unloaded edges. Relatively simple expressions governing the plate buckling problem can then be obtained.

The variation of K with e for a range of different R values is shown in Fig. 2 for a typical unidirectional GRP plate. When $R_0 = R_b = 0$ the plate is

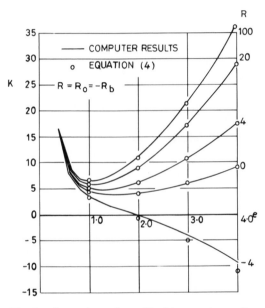

FIG. 2. Comparison of eqn. (4) with computer results.

simply supported, while for high positive values of R_0, the plate can be considered as fully fixed. Note that for negative values of R_0 the plate buckling is being assisted and hence the buckling coefficient is lower than that for a simply supported plate.

The variation of K with e was obtained, as indicated in reference 2, in the form

$$K = \frac{K_0 + RQ_1K_\infty}{1 + RQ_1} \tag{4}$$

where K_0 is the coefficient for a simply supported orthotropic plate and is given by

$$K_0 = 1.753 + 0.4471e^2 + \frac{2.236}{e^2} \tag{5}$$

and K_∞ is the coefficient for a clamped plate and is given by

$$K_\infty = 2.204 + 2.270e^2 + \frac{2.098}{e^2} \tag{6}$$

and Q_1 is a function of e given by

$$Q_1 = \frac{1 + 5.902e}{22.01 + 55.29e} \tag{7}$$

Thus, knowing the value of e, the value of K could be obtained from eqn. (4) for any particular value of R. Figure 2 also gives a comparison of results obtained using eqn. (4) with those obtained directly from the computer. The agreement is seen to be excellent and eqn. (4) can therefore be claimed to describe accurately the buckling coefficient for this type of plate.

The Postbuckling of Reinforced Plastic Plates

It is necessary in the postbuckling range to obtain expressions for both relative stiffness at buckling and the edge slope coefficient.

Relative stiffness at buckling

Expressions for the ratio of postbuckling to prebuckling stiffness E^*/E were obtained in a similar fashion to that described above for the buckling coefficients. The variation of E^*/E with e for a range of R is given in Fig. 3. The full lines given on this graph are those obtained from the derived equations. Values from the computer output are given at selected points. The form of the expression for the simply supported plate, i.e. $R = 0$ was taken as

$$\frac{E_0^*}{E} = 1 - \frac{2}{3 + \dfrac{Ae^3}{e^3 + B}} \tag{8}$$

By a curve fitting process, A was found to be 1.87 and B to be 13.47. For the fixed plate the form of the expression was altered slightly and was found to be of the form

$$\frac{E_\infty^*}{E} = 1 - \frac{2}{3.448 + \dfrac{4.885e^3}{e^3 + 10.22}} \tag{9}$$

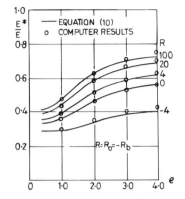
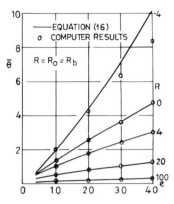

FIG. 3.   Variation of relative stiffness with R and e showing both computer results and those from eqn. (10).

FIG. 4.   Variation of edge slope with R and e showing both computer results and those from eqn. (16).

To take account of its variation with R the general form

$$\frac{E^*}{E} = \frac{\dfrac{E_0^*}{E} + Q_2 R \dfrac{E_\infty^*}{E}}{1 + Q_2 R} \tag{10}$$

was used. To obtain the results shown in Fig. 4, Q_2 was taken in the form

$$Q_2 = 0\cdot08 + 0\cdot01e \tag{11}$$

Edge slope coefficient

To satisfy compatibility conditions between linked plates it is necessary to know how the edge slope varies after buckling. It has been shown,[3,4] that by suitable manipulation of the theoretical equations an expression relating maximum deflection, load and applied end displacement can be obtained in the form

$$\bar{w} = H(\bar{u} - K)^{1/2} \tag{12}$$

where H is a constant and $\bar{w}$ is the nondimensional maximum deflection which occurs at the plate centre. This equation assumes that the deflected form at buckling does not vary. Also the deflection may be related to the edge slope by

$$\bar{w} = G\theta \frac{b}{t} \tag{13}$$

where G is a constant. Hence, the edge slope can be written in the general form

$$\theta \frac{b}{t} = \bar{\theta}(\bar{u} - K)^{1/2} \tag{14}$$

where $\bar{\theta}$ is the edge slope coefficient and depends on e and R. As before the values of $\bar{\theta}$ were plotted for various e and R values. The results are shown in Fig. 4.

For the simply supported case curve fitting produced an equation for $\bar{\theta}_0$ of the form

$$\bar{\theta}_0 = 0.5 + 1.06e - \frac{0.21}{e + 0.22} \tag{15}$$

As the restraint increases, the value of $\bar{\theta}_0$ reduces. This led to an equation for the general expression of the form

$$\bar{\theta} = \frac{\bar{\theta}_0}{1 + Q_3 R} \tag{16}$$

with Q_3 given as

$$Q_3 = \frac{0.1835e}{1.474 + e} \tag{17}$$

A comparison of the computer and derived results are given in Fig. 4.

SECTION ANALYSIS

A reinforced plastic section such as that shown in Fig. 5 can be considered as a series of long flat plates connected at the edges. If these plates were fabricated using unidirectionally oriented glass fibre, for example, then they could be treated as orthotropic plates and the above plate analysis could be used as a basis for evaluating the section behaviour.

Short lengths of such sections can suffer local instability when the plate elements buckle. When this happens the common edges of the component plates normally remain straight and the original angle between the plates is maintained during buckling. In addition the wavelength of the buckles which occur in all plates simultaneously are the same.

The method of analysis can be indicated by considering two adjacent

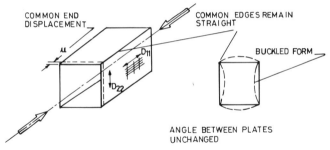

FIG. 5. Concepts involved in buckling of sections.

plate elements as shown in Fig. 6. Due to buckling the plates have out-of-plane deflections as indicated. The compatibility and equilibrium conditions for the plate edges require that

$$\theta_1 = \theta_2 \quad \text{and} \quad M_1 = M_2 \tag{18}$$

In addition the buckle wavelength for each plate is the same, i.e.

$$e_1 b_1 = e_2 b_2 \tag{19}$$

Introducing the coefficient of restraint defined earlier in eqn. (3) and remembering that α is given by

$$\alpha = \frac{M}{\theta} \tag{20}$$

gives the ratio of restraint on plates 1 and 2 using eqn. (18) as

$$R_2 = -R_1 \frac{b_2}{b_1} \left(\frac{t_1}{t_2} \right)^3 \tag{21}$$

The difference in signs of R_1 and R_2 arise because moment M_1 tends to reduce the rotation θ_1 while moment M_2 tends to increase the rotation θ_2.

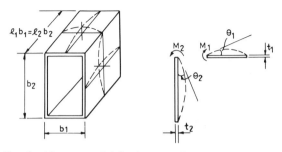

FIG. 6. Moments and deflections on adjacent plate elements.

Buckling Analysis of Sections

Assuming that the plates buckle simultaneously and that the critical stresses for each plate are thus equal, gives the relation between the buckling coefficients as

$$K_2 = K_1 \left(\frac{t_1 b_2}{t_2 b_1} \right)^2 \tag{22}$$

Thus, the stability of two plates with a common edge can be written in terms of the stability of one plate. It has already been shown that a general expression for K exists (eqn. (4)) in terms of the rotational restraint and the buckle wavelength. Substituting from eqns (4) and (21) into eqn. (22) gives the modified equation

$$\frac{(K_0)_1 + R_1(Q_1)_1(K_\infty)_1}{1 + R_1(Q_1)_1} = \left(\frac{b_1 t_2}{b_2 t_1} \right)^2 \left[\frac{(K_0)_2 - R_1 \dfrac{b_2}{b_1} \left(\dfrac{t_1}{t_2} \right)^3 (Q_1)_2 (K_\infty)_2}{1 - R_1 \dfrac{b_2}{b_1} \left(\dfrac{t_1}{t_2} \right)^3 (Q_1)_2} \right] \tag{23}$$

For any value of buckle half wavelength, e, this equation can be solved to obtain the value of R_1 and hence K_1 at the buckling of the section. To obtain the minimum value of K_1 it is necessary to examine a range of buckle lengths. Once K_1 has been obtained the critical buckling stress for the section can be evaluated from eqn. (2).

Postbuckling Analysis of Sections

In the postbuckling analysis an additional restraint is that each plate in the section is compressed by the same amount, i.e.

$$u_1 = u_2 \tag{24}$$

The nondimensional end displacement is given by

$$\bar{u} = \frac{ub^2}{at^2} \tag{25}$$

Substituting in eqn. (24) gives the relationship between the nondimensional end displacements as

$$\bar{u}_2 = \bar{u}_1 \left(\frac{b_2 t_1}{b_1 t_2} \right)^2 \tag{26}$$

In addition, using the compatibility condition as given in eqn. (18) and substituting from eqn. (14) gives

$$(\bar{u}_1 - K_1)\bar{\theta}_1^2 = \left(\frac{b_1 t^2}{b_2 t_1}\right)^2 \bar{\theta}_2^2 (\bar{u}_2 - K_2) \tag{27}$$

Substituting for $\bar{u}_2$ this equation can be written as

$$\bar{u}_1 = \frac{K_1 - \left(\dfrac{\bar{\theta}_2}{\bar{\theta}_1}\right)^2 \left(\dfrac{b_1 t_2}{b_2 t_1}\right)^2 K_2}{1 - (\bar{\theta}_2/\bar{\theta}_1)^2} \tag{28}$$

The values of K_1, K_2, R_1 and R_2 are thus related to the nondimensional end compression $\bar{u}_1$. As $\bar{u}_1$ is increased from its critical value, R will alter and the K values corresponding to these R values will also change. If an arbitrary value of R_1 is then chosen, K_1, K_2, $\bar{\theta}_1$ and $\bar{\theta}_2$ can be evaluated and substituted into eqn. (28) to obtain the corresponding value of $\bar{u}_1$. The procedure can be repeated to obtain the complete variation for the section. The load corresponding to this value of $\bar{u}_1$ can be obtained as follows.

The values of E^*/E and K can be obtained for each plate from eqns (10) and (4) for a given end compression knowing the values of R and e. The load on each plate can then be obtained from

$$\bar{P}_1 = K_1 + \left(\frac{E_1^*}{E}\right)(\bar{u}_1 - K_1)$$

$$\bar{P}_2 = \left[K_2 + \left(\frac{E_2^*}{E}\right)(\bar{u}_1 - K_2)\right]\frac{b_2 t_2}{b_1 t_1} \tag{29}$$

The total nondimensional load on the given rectangular box section considered is then

$$\bar{P} = 2(\bar{P}_1 + \bar{P}_2) \tag{30}$$

The postbuckling load/end displacement graphs can thus be obtained.

APPLICATION TO A SPECIFIC CASE

The above equations were derived for a glass reinforced plastic section with the following general properties.

$E_{11} = 30 \text{ GN/m}^2 \qquad E_{22} = 6 \text{ GN/m}^2 \qquad G_{12} = 5 \text{ GN/m}^2 \qquad \nu_{12} = 0.33$

The results of course could be derived for any reinforced plastic composite provided the fundamental mechanical properties of the material were known.

Buckling of Connected Plates or Sections

The application of the buckling analysis permitted Fig. 7 to be drawn. This shows the variation in the minimum buckling coefficient with b_2/b_1 for various thickness ratios.

It is seen that as b_2 increases with the other dimensions held constant, the buckling stress of the section reduces as expected. For the lower values of the t_1/t_2 ratio (i.e. t_2 thicker) the buckling stress remains fairly constant for increasing b_2. Alternatively as t_2 reduces in thickness the buckling stress for the section again reduces as expected.

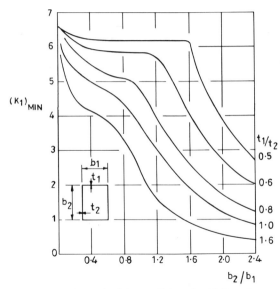

FIG. 7.   Minimum buckling coefficient for GRP sections.

Postbuckling of Connected Plates or Sections

The postbuckling analysis was applied in the first case to a square box section with length equal to wall width. The results for a thickness ratio of $t_2/t_1 = 2$ are shown in Figs 8 and 9. In Fig. 8 the variation of end displacement with load is given. Figure 9 shows the variation of ΔP_1 and ΔP_2 with increasing load. ΔP is the reduction in load of the plate compared with that for an unbuckled plate with the same deflection. It can be seen

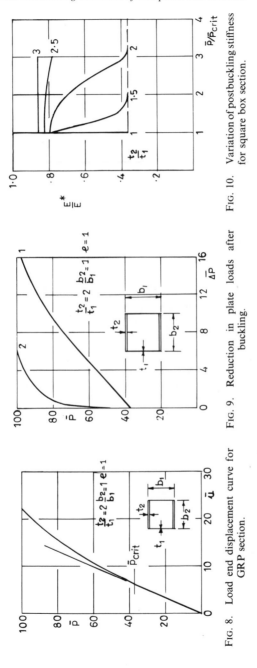

Fig. 8. Load end displacement curve for GRP section.

Fig. 9. Reduction in plate loads after buckling.

Fig. 10. Variation of postbuckling stiffness for square box section.

from this figure as expected, that plate 1 buckles first with a corresponding reduction in load. At first the load reduction in plate 2 is small but with increasing load the effects of buckling in plate 2 become apparent.

The variation in postbuckling stiffness for the section was obtained from Fig. 8 and is plotted in Fig. 10. It can be seen from this figure that after a sudden reduction in stiffness, there is a more gradual fall off. For comparison this figure also shows the postbuckling stiffness for a thickness ratio of $t_2/t_1 = 1\cdot5$, $2\cdot5$ and 3. The values become asymptotic to that for a simply supported plate at this aspect ratio viz. $0\cdot361$. Application to other sections was not given because of space limitations.

SUMMARY AND CONCLUSION

The paper extends earlier work on glass reinforced plastic plates to examine theoretically the buckling and postbuckling behaviour of GRP sections. The sections are considered as a series of linked orthotropic plates. Application is made to a box section in particular and the critical loads and postcritical behaviour predicted for particular geometries. The earlier work covered additional boundary conditions to those considered here and it is anticipated that the work will be extended to cover alternative section geometries, e.g. channels.

The market for reinforced plastic products is continually expanding. This is leading to the structural application of composites in, for example, the aircraft industry. This in turn means that problems solved earlier for isotropic systems need to be re-examined and analysed for the new materials. The work presented in this paper is a contribution in that direction.

REFERENCES

1. BANKS, W. M., HARVEY, J. M. and RHODES, J. The non-linear behaviour of composite panels with alternative membrane boundary conditions on the unloaded edges, *Proc. 2nd Int. Conf. on Composite Materials*, Toronto, April 1978, 316–36.
2. BANKS, W. M. and RHODES, J., The buckling behaviour of reinforced plastic box sections, *The Reinforced Plastics Congress '80*, Brighton, November 1980, 85–8.
3. HAMID, A. B. A., The examination of the behaviour of connected plates, MSc Thesis, University of Strathclyde, 1979.
4. RHODES, J., Secondary local buckling in thin-walled sections, *Hungarian Academy of Science*, **87**(1–2), 1978, 143–53.

27

The Effect of Thermal Strains on the Microcracking and Stress Corrosion Behaviour of GRP

F. R. Jones, A. R. Wheatley and J. E. Bailey

Department of Metallurgy and Materials Technology,
University of Surrey, Guildford, Surrey GU2 5XH, England

ABSTRACT

The tensile mechanical properties of two polyester/glass crossply laminates cured under different conditions have been investigated. Large thermal strains are generated which severely reduce the strain at which the cracking of the plies is initiated. The two chemically similar resins produced laminates which behaved differently.

The rapid stress corrosion failure of the outer 0° plies of a crossply laminate is shown to be associated with transverse cracks in the 90° ply.

INTRODUCTION

The design of many GRP structures is limited by the loss in integrity which occurs with the onset of microcracking at low composite strains, occurring particularly in those areas where the principal fibre direction is perpendicular to the direction of an applied stress. Loss of composite integrity may be important in applications such as containers and pipes where any flaws or microcracks may give rise to weepage or expose the fibres to corrosive attack.

This microcracking phenomenon is most readily studied in model crossply composites where fibres are laid up at right angles to each other. Sandwiched between the two outer 0° plies (*y* direction) is a transverse 90° layer designed to give strength and stiffness in the *z* direction. Tensile loading is applied parallel to the outer 0° fibres. On cooling down from the postcure temperature the inner transverse ply is prevented from contracting in the y direction by the high modulus $0°$ plies. Similarly the outer $0°$ plies are prevented from contracting in the z direction by the inner transverse ply. This results in tensile stresses being built into each lamina in a direction perpendicular to the direction of its fibres. It has been shown that the failure strain of the transverse ply can be drastically reduced by the existence of these large thermal strains. In a previous paper[1] it was suggested that the large thermal strains in polyester laminates were responsible for the absence of the 'whitening' effect seen in epoxy laminates made with the same glass fibres. We wish to report further examination of the microcracking behaviour of laminates made from two similar polyester resins with and without formal postcure. The effect of microcracking and transverse cracking on the stress corrosion failure of the longitudinal plies is also reported.

Under tensile loading, the Poisson's shrinkage of the $0°$ plies in the z direction is restricted by the transverse ply. Splitting of the $0°$ laminae in the fibre direction occurs under the influence of an applied stress. It has been shown that the applied stress necessary to induce longitudinal splitting of the $0°$ plies is strongly dependent upon the magnitude of the inbuilt thermal strains.

EXPERIMENTAL

Crossply laminates have been prepared from isophthalic polyester resins reinforced with Silenka 1200 tex 'E' glass fibres with a silane finish compatible with polyester and epoxy resins. The resins used were Crystic 272 and D3061 (Scott-Bader and Co. Ltd) which differ mainly in the chemistry of their formation. Crystic 272 is an isophthalic/fumaric acid based resin whereas Crystic D3061 is based on maleic anhydride. The resins are similar in that isomerisation of the maleate to fumarate links occurs during polyesterification.

Laminates were made by machine winding individual plies of glass rovings onto metal frames, followed by impregnation of the fibres with resin. In order to maintain a relatively long working time and prevent excessive temperature build-up a relatively small quantity of accelerator was used. The curing system was 2 % of a 50 % methyl ethyl ketone peroxide solution (catalyst M) and 0·25 % cobalt naphthenate solution (accelerator E) (Scott-Bader Co. Ltd).

In the preparation of $0°/90°/0°$ and $0°/90°$ crossply laminates the $90°$

ply was laid up first and allowed to gel for 24 h prior to the laying on of the outer $0°$ plies. This method of laminate preparation has proved extremely successful in the production of high quality laminates with volume fractions of glass in the range $V_f = 0.35 \pm 0.03$. Attempts to fabricate crossply laminates in a single process proved much less successful due to a tendency of the laminates to be of a much poorer finished quality with a large degree of cracking becoming evident on curing. Figure 1 compares edge damage in $0°/90°$ coupon samples prepared by both methods. Sample (a) was prepared in the conventional manner by laying the $0°$ ply onto the pre-gelled $90°$ ply. Sample (b) was prepared in a single process with both plies being wetted-out simultaneously. Sample damage in the former is considerably lessened.

Figure 1 also illustrates the method by which thermal strains are measured. Asymmetric $0°/90°$ beams have been prepared which behave like 'bimetallic strips' whereby the contraction of the $90°$ ply on cooling from the postcure temperature produces a strip of a fixed radius of curvature. Measurement of this radius of curvature may be used to determine the

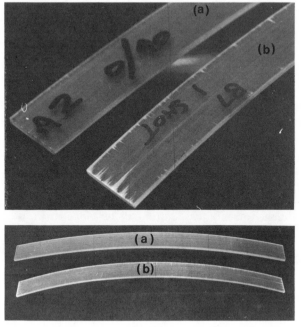

FIG. 1. $0°/90°$ coupon specimens prepared (a) by a two-stage process and (b) by a one-stage process.

magnitudes of thermal strains built into the plies of a crossply laminate on postcuring.[2]

The tensile mechanical properties of $0°/90°/0°$ laminates were determined on INSTRON 1195 and 1196 tensile testing machines. Coupon specimens were tested at a strain rate of 0.005 min^{-1}. The values of stress and strain required to produce the first transverse crack ($\sigma_{tlu}, \varepsilon_{tlu}$) and Poisson-type splitting of the longitudinal plies ($\sigma_{lls}, \varepsilon_{lls}$) have been recorded. The Young's modulus (E_{cl}) and ultimate tensile strength (σ_{uc}) of the laminates were also determined.

Optical and scanning electron microscope studies have been carried out in conjunction with a mechanical straining stage in order to study the cracking characteristics of strained laminates.

The stress corrosion behaviour of $0°/90°/0°$ laminates in four-point bending has been investigated. The force–extension and cracking characteristics of each laminate in four-point bending were determined by Instron testing. Tests in a four-point bending jig were then carried out on coupon specimens (15 mm wide, 3·2–4·0 mm thick) immersed in 1 M aqueous sulphuric acid. Loads both above and below the first transverse cracking force, F_{tlu}, were applied.

RESULTS AND DISCUSSION

The mechanical properties of laminates prepared from Crystic 272 and D3061 are presented in Table 1. The data refer to laminates prepared such that the thickness of the inner ply is twice that of the outer plies. Under these circumstances the thermal strain in the transverse ply in the longitudinal direction ε_{tl}^{th} is equal to the thermal strain in the longitudinal ply in the transverse direction ε_{lt}^{th}.

It may be seen from Table 1 that considerable internal stresses are developed in the transverse ply due to postcuring at elevated temperatures. For instance, the values of tensile strain, ε_{tl}^{th}, developed in the transverse ply on cooling from a postcure temperature of 80 °C in laminates prepared from Crystic 272 and D3061 are 0·40 % and 0·47 % respectively. This inbuilt thermal strain reduces the strain at which the first transverse crack appears ε_{tlu} (representing failure of the inner ply) from 0·6 % to 0·17 % and 0·13 % respectively. The transverse cracking strains of laminates postcured at 50 °C are also reduced, though to a lesser extent.

Similarly an equivalent transverse thermal strain is established in the longitudinal plies and this is responsible for a reduction in the strain at

TABLE 1

Effect of curing conditions on the tensile properties of crossply ($b = d$) glass fibre laminates from two polyester resins

Postcuring schedule	ε_{tl}^{th} (%)	ε_{tlu} (%)	$\varepsilon_{tl}^{th} + \varepsilon_{tlu}$ (%)	ε_{lls} (%)	E_{cl} (GPa)
Crystic 272					
170 h at 18 °C	0	0.59 ± 0.03	0.59 ± 0.03	1.77 ± 0.1	16.0
15 h at 50 °C	0.34 ± 0.01	0.24 ± 0.03	0.58 ± 0.04	0.84 ± 0.1	16.8
3 h at 80 °C	0.40 ± 0.01	0.17 ± 0.03	0.57 ± 0.04	0.64 ± 0.1	17.5
1.5 h at 130 °C	0.45 ± 0.01	0.13 ± 0.03	0.58 ± 0.04	0.66 ± 0.1	17.4
Crystic D3061					
170 h at 20 °C	0	0.32 ± 0.04	0.32 ± 0.04	1.62 ± 0.1	14.1
15 h at 50 °C	0.42 ± 0.02	0.15 ± 0.02	0.57 ± 0.04	0.72 ± 0.1	14.8
3 h at 80 °C	0.47 ± 0.02	0.13 ± 0.03	0.60 ± 0.05	0.56 ± 0.1	14.7

b = outer ply thickness

d = semi-inner ply thickness

ε_{tl}^{th} = thermal strain (calculated from the curvature of 0/90 laminates) in the longitudinal direction of the transverse ply. The cold-cured specimens have not experienced a temperature excursion and are considered to have zero thermal strain although a slight curvature is observed.

ε_{tlu} = experimental transverse ply cracking strain.

ε_{lls} = longitudinal splitting strain.

E_{cl} = Young's modulus of the 0°/90°/0° crossply laminate.

which Poisson's splitting of the longitudinal plies occurs. For Crystic 272 and D3061 laminates postcured at 80 °C, ε_{lls} falls from 1·77 to 0·64% and from 1·62 to 0·56% respectively.

Therefore the failure strain of the transverse ply can be computed from the measured thermal strain ε_{tl}^{th} and the experimental transverse cracking strain ε_{tlu} (Fig. 2). Figure 3 shows the variation of the transverse failure strain with inner ply thickness for laminates postcured at different temperatures. The outer ply thickness was kept constant at 1·0 mm (Crystic 272) and 1·1 mm (Crystic D3061). The volume fraction of glass fibres, V_f, was kept constant at 0.35 ± 0.3.

We have also consistently observed that 0°/90° laminates which have not experienced formal postcuring have a slight curvature. The curing conditions are such that the laminate temperature has remained constant and therefore we consider that strains can be built into the plies during gelation. The measured curing strains ε_{tl}^{c} in the transverse plies of Crystic 272 crossply laminates are given in Fig. 3. When this component is

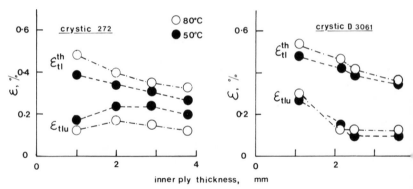

FIG. 2. The effect of transverse ply thickness $(2d)$ on ε_{tl}^{th} and ε_{tlu} developing in $0°/90°/0°$ GRP laminates from Crystic 272 and D3061 under different postcure temperatures (see Table 1 for definitions).

included, the transverse ply failure strain for the cold-cured laminates is increased, with the result that ε_{tu} for the Crystic 272 laminates is apparently independent of variations in thickness or postcuring schedule. This result is in contrast to Crystic D3061 laminates in which ε_{tu} increases with postcuring and is larger for the thinner transverse plies.

The transverse cracking behaviour of epoxy/glass laminates has been studied.[3] With inner ply thicknesses less than 0·4 mm transverse cracking

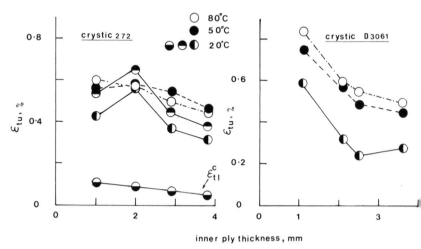

FIG. 3. The effect of transverse ply thickness $(2d)$ on the $90°$ ply failure strain (ε_{tu}) of crossply laminates with different postcure temperatures. For $\bigcirc$ $\bullet$ $\leftmoon$, $\varepsilon_{tu} = (\varepsilon_{tl}^{th} + \varepsilon_{tlu})$. $\ominus$ are internal strains in gelled laminates ε_{tl}^{c}, $\ominus$ $\varepsilon_{tu} = (\varepsilon_{tl}^{c} + \varepsilon_{tlu})$. ε_{tl}^{c} is the curing strain in cold-cured laminates.

was found to be constrained. Thus crack opening is restrained by the non-cracking phase so that a greater energy is required to propagate a crack. Therefore the cracks run in a controlled manner at higher strains. Examinations of the transverse ply of epoxy/glass laminates during loading have been made and have shown that the mechanism of crack initiation occurs by debonding of the resin/glass interface.[1] Both Crystic 272 and D3061 have similar properties; the latter has a slightly lower modulus and increased strain to failure, as expected of resins identical except that the latter relies on isomerisation of the maleate to fumarate unsaturations. Thus Crystic D3061 is slightly less reactive and forms a slightly more flexible resin. However, Fig. 3 shows that the crossply laminates behave differently and that transverse cracking is constrained in Crystic D3061 laminates, ε_{tu} increases markedly with decreasing inner ply thickness even when this is as large as 2 mm. This is absent in Crystic 272 laminates and ε_{tu} appears to be independent of inner ply thickness down to 1 mm (Fig. 3). The failure strain of the transverse plies in these laminates are dependent on the interfacial glass/resin bond. Therefore, since postcuring increases the value of ε_{tu} for laminates from Crystic D3061 and not for Crystic 272 we conclude that the molecular variations in the resin have affected the interfacial bond and the resin in the locality of the fibres. More important however are the larger thermal strains in the former which more than offset the improvement in ε_{tu}. The need for formal postcuring of Crystic 272 laminates is therefore brought into question.

Recent[4] measurements of the expansion coefficients of Crystic 272 resins and laminates have indicated an anomalous behaviour of the resin matrix within the laminate. This result, together with the high thermal strains reported here, suggests that the resin structure is modified by the glass fibres and therefore the interfacial resin properties may vary. These results are the subject of a further study and will be reported elsewhere.

MICROCRACKING IN POLYESTER/GLASS COMPOSITES

Transverse cracking in these model composites has been observed when the applied strain is as low as 0·13%. The appearance of the first transverse crack under loading represents the first total failure of the 90° ply. Further systematic cracking of the 90° ply is observed as the applied composite strain is increased such that shear transfer between the plies allows the strain elsewhere in the transverse ply to reach ε_{tu}. This effect has been described previously.[5] Considerable sample damage may, however, arise prior to ply failure.

As indicated above debonding has been observed in constrained transverse plies because the crack forms by the coalescence of debonds. Therefore in the stressed Crystic 272 laminates which we have studied, microcracks form. Laminates with and without formal postcure have been tensile loaded on the straining stage within a scanning electron microscope. Microcracks (Fig. 4) can also be seen by optical microscopy in postcured laminates in which a thermal strain has developed. These microcracks are absent in laminates prepared without formal postcure (Fig. 4). On cooling from the postcure temperature microcracking of the 90° ply takes place spontaneously (Fig. 4). The extent of the microcracking is dependent upon the postcure temperature, which in turn determines the level of thermal strain. These microcracks are absent in laminates with internal strains less than 0·1 % and appear when the thermal strains reach approximately 0·3 %. Microcracking occurs invariably in regions of high volume fractions of glass. The growth of the cracks across the 90° ply is inhibited by the resin rich areas both within the transverse ply and at the 0°/90° interface (Fig. 4). Cracks may be blunted by the 0° fibres at the 0°/90° interface and continue to propagate along this interface, but at applied strains of magnitude about 0·6 % for an 80°C postcured Crystic 272 laminate the transverse cracks penetrate the longitudinal plies (Fig. 5).

Figure 5 shows the transverse cracks produced under tensile loading and in four-point bend loading. It is seen that the latter only span the tensile loaded region of the 90° ply and that the first transverse cracks also penetrate the longitudinal 0°. This difference is further shown in the stress corrosion results reported below.

THE APPLICATION OF MICROSTRUCTURAL STUDIES TO ENVIRONMENTAL STRESS CORROSION CRACKING

One area in which the cracking and microcracking characteristics of polyester/glass laminates are of paramount importance is in the field of stress corrosion. Failure of the load bearing elements of GRP structures subjected to stress and a corrosive environment proceeds via attack on the principal load-bearing elements themselves, e.g. the fibres. Loss of integrity of structures through microcracking and transverse cracking provides a route through which corrosive attack of the fibres may proceed. Recent work has shown that a direct link exists between rapid stress corrosion failure of GRP laminates under a four-point bend and the appearance of transverse cracking in the system.

A comparison of Figs 6(a) and (b) shows that at the same applied loads

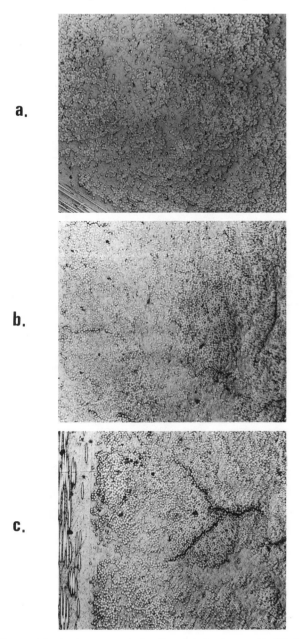

FIG. 4.   Microcracking in the transverse ply of crossply laminates with differing levels of internal strain developed from curing at (a) 20 °C, (b) 50 °C and (c) 80 °C.

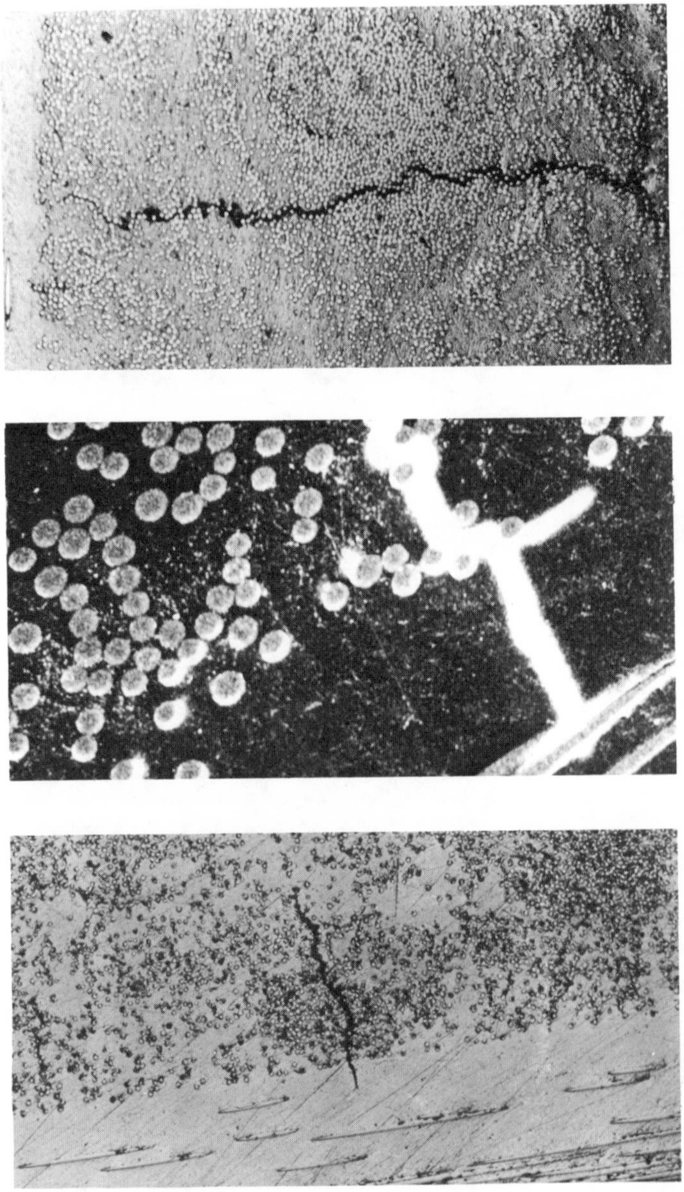

FIG. 5.   Transverse cracks produced in tension under (a) low strain, (b) high strain and (c) four-point bend.

the postcured laminate is more susceptible to stress corrosion attack. Furthermore none of those samples loaded below the transverse cracking force F_{tlu} failed, even though loads of $0.9 F_{tlu}$ have been employed. Loading above F_{tlu} resulted in stress corrosion failure of all but one sample within 12 h. Examination of the failed specimens showed that transverse cracking had occurred in all cases and that the characteristic smooth stress corrosion cracks had been formed in the tensile $0°$ ply immediately adjacent to a transverse crack. This is illustrated in Fig. 7. Failure is not always caused by the growth of a particular stress corrosion crack. Several stress corrosion cracks have been observed in the longitudinal ply, all of which are associated with a transverse crack in the $90°$ ply, and are usually initiated in the middle of the longitudinal ply face.

Figure 6(c) confirms that the stress corrosion failure is linked to transverse cracking since unidirectional $0°$ laminates loaded over the same strain range have not failed within the time scale of the experiment. Also, no significant reduction in mechanical properties of the unidirectional samples was observed after six weeks immersion in the acid. Recently,[6] however, notched longitudinal specimens have been seen to crack readily under the influence of stress corrosion.

The transverse ply of cut-coupons from crossply laminates has exposed fibre ends in contact with the corrodant. In order to overcome this problem, the edges of laminates gelled at 20 °C have been coated with the same resin and allowed to gel for a further 24 h. Specimens with and without an edge coating were subjected to the same four-point bending loads whilst immersed in 1 M aqueous sulphuric acid. Figure 6(d) shows that, as before, failure only occurred in samples loaded above F_{tlu}, and within 12 h. The quoted value of F_{tlu} is an average value and in those cases where samples which appear to be loaded above F_{tlu} have not failed, subsequent examination showed that no transverse cracking was present. No differences between the failure characteristics of the edge-coated and non-edge-coated samples were observed. Transverse failure of the $90°$ ply was accompanied by failure of the edge coating. However, differences in the microcracking behaviour in acid were observed. No transverse cracking was apparent but samples without edge coatings had developed diffuse edge cracks. This type of damage[7] did not cause catastrophic failure of the $0°$ plies but is much reduced by edge coating.

The probable explanation as to why no stress corrosion failure of extensively edge-cracked specimens has occurred is that no path to the outer $0°$ plies is created by edge cracking through the transverse layer. A true transverse crack traverses the whole thickness of the inner ply and

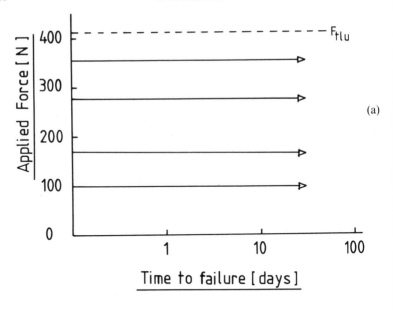

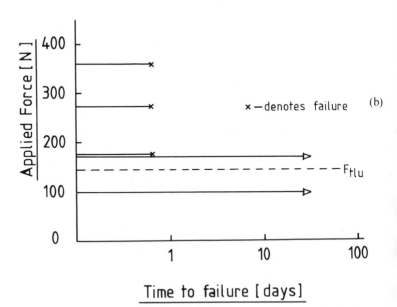

FIG. 6. Stress corrosion of Crystic 272 GRP laminates in 1 M aqueous sulphuric acid in four-point bend. (a) 0°/90°/0° cured at 20°C, (b) 0°/90°/0° postcured at 80°C, (c) unidirectional 0° postcured at 80°C, (d) 0°/90°/0° cured at 20°C with and without an edge coating.

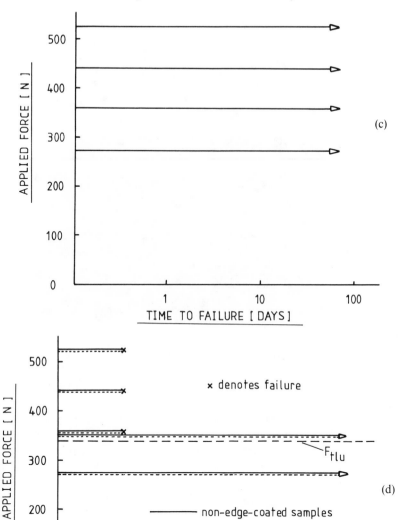

FIG. 6—*contd.*

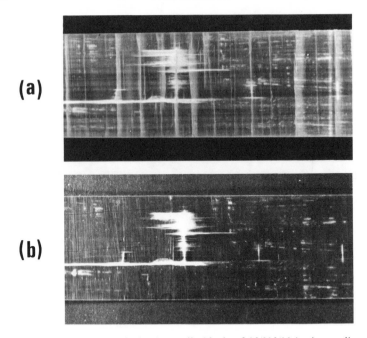

(a)

(b)

Fig. 7. Stress corrosion cracks in the tensile 0° ply of 0°/90°/0° laminate adjacent to transverse cracks (see Fig. 6). (a) Failed specimen and (b) the delaminated 0° ply.

under circumstances described above can extend slightly into the outer 0° plies, hence providing a direct, stress enhanced, route into the main load-bearing lamina. In no cases to date has a stress corrosion failure been observed in the compressive face of the laminate.

Figure 5(c) indicates that in bend the transverse crack penetrates the longitudinal ply, whereas in tension penetration only occurs at higher strains. This has a large influence on the stress corrosion cracking since tensile loaded specimens have not failed catastrophically at strains just above ε_{tlu} when transverse cracks are present,[7] in contrast to the results in Fig. 6.

CONCLUSIONS

The appearance of large thermal strains in crossply laminates reduces the transverse cracking strain. Since the strain at which stress corrosion cracking of the longitudinal plies occurs is strongly dependent on the formation of a transverse crack which penetrates the longitudinal ply, the knowledge of the inherent strain levels and the careful control of the lamination procedure are of prime importance to the design strains for these materials. It is also insufficient to rely on the ultimate failure strains of resins and composites, so that a detailed knowledge of the microstructural behaviour of the laminates is required.

ACKNOWLEDGEMENTS

We wish to thank the Polymer Engineering Directorate for a research award (ARW), the Science Research Council for an equipment grant and Scott-Bader and Co. Ltd for the resins.

REFERENCES

1. BAILEY, J. E. and PARVIZI, A., On fibre debonding effects and the mechanism of transverse ply failure in cross-ply laminates of glass fibre/thermosets composites, *J. Mat. Sci.*, **16** (1981) 649–59.
2. BAILEY, J. E., CURTIS, P. T. and PARVIZI, A., On the transverse|cracking and longitudinal splitting behaviour of glass and carbon fibre reinforced epoxy cross-ply laminates and the effect of Poisson and thermally generated strain, *Proc. R. Soc. Lond. A.*, **366** (1979) 599–623.
3. PARVIZI, A., GARRETT, K. W. and BAILEY, J. E., Constrained cracking in glass fibre-reinforced epoxy cross-ply laminates, *J. Mat. Sci.*, **13** (1978) 195–201.
4. (a) JONES, F. R., MULHERON, M. J., WHEATLEY, A. R. and BAILEY, J. E., The effect of curing conditions on the properties of the matrix and interfacial bond in glass fibre reinforced polyesters, *Interfaces in composites materials*, PRI meeting, Liverpool 1981, paper 6.
   (b) JONES, F. R. and MULHERON, M. J., unpublished results.
5. PARVIZI, A. and BAILEY, J. E., On multiple transverse cracking in glass fibre epoxy cross-ply laminates, *J. Mat. Sci.*, **13** (1978) 2131–6.
6. JONES, F. R. and WHEATLEY, A. R., in preparation.
7. BAILEY, J. E., FRYER, T. M. W. and JONES, F. R., Environmental stress-corrosion edge cracking of glass reinforced polyesters. In: *Advances in composite materials (ICCM 3)*, Vol. I, Bunsell, A. R. *et al.* (eds), Paris, Pergamon Press, 1980, 514–28.

28

Electrically Conductive Prepreg Systems

Gary L. Patz

*Hexcel Corporation, 11711 Dublin Boulevard,
Dublin, California 94566, USA*

ABSTRACT

In efforts to reduce weight and potentially reduce costs of aircraft, airframe manufacturers have been increasing their use of composite materials. A major problem with composite structures is that they lack the electrical and thermal characteristics of the metal structures they are replacing. Low cost, structural conductive prepreg materials are now being developed by Hexcel to offer solutions to these very difficult problems. These materials, which are being marketed under the Thorstrand^{tm} trade mark, are based on MBAssociate's aluminum coated E-glass fiber.

This paper will provide the background on the basic fiber, and detail how this fiber has been designed into systems that provide an airframe structure with the electrical and thermal conductivity that is necessary for the satisfactory electrical and thermal performance of the structure.

INTRODUCTION

Because of their weight, stiffness and strength characteristics and potential cost savings, composite materials are quickly replacing metal structures in aircraft design and construction. A main problem in the use of composite materials has been in achieving even a portion of the electrical and thermal conductivities of the metals required for functional purposes such as lightning strike protection, static dissipation, etc. Several methods are now being employed to provide surface conductivity to composite structures.

430

These include conductive paints, encapsulated wire screens and metal foils, and flame-sprayed aluminum. Unfortunately, the methods are difficult to employ as they require special processing by highly skilled operators. Furthermore, they add weight to the composite structure without any appreciable increase in strength.

To help solve some of these problems Hexcel has developed a family of thermally and electrically conductive prepreg systems with drapability and processability very similar to standard fiberglass prepregs. These systems are now being marketed by Hexcel under the Thorstrand™ trademark.

BACKGROUND ON THE THORSTRAND SYSTEMS

The Thorstrand family of prepreg systems is based on a unique aluminum coated E-glass fiber produced by MBAssociates of San Ramon, California. MBAs fiber, Metafil® G, possesses a continuous, uniform coating of pure aluminum that is chemically bonded to an E-glass filament (Fig. 1). Presently all Thorstrand systems use the half (surface)-coated fiber.

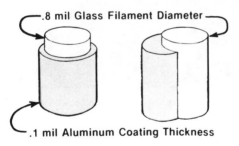

Fig. 1.   The two types of aluminum coated glass fibers. V/O glass 0·60; aluminum 0·40.

Photomicrographs of Thorstrand composites after tensile and compression failure have shown that the aluminum and glass behave as a single fiber (Fig. 2). Separation of the aluminum from the glass has never been observed.

Pure aluminum has very predictable electrical and thermal properties. Each Thorstrand system was designed to meet requirements of specific applications based on these very predictable properties. If so desired, E-glass, S-glass or Kevlar® can be combined with the Metafil yarns to provide increased structural properties.

Hexcel presently has three Thorstrand fabrics in full production. These are TEF7, TEFA and TSF181. Two other systems TEF5 and TKF7 are being produced in prototype quantities. The physical properties of these fabrics are found in Table 1.

TEF7 was designed specifically to provide protection to the composite substructure from the direct effects of a 200 000 amp lightning strike.

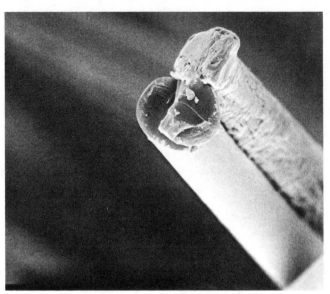

FIG. 2.   Photomicrograph of the Metafil® G fiber.

TEFA, originally developed for replacing flame spray in composite parabolic antennas, is also being used in aerospace applications to replace flame-sprayed aluminum for use as an antenna ground plane and for zone 2 lightning protection.[1]

TSF181 is being used for static electricity dissipation on fiberglass structures. It replaces a structural ply of 1581 fiberglass as well as the conductive paint.

TEF5 is a tightly woven fabric of about the same weight as TEFA. It allows for improved cure part surface characteristics.

TKF7 is a Thorstrand/Kevlar hybrid that incorporates the light weight and good tensile and modulus properties of Kevlar with the conductive and good compressive properties of Metafil G.

TABLE 1

	TEF7	TEFA	TSF181	TEF5	TKF7
Oz/sq yard	7·0	5·4	7·8	4·9	6·8
Grams/sq meter	237	183	265	166	230
Weave	5HS	plain	8HS	5HS	5HS
Construction	32 × 30	24 × 20	57 × 54	40 × 40	30 × 28
warp	2(45/45)/150-1/0	2(45/45)/150-1/0	(22/45)/150-1/0	(45/45)/900-1/0	2(45/45)/Kevlar 195
fill	2(45/45)/450-1/0	2(45/45)/450-1/0	(22/45)/150-1/0	(45/45)/900-1/0	2(45/45)/Kevlar 195
Number of Metafil fibers/inch	2 880 × 2 700	2 160 × 1 800	1 254 × 1 188	1 800 × 1 800	2 700 × 2 520

DEVELOPING THORSTRAND SYSTEMS TO SATISFY
ELECTRICAL REQUIREMENTS

Each individual Metafil G fiber is capable of conducting electricity up to that point where the aluminum melts and separates, thus halting the current flow. The Metafil G fiber has a nominal resistance of 2 ohm/cm length. Each fiber has an electrical current carrying capability as a function of duration and environment as follows:

Duration	Environment	I max
1 μsec	air	50 amps
10 μsec	plastic	25 amps
1 msec	air	0·5 amps
constant	air	0·3 amps
constant	plastic	0·6 amps

The fibers have an extraordinary current carrying capability in the region of 10 μsec to 1 msec because of the excellent coupling of the aluminum to the glass. The glass serves as a heat sink with a heat capacity of twice that of aluminum up to the melting point of aluminum.

A worst-case zone 1 (the primary lightning attachment zone on an aircraft) lightning strike is characterized by a peak amplitude of 200 000 amps and an action integral of 2×10^6 amp^2 sec. The zone 1 structure will then need to carry these high currents for about 10 μsec. For durations of approximately 10 μsec each Thorstrand fiber can conduct a maximum of 25 amps. Since 200 000 amps need to be dissipated, then 200 000/25 or 8000 fibers are required in the area of direct attachment. On a painted structure the area of attachment is estimated to be slightly less than two square inches. Using these simple calculations, the TEF7 fabric construction was developed.

Lightning Technologies, Inc. of Pittsfield, Massachusetts performed a series of tests on two separate sets of panels manufactured by Vought Corporation and Northrop Aircraft. The panels were both sandwich and laminate structures of fiberglass, Kevlar and graphite reinforced epoxy. TEF7 and flame-sprayed aluminum were the two protection systems employed. Lightning Technologies followed the testing parameters prescribed in the SAE report *Lightning Test Waveforms and Techniques for Aerospace Vehicles and Hardware* (June 1978).

The tests confirmed the results of earlier tests which demonstrated that a single ply of Thorstrand TEF7 aluminized glass-fabric will protect graphite/E and Kevlar/E laminates or sandwich skins typical of aircraft surfaces located within lightning strike zones 1A or 2A, for example, fuselages, engine nacelles, wing tips and empennage surfaces subject to direct lightning strikes with a low probability of flash hang-on. Trailing edges located in zone 1B must be expected to receive multiple strokes to the same spot and may have to be protected with several layers of Thorstrand or other means if structural damage to the laminate is to be avoided.

Thorstrand TEF7 and flame-sprayed aluminum appear to provide equivalent protection to laminates made of a combination of graphite/E and Kevlar/E plies. Thorstrand is also effective for protection of laminates made of Kevlar/E only. Flame-sprayed aluminum may be less effective than Thorstrand for protection of Kevlar/E laminates based on the significantly greater 'affected zone' after a lightning strike test, i.e. the greater area of flame spray burned away from the laminate about the heat affected attachment point.

In many applications the lightning protection systems must do more than protect the aircraft from the direct effects of lightning. The indirect effects can be just as hazardous. Low frequency magnetic field energy is emitted from a lightning strike. This energy can upset many of the electrical components with which it comes in contact. Thorstrand systems are now being tested for their effectiveness in providing protection against these indirect hazards.

It should be noted here that each structure possesses its own level of susceptibility to damage by lightning. Lightning protection of a composite structure is a problem that must be addressed early in the design stages. Once a part has been placed in production, it is very difficult to provide that structure with an efficient lightning protection system.

In a fashion analogous to the example cited above for lightning strike protection, the intrinsic electrical or thermal conductivity characteristics of the Metafil G fiber (used as the basic building block) provide the basis for the development of fabric constructions for specific applications such as: EMI (electromagnetic interference), static dissipation, parabolic dish antennas, and thermal dissipation.

Simulated tests are subsequently employed to confirm the effectiveness of the system design.

MANUFACTURING THORSTRAND PROTECTED STRUCTURES

Thorstrand prepregs possess the same drape, handling and processing characteristics as fiberglass prepreg systems, so the same processing techniques can be employed. After the Thorstrand prepreg is formed onto the mold as the exterior ply and then cured with the glass, Kevlar or graphite structural plies, the electrical and structural properties of the Thorstrand ply become an integral part of that structure. No secondary manufacturing step is necessary.

Over Kevlar structures the Thorstrand ply aids in minimizing the fraying of the Kevlar structure during subsequent cutting and drilling.

An adequate means of conducting electrical currents into and away from the Thorstrand-protected structure must be provided. This can be accomplished by direct contact between adjacent Thorstrand plies, between Thorstrand and other metallic elements, or by means of metallic fasteners. If metallic fasteners are utilized to conduct the currents from Thorstrand-protected structures to other conducting structures, direct contact should be made between the Thorstrand and the fasteners to minimize electrical sparking at these interfaces.

Lightning Technologies, Inc. recently evaluated a series of panels manufactured by Vought Corporation for the purpose of determining satisfactory metallic fastening methods for conducting current from the Thorstrand-protected panel to the grounded aircraft substructure. They reported that 'both the removable fasteners, which were spaced five inches apart, and the aluminum and monel rivets, which were spaced 0·75 to 1·5 inches apart, permitted sparking to occur in their immediate vicinity during the severe 200 kA stroke, but very little sparking was evident during the 50 kA stroke. Since current densities (i.e. current per rivet or fastener) in most aircraft applications would be lower than those resulting during these tests, the current per fastener would be less. Thus, any of the configurations tested here would appear to be adequate to transfer current to interior structural members without damage.'[3] Individual designs should be evaluated by simulated lightning tests if doubt exists as to their effectiveness.

SUMMARY

Because of the uniformity and predictability of the basic fiber, Thorstrand systems can be developed to meet the electrical, thermal and mechanical requirements for almost any application. Thorstrand systems have been successfully developed for lightning protection, static electricity dissipation, EMI shielding, parabolic antenna dish surfaces and thermal dissipation. Work is continuing at Hexcel in the areas of developing good fastening techniques and producing maximum strength Thorstrand systems with the lowest weight.

REFERENCES

1. Report of SAE Committee AE4 Special Task F, *Lightning Test Waveforms and Techniques for Aerospace Vehicles and Hardware*, May 5, 1976.
2. PLUMER, J. A., *Simulated Lightning Tests on Graphite Laminates Protected with Thorstrand Aluminized Glass Cloth*, Lightning Technologies, Inc., August 10, 1979.
3. PLUMER, J. A., *Simulated Lightning Tests on Kevlar and Graphite Laminates Protected with Thorstrand Aluminized Glass Cloth and Flame-Sprayed Aluminum*, Lightning Technologies, Inc., March 1981.

29

Analysis of Composite Materials by Dynamic Thermomechanometry (Dynamic Mechanical Analysis)

P. Burroughs and J. N. Leckenby

*Du Pont (UK) Ltd, Wedgwood Way,
Stevenage, Herts SG1 4QN, England*

ABSTRACT

Epoxy–fibre composite materials have been analysed by dynamic mechanical analysis. The technique has been used to detect the glass transition of highly cured composite materials and to investigate the curing reaction itself for normally formulated and experimental materials. Both isothermal and scanning temperature programs have been used, individually, and combined in more complex cure cycles.

Differences in mechanical and curing properties brought about by a variation in the composition of the reinforcing fibres have also been detected

INTRODUCTION

Composite materials consisting of a thermosetting polymer and a reinforcing agent are becoming increasingly important for applications in, for example, the aerospace industry, the automotive industry, building and electronics. In order to achieve maximum dimensional and high temperature stability from such materials it is necessary to characterise the curing (crosslinking) reaction in the polymer and to be able to measure the degree of cure in the final product. Furthermore, changes brought about by small differences in the product formulation and the reinforcing agent must also be monitored.

Thermal analysis techniques have been used for some years to study thermosetting polymers and, more recently, the same techniques have been

438

applied to commercially available reinforced thermosets. Data has been presented at two recent meetings[1,2] on the study of epoxy–graphite composites by several TA techniques, and this paper reports a further extension of that work with particular emphasis on the results obtained by dynamic thermomechanometry or dynamic mechanical analysis (DMA) which, for a number of reasons, seems to be the most successful TA technique for these materials.

MATERIALS AND INSTRUMENTATION

A variety of reinforced epoxy resins have been used in this study from both commercial and experimental sources, but the primary material investigated was Fibredux* 914 C which was supplied in the form of a unidirectional carbon fibre prepreg from Ciba-Geigy. The polymeric phase of this material is a proprietary mixture of epoxies that cure primarily in reactions promoted by a solid curing agent or hardener which is present in the prepreg in low quantities. The prepreg contains about 58 % by weight of carbon fibre.

Du Pont thermal analysis equipment was used throughout this work, together with either a model 990 thermal analyser or a model 1090 thermal analyser/data system. The TA modules used were a 910 Differential Scanning Calorimeter, a 981 Dynamic Mechanical Analyser and a 943 Thermomechanical Analyser.

RESULTS AND DISCUSSION

The study of the curing of composite epoxy systems by differential scanning calorimetry (DSC) and the calculation of efficient cure cycles from the kinetic parameters of the reaction is the subject of continuing work and will be reported in a future communication. Whilst the curing of a prepreg can be successfully followed at various pressures by DSC, as can be seen from the data for Fibredux given in Fig. 1, the glass transition (and hence degree of cure) for fully or nearly fully cured composites is sometimes not so straightforward to detect. Figure 2 is the DSC data for Fibredux samples cured under various conditions and shows nothing in the expected glass transition region of 180 to 200 °C, and is complicated by the exothermic

---

* Fibredux is a Ciba-Geigy registered trademark.

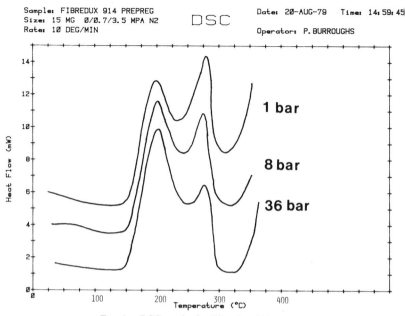

FIG. 1.   DSC results for Fibredux 914 prepreg.

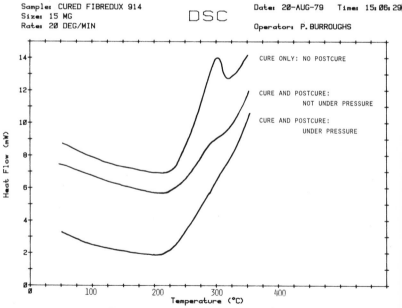

FIG. 2.   DSC results for cured Fibredux.

degradation reaction at higher temperatures. It has been found that the higher the percentage of reinforcing agent present in the composite, the more difficult it is to detect a glass transition by DSC at high levels of cure.

Thermomechanical analysis (TMA) or dilatometry is sometimes more successful. Hassel and Blaine[1] have reported measuring expansion coefficients for epoxy composites (glass fibre and graphite fibre reinforced) and noted changes in values on going through the glass transition. Using the thermomechanical analyser in its penetration mode (fitted with a loaded, pointed sample probe) may at times give a clear indication of the glass transition, but where the tip of the probe comes to rest directly on a reinforcing fibre the experiment may produce no useful data.

Using sample blocks of approximately 0·2 cm thickness in the direction of measurement, the expansion coefficient of cured Fibredux 914 unidirectional carbon fibre composite has been measured as 30 to 40 μm/m K in the axis at right angles to the carbon fibres. Figure 3 shows TMA data for a small strip of cured Fibredux, 0·6 cm square and 0·04 cm thick, standing on edge and with a flat sample probe loaded with 1 g resting on the top edge of the sample. This mode of operation allows the sample to buckle slightly as it softens at higher temperatures under the small compressive force of the probe, and is less affected by inconsistencies in the

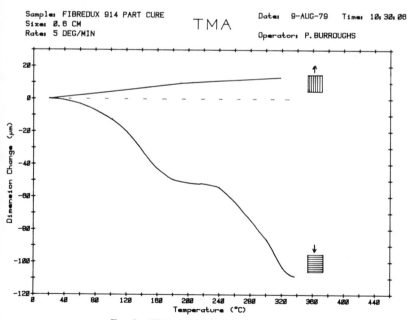

FIG. 3.   TMA results for cured Fibredux.

surface of the sample than operating the TMA in the penetration mode. The first break in the lower curve at 176 °C correlates well with the dynamic mechanical analysis data for the glass transition region. The second break at 240 °C is probably due to degradation.

By comparison with the data so far reported, the results obtained by studying composite samples by dynamic mechanical analysis are consistently clearer and potentially more useful. The use of a small visco-elastic tester for studying epoxy laminates has been reported by Hassel[3] and the instrument has been described in detail elsewhere.[4] Favre[5] has also reported the use of torsional braid analysis (TBA) to observe the curing of Fibredux 914. Figure 4 is a survey DMA scan of Fibredux 914 prepreg from a subambient temperature and shows the glass transition (as measured by a peak in the energy damping curve at 0 °C) and the curing reaction above 163 °C which is indicated by an increase in the sample's resonant frequency. This frequency may be used to calculate the Young's modulus of the sample when the sample is clamped vertically in the instrument, and the damping value, given here in mV, may be used to calculate the loss modulus or tan δ. These calculations are most conveniently performed using an on-line data reduction system.[6] Figure 5 shows DMA data for two Fibredux samples, one cured under high pressure and the other cured under only one atmosphere of pressure. Not only does the data show the glass transition clearly by a decrease in frequency and a peak in the damping, but the lower

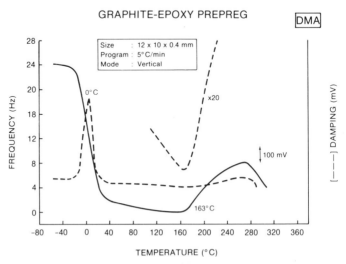

FIG. 4.   DMA results for Fibredux 914 prepreg.

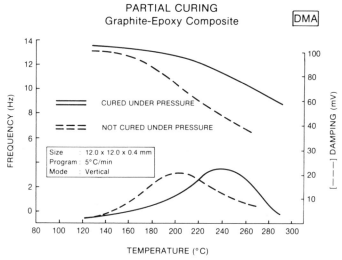

FIG. 5. DMA results for cured Fibredux.

high temperature mechanical strength of the sample which was not cured under pressure can be clearly seen. In the industrial curing process pressure is always applied to epoxy–graphite prepregs to enable the resin to fill the voids between the fibres and so give better mechanical properties in the final product.

The curing reaction itself may also be studied by DMA as the following data will show. Figure 6 shows the increase in frequency values of samples of Fibredux prepreg curing under isothermal conditions (the temperatures are noted on the curves). Gel times can be measured from the first upward turn in the frequency and are as follows.

170 °C	Gel point =	7·0 min
193 °C		4·5 min
205 °C		3·5 min
217 °C		2·8 min

Whilst the gel time is an important parameter for a thermosetting material it is felt that, in this particular system, the gel time is governed by (i) the curing agent melting and going into solution and (ii) the conduction of heat through the predominantly graphite matrix and the consequent slow time for the whole sample to achieve thermal equilibrium. However, once the curing reaction has started and been in progress for some minutes these two factors become less important. The initial rate of reaction can be

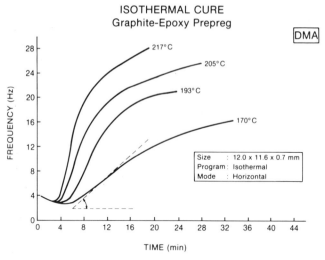

FIG. 6. Isothermal curing studied by DMA.

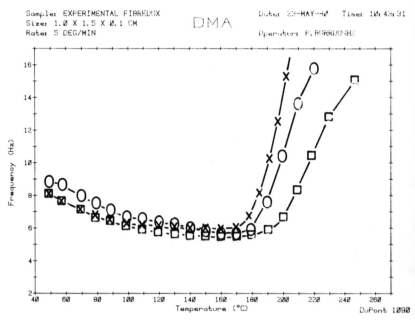

FIG. 7.   Variation of resonance frequency during cure: ○, normal formulation, ×, +20%
curing agent, □, −20% curing agent.

estimated from the tangent of the angle marked on the Figure and preliminary calculations show that this may be a more useful parameter for characterising the curing reaction than the gel time.

The dynamic mechanical analyser used for this study has the ability to take small samples (for example $1.0 \times 1.5 \times 0.1$ cm strips) and can thermally program such samples at the relatively high rate of 5 °C/min. This has been done on three experimental Fibredux 914 C epoxy–graphite composites, two with non-standard hardener contents ($+20\%$ and -20% of the normal values) and one correctly formulated. Figure 7 shows the frequency curves for the three samples (each point being the mean value of the corresponding frequency from three separate runs on the same type of sample). Even though it is not possible to calculate accurate elastic modulus values for these samples which tend to separate and distort on curing, the data shows that the sample low in hardener content requires a higher temperature to cure to the same degree as the other two samples. The curve needs to be shifted by approximately 20 degrees to overlap the other two. The hardener rich sample and the correctly formulated sample are much closer together and difficult to distinguish. However, Fig. 8 compares the

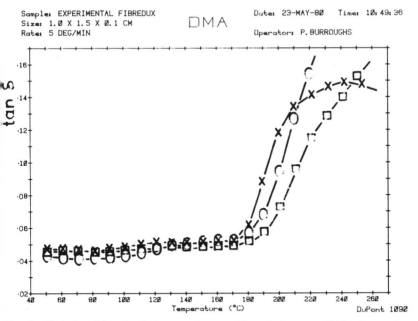

Sample: EXPERIMENTAL FIBREDUX   DMA   Date: 23-MAY-80 Time: 10: 49: 36
Size: 1.0 X 1.5 X 0.1 CM
Rate: 5 DEG/MIN   Operator: P. BURROUGHS

DuPont 1090

FIG. 8.   Variation of damping during cure: ○, normal formulation, ×, $+20\%$ curing agent, □, -20% curing agent.

calculated tan δ values of the same three samples (once again each point is a mean of three runs) and differences between samples are more obvious. As might be expected, the data from the normally formulated sample falls between that from the hardener rich and deficient samples. The tan δ curve for the hardener rich sample also appears to be peaking at about 240 °C, suggesting that the curing reaction is nearer completion than the other two samples at this temperature.

During the industrial curing process for epoxy composites the thermal profile is not normally a linear increase in temperature up to a preset maximum level, but is usually a cure cycle in which the sample is raised to an intermediate temperature to decrease the viscosity of the resin, and held for a period of time at this temperature (the high pressure is often applied at this point). The temperature is then raised again to initiate the cure, and the sample held for some time at the higher temperature. The effectiveness of such cure cycles may be tested using DMA by simulating the thermal profile used industrially. A typical cure cycle for a graphite–epoxy composite consists of raising the temperature at 5 °C/min, holding at 175 °C for 1 h, then raising the temperature to 195 °C and holding it there for 4 h. Figure 9 shows how the resonant frequency (elastic modulus) of a normal Fibredux graphite–epoxy prepreg changes on undergoing this cure cycle. The importance of the 195 ° post-cure period can be seen by the increase in frequency from 35 Hz to 50 Hz during this time, corresponding to a

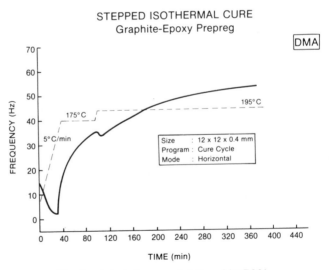

Fig. 9. A typical cure cycle followed by DMA.

doubling of the modulus. The use of high pressure probably causes an even greater increase in modulus during this postcure period, but the 4-h limit was set since it is not economically worth while to cure the sample for longer periods. Only a small further increase in modulus would take place after this time.

The final series of Figures show how DMA can easily reveal differences in the mechanical properties of the cured composites formed from a single epoxy resin reinforced with different fibre materials. A graphite reinforced system (Fig. 10) shows approximately a 70 % drop in modulus in going through the glass transition, whereas a glass-fibre reinforced composite (Fig. 11) shows a slightly lower drop in modulus, approximately 60 % over the same temperature range. The damping peak maxima at the glass transitions for these two samples are both around 200 °C. By comparison, an epoxy system reinforced with aramid fibres (Fig. 12) also shows a modulus drop of about 70 %, but at a temperature some twenty degrees lower, suggesting that the degree of cure is lower in this composite than in the other two samples. This may be due to a blocking effect on the crosslinking reaction caused by reaction with the active sites on the aramid fibre. The glass transition of these three materials could not be detected easily using DSC, and only DMA was able to show the difference in mechanical properties in the region on the glass transition.

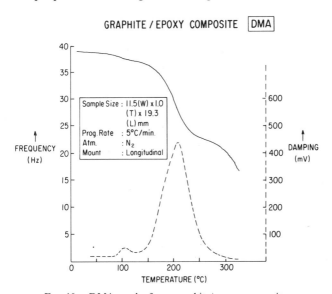

FIG. 10.   DMA results for a graphite/epoxy composite.

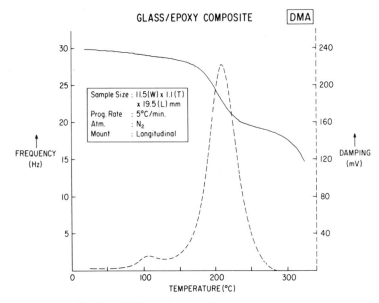

F<small>IG</small>. 11.   DMA results for a glass/epoxy composite.

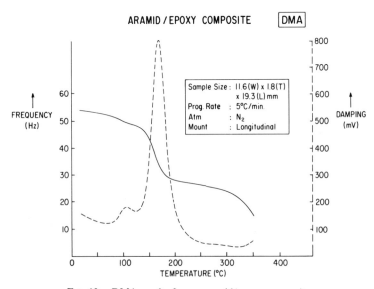

F<small>IG</small>. 12.   DMA results for an aramid/epoxy composite.

ACKNOWLEDGEMENTS

The authors wish to thank the staff of Ciba-Geigy Ltd, Duxford, UK, for several helpful discussions and for providing fresh normal and experimental Fibredux 914 C prepreg samples.

REFERENCES

1. HASSEL, R. L. and BLAINE, R. L., *Proceedings of the SAMPE Symposium, San Francisco*, 1979.
2. BURROUGHS, P., BLAINE, R. L. and GILL, P. S., *Proceedings of the NATAS Meeting, Chicago*, 1979.
3. HASSEL, R. L., Quality control of thermosets, *Industrial Research/Developments*, **20** (1978).
4. LOFTHOUSE, M. G. and BURROUGHS, P., *J. Thermal Anal.*, **13** (1978) 437–53.
5. FAVRE, J. P., *Proceedings of the '2ᵉ Conférence International sur les Materiaux Composites', Toronto*, 1978.
6. LEVY, P. F., BLAINE, R. L., GILL, P. S. and LEAR, J. D., Thermal analysis—Advances in instrumentation, *International Laboratory*, **9** (1979) 53–60.

30

Evaluation of Composite Structures by Stress-Wave-Factor and Acoustic Emission

ALLEN T. GREEN

Acoustic Emission Technology Corporation, 1812J Tribute Road, Sacramento, California 95815, USA

ABSTRACT

The development of acoustic emission and stress-wave-factor methods for the evaluation of composite materials and structures is discussed. Reference to existing work and on-going programs shows the direction for the use of both methods. Acoustic emission measurements rely on the self-originating signals created during deformation whilst the stress-wave-factor method utilizes an externally stimulated impulse and properties within the material which perturb the signal.

Promising results are shown for both methods.

INTRODUCTION

It is well known that the analysis and inspection of composite materials and structures is currently a difficult and imprecise science. Two methods which are finding greater use in both these areas are the stress-wave-factor and acoustic emission. The stress-wave-factor is basically a measure of the efficiency of stress-wave energy transmission and acoustic emissions are self-generated stress waves created by deformation of the material.

While both stress-wave-factor (SWF) and acoustic emission (AE) inspections (or examinations) are relatively easy to perform and permit one to obtain data, they are distinctly different in many aspects. The major differences are that the stress-wave-factor is a dynamic method and acoustic emission is a passive method, and that AE is generated by actual deformation of the material while the SWF is a measure of the transmission of an artificially generated stress-wave. Additionally, AE[1] methodology in composites has been developing since its application to E and S glass 'Polaris' filament wound rocket motor cases[1] in the early 1960s to where, today, we find nearly a dozen companies producing AE equipment for commercial sale. Stress-wave-factor, also referred to as acousto-ultrasound, on the other hand, is of more recent origin[2]—the late 1970s—with only a single company producing commercial equipment.

Acoustic emission examinations of composite materials and structures are generally conceived with the normal non-destructive examination objectives of the detection and location of defects. Stress-wave-factor examinations, while capable of detection and location, are also suitable for ranking composite structures according to strength.

BACKGROUND AND DISCUSSION

(A) Acoustic emission methods in composite structures began in the early 1960s with the 'Polaris' rocket motor case project as a research and development program.[3] In all AE methods the microscope signals are amplified and frequency filtered and then processed by two primary methods. One is counting threshold crossings (each AE event creates multiple threshold crossings due to the resonant decay of the signal from the sensor and/or structure), and the other is measuring the total signal level. In the first method, simple counters may be utilized to establish total counts or count rates, and the second method may utilize RMS or TMS (Root-Mean-Squared or True-Mean-Squared) voltage measuring circuits. Within threshold measurement methods other characteristics of the AE event, such as peak amplitude, event duration and rise-time, can be established. Other parameters can be associated and tracked for additional correlation to the AE data. Filtering, based upon characteristics of the AE event and/or external parameters and/or spatial notations, enables a wide range of discriminatory levels to be applied. This feature enables portions of structures to be isolated while being monitored and certain material performance levels, such as resin (matrix) cracking, to be separated from filament or interlaminar failures.

For example, in composite materials, the multitude of matrix failures can best be followed by RMS/TMS measurements, while filament failures or delaminations can be defined by the threshold counting methods.

During the early AE effort, we observed that not only could we distinguish good structure from bad (i.e. unable to meet minimum pressure test levels on a second cycle), but we could also: (1) measure the consistency of the production process and (2) define at least three modes of structural degradation—(a) matrix crazing, (b) filament fracture and (c) interlaminar shear.

Figure 1 illustrates the amplitude analysis performed on the analog recorded AE data, which defined filament failures and interlaminar shear failures. Similar results are shown in Fig. 2, obtained from a frequency spectral analysis. These analyses were confirmed by other experimental destructive and non-destructive methods, including special specimen testing, interlaminar strain gages and destructive sectioning.

In most materials, an effect named after one of the original researchers in the field of AE is very demonstrative and has been utilized in helping to diagnose the remaining structural integrity. The effect is the short-term irreversibility of AE, known as the Kaiser Effect. The Kaiser Effect is illustrated by stressing a material to a predetermined level, reducing the stress, and then returning to a level beyond that originally reached. On the first cycle, acoustic emission data will be noted whilst, on the second cycle, no AE data will be detected until the first level is exceeded or, as might

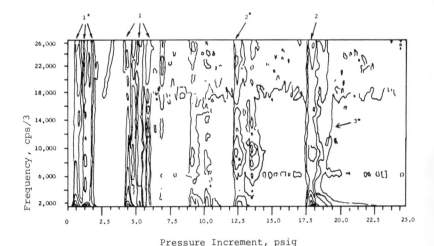

Pressure Increment, psig

*Notes: 1. Filament Failures
          2. Interlaminar Shear Failures
          3. Amplitude Shown as 6 dB (Acceleration Ratio of 2)
             Between Contours

FIG. 1.   Amplitude analysis (composite pressure vessel).

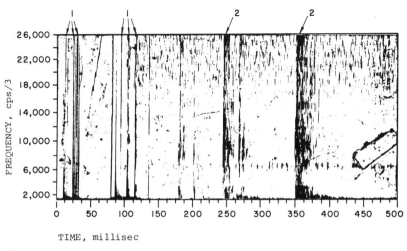

FIG. 2. Frequency analysis. Spectrograph of hydrotest of 'Polaris' A3 filament wound chamber.

happen under fatigue conditions, additional damage is sustained. In instances where sufficient background information regarding the AE performance of a material has been previously established, the Kaiser Effect, or the lack of it (sometimes referred to as the Felicity Effect), may be used to establish the deformation or damage level previously sustained.

The classic example of using the Kaiser Effect in composites was accomplished on the 'Trident' rocket motor cases, made from Kevlar-49/ epoxy as a filament wound structure.[4] A hydropressure test schedule was established for these structures so as to repeat a stress level range once during each test. During the test, little or no acoustic emission is expected while the motor case pressurization passes through this zone. The pressure schedule and data obtained from two sensors are shown in Fig. 3. On chambers which failed prematurely at low stress levels, high amounts of AE data were recorded in this pressure increment. Conversely, good structures showed little or no emission while passing through this pressure range. Refinement of the data enabled a quantitative measurement of performance for these limited use structures to be obtained, as is shown in the correlation of the data points and the 95 % confidence limits of Fig. 4.

The most widespread use of AE has been in the chemical industry under the auspices of the Society of the Plastics Industry (SPI) and their Committee on AE for Reinforced Plastics (CARP) group. CARP has developed a Recommended Practice for Acoustic Emission Testing of

Allen T. Green

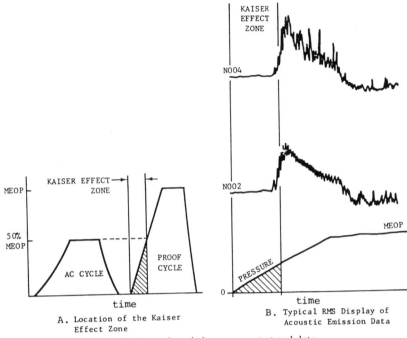

KAISER
EFFECT
ZONE

NO04

KAISER EFFECT → ZONE

MEOP

50%
MEOP

KAISER EFFECT →
ZONE

NO02

MEOP

PROOF
CYCLE

AC CYCLE

PRESSURE

time

0

time

A. Location of the Kaiser
Effect Zone

B. Typical RMS Display of
Acoustic Emission Data

FIG. 3.   Acoustic emission pressure test and data.

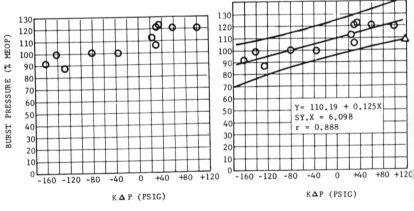

$$Y = 110.19 + 0.125X$$
$$SY.X = 6.098$$
$$r = 0.888$$

A. Scatter Diagram for First Stage
Trident-I (C-4) Chambers

B. Regression for First Stage
Trident-I (C-4) Chambers

FIG. 4.   Kaiser Effect zone acoustic emission pressure increment versus burst pressure
(Kevlar chambers).

Fiberglass Reinforced Plastic Resin (FRP) Tanks/Vessels.[5] This recommended practice is for application on new and in-service equipment, limited to tests not to exceed 65 psia above hydrostatic test pressure and a vacuum of less than 9 psia. The fact that FRP exhibit the Kaiser Effect only up to a percentage of ultimate load, beyond which emission will begin at loads lower than the previously attained maximum, is strongly used in the recommended practice (Felicity Effect). Use of this effect makes it possible to test in-service structures without exceeding their maximum operating stresses. The chance of permanent damage is eliminated. The ratio of the load (stress) at the onset of AE to the previously attained maximum load has been defined as the Felicity Ratio. The lower the number, the greater the damage. While much good use has been made of the Kaiser Effect—or the lack of it (Felicity Ratio)—it is an area where research must continue primarily because of the need to understand the effect of time on this phenomenon.[6] Table 1 presents the acceptance criteria developed by CARP for new and in-service vessels. While the criteria may appear circumspect,

TABLE 1
Acceptance criteria

	Criteria I. New vessels first filling	Criteria II. All other vessels	Significance of criterion
Emissions during hold	*None beyond 2 min*	*None beyond 2 min*	Measure of continuing permanent damage caused by creep of the matrix
Felicity Ratio	Greater than 0·95[a]	*Greater than 0·95*	Measure of severity of previously induced damage
Total counts	*Less than N_1[b]*	Less than N_2[b]	Measure of overall damage during a load cycle
Number of events greater than reference amplitude threshold[b]	*Less than 10*	Less than 10[a]	Measure of high-energy micro-structural failures. This criterion is often associated with fiber breakage

NOTES: Criteria in italics are the most significant.
[a] This criteria will seldom govern.
[b] Varies with instrumentation manufacturer.

they have been based upon a relatively large number (over 400) of actual tests of FRP structures.[7] Further data continue to be acquired regularly.

Similar use of AE testing is widespread among 'boom' truck industrial users. Here again, structural loading over a predetermined program range and use of the Kaiser Effect have become accepted methods of establishing the structural adequacy of these booms.[8]

(B) The stress-wave-factor method of evaluating composite materials and structures involves injecting a low amplitude stress pulse into a specimen and processing the received signal with acoustic emission methods. The processed data have been found to characterize the factors attributed to composite failures, such as constituent strengths and moduli lamination orientations, flaw populations and distributions, applied and residual stresses, energy dissipation dynamics and fracture propagation paths. They have also been found to assess the integrated effect of flaw populations as well as the more subtle effects such as fiber–resin bonding and ratios.[9] In the method the pulses are generated at a repetition rate, g, with each successive pulse identical to its predecessors. After amplification, the received signals are sent to a electronic counter and a root-mean-squared voltmeter. The counter determines the number of oscillations, n, received in each time window which exceed a fixed threshold value. The time window of the counter is reset after each interval, r. The displayed count assumes a constant value soon after the sensors are coupled to the specimen. The number is described as the stress-wave-factor, e, where $e = grn$.

Since the number e is arbitrary and depends on factors such as sensor pressure, coupling, gain, reset time (r), threshold voltage, repetition rate

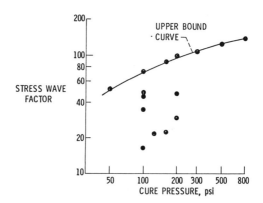

FIG. 5.   Stress wave factor versus core pressure (Vary[10]).

(g), and so forth, it will reflect only material variation of the specimen tested if all these factors are kept constant for any series of measurements.

Figure 5 shows the stress-wave-factor variation with cure pressure for graphite polyimide composite panels of AS/PMR-15 12 ply unidirectional construction. According to the author, prior tests had indicated that the range of cure pressure would produce a significant range of void contents and fiber/resin ratios. The material strength increases directly with cure pressure, the higher values of SWF (e) corresponding to greater interlaminar shear strength (see Fig. 6).

The curve in Fig. 5 is actually an upper bound that represents an apparent optimum condition for a given cure pressure. Ultrasonic C scan, performed on one of the low-cure pressure and e value panels, has shown the SWF to distinguish inferior material that can arise even when a key processing parameter, cure pressure, is controlled.

As with all stress wave measurements, the C values are related to positions along the specimen length. The C value for the specimens subjected to interrupted loadings showed that the point of specimen fracture coincided with minimal values. The data shown in Fig. 7 demonstrate that the stress-wave-factor correctly ranked the ultimate strengths despite porosity variations.[10]

Two $\frac{1}{2}$ in-thick graphite–epoxy panels, purposely damaged by different energy impacts, were scanned by acousto-ultrasound and SWF readings

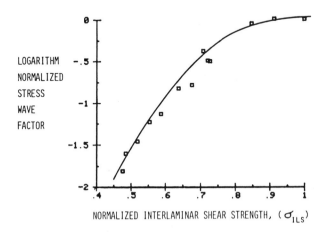

FIG. 6. Stress wave factor versus interlaminar shear strength in graphite/polyimide laminate (Vary and Bowles[2]).

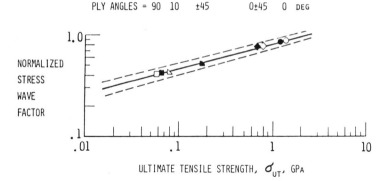

FIG. 7.   Stress wave factor versus tensile strength for graphite/epoxy laminates. (Vary and Lark[9]).

were developed. Figure 8 shows the panels, as marked off in $\frac{1}{2}$-in squares, with the associated SWF readings noted graphically. The darker the color, the lower the SWF.[11]

Panel No. 1 had three impacted areas at locations (1) F–G/12–13, (2) F–G/24–25 and (3) F–G/35–36. Locations (1) and (2) showed de-laminations on the rear surface of the panel while location (3) was only casually visible on the front surface. Panel No. 2 had a total of ten impacted areas, three of which showed no visible surface marks while five penetrated the panel thickness, causing rear surface delaminations. The SWF examination was conducted only from the front surface. Through thickness readings could also have been obtained in this case, but were not.

Figure 9 shows a computer-derived three-dimensional representation of the SWF data for Panel 1. The three impact points are clearly delineated. The SWF data for Panel 2 are shown in three-dimensional projection in Fig. 10. The heavily damaged (dark areas in Fig. 8) areas are clearly shown as peaks in this display. Similar representation of this information can be obtained as a grey scale for reference to conventional NDE methods.

Williams and Lambert[12] have reported SWF measurements, also conducted on unidirectional AS/3501-1 continuous graphite fiber–epoxy composites. Specimens damaged by impact and then tested for residual tensile strength were SWF inspected. The residual tensile strength was correlated with the SWF. The authors concluded that the results suggest that impact damage in graphite–fiber composites can be non-destructively assessed quantitatively using either through thickness attenuation (not always possible) or the SWF.

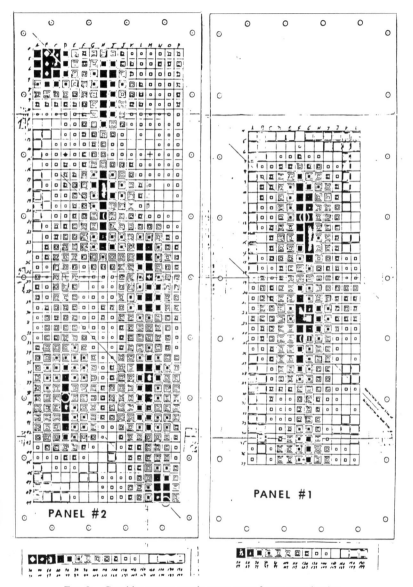

FIG. 8. Graphite–epoxy panels stress wave factor examination.

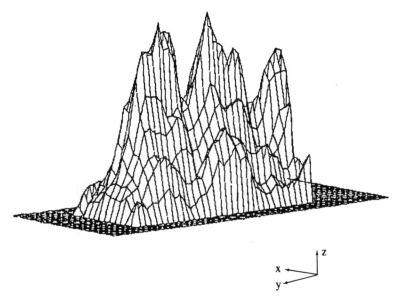

FIG. 9.   Panel 1 with $\theta = -10$ and $\phi = 45$.

CONCLUSIONS

In summary, the following conclusions may be drawn regarding AE and SWF inspections of composite materials and structures.

(1) AE tests, when properly planned and conducted, can provide information regarding the serviceability of the part and may be capable of establishing the mode of degradation.

(2) AE testing requires some means of actively stressing the part under investigation. While this might be done in-service, usually it is through an additional test procedure.

(3) SWF is a sensitive indicator of composite strength variations that accompany various fiber orientations relative to the load axis.

(4) For a given fiber orientation, SWF is also sensitive to strength variations associated with differences in fiber–resin bonding, voids and fiber–resin ratios. For the composite tensile specimens studied, the SWF decreased proportionally with fractional powers of ultimate strength.

(5) The SWF may be a useful aid in predicting potential failure locations in composite laminates.

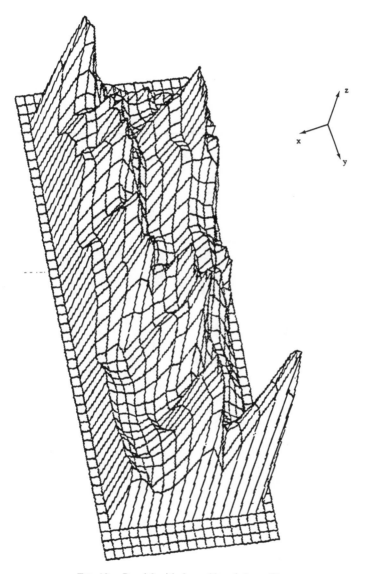

FIG. 10.   Panel 2 with $\theta = -80$ and $\phi = -30$.

REFERENCES

1. GREEN, A. T., LOCKMAN, C. S. and HAINES, H. K., *Acoustic analysis of filament-wound 'Polaris' chambers*, Aerojet-General Corp. Report 0672-01F, 16 September, 1963.
2. VARY, A. and BOWLES, K. J., *Use of an ultrasonic-acoustic technique for nondestructive evaluation of fiber composite strength*, NASA Technical Memorandum TM-73813, February, 1978.
3. GREEN, A. T., LOCKMAN, C. S. and STEELE, R. K., *Acoustic verification of structural integrity of 'Polaris' chambers*, Society for Plastic Engineers, 27–30 January, 1964.
4. JESSEN, E. C., SPANHEIMER, H. and DeHERRA, A. J., Prediction of composite pressure vessel performance by application of the Kaiser Effect in acoustic emissions, ASME Paper H3000-12-2-037, June, 1975.
5. Recommended Practice For Acoustic Emission Testing of Fiberglass Reinforced Plastic Resin (FRP) Tanks/Vessels, Committee on Acoustic Emission from Reinforced Plastics (CARP). A Working Group of the Society of the Plastics Industry, 2–3 December, 1980.
6. HAMSTAD, M. A., *Testing fiber composites with acoustic emission monitoring*, University of California, UCRL No. 85428, 23 January, 1981.
7. FOWLER, T. J., *Acoustic emission testing of plant components*, ETCE, New Orleans, Louisiana, 3–7 February, 1980.
8. SWAIN, W. E., *Nondestructive testing of aerial man-lift and digger derrick equipment using acoustic emission*, International Conference of Acoustic Emission, Anaheim, California, 13 September, 1979.
9. VARY, A. and LARK, R. F., *Correlation of fiber composite tensile strength with the ultrasonic stress-wave-factor*, NASA TM-78846, April, 1978.
10. VARY, A., *Recent advances in acousto-ultrasonic measurements of composite mechanical properties*, ASNT, St Louis, October, 1979.
11. THOMPSON, L. A., *Stress-wave-factor examination of impact-damaged graphite epoxy panels*, AET Memorandum, 17 October, 1980.
12. WILLIAMS, J. H. and LAMBERT, N. R., Ultrasonic evaluation of impact-damaged graphite fiber composite, *Materials Evaluation* (December, 1980), 68–72.

31

Vibration Testing of Composite Materials

J. L. Wearing and C. Patterson

*Department of Mechanical Engineering, University of Sheffield,
Mappin Street, Sheffield S1 3JD, England*

ABSTRACT

*The use of composite materials for engineering applications requires the
determination of their properties for the analysis, manufacture and quality
control of the materials. Conventional static tests are normally unsuitable
for composites because of the problems associated with clamping the
specimens. Suitable alternative experimental techniques must be employed,
therefore, if reliable results are to be obtained for the properties of these
composite materials.*

*In this paper a vibrational technique for the determination of the
properties of thin laminated plates is discussed. An expression is developed
for the determination of the natural frequencies of orthotropic rectangular
plates. Experimentally determined natural frequencies and mode shapes of
appropriately manufactured specimens are used in conjunction with the
developed expression to determine the properties of the laminate under
consideration.*

*In this initial investigation, to test the validity of the proposed approach,
industrially derived specimens have been tested. The results are presented and
their accuracy is discussed.*

NOTATION

a, *b*, *h*	length, breadth and thickness of rectangular plate
m, *n*	integers
t	time

w	deflection of plate
W	maximum deflection of plate
x, y	co-ordinates directions
$\left.\begin{array}{l} C_{11}, C_{22}, C_{33} \\ C_{12}, C_{21} \end{array}\right\}$	constants related to material properties
A_{mn}	constant in assumed deflected form of plate
$\left.\begin{array}{l} D_x D_y \\ D_{xy}, D_1 \end{array}\right\}$	constants related to material properties and plate thickness
E_x, E_y	Young's moduli in the x and y directions
G_{xy}	torsional rigidity in the xy plane
$\left.\begin{array}{l} H_1^{(mn)}, H_2^{(mn)} \\ H_3^{(mn)}, H_4^{(mn)} \end{array}\right\}$	constants related to derivatives of plate's deflected form
K_{mn}	element in matrix obtained from maximum kinetic energy
M_{mn}	element in matrix obtained from maximum strain energy
T_{max}	maximum kinetic energy
U_{max}	maximum strain energy
σ_x, σ_y	normal stresses in the x and y directions
τ_{xy}	shear stress in xy plane
$\varepsilon_x, \varepsilon_y$	normal strains in the x and y directions
γ_{xy}	shear strain in xy plane
v_x, v_y	Poisson's ratios in the x and y directions
$\phi_m(x), \theta_n(y)$	modal shapes of beams used in assumed deflected form of plate
ω_n	natural frequency of plate
λ	Eigenvalue of plate
ρ	material density

INTRODUCTION

The demands imposed on present day structures have led engineers to investigate the possibility of using new materials for the manufacture of engineering components and structures. For example, engineering applications in the use of stronger and lighter composite materials can be found in the design and manufacture of space vehicles, heat shields and deep submergence vessels.

The use of composite materials for engineering applications requires, however, the determination of their properties for the analysis, manufacture and quality control of the components which are to be produced with such materials. One of the prerequisites in any experimental programme for the determination of material properties is to use specimens which are manufactured in the same way as the intended end product. If that requirement is not met, the experimental effort is wasted. Previous methods of determining the elastic constants of composites have been based mainly on ultrasonic pulse propagation techniques and static tests.[1,2]

The ultrasonic technique is based on the measurement of the velocities of high frequency elastic waves propagating along known directions in the composite. The velocity of the waves is dependent on the nature of the wave, the material density and the elastic stiffness components. Ultrasonic techniques are normally used to determine the properties of viscoelastic composites or polymeric materials. These techniques have been used successfully by Dean and Turner[3] for the characterization of elastic composite materials.

Static tests require simple tensile compressive or bending specimens of the material to be investigated, but there are inherent difficulties in such tests. The simple tensile test requires undirectional samples and is therefore unsuitable for cross-ply laminates. The formulae, which are used to determine the experimental data, are based on linear stress–strain relationships. Problems arise, therefore, in maintaining the deflections in static tests within the linear stress–strain regime. Additionally, information obtained from static tests may be subject to errors arising from local stress concentrations, frictional effects, deformations around the loading and support points and creep effects.

The industrially derived specimens tested for the work outlined in this chapter were composites in the form of thin laminates and a vibrational technique has been used for the determination of the material properties. Thin rectangular plate specimens of the laminates were employed and the Rayleigh–Ritz method was used to develop an expression for the calculation of the natural frequencies of thin orthotropic plates. Experimentally determined natural frequencies of appropriately manufactured specimens were used, in conjunction with the developed expression, to determine the properties of the laminates and the results of this initial investigation are presented here.

STRESS–STRAIN RELATIONSHIPS AND MATERIAL PROPERTIES

It is necessary to examine the basic stress–strain relationships to ascertain the number of independent material properties required to completely characterize any material. The thin rectangular plate specimens tested were comprised of bonded layers of fibre such that the fibre directions of alternate layers were orientated in the same direction with the fibre directions of the intervening layers being in a perpendicular direction (i.e. the fibre formation was $0°-90°-0°$). They were, therefore, assumed to be orthotropic. As the normal stress components of thin rectangular plates are assumed to be negligible, the stress–stress relationships for a thin orthotropic rectangular plate lying in the xy plane are given by the matrix expression:

$$\begin{Bmatrix} \sigma_x \\ \sigma_y \\ \tau_{xy} \end{Bmatrix} = \begin{bmatrix} C_{11} & C_{12} & 0 \\ C_{21} & C_{22} & 0 \\ 0 & 0 & C_{33} \end{bmatrix} \begin{Bmatrix} \varepsilon_x \\ \varepsilon_y \\ \gamma_{xy} \end{Bmatrix} \tag{1}$$

The coefficients, C_{ij}, in eqn. (1) are related to the plate's material properties by the expressions:

$$C_{11} = \frac{E_x}{1 - v_x v_y} \tag{2}$$

$$C_{22} = \frac{E_y}{1 - v_x v_y} \tag{3}$$

$$C_{12} = \frac{v_y E_x}{1 - v_x v_y} \tag{4}$$

$$C_{21} = \frac{v_x E_y}{1 - v_x v_y} \tag{5}$$

$$C_{33} = G_{xy} \tag{6}$$

Of the five independent elastic constants given in the above equations, only four are independent. As the matrix of coefficients in eqn. (1) is symmetrical, the following relationships can therefore be obtained from eqns (4) and (5):

$$\frac{E_x}{E_y} = \frac{v_x}{v_y} \tag{7}$$

leading to the determination of the fifth property.

NATURAL FREQUENCIES AND MODE SHAPES OF RECTANGULAR PLATES

The equation of motion of free vibrations of a thin orthotropic rectangular plate is:

$$D_x \frac{\partial^4 w}{\partial x^4} + 2(D_1 + 2D_{xy}) \frac{\partial^4 w}{\partial x^2 \partial y^2} + D_y \frac{\partial^4 w}{\partial y^4} = \rho h \omega_n^2 \frac{\partial^2 w}{\partial t^2} \tag{8}$$

in which the constants, D_x, D_1, D_{xy} and D_y, are given by the expressions:

$$\left. \begin{aligned} D_x &= \frac{C_{11} h^3}{12} & D_1 &= \frac{C_{12} h^3}{12} \\ D_y &= \frac{C_{22} h^3}{12} & D_{xy} &= \frac{C_{33} h^3}{12} \end{aligned} \right\} \tag{9}$$

Equation (8) cannot normally be solved to determine the natural frequencies and mode shapes of a rectangular plate with any combination of boundary conditions. An approximate technique must, therefore, be used for the calculation of the natural frequencies. Of the approximate techniques which may be used the Rayleigh–Ritz method is one of the better known. To determine the natural frequencies using the method, the maximum strain and kinetic energies of the vibrating plate are equated. These are given respectively by the expressions:

$$U_{max} = \frac{1}{2} \int_0^a \int_0^b \left[D_x \left(\frac{\partial^2 W}{\partial x^2} \right)^2 + D_y \left(\frac{\partial^2 W}{\partial y^2} \right)^2 \right.$$
$$\left. + 2D_1 \left(\frac{\partial^2 W}{\partial x^2} \right) \left(\frac{\partial^2 W}{\partial y^2} \right) + 4D_{xy} \left(\frac{\partial^2 W}{\partial x \partial y} \right)^2 \right] dy \, dx \tag{10}$$

and:

$$T_{max} = \frac{1}{2} \rho h \omega_n^2 \int_0^a \int_0^b W^2 \, dy \, dx \tag{11}$$

Equating eqns (10) and (11) results in the following expression:

$$U_{max} - T_{max} = 0 \tag{12}$$

The natural frequencies of the vibrating plate are obtained from eqn. (12) by assuming, initially, that the deflected form of the vibrating plate is a series solution of the type indicated by eqn. (13):

$$W = \sum \sum A_{mn} \phi_m(x) \theta_n(y) \tag{13}$$

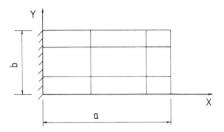

FIG. 1.   Nodal pattern 3/2 of a cantilevered rectangular plate.

The assumed deflected form is substituted into eqn. (12) and the constants, A_{mn}, adjusted to make it a minimum. This is achieved by differentiating the resulting expression with respect to each of the constants, A_{mn}, to give a series of equations of the form:

$$\frac{\partial U_{max}}{\partial A_{mn}} - \frac{\partial T_{max}}{\partial A_{mn}} = 0 \qquad (14)$$

from which the natural frequencies of the vibrating plate are obtained.

The functions $\phi_m(x)$ and $\theta_n(y)$ which have been chosen to represent the deflected form (eqn. (13)) of the vibrating plate are those which represent the normal modes of vibration of uniform beams. Beam functions have been used by Warburton[4] and Young[5] for the determination of the natural frequencies of isotropic plates. They were chosen because the nodal patterns which correspond to the natural frequencies of rectangular plates take the form of lines which are approximately parallel with the edges of the plate. The nodal patterns can be defined, therefore, by the notation m/n, in which m is the number of nodal lines in the x direction and n is the number of nodal lines in the y direction. When an edge is simply supported or fixed it is assumed to be a nodal line. Using that notation, the nodal pattern, 3/2, of a cantilevered rectangular plate is indicated in Fig. 1.

A typical term, $\phi_4(x)\theta_3(y)$, for example, in eqn. (13), for the deflected form of a cantilevered plate is comprised of the deflected form $\phi_4(x)$ of the fourth natural frequency of a cantilevered beam and the deflected form $\theta_3(y)$ of the second non-rigid body mode of a free–free beam.

FREQUENCY EXPRESSION FOR ORTHOTROPIC RECTANGULAR PLATES

When eqn. (13) is substituted into eqn. (14) for the determination of the natural frequencies of orthotropic rectangular plates, the equations which are obtained, when the appropriate differentiations and integrations have been performed, may be expressed in matrix form, as indicated by eqn. (15):

$$([M_{mn}] - \lambda[K_{mn}])\{A_{mn}\} = 0 \tag{15}$$

The elements in the matrix $[M_{mn}]$ are obtained from the strain energy expression and the values of the appropriate integrals in that integral are tabulated by Young[5] for various values of m and n. The elements in the matrix $[K_{mn}]$ are obtained from the kinetic energy integral and are equal to zero, apart from one principal term which is equal to unity in each equation. A typical equation of those specified by eqns (15) would therefore be:

$$M_{11}A_{11} + M_{12}A_{12} + \cdots + (M_{mn} - \lambda)A_{mn} + \cdots = 0 \tag{16}$$

Normally the natural frequencies are obtained by equating the determinant of the matrix $([M_{mn}] - \lambda[K_{mn}])$ in eqn. (15) to zero. In the work discussed in this chapter it was observed that the term $(M_{mn} - \lambda)$ in each equation was large compared with the others which were, therefore, neglected. This means that each equation is reduced to one term, as indicated by eqn. (17), and is similar to the approach adopted by Warburton:[4]

$$(M_{mn} - \lambda)A_{mn} = 0 \tag{17}$$

The term M_{mn} in eqn. (17) is given by the expression:

$$M_{mn} = D_x H_1^{(mn)} + 2D_1 H_2^{(mn)} + D_y H_3^{(mn)} + D_{xy} H_4^{(mn)} \tag{18}$$

and λ is obtained from:

$$\lambda = \rho h \omega_n^2 \tag{19}$$

Hence, if eqns (18) and (19) are substituted into eqn. (17) the natural frequency of mode m/n is obtained from the following expression.

$$\rho h \omega_n^2 = D_x H_1^{(mn)} + 2D_1 H_2^{(mn)} + D_y H_3^{(mn)} + D_{xy} H_4^{(mn)} \tag{20}$$

in which the constants $H_1^{(mn)}$, $H_2^{(mn)}$, $H_3^{(mn)}$ and $H_4^{(mn)}$ are obtained from the following derivatives in the strain energy integral:

$$\left.\begin{aligned}
H_1^{(mn)} &= \int_0^a \int_0^b \left(\frac{\partial^2 \phi_m(x)}{\partial x^2}\right)^2 (\theta_n(y))^2 \, dy \, dx \\[2mm]
H_2^{(mn)} &= \int_0^a \int_0^b \left(\frac{\partial^2 \phi_m(x)}{\partial x^2}\right)\left(\frac{\partial^2 \theta_n(y)}{\partial y^2}\right) dy \, dx \\[2mm]
H_3^{(mn)} &= \int_0^a \int_0^b (\phi_m(x))^2 \left(\frac{\partial^2 \theta_n(y)}{\partial y^2}\right)^2 dy \, dx \\[2mm]
H_4^{(mn)} &= \int_0^a \int_0^b \left(\frac{\partial \phi_m(x)}{\partial x}\frac{\partial \theta_n(y)}{\partial y}\right)^2 dy \, dx
\end{aligned}\right\} \tag{21}$$

The constants, D_x, D_y and D_{xy} and D_1, in eqn. (20) are related to the material properties of the plate through eqns (9). It follows, therefore, that if four natural frequencies and their corresponding mode shapes are known, four simultaneous equations can be obtained from eqn. (20) for the determination of D_x, D_y, D_{xy} and D_1, which can then be used for the determination of the plate's material properties.

EXPERIMENTAL PROCEDURE

The objective of the experimental work was the determination of the natural frequencies and corresponding mode shapes of rectangular plates manufactured from the laminate under consideration. The carbon epoxy resin specimen which was tested was laminated with three layers of cloth with the main fibre directions orientated at 0°–90°–0°. A cantilevered rectangular plate and a rectangular plate with two opposite edges fixed and the other two free were tested. The apparatus used for the experimental determination of the natural frequencies and nodal patterns of the plate is shown in block diagram form in Fig. 2.

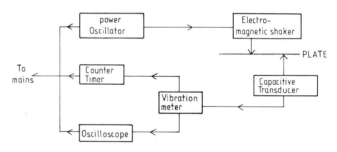

Fig. 2. Block diagram of experimental apparatus.

The natural frequencies of the plate were obtained by sweeping through a range of frequencies and noting the maximum deflections on the oscilloscope. The nodal patterns were found by scattering fine dry sand on the plate and when it vibrated in a natural frequency the sand settled along the nodal lines. Having found the natural frequencies and mode shapes, the experimental results were used in conjunction with eqn. (2) for the determination of the constants D_x, D_y, D_{xy} and D_1, from which the plate's material properties were obtained.

RESULTS AND CALCULATIONS

Experimental natural frequencies for a cantilevered rectangular plate with a length of $0 \cdot 25$ m, a breadth of $0 \cdot 15$ m and a thickness of $0 \cdot 85 \times 10^{-3}$ m, are presented in Table 1. The density of the material from which the plate was manufactured is $1568 \, \text{kg/m}^3$. In the table the constants m and n are the number of nodal lines in x and y directions respectively. Hence the natural frequency corresponding to any particular nodal pattern m/n can be found directly from Table 1.

TABLE 1

Experimental natural frequencies of cantilevered plate ($a = 0 \cdot 25 \, m$, $b = 0 \cdot 15 \, m$ and $h = 0 \cdot 85 \times 10^{-3} \, m$)

n \ m	1	2	3	4
0	15·000	88·000	241·000	489·000
1	38·397	120·644	279·500	509·560
2	165·734	254·324	400·107	1 017·235
3	554·120	667·067	676·000	
4	1 080·45			1 494·340
5		1 850·945		1 980·187

The experimental results presented in Table 1 can be substituted into eqn. (20) for the determination of the plate's material properties. If a natural frequency corresponding to any of the modes $m/0$ is used, D_x can be calculated directly. This occurs because the integrals in the strain energy expression (eqn. (10)) are zero for $n = 0$ when both boundaries which are parallel to the y axis are free (Fig. 1), making $H_2^{(mn)}$, $H_3^{(mn)}$ and $H_4^{(mn)}$ equal to zero in eqn. (20). Similarly the constants $H_2^{(mn)}$ and $H_3^{(mn)}$ are zero in eqn. (20) for $n = 1$ giving an equation with D_x and D_{xy} as unknowns. For values of $n > 2$ all integrals in the strain energy expression are activated for free–free boundaries, leading to four unknowns in eqn. (20).

It is therefore possible to calculate D_x using one of the frequencies in Table 1 having a nodal pattern of the form $m/0$. Having found D_x, a natural frequency having a nodal pattern $m/1$ is used to calculate D_{xy}. The other constants, D_1 and D_y, can then be found using two of the natural frequencies from Table 1 for which $n \geq 2$.

The detailed procedure discussed above for the determination of the constants D_x, D_y, D_1 and D_{xy} applies to plates with two free edges parallel to

TABLE 2

Experimental natural frequencies of a plate with two parallel edges fixed and two parallel edges free ($a = 0.21\,m$, $b = 0.15\,m$ and $h = 0.85 \times 10^{-3}\,m$)

n \ m	1	2	3	4
0	140·033	355·160	683·983	
1	151·016	390·188		
2	283·400	466·905	831·214	1 296·173
3	606·665	776·550	969·915	1 693·637
4		1 234·916	1 370·380	
5	1 826·257	1 909·267	2 118·210	

the x axis. In addition to the results presented in Table 1, natural frequencies and mode shapes were obtained experimentally for a rectangular plate which was fixed on the edges parallel to the y direction and free on the edges parallel to the x direction. The length and breadth of that plate were 0·21 m and 0·15 m, respectively and the results are shown in Table 2.

The experimental results presented in Tables 1 and 2 were used independently to determine the constants D_x, D_y, D_{xy} and D_1 from eqn. (20). These values were then substituted into eqn. (9) for the determination of the plate's material properties. The average values of these properties were computed from the results of several calculations and are shown in Table 3.

TABLE 3

Material properties of three ply laminate

$$E_x = 92.53 \times 10^9 \, \text{N/m}^2$$
$$E_y = 6.049 \times 10^9 \, \text{N/m}^3$$
$$G_{12} = 10.71 \times 10^9 \, \text{N/m}^3$$
$$v_x = 0.0371$$
$$v_y = 0.5682$$

DISCUSSION OF RESULTS AND CONCLUSIONS

In this preliminary investigation it is important to make some assessment of the accuracy of the values which have been obtained for the material properties of the laminate under consideration. Inaccuracies can occur in two main areas. The analytical expression which was derived for the calculation of the constants D_x, D_y, D_{xy} and D_1 is based on the Rayleigh–Ritz method for the calculation of the natural frequencies of rectangular plates and it is well known that the results from the technique are normally higher than the true value. A further approximation was introduced by considering only the principal term in any equation for the determination of the natural frequencies. Warburton[4] has shown that, when using a single term approximation for the deflected form of the vibrating plate, the discrepancies occurring in the calculated natural frequencies are worse for plates with two 'parallel' free boundaries than for plates with other combinations of boundary conditions, with the worst cases occurring at the modes 1/1 and 2/1 of cantilevered plates. For example, the natural frequency of mode 1/1 of a cantilevered plate with a length to breadth ratio of 5 is 27% high and for mode 2/1 it is 12% high. The results for the other modes are usually considered to be about 4% to 7% high.

There are also sources of error in the experimentally derived natural frequencies. Apart from normal experimental problems when using simple apparatus, such as ensuring that the noted frequency is an accurate value for the natural frequency, the main source of error arises from the method of obtaining the fixed edge which is very difficult to achieve experimentally. In the work discussed in this chapter, the fixed edge was simulated by clamping the plate between two heavy blocks of steel using heavy clamps. It is unlikely, therefore, that the fixed boundary conditions would be completely satisfied using that technique. Hence, the experimental natural frequencies are probably lower than those of a true cantilever.

The constant, D_x, and hence Young's modulus, E_x, in the x direction was calculated using a natural frequency from the family $m/0$. The constant, D_{xy}, and hence the shear modulus, G_{xy}, was calculated using a natural frequency from the family $m/1$. Finally, the constants D_y and D_1 were calculated using two natural frequencies from the families m/n, where $n \geq 2$, and the previously calculated values of D_x and D_{xy}. These results lead ultimately to the determination of Young's modulus, E_y, in the y direction and the Poisson's ratios v_x and v_y in the x and y directions.

Considering the possible sources of inaccuracy in the analytical approach and in the experimental results, which were used, of the constants D_x, D_y, D_{xy} and D_1, which were calculated, D_{xy} is likely to provide the most accurate results. Of the material properties which were obtained from these constants G_{xy} is probably the most accurate at around 20% higher than the true value, with v_x and v_y being the least accurate at around 60% higher than the true value and E_x and E_y lying between these.

This investigation was undertaken to establish the feasibility of using a vibrational approach to determine the material properties of thin laminates. Although the best results which were obtained are in error by around 20%, the technique is worthy of further investigation. The trial functions for the plate deflected form were simple approximations. If these functions were improved, together with improvements in the experimental approach, the results would be much more accurate, making this simple, non-destructive approach attractive.

REFERENCES

1. TSAI, S. W. and SPRINGER, G. S., The determination of moduli of anisotropic plates, *ASME Trans., J. App. Mech.*, **30** (1963) 467–8.
2. MEGSON, T. H. G., *Aircraft structures for engineering students*, New York, Crane Rusak Co., 1972.
3. DEAN, G. D. and TURNER, P., The elastic properties of carbon fibres and their composites, *Composites*, **4** (1973) 174–80.
4. WARBURTON, G. B., The vibration of rectangular plates, *Proc. Instn. Mech. Engrs.*, **168** (1954) 371–84.
5. YOUNG, D., Vibration of rectangular plates by the Ritz method, *ASME Trans., J. App. Mech.*, **17** (1950) 448–53.

32

A Minimum Energy Composite Automobile

RICHARD W. MCLAY

COMtech Inc., 18 Athens Drive, Essex Junction,
Vermont 05452, USA

AND

JAMES BUCKLEY, THOMAS FLOYD AND DANIEL VIENS

Mechanical Engineering, Votey Building,
The University of Vermont, Burlington, Vermont 05405, USA

ABSTRACT

A two-passenger, minimum energy vehicle is discussed as it is based on the concept of a four-wheeled moped. Power requirements are explored. The results of a student competition for the building of a two-passenger model airplane engined car point toward a minimum energy vehicle (MEV). The preliminary design of the MEV is shown as a systems study of a 70 cc powerplant and drive train combined with a composite/sandwich unitized body. A PVC core/fiberglass skin sandwich is concluded to combine the attributes of low weight and high impact toughness necessary for the MEV. The mechanical/sandwich connections are explored.

INTRODUCTION

In recent years mopeds have become common on American roads. Based on minimum energy concepts, the moped is designed to have a minimum weight combined with a powerplant that slowly accelerates it to a speed just under the threshold of aerodynamic drag, as shown in Fig. 1. The resulting vehicle is highly efficient for transportation over a short distance. For higher speeds, the power requirements increase roughly proportionate to v^3, which means that high speed, long distance travel requires higher energy

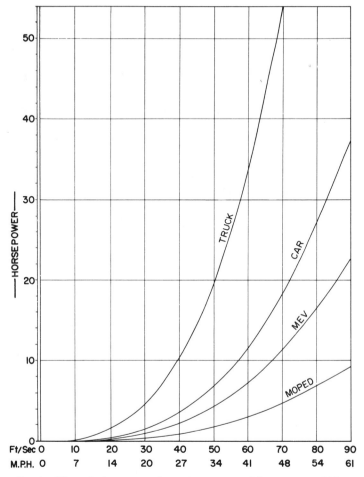

Fig. 1.   Theoretical power requirements versus speed for common vehicles.

outlays per passenger mile. Figure 1 does imply, however, that small, short-ranged vehicles can be built that will be similar to the moped in economy. One such vehicle is the subject of this paper. Called a minimum energy vehicle (MEV), the two-passenger car appears as the curve labelled MEV in the figure.

The following sections of the paper present the studies leading up to the construction of an MEV powered by a 70 cc engine. First, a preliminary project is reviewed in which two groups of students built competing cars powered by 0·21 in³ (3·44 cm³) model airplane engines. The insight gained from this project is shown to point toward the MEV. The design studies for the MEV are next presented, the principal problem being the building of a lightweight unitized body to accommodate the air-cooled engine and two passengers. Finally, the mechanical/composite connections are discussed.

A SMALL-ENGINED AUTO COMPETITION

A competition was recently held in the Mechanical Engineering Department of the University of Vermont for the construction of a two-passenger automobile powered by a 0·21 in³ (3·44 cm³) model airplane engine. The guidelines of the competition required that the engine be combined with four lightweight bicycle wheels to result in an automobile capable of:

(1)   Carrying two passengers over a 100 foot (30·48 m) straight-and-level course for time trials.
(2)   Racing from a Le Mans start with only a driver around a closed circuit of 250 feet (76 m).

The four students of each group of senior mechanical engineering students were assigned the following respective tasks in the design, construction and testing of the cars: (1) the frame; (2) the suspension and steering; (3) the engine run-in, controls, cooling system and general operation; (4) the transmission (gearbox).

The overall winner in this competition is shown in Fig. 2. Briefly, the attributes of this vehicle are as follows:

(1)   The tubular aluminum frame weighs 8 lbs (3·63 kg).
(2)   The entire vehicle weighs 32 lbs (14·52 kg).
(3)   The engine produces an estimated 1/2 hp at 26 000 rev/min.
(4)   The transmission provides power to a Lexan final drive wheel through a series of belt and chain reducers and a bicycle type sprocket/chain transmission.
(5)   The speed of the vehicle is 2 mph (3·22 km/h).

This project offered a series of challenges that were both interesting and technically valuable to the students' professional development. However, the insight gained in the design and construction of this first car was most valuable in the preliminary design of the minimum energy vehicle that is the

FIG. 2.   An auto powered by a 0.21 in^3 (3.44 cm^3) methanol, castor oil, nitromethane engine.

subject of this paper. The following conclusions were drawn from this initial work:

(1)   The vehicle must have a minimum weight.
(2)   The engine and drive train should be a standard unit to ensure reliability.
(3)   The frontal area should be kept at a minimum to minimize the power requirements.

THE PRELIMINARY DESIGN

The direction for the MEV design starts with the experience of the moped as shown on the graphs of Fig. 1. Because the speed of the moped is less than 35 mph (56 km/h), its power requirements are minimal. If we take the moped design and extend it to a two-passenger vehicle, we are faced with the following problems as compared with the moped:

(1)   The weight of the vehicle will increase by a factor of eight with a unitized metal body.
(2)   The engine and drive train weight will increase by a factor of six for a water-cooled design.
(3)   The frontal area will increase by a factor of two for a two-passenger, side-by-side configuration.

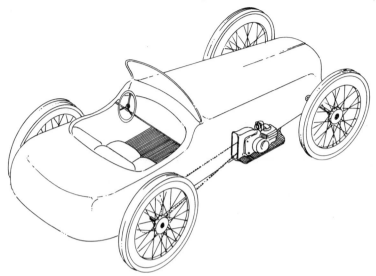

FIG. 3. The MEV preliminary design.

Thus, the design of the two-passenger MEV is a systems study that must be undertaken with a view toward the constraints of weight, frontal area and the engines available.

The concepts for the MEV design start with the choice of a 70 cc motorcycle engine and drive train with a power output of 5 hp (3·73 kW). With a frontal area of 12 ft² (1·12 m²), the theoretical curve for the MEV falls on the graph as shown in Fig. 1. Since the engine is air-cooled, the increase in weight over that of the moped is minimal; the weight of the engine and drive train is 40 lbs (18 kg). However, since the engine is air-cooled, it will have to be mounted externally to the body to facilitate air flow. Finally, the problem of the increased body weight must be solved in order to give the MEV a performance adequate for local driving. To solve this problem, which is the principal problem in the design, the weight of the unitized body must be minimized through the new technology of the composite/sandwich,[1] the subject of this paper.

Figure 3 illustrates the general configuration of the composite/sandwich unitized body. Molded in one piece, the body contains room for two passengers side-by-side. The engine and transmission unit is mounted externally and drives the right rear wheel through a chain. Cooling is accomplished by natural air flow in the same way as on the motorcycle. All body parts are fiberglass/PVC sandwich. The only metal parts are in the suspension, steering and running gear.

THE COMPOSITE/SANDWICH UNITIZED BODY

The detailed design for the unitized body was done through the use of polyvinyl chloride foam[2] for the core combined with a laminate of polyester resin and 0·040 inch (1 mm) glass fiber fabric of 16 by 14 weave. The measured breaking strength of the fabric was found to be 410 lbs/in (718 N/cm).

The PVC core material was found from experiments to be the best overall for use in the MEV body. It is a rigid PVC foam that is serviceable in a range of temperatures between $-40°$ and $+75°C$. At approximately 75 °C, the foam loses its structural strength; thus, it can be molded with a return of the properties as the temperature falls again to room temperature. The foam is also compatible with the polyester resin used with fiberglass and forms a strong mechanical bond with the glass fiber laminate. The main advantage in the use of the PVC foam, other than its light weight, is its deformation characteristics at ultimate load on the sandwich. In contrast to the more brittle materials, the foam does not fracture at ultimate load; instead, it compresses without debonding at the foam/composite interface. With a release in load, the sandwich returns to its original shape with essentially no loss of properties. Thus, by the definition of toughness, the energy of failure, the sandwich absorbs a very large amount of energy as the core compresses and the laminate debonds from its surfaces, but only after several cycles of loading have occurred. The same properties make it very tough from the standpoint of impact. In addition, the core has very good damping characteristics from the standpoint of vibration, so that the small, light engine may be used without fear of fatigue failure in the body components adjacent to the motor mount platform.

Chemically, the PVC offers further advantages. It is naturally inert to most chemicals in the environment including gasoline and lubricating oils. It is self extinguishing in a fire. Finally, it is compatible with the polyester resin in the laminate.

The production process for the glass fiber laminate/PVC foam sandwich can be quite flexible. The foam and the finished sandwich can be cut, sawed, and drilled as well as adhesively bonded to other materials. But, the major advantage of the foam is its weight of 5 ounces/ft^2 (0·47 kg/m^2) in the $\frac{3}{4}$ inch (20 mm) thickness, which was ultimately chosen for the design. It is

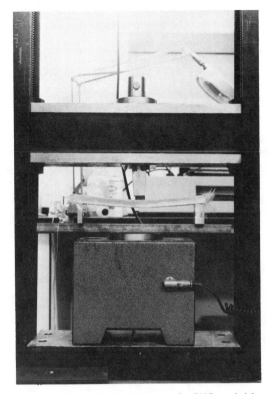

Fig. 4. Three-point bending test of a PVC sandwich.

this property that allows the weight of the unitized body to be kept to an absolute minimum while maintaining the stiffness, strength, and toughness.

The study of the PVC sandwich included a series of three-point bending tests, one of which is shown in Fig. 4. In the figure, a beam consisting of a $\frac{7}{8}$ inch (22 mm) PVC core with a two-layer fiberglass laminate top and bottom is loaded at the center by a fixture on the crosshead of the testing machine. In this test, the sandwich was observed to suffer a deflection that changed the geometry of the beam before the core compressed to cause a loss of structural stiffness. On recovery, a crack was found in the top laminate at the vicinity of the loading block. The crack was a shear failure, illustrated by broken fibers and a locally debonded laminate. The breaking load for this laminate, a 4 inch by 12 inch beam, was 275 lbs (1224 N). The insight gained from these tests indicated adequate strength in the sandwich as long as the mechanical connections were properly designed to distribute the loads from the engine and running gear.

THE MECHANICAL/SANDWICH CONNECTIONS

The major design problem in the use of the sandwich is to avoid high local loads that will cause debonding of the laminate from the core. This is especially important in the MEV since the laminate consists of two layers of cloth over the majority of the body panels. Several methods were used in the detailed design to distribute the loads. Most of them consisted of careful bracing with redundant supports such as is found in the running gear connection. But one common design feature is found in all of the connections: the PVC core is replaced locally by a plywood insert to distribute the large shear loads uniformly into the sandwich.

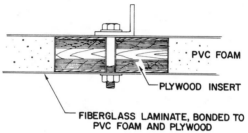

FIG. 5. Example of a mechanical/sandwich connection.

A schematic of a typical plywood insert connection is shown in Fig. 5. The fiberglass laminate is bonded straight across the joint between the PVC and plywood. All interior joints between the PVC and plywood are also bonded using a contact cement. The result is an incompressible connection that can be bolted and that will distribute the load uniformly to the rest of the sandwich.

CONCLUSIONS

We have presented work done on two phases of a minimum energy vehicle study. In the first phase a student competition on a model airplane engined car is reported. The experience gained in this first study is shown to point toward the MEV as a natural extension of a moped. The preliminary design of the MEV is discussed from the standpoint of a PVC/fiberglass laminate sandwich body. The advantages of the sandwich are shown to be low weight, high stiffness and strength, and high impact toughness. Tests on sandwich specimens give insight as to the design requirements of the unitized sandwich body.

At the time of writing this paper the project has progressed to the layout stage with various components of the running gear constructed to mate with the sandwich body. The body components have been sized. Construction and development should be completed by summer 1981. Field tests are expected to last through August 1981.

REFERENCES

1. DHARAN, C. K. H., Design of automotive components with advanced composites, In: *Proceedings of the 1978 International Conference on Composite Materials*, Warrendale, Pennsylvania, The Metallurgical Society of the AIME, 1978, 1446–61.
2. JOHANNSEN, T. J., Airex cored fiberglass for construction of larger recreational power boats, Symposium on design and construction of recreational power boats, Department of Naval Architecture and Marine Engineering, Ann Arbor, Michigan, 1979.

33

Structures in Reinforced Composites

W. S. CARSWELL

*National Engineering Laboratory, East Kilbride,
Glasgow G75 0QU, Scotland*

ABSTRACT

In the design of structures it is necessary to have accurate information on the stress distribution produced by the service-loading conditions and geometry of the structure and on the stress or strain limits for the material to avoid premature failure or permanent damage.

Even in apparently uniform structures in composite materials fabrication methods can introduce geometric and material variations which will significantly distort the stress distribution. The stress or strain limits for failure derived from 'ideal' testpieces and short-term tests may not be representative of the failure conditions in large structures.

INTRODUCTION

As fibre-reinforced structures become more widely used in more critical applications it becomes increasingly important that such structures are properly designed to ensure effective and safe use of the structure or system and of the materials used. It is an essential requirement in the design to have accurate and precise information on the service-loading to enable the stress distribution in the component to be obtained and to ensure that the critical values of stress or strain are not exceeded for any type of failure. In this context the term structure is intended to mean layered composite structures possibly with geometrical discontinuities and curvature, as distinct from uniform boards or strips prepared in ideal conditions. To a large extent the

484

form and geometry of such structures will be defined and dictated by the application and the system of which it is part. From a model of the structure the stress distribution can be obtained under any loading condition by considering the material as homogeneous with derived elastic constants. From this the stress distribution in each layer can be obtained. The critical values of stress or strain are obtained for each layer or each component of the structure by simple 'ideal' tests. The form of such data will depend on the appropriate method of failure, and the service-loading of the structure.

The purpose of the paper is to review briefly the relevance of the data obtained from simple ideal structures in a real structural application and the problems of translation from one component to another.

STRUCTURES AND MATERIALS

The structures examined are mainly curved structures, i.e. pipes formed by winding filaments or fabric with resin around a mandrel. The materials forming the composite are glass and polyester resin; the former may be in the form of continuous filaments as in filament winding or short fibres in a mat form.

The fabrication effects described may be typical of such a process as described but it is considered that similar effects will occur in other structures with other means of fabrication and with other constituent materials.

The loading is generally multi-axial which requires the composite structure to be designed accordingly.

FABRICATION

In the formation of large structures, particularly in the winding arrangements, surfaces are not fully covered on the first ply and there are often areas of overlap between layers. Such arrangements can cause regions of different fibre or resin density. Figure 1 shows a typical arrangement of filament winding. The effect of such variations is shown in Fig. 2 which shows the variation in distortion of such a construction under internal pressure.

Figure 3 shows the variation in circumferential strain around a pipe produced by overlapping bands of isotropic chopped strand mat. The variation in circumferential stress in a pipe produced by butting the bands

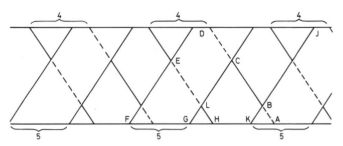

FIG. 1. Filament winding arrangement.

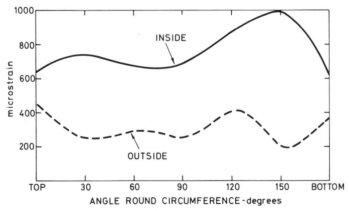

FIG. 2. Hologram of distortion of filament wound pipe under internal pressure.

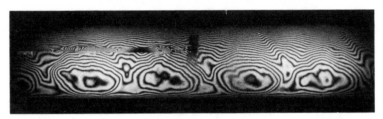

FIG. 3. Strain distribution around the circumference of a 100 mm diameter pipe, 5 mm thick with overlap, under 3·5 bar internal pressure.

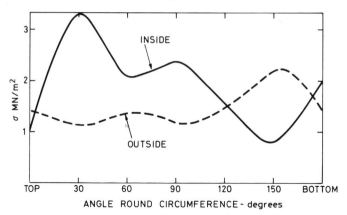

FIG. 4. Stress distribution around the circumference of CSM pipe, 100 mm diameter, 5 mm thick with no overlaps, under 3·5 bar internal pressure.

without overlap is shown in Fig. 4. The strain gradients and shell bending produced in these examples is a result of local material or geometrical variation around the circumference. The effect of a simulated thickness variation at three points around the circumference of such a pipe under pressure is shown in Fig. 5.

The effect of tabs on the ends of tubular testpieces on the axial stress distribution as shown in Fig. 6 has been shown by several workers.[1] The tabs are fixed to assist gripping and load transfer into such testpieces. They are very similar to overwrapping applied at joints for the same purpose.

It is well known that geometric changes in a structure introduce stress or

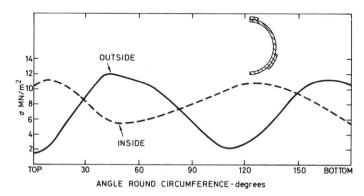

FIG. 5. Circumferential stress distribution in 100 mm diameter pipe, 5 mm thick isotropic material, with thickness variation at 0° (top) and 120°, under 7 bar internal pressure.

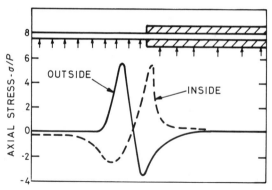

FIG. 6. Stress distribution along pipe with end tabs (from Rizzo and Vicario[6]).

strain concentrations. With curved structures and finite sizes such effects can be considerably increased due to shell bending. Anisotropy in the layers of such structures can also introduce non-uniformity or stress gradients into the structure. However, even in simple structures material variation and small geometric variations, introduced accidentally or by the manner of fabrication, can introduce stress gradients and local bending not generally expected. The magnitude of such effects will depend on the manner of loading. On the other hand, non-uniform stress distributions can be introduced into small testpieces under simple uniform loading due to the anisotropy of the material.[2]

Investigations have shown that residual stresses are present in some

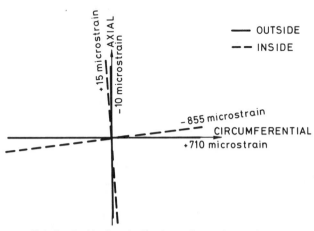

FIG. 7. Residual strain for chopped strand mat tube.

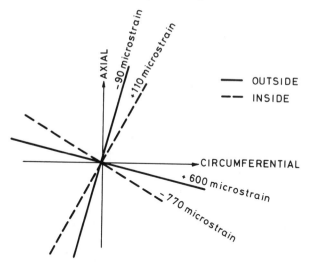

FIG. 8. Residual strain for ± 54° filament tube.

structures.[3] The magnitude of these stresses will be a function of the material, the geometry of the structure, etc., and are produced by the contraction of the resin during the curing cycle. These stresses are macrostructural as distinct from the microstress present between resin and glass fibre.

Examples of the residual stresses found in filament wound tubes are shown in Figs 7 and 8. They are essentially bending stresses and although the principal directions differ the magnitudes are all of the same order. Such stresses will be superimposed on the stress distribution resulting from the applied loads and will influence failure.

STRENGTH

Generally structures are not intended for uniaxial loading but have to withstand multiaxial loading and the composite material of the structure must be designed accordingly. This means that in composites the fibres are not generally unidirectionally aligned but distributed in multidirectional arrangements.

The strength of these constructions will therefore be anisotropic, and dependent on the direction and magnitude of the principal loads applied. A failure envelope is necessary to describe the behaviour over the full range of applied load. These can be obtained by tests on the actual structures but this approach is time-consuming and expensive. Alternatively, they can be derived using one of the many theories or interaction expressions available for multiaxial failure[4] and the failure characteristics of the plies or constituents making up the construction. In this latter method the failure parameters are obtained for each ply when subjected to a uniform stress distribution.

It is now generally accepted that higher strengths are obtained in the presence of stress gradients than in uniform stress distributions.[5] The reason proposed for this is that, since failure in a composite initiates at a defect or some point with different properties, the probability of failure is governed by the distribution of defects and the volume of the stressed area. Under these circumstances the failure envelope for a structure with a non-uniform stress distribution will have to be altered to obtain a realistic failure prediction.

In testing constructions of basic balanced plies with simple testpiece forms to ensure as uniform a stress distribution as possible, two features are often prominent:

(a)   the scatter in ultimate load values; and
(b)   the non-linearity of the stress/strain curve.

Scatter is due mainly to variability both in the construction of the laminate and in the properties of the constituents. Non-linearity is due to the progressive or multiple nature of the failure. For a material with

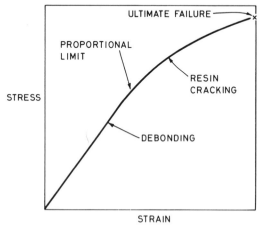

Fig. 9.   Tensile stress/strain curve for glass reinforced material (from Johnston[7]).

reinforcing fibres in more than one direction, the form of the stress/strain curve is often as shown in Fig. 9.[7] The curve can be divided into sections, each of which can be broadly associated with a form of damage within the material, e.g. debonding, resin cracking, and fibre failure. Various physical phenomena, e.g. acoustic emission and visual effects, can also be associated with such forms of damage.

In such tests the ultimate stress at failure, which is controlled by fibre strength, is a function of the volume fraction of the reinforcing fibres but the proportional limit (the point at which fibre debonding causes a significant reduction in materials stiffness) is little affected. The dependence on fibre volume fraction increases with higher levels of damage, e.g. gel-coat cracking, as shown in Fig. 10.[8]

The presence of cross fibres or angled fibres introduces strain magnifications which can lead to debonding and resin cracking as shown in

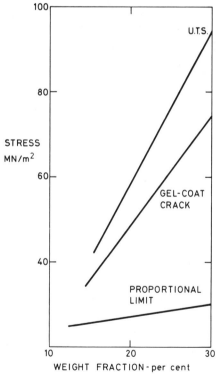

FIG. 10. Dependence of ultimate strength, gel-coat cracking and proportional limit on the volume fraction for CSM material (from Raymond[8]).

Fig. 11.   Resin crack from transverse fibre.

Fig. 11. This effect has been analysed by Kies[9] and the strain magnification factor (SMF) simply stated as

$$\text{SMF} = \frac{1}{\dfrac{2r}{s}\dfrac{E_r}{E_f} + \dfrac{b}{s}} \quad \text{or} \quad \frac{1}{1 - \left(\dfrac{4V_f}{\pi}\right)^{1/2}\left(1 - \dfrac{E_r}{E_f}\right)}$$

where b is the spacing between transverse fibres, r is the radius of fibres, $s = b + 2r$, E_r is the elastic modulus of resin, E_f is the elastic modulus of fibre and V_f is the volume fraction of fibres.

The strain magnification factor is a function of the volume fraction of reinforcing fibres because of the dependence on b and s. Thus, although the strain at fibre debonding and the stiffness of the material are functions of the volume fraction, the stress at this point will not be affected by the volume fraction to the same extent. Also, if there is variation in construction or local variations in fibre concentration any variations in bond strength, etc. will lead to scatter in the initiation and growth of damage. Thus a point on the stress/strain curve for any material is indicative of the level of damage within the material. This level of damage may be used to define a design limit as an alternative to the ultimate strength. This point with the chosen level of damage on the stress/strain curve or from another means of detection, e.g. acoustic emission, will be termed 'point of first cracking'. The form of construction and mixture of different construction, i.e. chopped strands, woven roving or mixtures, will influence the magnitude of the point of first cracking as shown in Table 1.

TABLE 1
Properties of mixed laminates with polyester resin

Construction	Modulus (GN/m^2)	First cracking stress (acoustic) (MN/m^2)	First cracking strain (acoustic) (%)	UTS MN/m^2
All chopped strand mat	6·96	32·0	0·46	81·5
Chopped strand mat with woven roving	10·10	28·0	0·28	96·4
Chopped strand mat with cross plied unwoven layers	11·3	37·3	0·33	116·7

The long-term strength in fatigue or creep, i.e. strength after a large number of cycles of stress or a sufficiently long period of time under sustained static loading, where strength is defined as complete separation of material or inability to sustain the load, appears to be related to first cracking strain for different materials (Fig. 12). The sensitivity of a material to an environment is demonstrated by the decrease in strength under sustained or cyclic loading conditions. The critical level of 'first cracking' in the short-term tests is thus decreased with more aggressive environments (Fig. 13). The strain at first cracking is therefore a better or more sensitive indication of performance in long-term conditions than ultimate strength. Other parameters such as overall structure size, which might influence

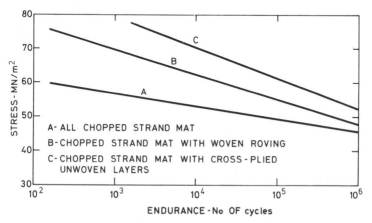

FIG. 12. Fatigue curves for mixed laminates.

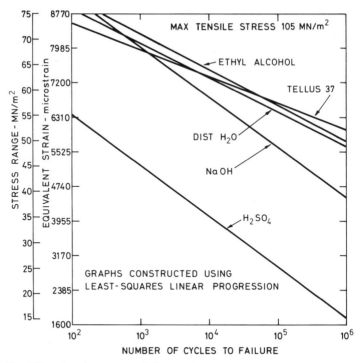

Fig. 13.   Effect of environment on fatigue performance of chopped strand mat/polyester
material.

ultimate strength, should not influence the point of first cracking in the
same way.

It is important to note that where stiffer materials are incorporated in a
structure with less stiff layers in attempting to improve overall stiffness for
the same material dimensions, corresponding proportional increase in
strength is not likely to be produced.

In design standards such as BS 4994[11] strength is related to the ultimate
strength of a material and various design factors are introduced to allow for
fabrication effects, long-term effects such as cyclic stress, or curing. A safety
factor of three is also introduced to ensure that, for any structure designed
to that standard, the design factor shall not be less than six. These designs
and safety factors can be shown to be dependent on the variability in
material and in stress distribution and on long-term effects. An alternative
approach is to limit the design strain, i.e. to 0·2 % as recommended in
BS 4994. This has been shown to be more relevant in terms of first cracking

but it must be remembered that the limit for failure can vary with material and with operating conditions.

ACKNOWLEDGEMENT

This paper is presented by permission of the Director, National Engineering Laboratory, Department of Industry. It is Crown copyright. The work described was supported by the Engineering Materials Requirements Board of the Department of Industry.

REFERENCES

1. RIZZO, R. R. and VICARIO, A. A. 'A Finite Element Analysis of Stress Distribution in Gripped Tubular Specimens', *Composite Materials (Testing and Design) 2nd Conf.*, STP497, ASTM, 1972, 68–88.
2. PAGANO, N. J. and BYRON PIPES, R. Some observations on the interlaminar strength of composite laminates, *J. Composites*, **15**, 1973, 679–88.
3. DANIEL, I. M. and LIBER, T. 'Lamination Residual Strains and Stresses in Hybrid Laminates', *Composite Materials (Testing and Design) 4th Conf.*, STP617, ASTM, 1977, 330–43.
4. SENDECKYJ, G. P. 'A Brief Survey of Empirical Multiaxial Strength Criteria for Composites', *Composite Materials (Testing and Design) 2nd Conf.*, STP497, ASTM, 1972, 41–51.
5. WHITNEY, J. M. and KNIGHT, M. The relationship between tensile strength and flexure strength in fibre-reinforced composites, *Exp. Mech.*, **20**(6), 1980, 211–16.
6. AVESTON, J. and KELLY, A. Tensile first cracking strain and strength of hybrid composites and laminates, *Phil. Trans. Roy. Soc.*, **A294**, 1980, 519–34.
7. JOHNSTON, A. F. Engineering design properties of GRP. Publication 215/26, British Plastics Federation, London, 1979.
8. RAYMOND, J. A. 'The Effect of Resin Content on the Mechanical Properties of Glass Reinforced Polyester Laminates', *BPF Reinforced Plastics Conf., Brighton, 1974.*
9. KIES, J. A. Maximum strains in the resin of fibreglass composites. Naval Research Laboratory Report No. 5752, AD-274560, 1962.
10. ROBERTS, R. C. and CARSWELL, W. S. Environmental fatigue stress failure mechanism for glass fibre mat reinforced polyester. *J. Composites*, **11**, 1980, 95–9.
11. BRITISH STANDARDS INSTITUTION. Specification for vessels and tanks in reinforced plastics, BS 4994: 1973.

34

Properties and Performance of GRC

B. A. PROCTOR

Pilkington Brothers Limited, Research and Development Laboratories, Lathom, Ormskirk, Lancashire L40 5UF, England

ABSTRACT

Glassfibre Reinforced Cement (GRC) is a new building and construction material which is rapidly gaining acceptance around the world. This paper outlines the test programmes which have been found necessary to provide design information and to engender the necessary confidence for its use. Material properties, long term strength predictions and some measurement difficulties are also discussed.

INTRODUCTION

The introduction of any new material into real engineering or constructional use is today dependent on the availability of a wide range of property data. This is especially important in construction and building applications since materials are used in a variety of conditions in different climates and lifetimes are generally expected to be quite long.

The invention, development and commercialisation of alkali-resistant glass fibres for cement reinforcement[1-3] launched a new material—Glassfibre Reinforced Cement (GRC)—and generated just such a need for new information. As the manufacturers of the alkali-resistant fibre Pilkington Brothers recognised that need and responded to it by establishing extensive research and development programmes devoted to the investigation and testing of GRC materials and components under a wide range of conditions. That work has led to the establishing of design

496

procedures, the development of appropriate test methods and the definition of quality control methods for GRC which are all described in a number of privately published brochures[4,5] available to Pilkington Licencees and to major GRC specifiers and designers. Some of this work has also been published in the open literature.[6-9]

In principle, as with all fibre reinforced materials, even a combination of one type of alkali-resistant (AR) fibre with only one basic cement matrix can yield an infinite number of composites depending on fibre content, orientation, method of manufacture, etc. The problems of information gathering are dauntingly impossible.

In practice, again as with most fibre reinforced materials, very few materials turn out to be economic, practicable and useful. In the case of GRC the requirements of most applications have been met by the use of a relatively standard sprayed material containing about 5% wt of AR glassfibre. This has made the data gathering more manageable and most of the information given below refers to that type of material. From a knowledge and understanding of the properties of that standard composite, and of the different roles played by fibre and matrix, the properties of other types of composite can generally be predicted.

These predictions have been supported by very much more limited experimental programmes on 'special' composites of varying glass content, differing cement matrices, etc. In general it is true to say that the Young's modulus, Limit of Proportionality (LOP) and creep behaviour of GRC are strongly dependent on matrix properties and are therefore influenced by: (a) the type of matrix being used (neat cement, mortar, etc.), and (b) the degree of cure of the cement.

The ultimate strength of GRC, whether it be bending, tensile or impact strength is essentially determined by the presence of the glass fibres and is therefore dependent on:

(1) glassfibre content and form,
(2) orientation and disposition of the fibres,
(3) the bonding between fibre and matrix (a function of matrix type, porosity and curing of the matrix).

Another factor which has eased the problem of data gathering and encouraged more acceptable application of GRC is that the 'standard' material has been based on an RHPC or OPC mortar matrix. Composites based on other cements (aluminous and supersulphate) are possible and have been studied—but practical considerations have almost always led

back to the *use* of composites based on RHPC/OPC. The wide knowledge
experience and confidence in this cement system, which forms 95 % of GRC
have been important factors in the ready acceptance of GRC as a
construction material.

WEATHERING STUDIES

The first GRC composites were developed and made at the Building
Research Establishment[1] where the need to set up long term test
programmes to establish the durability of this new material was
immediately recognised. Samples of spray dewatered GRC initially with a
neat cement paste matrix and containing 5 % wt of AR glassfibre, were laid
down in three basic storage conditions for subsequent testing in a series of
programmes beginning in 1968. Results from the earliest series were
reviewed by a BRE/Pilkington Working Party after the 5-year results were
available.[10] Ten-year results have recently been reported by BRE[11] and are
commented on by Majumdar.[12]

The three basic storage conditions used in the BRE experiments were:

(1)   dry, indoors (40 % RH, 20 °C),
(2)   water immersion at 20 °C,
(3)   UK weather exposure (at BRE Garston).

As Pilkington became involved in the commercial development of GRC
in the early 1970s it was realised that weathering information was needed
over a much wider range of climatic conditions. Also it became clear that,
even with the 5 % glass content of the standard material held constant, there
would be detailed variants of formulation and manufacturing method—in
particular sand was introduced early on to reduce moisture movement and
many components were made by a hand spraying method which did not
include dewatering.[13] Consequently in the years following 1972 a number
of Pilkington weathering programmes were set up to supplement the BRE
experiments and provide additional information on the effects of different
climates on a variety of GRC materials. The main variables covered in this
work are shown in Tables 1 and 2. Early programmes were planned for 10
years. As the work has developed the amount of material exposed has been
increased to allow for longer exposure times (20 years) and a wider range of
tests including shear strengths, density, porosity and permeability
measurements.

TABLE 1
GRC weathering stations

UK, Lathom	Temperate, low rainfall, rural
UK, Fort William	Temperate, high rainfall, rural
UK, St Helens	Temperate, low rainfall, industrial
UK, Nottingham University	Controlled soil burial (pH 4–8)
UK, Anglesey	Sea water immersion/spray
Singapore ⎫ Nigeria, Lagos ⎬	Tropical, high rainfall
Canada, Toronto	Hot summer/freezing winter
India, Bombay	Monsoon
Australia, Cloncurry ⎫ USA, Arizona ⎬	Hot/dry, desert site
Australia, Innisfail	Hot/wet, jungle site

It will be appreciated that the amount of test data available over the years from these BRE and Pilkington weathering programmes could present severe problems of assimilation and analysis. In common with many cement based or composite materials properties may vary from sample to sample and the ability to summarise the total range of results from many similar experiments is invaluable. This has been achieved by the establishment of a computer data bank for GRC. Pilkington weathering data, BRE data available under the Pilkington–NRDC licence and some of the results from accelerated tests referred to below are all included, and it is possible to produce comprehensive summaries of all available information. For example, Fig. 1 shows all available bending strength results in UK weather for spray dewatered GRC containing 4·8 % to 6·0 % (wt) of Cem-FIL fibre in Portland Cement matrices, with and without sand up to sand/

TABLE 2
Main materials variants in weathering experiments

1. Spray dewatered material	OPC and OPC/pfa matrices
2. Spray dewatered ⎫ Spray non-dewatered ⎬ materials	RHPC and HAC matrices
3. Spray dewatered materials	Swiftcrete, white OPC, supersulphate cement, Aquacrete, Revinex addition
4. Spray dewatered materials	RHPC, sand and limestone flour matrix RHPC and OPC with and without sand
5. Spray dewatered ⎫ Spray non-dewatered ⎬ materials	RHPC and sand matrix, new series with larger panels (tests include shear, shrinkage, permeability)

B. A. Proctor

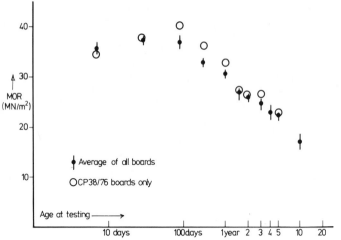

FIG. 1.   Bending strengths of spray dewatered GRC in UK weather (longitudinal samples).

cement ratios of 0·5, and made with water/cement ratios less than 0·4. Table 3 indicates that up to 5 years the data points are mainly the averages of tests on samples from 10 to 40 separate experiments. This type of study has shown that sand-containing GRC tends to have slightly lower early life strengths than the prototype neat cement paste material studied and reported in CP 38.[10] However after 18 months the overall average strengths

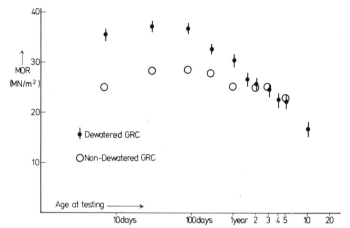

FIG. 2.   Bending strengths of dewatered and non-dewatered sprayed GRC in UK weather (5 % wt Cem-FIL fibres, longitudinal specimens).

TABLE 3

Bending strength results for longitudinal specimens of spray dewatered GRC containing from 4·8 to 6·0 wt % Cem-FIL AR fibre in UK weather for up to 5 years

Age	MOR (MN/m²)	Standard deviation (MN/m²)	No. of sets of results	LOP (MN/m²)	Standard deviation (MN/m²)	No. of sets of results
Initial strength 7 days	35·6	5·4	51	12·4	2·1	29
Initial strength 28 days	37·2	5·3	129	12·5	2·9	100
3 months	36·7	3·9	21	13·2	2·4	18
6 months	32·6	2·6	29	13·6	2·5	24
1 year	30·4	3·1	41	14·1	2·1	31
1½ years	26·8	3·3	12	14·5	1·0	6
2 years	25·7	2·5	27	14·2	2·0	22
3 years	24·4	2·4	12	14·7	2·5	11
4 years	22·7	1·3	3	15·5	2·4	2
5 years	22·4	3·5	16	16·2	2·0	13

for all materials tested are very similar to the original neat cement paste based samples and clearly indicate that the long term strength predictions from the earliest weathering programmes may be applied to newer sand-containing materials.

Another important comparison is shown in Fig. 2, which shows weathering results for dewatered and non-dewatered materials of similar fibre content and density. Although the initial values of the non-dewatered material are significantly lower, the strengths after 2 years are indistinguishable—and are expected to remain so thereafter.

ACCELERATED DURABILITY TESTING

Alkali-resistant glass fibres represent an enormous step forward over non-alkali resistant E-glass fibres, making possible the manufacture of useful reinforced composites. Nonetheless the highly alkaline environment in wet cement still leads to some attack on the fibres and this is shown up as strength loss in the weathering results given (for example) in Figs 1 and 2 and discussed in more detail in Refs 3, 6, 10, 11 and 12. Design and use of a new material such as GRC must be based on reliable estimates of long term strength values which can be regarded as minimum values over the lifetime of the components. One way of making such predictions is to carry out weathering studies over a number of years (as described above) and then attempt to extrapolate from observed trends after (say) 5 or 10 years.[10-12] An alternative and valuable complementary approach is to find some way of accelerating these changes, to attempt to relate the time scales of the accelerated and real changes, and then to use the accelerated test results to indicate expected long term behaviour.

Early work at BRE had indicated that GRC composites lost strength more rapidly when immersed in hot water; subsequent Pilkington work confirmed this and indicated further that there was no additional strength loss induced by cycling between hot and cold conditions. Separate fundamental studies of the tensile strengths of glass strands set into small blocks of cement and immersed in water at various temperatures—as distinct from the more complex behaviour of GRC composites—also indicated a general pattern of early strength loss followed by a continually reducing rate of change. The pattern of behaviour was similar at all temperatures. Strength losses occurred more rapidly at high temperatures and it was possible to simulate the changes which would occur over many years—even tens of years—at lower practical working temperatures, by tests lasting days or weeks at higher temperatures.

Thus both the work on GRC composites and on glass strands led to the concept of accelerated testing of GRC based on immersion in hot water. With the advantage of several years' real weathering data also available it has been possible to establish relationships between the accelerated testing times and real weathering. This has led to the ability to make strength predictions over tens of years covering most practicable lifetimes. One such approach for UK weathering, based on a summary of all relevant data in the computer data bank, is shown in Fig. 3. Composite strengths from accelerated tests at 50 °C, 60 °C and 80 °C have been combined with actual weathering results for up to 10 years to give an indication of strengths expected over 60 years or more. To achieve this the strength plots for the higher temperature accelerated results have been displaced along the log (time) axis until the early results overlap with the measurements made in real weathering conditions. It can be seen that the results for the different acceleration temperatures then form a common curve, extending forward for many years in real weathering, and it is believed this provides a sound basis for extrapolation to indicate patterns of expected future strength values.

Of course Fig. 3 deals with one particular type of spray dewatered material in one climate but similar treatments can be used for other

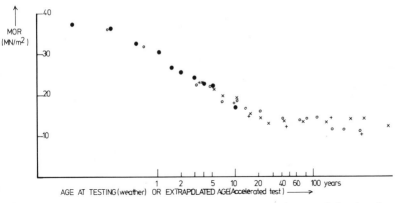

Fig. 3. Bending strengths of dewatered GRC in UK weather, extrapolation based on accelerated test (5% wt Cem-FIL glassfibre). ● UK weather accelerated test at: ○, 50 °C; +, 60 °C; ×, 80 °C.

conditions. The more fundamental approach based on glass strand strength measurements will be described in detail elsewhere. The results provide further justification for the approach outlined above and give very similar extrapolations from somewhat different types of calculation. This approach is more general and can be used to predict the behaviour of a wider range of composites with, for example, different glass contents, orientations, and even different types of cement.

BEHAVIOUR UNDER CONTINUOUS LONG TERM LOADING

So far GRC has not been used to any great extent in situations in which it would be subjected to significant permanently applied or continuous stresses. Thus properties such as creep, stress-rupture and fatigue have not been critical. The fact that creep strains are generally less than moisture movements creates certain difficulties of measurement. Even in a controlled humidity laboratory direct tensile creep measurements can be obscured by small changes in the moisture level of the environment. By measuring deflections in bending tests on simple strip specimens the moisture movement effects are suppressed and creep may be observed and measured in normal laboratory conditions. The general form of creep strain variation with time for GRC materials is shown in Fig. 4 which gives results for neat cement paste based samples tested in bending. At bending stresses below the Limit of Proportionality, LOP (i.e. in the design stress range, see below) the

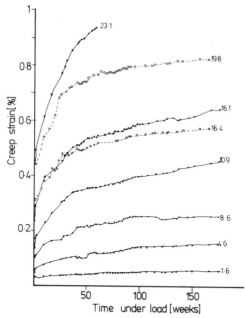

Fɪɢ. 4. Creep of spray dewatered GRC with cement paste matrix in dry conditions (5 % wt glass; water/cement ratio, 0·3; 1 month old at loading). Numbers on curves are nominal applied bending stresses in MN/m².

creep behaviour of sprayed GRC is not influenced by the fibres and is identical with that of the matrix material made in the same way. Both water content and aggregate content have a big effect on creep rates—as they do for mortars—and creep coefficients for a neat cement paste GRC and a sand-containing GRC in wet conditions are shown in Fig. 5. In dry conditions creep is initially somewhat greater but approaches the 'wet creep' at the longer times.

Recently some additional experiments in natural weather conditions have been started with hollow box beams 200 mm wide × 100 mm deep with 10 mm thick skins loaded in four-point bending over a 1·3 mm span to stresses of nominally 3 MN/m² and 4·5 MN/m². Because of the hollow sections the stress condition is much nearer to pure tension, rather than bending. Over a period of 18 months any creep which may have occurred has been indistinguishable from the variations due to moisture and temperature changes.

Stress rupture tests in bending have been in progress with spray dewatered GRC for between 3 and 5 years in dry indoor conditions, water

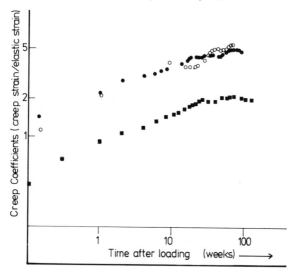

FIG. 5. Effect of matrix type on creep of GRC loaded in flexure at 1 month (creep under water). ○, ●, Neat cement paste, water/cement ratio 0·3; ■, sand/cement mortar, water/cement ratio 0·29.

immersion and UK weather. No stress rupture failures have been observed in any of these conditions at stresses below twice the normally recommended design stress and overall no additional strength loss due to a stress corrosion effect has been observed so far. This is illustrated in Fig. 6 for the water immersion conditions where stress rupture failures (points) are compared with the normal strength changes with age for similar material

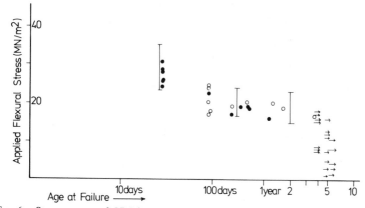

FIG. 6. Stress rupture of GRC in water. Age when loaded: ●, 1 month; ○, 3 months.

B. A. Proctor

(bars). All the stress rupture failures lie within the normal (unstressed storage) scatter bands and several samples survive after 5 years under stress in these wet conditions.

Fatigue tests have been carried out under conditions of non-reversed flexural loading in four-point bending[14] and in direct tension under zero load–tension and tension–compression cycling.[4] While more work is certainly needed, and is planned, particularly in fully reversed flexure, the present results indicate fatigue lives in excess of 10^6–10^7 cycles at the recommended design stress levels, showing that fatigue is not a critical consideration in the uses so far considered for GRC.

COMPRESSIVE AND SHEAR STRENGTHS

The mode of failure and compressive strength of GRC depend on the orientation of the test relative to the fibres. The in-plane compressive strength (stress parallel to fibres) is about 70 % of the across-plane strength (stress perpendicular to fibres). Both strengths are strongly dependent on the type of matrix and the latter strength is in fact close to that of the unreinforced matrix material.

Initially values were obtained from 25 mm cubes cut from extra thick sprayed sheet. Because GRC is so frequently used as a much thinner sheet material we wished to test coupons of representative thickness cut from normal sheets. A recent study[15] has indicated that consistent results for in-plane strength could be obtained with a specimen size of $60 \times 50 \times 8$–10 mm. Compressive strength values for a spray dewatered material containing 5 % Cem-FIL fibre in a matrix with a sand:cement ratio of 1:3 lay in the range 64–83 MN/m^2. Stress–strain curves were linear to about 45–55 MN/m^2 with Young's moduli in the range 27 to 31 MN/m^2 and Poisson's ratios of 0·12 to 0·22. Tests on samples stored in natural weather for 5 years indicate a tendency for compressive strengths to increase on ageing, and for the in-plane strength to approach the across-plane strength values.

Work on the shear strength testing of GRC has been described by Oakley and Unsworth.[8] Because of the anisotropy of sprayed sheet materials there are three different types of shear strength:

(1) Interlaminar shear with failure plane and shear direction parallel to the fibre plane. This is essentially a matrix-controlled property,

strength values being about 3 to 5 MN/m^2 and tending to increase slightly on wet ageing.

(2) In-plane shear with failure plane perpendicular to fibres and shear direction in the fibre plane. This tends to be a fibre-controlled property and Oakley and Unsworth have pointed out that it is numerically equal to tensile strength. Values fall on wet ageing from about 15 MN/m^2 to about 5 MN/m^2.

(3) Punch-through shear with both failure plane and shear direction perpendicular to the fibre plane. This again is a matrix-controlled property and is thought to be related to the compressive strength of the materials. Values tend to lie in the range 30 to 50 MN/m^2.

Tests on samples aged for up to 3 and 5 years in UK weather and indoors respectively indicate no reduction in punch through shear strengths. Tests on samples aged in water at 20 °C for 4 years show a possible slight fall from 35 to 40 MN/m^2 to rather less than 30 MN/m^2.

FREEZE–THAW AND LOW TEMPERATURE BEHAVIOUR

Construction materials are often exposed to prolonged freezing conditions and to cycling between normal and sub-zero temperatures. Part of the GRC weathering programme involves exposure of material on a roof site in Toronto which is subject to prolonged cold winters with hot summers. Samples of spray dewatered and non-dewatered GRC with a neat cement paste matrix have recently been examined after 5 years exposure at this site and compared with similar material weathered in the UK. The Canadian samples showed no deterioration over the UK samples, visually the surfaces were free from pitting or cracking and the strengths were if anything slightly higher than the UK weathered samples.

In laboratory tests based on BS 4624, i.e. cycling wet samples 25 times (as in BS 4624) or 50 times between $+20$ °C and -20 °C in air, no loss of strength could be detected with dewatered or non-dewatered GRC in either a freshly made or artificially aged condition. When tested at low temperatures (e.g. -55 °C) GRC actually shows an increase in strength over room temperature values; this is quite a slight effect with dry material but is quite marked for wet material.

ASTM C666 (Procedure A), freezing and thawing immersed in water, provides a very severe freeze–thaw condition and most materials show some degradation. GRC is no exception but it still compares favourably with

B. A. Proctor

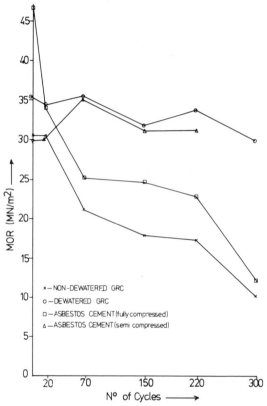

F<small>IG</small>. 7.   Effect of freezing and thawing in water on the strength of GRC and asbestos cement.

older established construction materials as shown in Fig. 7 and the conclusion is that freeze–thaw conditions do not pose a significant problem in the use of GRC.

DESIGN AND WORKING STRESS LEVELS

In selecting design stress values a number of factors must be considered. They include:

(1)   the type of stress system and relevant material strength,
(2)   the retention of that strength over product lifetimes in given conditions,

(3) the type of stress–strain curve for the material,
(4) material quality and probability of providing safety margins.

The main reason for adding alkali-resistant glass fibres to cement is to provide useable and reliable flexural and tensile strengths. The flexural strength of GRC is generally 2 to 3 times the tensile strength,[6,16–19] both decrease with wet ageing but change little in dry indoor conditions.[4,6,10–12,16] The stress–strain curve for GRC in bending or tension has a linear initial region and then bends over at an apparent yield point or Limit of Proportionality (LOP) to a region of lower slope. The yielding is in fact due to very fine distributed cracking in the material[6,20] and cycling above the LOP leads to reductions in modulus and, eventually, to fatigue failures.

The approach which Pilkington has taken in recommending design or working stress levels has been to select different values for the direct tension and bending stress conditions such that a considerable margin exists between very long term strength predictions (say 60 years based on extrapolations and/or accelerated tests) and the design values. These stress values were also chosen to be below the LOP so as to avoid matrix cracking or stiffness changes, and to ensure that fatigue, if it occurred, would not be a problem.

With standard types of GRC material, over a wide range of uses and conditions, it has been convenient and practical to choose standard nominal values of $3 MN/m^2$ (tension) and $6 MN/m^2$ (bending) for these design stresses. Material testing and quality control procedures[5] have then been recommended to ensure that material quality and performance is adequate. In rare cases of use in severe conditions these design recommendations have been lowered, in other instances where lifetimes and use conditions were precisely known, they have been raised somewhat.

This approach to the design and use of GRC may appear to be cautious or conservative but it is desirable to introduce a new construction material carefully. As a consequence, GRC components when first manufactured, and during their early lifetimes—when they are handled, transported and installed—possess a considerable reserve of strength over and above their working requirements.

CONCLUDING REMARKS

The object of this paper has been to draw attention to the enormous amount of information that is required in order to establish a new construction material and provide enough confidence for it to be used, and also to indicate how much of this has so far been achieved for Glassfibre Reinforced Cement based on OPC/RHPC and Cem-FIL AR glassfibre.

It is impossible to list all the properties and measurements in a single paper. In addition to the work outlined above considerable effort has been devoted to the study and measurement of shrinkage and moisture movement, thermal movement, fire testing, erosion, permeability, etc.

The incorporation of AR glassfibre into Portland Cement has provided a new cement based sheet material. Non-combustible, capable of manufacture in thin and lightweight sections, it is extremely tough when made. Despite a loss of strength and toughness over the years it retains useful, predictable properties over long practical lifetimes and is already extensively used in building and construction all around the world.

ACKNOWLEDGEMENT

I gratefully acknowledge the contributions of many of my colleagues in the GRC Research Department of Pilkington Brothers Research Laboratory.

This paper is published with the permission of the Directors of Pilkington Brothers Ltd and Mr A. S. Robinson, Director of Group Research and Development.

REFERENCES

1. MAJUMDAR, A. J. and RYDER, J. F., Glass fibre reinforcement of cement products, *Glass Technology*, 9(3) (June, 1968) 78–84.
2. PROCTOR, B. A. and YALE, B., *Glass fibres for cement reinforcement*, Royal Society Discussion Meeting on New Fibres and Their Composites, London, May, 1978. (To be published in *The Proceedings of the Royal Society*.)
3. BLACKMAN, L. C. F., PROCTOR, B. A., SMITH, J. W. and TAYLOR, J. W., Glass fibre reinforced cement, *The Chartered Mechanical Engineer* (January, 1977) 45–51.
4. 'Design guide—glassfibre reinforced cement', issued by Pilkington Brothers Ltd to all Cem-FIL Licensees and available on request to GRC Specifiers and Designers.
5. 'Application data for use with Cem-FIL fibre—quality control test booklet', issued by Pilkington Brothers Ltd to all Cem-FIL Licensees and available on request to GRC Specifiers and Designers.
6. PROCTOR, B. A., Principles and practice of GRC. A review, *Composites* (January, 1978) 44–8 (also in *Proceedings of the International Congress on*

Glass Fibre Reinforced Cement (Brighton, 1977), Cross, S. H. (ed.), The Glassfibre Reinforced Cement Association, 51–67).

7. WARD, D. and PROCTOR, B. A., 'Quality control test methods for glassfibre reinforced cement'. In: *RILEM Symposium: Testing and Test Methods of Fibre Cement Composites* (1978), Swamy, R. N. (ed.), The Construction Press, 35–44.

8. OAKLEY, D. R. and UNSWORTH, M. A., 'Shear testing for glass reinforced cement'. In: *RILEM Symposium: Testing and Test Methods of Fibre Cement Composites* (1978), Swamy, R. N. (ed.), The Construction Press, 233–41.

9. LEE, J. A. and WEST, T. R., 'Measurement of drying shrinkage of glass reinforced cement composites'. In: *RILEM Symposium: Testing and Test Methods of Fibre Cement Composites* (1978), Swamy, R. N. (ed.), The Construction Press, 149–57.

10. 'A study of the properties of Cem-FIL/OPC composites', Building Research Establishment Current Paper, CP 38/76, 1976.

11. 'Properties of GRC:10 year results', Building Research Establishment Information Paper, IP 36/79.

12. MAJUMDAR, A. J., 'Properties of GRC'. In: *Concrete International: Symposium on Fibrous Concrete* (April, 1980), The Construction Press, 48–58.

13. LEE, J. A., 'Modifiers and additives for GRC. A review'. In: *Proceedings of the International Congress on Glassfibre Reinforced Cement* (Brighton, 1977), Cross, S. H. (ed.), The Glassfibre Reinforced Cement Association, 39–49.

14. HIBBERT, A. P. and GRIMER, F. J., 'Flexural fatigue of glass fibre reinforced cement', Building Research Establishment Current Paper, CP 12/76, 1976.

15. MCKENZIE, H. W. and CHOONG, C., Private communication, 1979.

16. PROCTOR, B. A., *Fibre reinforcement of cement and concrete*, Fourth South African Building Research Congress, Capetown, May, 1979.

17. NAIR, N. G., Private communication (Pilkington R&D Report, 1973).

18. ALLEN, H. G., Stiffness and strength of two glass-fibre reinforced cement laminates, *J. Comp. Mat.*, **5** (April, 1971) 194–207.

19. AVESTON, J., MERCER, R. A. and SILLWOOD, J. M., 'Fibre reinforced cements—Scientific foundations for specifications'. In: *Composites—Standards, Testing and Design, Conference Proceedings* (April, 1974), IPC Science and Technology Press, 93–103.

20. OAKLEY, D. R. and PROCTOR, B. A., 'Tensile stress–strain behaviour of glass fibre reinforced cement composites'. In: *RILEM Symposium: Fibre Reinforced Cement and Concrete* (1975), Neville, A. (ed.), The Construction Press, 347–59.

35

Buckling of Plate Strips—An Evaluation of Six Carbon-Epoxy Laminates

J. F. M. WIGGENRAAD

Department of Structures and Materials,
National Aerospace Laboratory, NLR, Anthony Fokkenweg 2,
1059 CM Amsterdam, The Netherlands

ABSTRACT

Buckling tests have been performed on 6 different laminates. The objectives were to select the best configuration for application in hat-stiffened panels and to investigate how close the test results could be estimated with classical plate theory. A description of test specimens, testing equipment and instrumentation is given. The problem of deriving the buckling load from the test data is discussed, as well as the difficulties encountered in obtaining stiffness matrices with the lamination theory for the specimens at hand. A comparison between experimental results and theory is made for Young's modulus, buckling load and post-buckling behavior of the laminates.

NOTATION

A	coefficient of deflection function
A_{ij}	membrane stiffness matrix elements
a	half wavelength
b	plate strip width between supports
D_{ij}	bending stiffness matrix elements
E_1, E_2	modulus of elasticity of the plate in the x and y directions respectively
G_{12}	elastic shear modulus of the plate with respect to the x and y directions

H   elastic constant defined by

$$H = \frac{1}{G_{12}} - 2\frac{v_{12}}{E_1}$$

P, P_{cr} applied load and critical load in x direction

t plate thickness

u, u_{cr} in-plane displacement and critical value of in-plane displacement of plate middle surface in x direction

w out-of-plane deflection

v_{12}, v_{21} Poisson's ratio in the x and y directions

σ_{xb}, σ_{yb} bending stress in the x and y directions

σ_{xm}, σ_{ym} membrane stress in the x and y directions

INTRODUCTION

As a first stage in designing and testing carbon-epoxy hat-stiffened panels, 6 different laminates were evaluated in buckling tests on plate strips. Tests were performed on long plate strips as these form the basic elements of a panel. In this respect four laminates were selected which could serve as panel skins ((I)–(IV)) while the other two laminates ((V), (VI)) were to be applied in the stiffeners. The objectives of the tests were to select the best skin configuration and to investigate how close the compressive pre-buckling stiffness and buckling load could be estimated with classical plate theory. As the buckling load was assumed to be the design load of the panels, post-buckling calculations were not considered at first. However, in order to derive the buckling load from the test data, such calculations turned out to be essential, as they gave a better understanding of the plate strip behavior. Results from a simple approach and from computer calculations are compared.

TEST SPECIMENS

The plate strips were manufactured at Fokker-Schiphol's Technological Center (Manufacturing and Product Development Department). They were assembled from uni-directional and fabric layers. The fibres were T300-Torayca and the epoxy was Narmco-550. The fabric was coded No. 6141 for thin layers ($t = \pm 0.25$ mm) and No. 6341 for thick layers ($t = \pm 0.5$ mm). It contained 3000 fibres per bundle in an 8 harness satin

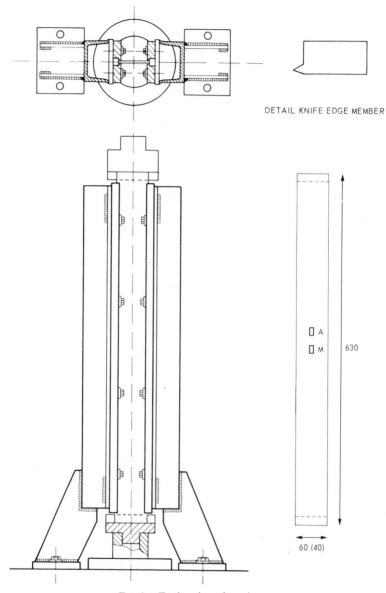

DETAIL KNIFE EDGE MEMBER

A
M
630

60 (40)

Fig. 1.   Testing rig and specimen.

TABLE 1
Laminate configurations

No.	Configuration	Average layer thickness (mm)	Total thickness (mm)
(I)	$\pm 45°$F, $\pm 45°$F	0·505, 0·505	1·01
(II)	$\pm 45°$F, 90°T, 90°T, $\pm 45°$F	0·225, 0·3, 0·3, 0·225	1·05
(III)	$\pm 45°$F, 0°T, 0°T, $\pm 45°$F	0·225, 0·3, 0·3, 0·225	1·05
(IV)	$\pm 45°$F, 0°/90°F, $\pm 45°$F	0·29, 0·55, 0·29	1·13
(V)	0°/90°F, 0°/90°F	0·49, 0·49	0·98
(VI)	0°/90°F, 0°T, 0°/90°F	0·24, 0·31, 0·24	0·79

F = fabric, T = uni-directional tape. 0° is the loading direction.

weave. Further details are given in reference 1. Fabric has been used when possible, because it is easier to handle than uni-directional tape. A disadvantage of fabric is the decrease in stiffness due to waving of the fibre bundles. The configuration of each of the 6 laminates is given in Table 1. The plate strips had a length of 630 mm and a width of 60 mm, except for laminate (VI) which had a width of 40 mm (see Fig. 1). For each configuration a number of specimens were tested (up to 6).

TESTING EQUIPMENT

The plate strips were tested in a rig that was attached to an Instron-1122 machine with a maximum capacity of 5000 N. The loading was displacement-controlled. A general impression of the rig is given in Fig. 1. It was designed and built at the NLR. Essentially it consists of 2 supports which are bolted to the base of the testing machine. To give simple support conditions at the unloaded plate strip edges each support has 2 adjustable knife edge members, in between which the plate strip is guided. The unrounded knife edges are made of steel. In combination with some grease this provided a minimum of friction. This was essential as the lengthwise edge displacements of the plate strips were quite substantial due to the length of the strips. The line-contact between the plate strip and the knife edges was situated at 1 mm from the plate edges, which left a buckling width of 58 (38) mm between the knife edges. The short ends of the plate strips fitted into slotted adapters, attached to the heads of the testing machine, to give approximately clamped end conditions. The free length of the plate strips between the heads was 600 mm. Test data (outputs of strain gauges and displacement transducers and end load) were scanned during the test and recorded by a 'programmable apparatus for data acquisition and procedures' (PADAP). During a test run 13 data channels were scanned between 100 and 200 times each at a speed of 25 channels per second. Hence, there was no need to stop the test in order to take the readings. When a test run was completed the recorded data were plotted or printed. It was also possible to manipulate the data before plotting. This was done to obtain bending and membrane strains from the recorded surface strains. The buckling pattern was plotted at certain load levels by means of a displacement transducer (LVDT) which was guided along the centerline of the plate strip. The output of a linear potentiometer determined the position of the transducer along the traverse and was used, together with the output of the LVDT to drive an x–y plotter. In order not to disturb the readings of the test data a second test run was made to obtain the buckling patterns. This was allowed because the actual buckling tests were stopped before failure occurred, while the material remained elastic up to the maximum applied loads.

INSTRUMENTATION

Strain gauges were applied on both sides of the plate strips in the loading direction, one pair exactly at the centre of the strip, another 30 mm further along the central axis of the plate strip (see Fig. 1). In this way at least one pair of strain gauges was positioned at or close to the crest of a buckle. Displacement transducers recorded the load-end displacement and the out-of-plane deflection at a crest of the buckling pattern. At the time when the experiments were carried out a post-buckling analysis was not considered. Therefore the load-end displacement was recorded to give an indication of the buckling load. For a post-buckling study this is not a useful parameter because it records the displacement of the moving head rather than the end displacement of one or a number of regular waves (away from the clamped ends).

BUCKLING LOAD FROM TEST DATA

Results of one buckling test are given in Fig. 2. Figures 2a and 2b show the output of the two strain gauge pairs as well as the membrane and bending strains which were calculated with:

$$\varepsilon_m = \frac{\varepsilon_1 + \varepsilon_2}{2} = \frac{(\varepsilon_m + \varepsilon_b) + (\varepsilon_m - \varepsilon_b)}{2}$$

$$\varepsilon_b = \frac{\varepsilon_1 - \varepsilon_2}{2} = \frac{(\varepsilon_m + \varepsilon_b) - (\varepsilon_m - \varepsilon_b)}{2}$$

Figure 2c gives the deflection w versus end load, Fig. 2d gives the end displacement versus end load and Fig. 2e shows the buckling pattern at various load levels. From these data the buckling load has to be derived. This can be done in various ways:

(a) The Southwell-plot[2] (Fig. 2c). From the deflection-load plot a new plot can be made, where w/P is plotted against w. A straight line should then be found in the region where w is not too small, but less than the plate thickness. The inverse of the slope of this line gives the buckling load. This method is not very accurate for these configurations, as the line is not very straight in most cases, while small changes in the slope give large variations in the buckling load.

(b) The load versus end displacement plot (Fig. 2d) shows a change of slope close to the buckling load. The pre-buckling part of this line is straight and if the post-buckling part of this line forms a straight line too, the point of intersection can be taken as the buckling load. However, as appears from the post-buckling analysis and also from test data, in many cases this line is not straight so an accurate point of intersection cannot be found.

(c) The sliding transducer produces useful information about the behavior of the plate strips during the test. It appears that some waves are developed earlier and faster than others, possibly because of variations in the plate thickness, but when the buckling load has been exceeded the amplitudes are all of the same magnitude. An exact point where the buckling load has been reached cannot be found, however.

(d) The load versus strain plots (Figs 2a and b) show the strain reversal behavior near the buckling load. Imperfections have a great influence on this behavior. The test data show in many cases a large transition area between the pre- and post-buckling situation so a well defined buckling point cannot be given.

(e) The membrane strains were plotted versus end load (Figs 2a and b) and it appeared that the post-buckling part of these lines was almost

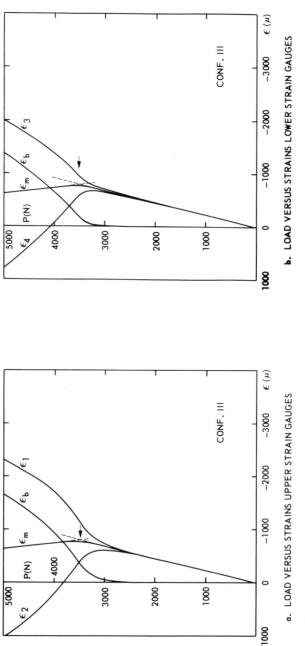

FIG. 2. Test results for configuration (III).

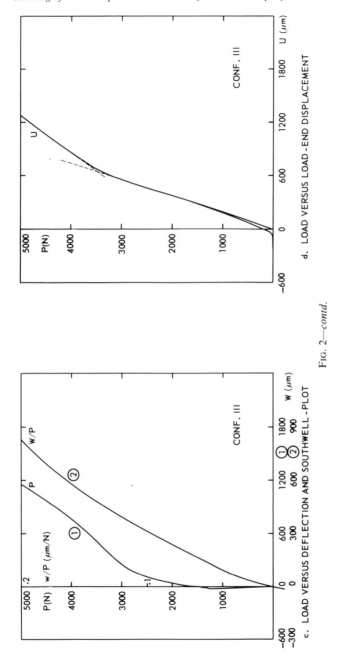

FIG. 2—*contd.*

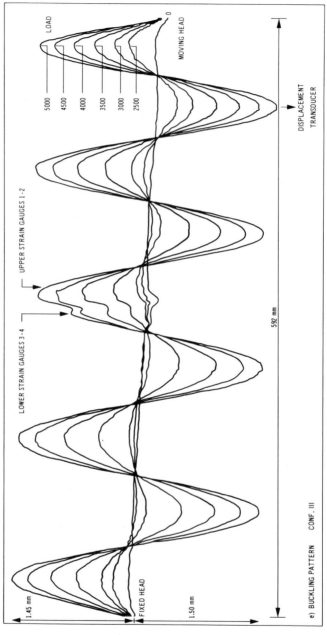

FIG. 2—contd.

linear far into the post-buckling region. This was found even for strain gauges which were positioned on a node in the buckling pattern, so the point of intersection of these lines gives a well-defined buckling point. This approximate linear behavior is explained by the post-buckling theory discussed in the penultimate section of the paper.

As the last method gives the best defined buckling point, this method was used in the analysis of the test data.

THEORETICAL BUCKLING LOAD AND YOUNG'S MODULUS

As can be seen from Table 1 all laminates are built up (symmetrically) from homogeneous orthotropic layers. This implies that for all layers ($0°$, $90°$, $0°/90°$, $\pm 45°$) the stress–strain equations with respect to the plate axes x–y reduce to

$$
\begin{vmatrix} \sigma_x \\ \sigma_y \\ \tau_{xy} \end{vmatrix} = \begin{vmatrix} Q_{11}^k & Q_{12}^k & 0 \\ Q_{12}^k & Q_{22}^k & 0 \\ 0 & 0 & Q_{66}^k \end{vmatrix} \begin{vmatrix} \varepsilon_x \\ \varepsilon_y \\ \gamma_{xy} \end{vmatrix}
\tag{1}
$$

where

$$
Q_{11}^k = \left(\frac{E_1}{1 - v_{12}v_{21}} \right)^k \qquad Q_{22}^k = \left(\frac{E_2}{1 - v_{12}v_{21}} \right)^k
$$
$$
Q_{12}^k = \left(\frac{v_{21}E_1}{1 - v_{12}v_{21}} \right)^k = \left(\frac{v_{12}E_2}{1 - v_{12}v_{21}} \right)^k \qquad Q_{66}^k = G_{12}^k
\tag{2}
$$

and k refers to the properties of layer k.

From these values the constitutive relations for the plate follow:

$$
\begin{vmatrix} N_x \\ N_y \\ N_{xy} \\ M_x \\ M_y \\ M_{xy} \end{vmatrix} = \begin{vmatrix} A_{11} & A_{12} & 0 & 0 & 0 & 0 \\ A_{12} & A_{22} & 0 & 0 & 0 & 0 \\ 0 & 0 & A_{66} & 0 & 0 & 0 \\ 0 & 0 & 0 & D_{11} & D_{12} & 0 \\ 0 & 0 & 0 & D_{12} & D_2 & 0 \\ 0 & 0 & 0 & 0 & 0 & D_{66} \end{vmatrix} \begin{vmatrix} \varepsilon_x^0 \\ \varepsilon_y^0 \\ \gamma_{xy}^0 \\ -w_{,xx} \\ -w_{,yy} \\ -2w_{,xy} \end{vmatrix}
\tag{3}
$$

where

$$A_{ij} = \int_{-t/2}^{t/2} Q_{ij}\,dz$$

$$D_{ij} = \int_{-t/2}^{t/2} Q_{ij}z^2\,dz \tag{4}$$

As far as in-plane behavior is concerned, the laminate is equivalent to a homogeneous plate with properties E_1, E_2, v_{12}, v_{21} and G_{12} which can be derived from:

$$v_{12} = \frac{A_{12}}{A_{22}} \qquad v_{21} = \frac{A_{12}}{A_{11}} \qquad G_{12} = \frac{A_{66}}{t}$$

$$E_1 = \frac{A_{11}}{t}(1 - v_{12}v_{21}) \qquad E_2 = \frac{A_{22}}{t}(1 - v_{12}v_{21}) \tag{5}$$

The buckling load of a simply supported rectangular plate with properties described by eqn. (3) is given by:

$$P_{cr} = \frac{\pi^2}{b}\left\{ D_{11}\left(\frac{b}{a}\right)^2 + 2(D_{12} + 2D_{66}) + D_{22}\left(\frac{a}{b}\right)^2 \right\} \tag{6}$$

The strips had clamped ends, but at some distance from these ends this influence vanishes. Apart from the first and last one or two buckles the buckling pattern is regular, so eqn. (6) gives the buckling load when the half wavelength of these buckles is substituted for a. The calculation of Young's modulus and the theoretical buckling load is thus based on the membrane and bending stiffness matrices A and D, defined by eqn. (4). These values can be determined when the lamina-thickness and properties in eqn. (2) are known. The fabrication process of the laminates is still such that thickness variations (leading to fibre volume percentage variations) are unavoidable, even more so when fabric is used. This results in a range of values for Young's modulus and the buckling load for each laminate, so it would be very desirable if these values could be determined more precisely from tests. Young's modulus is measured indirectly during the buckling tests (through the compressive strain recordings) but it is difficult to determine the bending stiffness coefficients by a test, because it appears that these values are very sensitive to small deviations of the displacement parameters. The development of a suitable test method is the subject of study at the NLR.

POST-BUCKLING CALCULATIONS

Post-buckling calculations have been made based on the theory as presented by Banks.[3] According to this method a suitable expression for the deflection function w is chosen, containing a number of independent parameters. Subsequently the stress function F is solved in terms of w, by means of compatibility equations. Finally the unknown parameters are determined by minimizing the total potential energy expression. With the deflection known, stresses and strains can be calculated.

In the case of the plate strips a calculation of post-buckling values of the deflection can be made by considering a plate with a length equal to one buckle, which is assumed to be simply supported along all edges. At the buckling load the end displacement is given as follows:

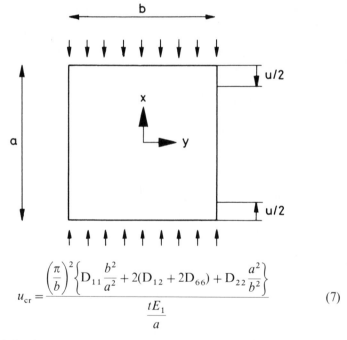

$$u_{cr} = \frac{\left(\dfrac{\pi}{b}\right)^2 \left\{ D_{11}\dfrac{b^2}{a^2} + 2(D_{12} + 2D_{66}) + D_{22}\dfrac{a^2}{b^2} \right\}}{\dfrac{tE_1}{a}} \tag{7}$$

while the deflection w takes exactly the form of:

$$w = A\cos\left(\frac{\pi x}{a}\right)\cos\left(\frac{\pi y}{b}\right) \quad \text{(with A indeterminate)} \tag{8}$$

Post-buckling calculations can be performed when assuming the deflection field to remain unchanged apart from the amplitude A, which is thus the only parameter to be calculated. Banks showed[3] that this approach gives good results for values not too far in excess of buckling ($\simeq P/P_{cr}$ less than 2). Such calculations can be done in closed form as shown hereafter.

Calculations with the program STAGS have been made as well, to show the limits to the accuracy of these post-buckling formulas. The amplitude of the deflection is determined as follows:

For plates with $E_1^2 H^2 > 4\dfrac{E_1}{E_2}$:

$$k_1 = \frac{2\pi}{a} \sqrt{\frac{E_1 H}{2} + \frac{1}{2}\sqrt{E_1^2 H^2 - 4\frac{E_1}{E_2}}}$$

$$k_2 = \frac{2\pi}{a} \sqrt{\frac{E_1 H}{2} - \frac{1}{2}\sqrt{E_1^2 H^2 - 4\frac{E_1}{E_2}}} \qquad (9)$$

$$s_1 = \sinh\left(k_1 \frac{b}{2}\right) \qquad s_2 = \sinh\left(k_2 \frac{b}{2}\right)$$

$$c_1 = \cosh\left(k_1 \frac{b}{2}\right) \qquad c_2 = \cosh\left(k_2 \frac{b}{2}\right) \qquad (10)$$

$$A = \sqrt{\frac{2E_1(u - u_{cr})}{\dfrac{\pi^2}{ab}\left\{E_1 \dfrac{3b}{8} + \dfrac{a^4}{b^4} \dfrac{E_2}{8}\left[\sqrt{E_1 E_2 H^2 - 4}\left\{\dfrac{2s_1 s_2}{k_2 s_2 c_1 - k_1 s_1 c_2}\right\} + b\right]\right\}}} \qquad (11)$$

(with u positive for compression)

For plates with $E_1^2 H^2 < 4\dfrac{E_1}{E_2}$:

$$p = \frac{2\pi}{a} \sqrt{\frac{1}{2}\sqrt{\frac{E_1}{E_2}} + \frac{E_1 H}{4}}$$

$$q = \frac{2\pi}{a} \sqrt{\frac{1}{2}\sqrt{\frac{E_1}{E_2}} - \frac{E_1 H}{4}} \qquad (12)$$

$$sp = \sinh\left(p \frac{b}{2}\right) \qquad sq = \sin\left(q \frac{b}{2}\right)$$

$$cp = \cosh\left(p \frac{b}{2}\right) \qquad cq = \cos\left(q \frac{b}{2}\right) \qquad (13)$$

$$A = \sqrt{\dfrac{2E_1(u-u_{cr})}{\dfrac{\pi^2}{ab}\left\{E_1\dfrac{3b}{8}+\dfrac{a^4 E_2}{b^4\,8}\left[-\sqrt{4-E_1 E_2 H^2}\left\{\dfrac{(cp)^2(sq)^2+(sp)^2(cq)^2}{(p)(sq)(cq)+(q)(sp)(cp)}\right\}+b\right]\right\}}}$$

$$(14)$$

For plates with $E_1^2 H^2 = 4(E_1/E_2)$, i.e. isotropic plates, one has to calculate A from the limiting case of either group: $\lim(k_1-k_2)\to 0$ or $\lim q\to 0$. with $E_1 = E_2 = E$, $H = 2/E$.

The total in-plane load is given by:

$$P = \frac{tE_1 b}{a}\left\{u-\frac{\pi^2}{8a}A^2\right\} \tag{15}$$

hence, because of eqn. (14), the relation between P and u is linear. The membrane stresses are found as follows:

$$\sigma_{x_m} = \frac{-uE_1}{a}+\frac{E_1\pi^2}{4a^2}A^2\cos^2\left(\frac{\pi y}{b}\right)+\frac{\pi^2}{2a^2 b^2}A^2\left(\cos\frac{2\pi x}{a}\right)(\phi'')$$

$$\sigma_{y_m} = -\frac{2\pi^4}{b^2 a^4}A^2\left(\cos\frac{2\pi x}{a}\right)(\phi)$$

$$(16)$$

The relation between the membrane stresses and u is linear.

For plates with $E_1^2 H^2 > 4\dfrac{E_1}{E_2}$:

$$\phi = -\frac{E_2 a^4}{16\pi^2}\left\{1-\frac{k_2 s_2\cosh k_1 y}{k_2 s_2 c_1-k_1 s_1 c_2}+\frac{k_1 s_1\cosh k_2 y}{k_2 s_2 c_1-k_1 s_1 c_2}\right\}$$

$$\phi'' = -\frac{E_2 a^4}{16\pi^2}\left\{-\frac{k_1^2 k_2 s_2\cosh k_1 y}{k_2 s_2 c_1-k_1 c_1 s_2}+\frac{k_1 k_2^2 s_1\cosh k_2 y}{k_2 s_2 c_1-k_1 s_1 c_2}\right\}$$

$$(17)$$

For plates with $E_1^2 H^2 < 4\dfrac{E_1}{E_2}$:

$$R = -(p)(cp)(sq)-(q)(sp)(cq)$$

$$S = (p)(sp)(cq)-(q)(cp)(sq) \tag{18}$$

$$N = (p)(sq)(cq)+(q)(sp)(cp)$$

$$\phi = -\frac{E_2 a^4}{16\pi^2}\left\{1+\frac{R}{N}\cosh py\cos qy+\frac{S}{N}\sinh(py)\sin(qy)\right\} \tag{19}$$

$$\phi'' = -\frac{E_2 a^4}{16\pi^2}\left\{\frac{(p^2 - q^2)R + 2pqS}{N}\cosh py \cos qy\right.$$

$$\left. + \frac{(p^2 - q^2)S - 2pqR}{N}\sinh py \sin qy\right\} \quad (20)$$

Membrane strains can be calculated with:

$$\begin{vmatrix} \varepsilon_{x_m} \\ \\ \varepsilon_{y_m} \end{vmatrix} = \begin{vmatrix} \dfrac{1}{E_1} & \dfrac{-v_{12}}{E_1} \\ \\ \dfrac{-v_{12}}{E_1} & \dfrac{1}{E_2} \end{vmatrix}\begin{vmatrix} \sigma_{x_m} \\ \\ \sigma_{y_m} \end{vmatrix} \quad (21)$$

The relationship between membrane strains and u is linear.

Bending strains at the plate surfaces can be derived with:

$$\varepsilon_{x_b} = -\frac{t}{2} w_{,xx} = \frac{t}{2}\left(\frac{\pi}{a}\right)^2 A\cos\frac{\pi x}{a}\cos\frac{\pi y}{b}$$

$$\varepsilon_{y_b} = -\frac{t}{2} w_{,yy} = \frac{t}{2}\left(\frac{\pi}{b}\right)^2 A\cos\frac{\pi x}{a}\cos\frac{\pi y}{b} \quad (22)$$

Bending stresses at the plate surfaces follow with:

$$\begin{vmatrix} \sigma_{x_b} \\ \\ \sigma_{y_b} \end{vmatrix} = \begin{vmatrix} \left(\dfrac{E_1}{1 - v_{12}v_{21}}\right)^s & \left(\dfrac{v_{21}E_1}{1 - v_{12}v_{21}}\right)^s \\ \\ \left(\dfrac{v_{21}E_1}{1 - v_{12}v_{21}}\right)^s & \left(\dfrac{E_2}{1 - v_{12}v_{21}}\right)^s \end{vmatrix}\begin{vmatrix} \varepsilon_x^b \\ \\ \varepsilon_y^b \end{vmatrix} \quad (23)$$

while E_1^s, E_2^s, v_{12}^s, v_{21}^s are constants for the surface layers.

It has been shown by Banks et al.[5] that an improved function for the displacement w, which allows for a flattening out of the centre of the plate, gives a deviation of the linear relationships mentioned before. These deviations were confirmed by the STAGS calculations.

EXPERIMENTAL RESULTS AND CONCLUSIONS

Figure 3 shows membrane and bending strains versus end load and Fig. 4 shows out-of-plane deflection and end-displacement versus end load for representative specimens of all 6 laminates, both theoretical and

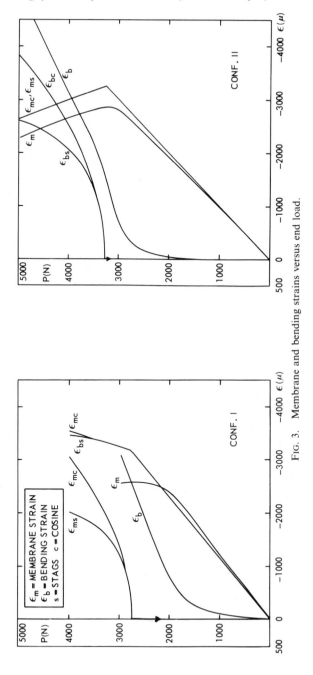

FIG. 3.   Membrane and bending strains versus end load.

J. F. M. Wiggenraad

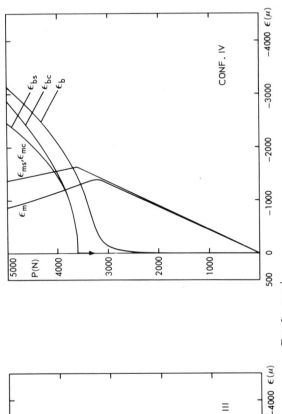

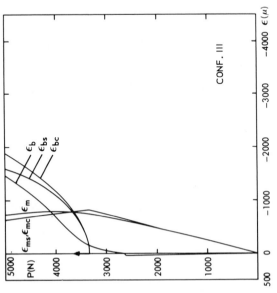

Fig. 3—*contd.*

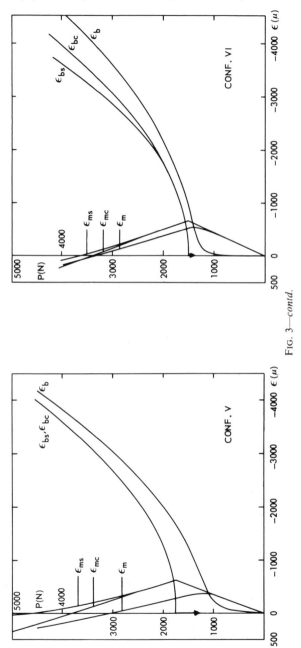

FIG. 3—*contd.*

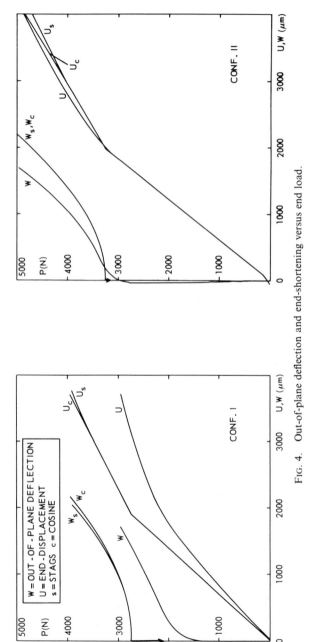

FIG. 4.   Out-of-plane deflection and end-shortening versus end load.

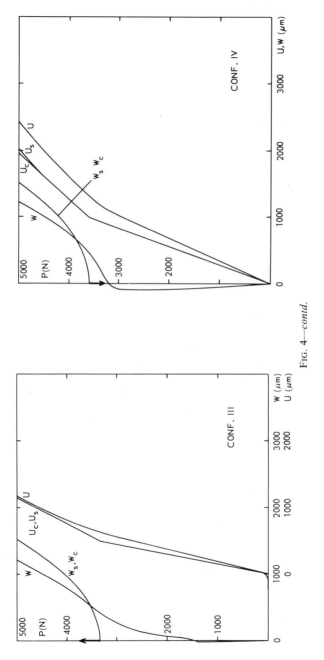

Fɪɢ. 4—*contd.*

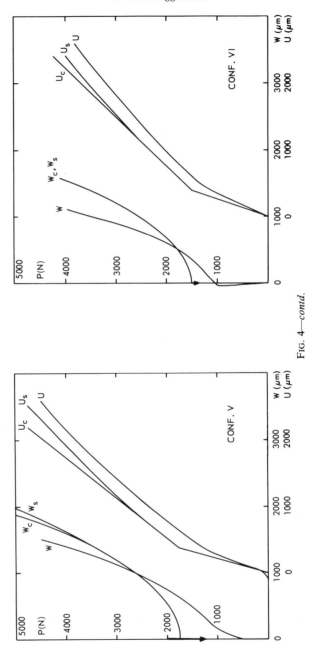

FIG. 4—*contd.*

TABLE 2
Theoretical and experimental results for Young's modulus and buckling load

Conf. no.	t (mm)	E_x (N/mm^2)			P_{cr} (N)		
		Theory	Experiment	%	Theory	Experiment	%
(I)	1·01	14 310	13 766	−4	2 750	2 100	−24
(II)	1·05	15 896	16 922	+6	3 270	3 125	−4
(III)	1·05	63 461	70 495	+11	3 350	3 675	+11
(IV)	1·13	32 369	32 954	+2	3 600	3 225	−10
(V)	0·98	47 377	49 724	+5	1 750	1 200	−31
(VI)	0·79	72 696	75 697	+4	1 500	1 350	−10

experimental results. The values of the end load as used in the calculations have been multiplied by 60/58 (40/38 for configuration (VI)) to account for the actual width (60 mm resp. 40 mm) of the plate strips compared to the free buckling width (58 mm resp. 38 mm) which was used in the calculations. Theoretical results are presented, obtained with STAGS calculations and with calculations based on the assumption that the post-buckling deflection form can be described by a simple cosine function in x and y direction of the plate.

Table 2 compares the theoretical values of Young's modulus and the buckling load with experimental results for all 6 laminates. A good agreement is found for Young's modulus; most theoretical values are slightly lower than the experimental values. Theoretical buckling loads are generally higher than experimental values, except for configuration (III) (for which the longitudinal stiffness was the most underestimated of all the laminates). The agreement is reasonable except for configurations (I) and (V). A possible reason for these deviations from theoretical results can be found in the nature of 8 H satin weave fabric. This material is not symmetrical with respect to its midplane, but shows some resemblance of a two-layer laminate: $(0°, 90°)$ or $(+45°, −45°)$. As laminates (I) and (V) consist purely of 2 layers of thick fabric these are most affected. For laminate (I) the effect is a considerable amount of bending torsional coupling (D16, D26 $\neq$ 0), which is known to reduce the buckling load with up to 30 %.[6] Laminate (V) resembles a $(90°, 0°, 0°, 90°)$ laminate with a lower longitudinal bending stiffness D_{11}, resulting in quite a shorter wavelength. Thus, the theoretical result is too high because the buckling load was calculated for the inadequately modelled material (0/90) forced into the buckling pattern of the actual material $(90°, 0°, 0°, 90°)$ as observed in the test.

Post-buckling Behavior

The agreement between membrane strains calculated with the 2 different methods is good with some deviations for configurations (V) and (VI), while the agreement between the calculated bending strains is good ((V)), reasonable ((III), (IV) and (VI)) and poor ((I) and (II)). A comparison of theoretical and experimental results reveals that the post-buckling load–strain plots have different starting points but they generally are in reasonable to good agreement otherwise. These starting points are determined by Young's modulus and the buckling load, so the differences between these points are caused by the difficulty in obtaining the actual membrane and bending stiffness matrix of a specimen with the lamination theory as discussed earlier. As a different choice from the range of stiffness values for these specimens gives merely a translation of the theoretical post-buckling curves it seems probable that a better set of stiffness values could lead to a closer fit of theoretical and experimental curves. The theoretical results for out-of-plane deflection and end-displacement as calculated with both methods are in good agreement, with some deviations of the end-displacements for configurations (II), (V) and (VI). The theoretical values for the out-of-plane deflection are generally larger than the experimental values, which was also noted by Banks.[7] Theoretical values of the end-displacement have been multiplied by $600/$(half wavelength) to give some conformity with the measured end-displacement of the moving head of the testing machine. It was mentioned before that this parameter is not suitable for a comparison with theoretical results.

It can be concluded that, for this class of laminates, hand calculations, based on the assumption that the post-buckling deflection form of a plate can be approximated by a cosine function, give generally good results compared to more accurate computer calculations, for loads not too far into the post-buckling region. To improve the agreement between theoretical and experimental results it should be possible to determine laminate stiffness parameters more accurately, for instance by a suitable bending test.

REFERENCES

1. SHIBATA, N., NISHIMURA, A. and NORITA, T., Graphite fibre's fabric design and composite properties, *SAMPE Quart.*, 1976, July, 25–33.
2. SOUTHWELL, R. V., On the analysis of experimental observations in problems of elastic stability, *Proceedings, Royal Society, London, Series A*, **135** (1932), 601–16.

3. BANKS, W. M., The post-buckling behaviour of composite panels, *1st International Conference on Composite Materials, Geneva, 1975*, **2**, 272–93.
4. ALMROTH, B. O., BROGAN, F. A. and MARLOWE, M. B., Collapse analysis for shells of general shape, Air Force Flight Dynamics Laboratory, TR-71-8, 1972.
5. BANKS, W. M., HARVEY, J. M. and RHODES, J., The non-linear behavior of composite panels with alternative membrane boundary conditions on the unloaded edges, *2nd International Conference on Composite Materials, Toronto, 1978*, 316–36.
6. WIGGENRAAD, J. F. M., The influence of bending-torsional coupling on the buckling load of general orthotropic, midplane symmetric and elastic plates, National Aerospace Laboratory, TR 77126 U (1977).
7. BANKS, W. M., Experimental study of the nonlinear behaviour of composite panels, *3rd International Conference on Composite Materials, Paris, 1980*.

36

The Damage Tolerance of High Performance Composites

R. J. Lee

*School of Materials Science, University of Bath,
Claverton Down, Bath, Avon BA2 7AY, England*

AND

D. C. Phillips

*Materials Development Division, AERE,
Harwell, Oxfordshire OX11 0RA, England*

ABSTRACT

A study has been carried out of the effects of severe stress concentrations on the strength of high performance laminates. The aims have been to establish the most appropriate strength criterion for engineering design and to determine the effects of microstructural variables on the damage tolerance. The development of damage prior to and during fracture has been observed by a variety of techniques including acoustic emission, optical microscopy, and fine focus x-ray radiography. The residual strength depends upon geometrical factors such as crack size, crack tip radius, plate dimensions and on the notch sensitivity of the material. The development of damage and notch sensitivity depend upon material parameters such as fibre type, bond strength, ply orientations and thickness, and lamination sequence. The influences of these effects on residual strength are outlined and guidelines for the development of tough laminates are suggested.

1. INTRODUCTION

Present understanding of the factors which control the strength of composites containing damage or severe stress concentrations is very limited and barely adequate for materials selection and component design. A better understanding is required of the usefulness and limitations of available predictive theories, such as linear elastic fracture mechanics

536

(LEFM), and of the effects of materials variables such as fibre type, ply orientation and stacking sequence. Although the mechanical properties of undamaged laminates can be calculated with reasonable accuracy using laminate theory no accurate means of predicting toughness are available. In contrast to metals where stress relaxation at a notch occurs by yielding, and failure usually involves the extension of a single macroscopic crack perpendicular to the applied stress, in laminates stress relaxation and fracture usually result from combinations of splitting parallel to the fibres, matrix microcracking, fibre failure, delamination, and pulling out of fibres and under some circumstances of complete plies. The formation of a stable zone of damage has been observed by several workers[1,2] using optical microscopy and laser interferometry and its size has been shown to be governed by such factors as the radius of curvature at the crack tip, the orientation and localised interactions between plies, fibre matrix bond strength, ply stacking sequence and thickness.[3,4] Any micromechanical mechanism which absorbs energy will increase the fracture resistance and a larger damage zone produced at a notch tip will result in a higher toughness and greater damage tolerance. Because of the complexity of the failure process and the number of parameters involved there is not yet any general analytical model that successfully predicts the toughness of laminates, and failure theories which have been established are all over-simplifications, usually resulting in two- or three-parameter formulae which ignore fine details. Most theories are essentially two-dimensional and are inadequate to describe the complex stress distribution near cracks and free-edges where three-dimensional effects occur, such as out-of-plane normal and shear stresses. Three-dimensional analyses performed so far for notched laminates using minimum complementary potential energy principles and finite element numerical analysis have not resulted in any immediate implications for designing a tough laminate.

This paper summarises some of the results of an investigation of the damage tolerance and residual strengths of notched and holed carbon fibre laminates. The applicability of a variety of failure theories has been considered, and some guidelines for the design of tougher laminates are suggested.

2. FRACTURE MODELS AND RESIDUAL STRENGTH CRITERIA FOR LAMINATES

Mathematical treatments describing the fracture of laminates can be classified essentially into three categories. At one extreme microstructural models, which treat the material as heterogeneous and anisotropic, have been useful in elucidating the physical processes of energy absorption such as fibre pull-out, debonding and post-debond friction. A good understanding exists of the relative importance of these models for unidirectional materials but their significance for laminates, in which delamination and splitting are important, has not been well established. A general micromechanical approach would require a detailed understanding of all the various material parameters and of the complex three-dimensional stress field surrounding the notch tip, which is not presently practical. At the other extreme macroscopic models which ignore heterogeneity are simpler to apply but are physically unrealistic. An intermediate approach[5] which treats the material as a homogeneous orthotropic continuum but incorporates local heterogeneous regions near crack tips has been used with some degree of success for unidirectional materials although extension to laminates of arbitrary lay-up have not been reported.

In this work five macroscopic models have been considered and compared with experimental data from a variety of laminates.

LEFM has been applied successfully to high strength, low ductility alloys but its relevance to composites which fail by a variety of complex modes appears *a priori* doubtful. However in practice it has been applied with some success to notched laminates when allowances were made for the formation of damage zones which effectively extend the notch[6,7]

$$\sigma_N = \frac{K_{Ic}}{Y\sqrt{(a + R_p)}} \tag{1}$$

where σ_N is the notched strength, K_{Ic} is the critical stress intensity factor, a is the semi-crack length, Y is a width correction factor and R_p is the effective increase in notch length, a characteristic of the laminate.

Waddoups *et al.*[8] introduced a two parameter fracture mechanics model which relied on a hypothetical 'intense energy' region of dimension α adjacent to the notch or hole modelled as an inherent flaw. The characteristic length was assumed to be a material property and the notched (σ_N) and unnotched (σ_0) strengths were related

$$\frac{\sigma_N}{\sigma_0} = \sqrt{\frac{\alpha}{a + \alpha}} \tag{2}$$

In rather similar approaches Nuismer and Whitney[9] introduced two different fracture models which both relied on characteristic dimensions which were material parameters. The 'point stress' failure criterion assumes that failure occurs when the stress at some distance, d_0, ahead of the notch reaches the unnotched strength. For a sharp central crack in an infinite width plate this results in

$$\frac{\sigma_N}{\sigma_0} = \sqrt{\left(1 - \frac{a}{a + d_0}\right)^2} \tag{3}$$

The 'average stress' criterion assumes that failure occurs when the stress averaged over a characteristic distance, a_0, ahead of the crack reaches the unnotched strength, resulting in the relationship for a sharp central crack

$$\frac{\sigma_N}{\sigma_0} = \sqrt{\frac{1 - (a/a + a_0)}{1 + (a/a + a_0)}} \tag{4}$$

Experimental investigations have shown that the average stress failure criterion can be used to produce predictions that are in reasonable agreement with experimental results for three-point bending[10] and for centrally cracked laminates in both uniaxial tension[11] and compression.[12] However, it has been recently reported[13] that a constant characteristic dimension is inappropriate for the notched behaviour of laminates and extension of the analysis by the use of additional parameters[14] has been suggested, although this is essentially a curve fitting procedure and does not lead to a better understanding of the physical processes during fracture.

Mandell *et al.*[15] have questioned the validity of trying to apply directly a LEFM approach where the main crack is blunted by growth of sub-cracks which tend to reduce the stress concentration and where the size of the damage zone near the notch tip, which is the same order of magnitude as the crack length, would be expected to destroy the classical stress singularity and render the calculated value of K_Q meaningless. They proposed an extension of the stress concentration approach where for an elliptical notch in an infinite uniaxially loaded specimen the maximum stress (σ_{max}) at the notch tip is given by:

$$\sigma_{max} = \bar{\sigma}\left(1 + 2\left[\frac{a}{\rho}\right]^{1/2}\right) \tag{5}$$

where ρ is the notch radius, a the semi-crack length and $\bar{\sigma}$ the remote applied stress. This expression has been modified to take into account the effects of material anisotropy[2,15]

$$\frac{\sigma_N}{\sigma_0} = \frac{1}{1 + q\left[\dfrac{a}{\rho_0}\right]^{1/2}} \tag{6}$$

where

$$q = \left(2\left(\alpha' E_{11} + \left[\frac{E_{11}}{E_{22}}\right]^{1/2}\right)\right)^{1/2}$$

$$\alpha' = \frac{1}{2G_{12}} - \frac{v_{12}}{E_{11}}$$

and where ρ_0 is the critical crack tip radius which is a characteristic of the laminate.

Comparison of eqns (1)–(6) shows that although the models are physically different a common result is that they all predict a notched strength which varies with the reciprocal of the square root of the notch length and contain a characteristic parameter which can be suitably adjusted to fit to notched strength data.

3. EXPERIMENTAL CONSIDERATIONS

A variety of carbon fibre laminates were manufactured from commercially available unidirectional pre-impregnated sheets of moulded ply, thickness 0·25 mm, which were laid up into a range of stacking sequences, representing extremes of elastic behaviour. The laminates were balanced and symmetrically stacked about the central plies in order to eliminate warping, and contained approximately 60 volume percent of high modulus carbon (Grafil EHM-S) or high strength carbon (Grafil EHT-S) fibres with two different levels of surface treatment. Straight sided tensile specimens were cut using a diamond tipped slitting wheel, and aluminium alloy end tabs were bonded to transfer the load into the specimen. Central notches with both sharp and rounded ends were cut in order to look at the effects on the notched strengths of the following parameters.

(a) Geometrical parameters
   (i) notch length [$2(a/W)0·2 \rightarrow 0·7$]
   (ii) specimen dimensions (plate width, laminate thickness)
   (iii) notch tip radius ('sharp', 0·5 mm, 1 mm)
(b) Material parameters
   (i) fibre–matrix bond strength
   (ii) fibre type
   (iii) ply orientations
   (iv) style of reinforcement (woven or plied)

Specimens were clamped in wedge action grips and aligned by universal joints and loaded uniaxially at a crosshead displacement of 0.5 mm min^{-1}. A piezo-electric transducer was attached to monitor acoustic emissions in order to indicate when irreversible processes were occurring and the surface deformations were examined visually under low magnification during loading. Some of the specimens were unloaded at pre-determined levels, selected on the basis of acoustic emission, to see what damage had occurred and in some cases were soaked in an x-ray opaque additive and radiographed.

4. RESULTS

Fracture mechanics measurements were made on the centre notched specimens where the Mode I K_{Ic} was calculated from the load at failure and crack length and adjusted for finite width by using an isotropic correction factor,[16]

$$K_{Ic} = Y\sigma_N\sqrt{a} \qquad (7)$$

For metals the candidate fracture toughness (K_Q) measured from a test has to satisfy certain requirements concerning the amount of inelastic deformation at the crack tip and specimen size before it is classified as the true fracture toughness, K_{Ic}. No such standardised procedures have been established for composites, consequently the candidate fracture toughness (K_Q) will be used. We recognise that LEFM may not be appropriate to these materials but the value of K_Q provides a useful comparison of the damage tolerance of these materials and is used as such in the following.

4.1. Effect of Notch Length
The results could be broadly classified into two main categories.

(a) Notch insensitive
This behaviour occurred in two classes of laminates, $[0_8]_s$ and $[\pm 45]_{2s}$, where the net-section stress measured at the notch section equalled the unnotched tensile strength as shown in Fig. 1. In this case linear elastic fracture mechanics is clearly not applicable. Failure of the $[\pm 45]_s$ laminates started by initiation and extension of sub-cracks from the notch tip parallel to fibres, spreading across the entire width of the plate to produce a large damage zone and eliminating the stress concentration. Final failure was by macroscopic shearing and was controlled and non-catastrophic. In the $[0_8]_s$

laminate, longitudinal shear cracks initiated and grew from the notch tip, thereby reducing the stress concentration. These subsequently grew the length of the specimen and eventually tensile failure of the specimens occurred catastrophically near the loading grips.

(b) Notch sensitive

The remaining laminates were $[0, \pm 45, 0]_s$, $[0, 90, 0, 90]_s$ and $[0, \pm 45, 90]_s$. These were sensitive to the presence of notches, their strengths falling below the notch insensitive line, Fig. 1. The notch sensitive behaviour could be divided into two distinct classes.

(i) *High notch sensitivity* ($K_Q < 20 \, MPa \sqrt{m}$). In our experiments this was exhibited by laminates containing high strength carbon fibres which had received high levels of fibre surface treatment resulting in strong fibre–matrix bond strengths and high interlaminar shear strengths ($> 100 \, MPa$).

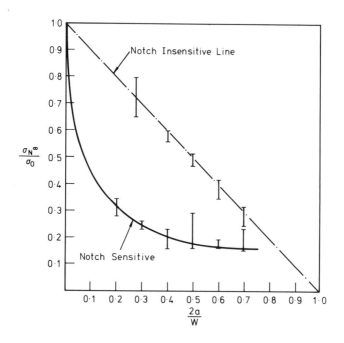

①   Notch insensitive – EHM–S/CODE 69 $[(\pm 45)_2]_s$

②   Notch sensitive – EHT–S/CODE 69 $[0, 90, 0, 90]_s$ (strong bond)

FIG. 1.   Residual strengths of two laminates with different notch sensitivities.

These laminates included the stacking sequences $[0, 90, 0, 90]_s$, $[0, \pm 45, 0]_s$ and $[\pm 45]_{2s}$. There was little acoustic emission prior to failure nor was there any visible splitting near the notch tip during loading. Because of the strong fibre–matrix bond the stress relaxation resulting from splitting and localised delamination was suppressed. Failure occurred in a catastrophic manner in which the main crack grew from the notch tip in a direction perpendicular to the applied stress and no delamination between plies took place. Measured values of K_Q were independent of notch length for a wide range of aspect ratios, as shown by case 1 in Fig. 2, indicating that LEFM may be appropriate in this case.

(*ii*) *Low notch sensitivity* ($K_Q > 20\ MPa\sqrt{m}$). The majority of laminates had K_Q values ranging from 30–50 MPa$\sqrt{m}$ and failure involved complex interactions between plies with the occurrence of large amounts of

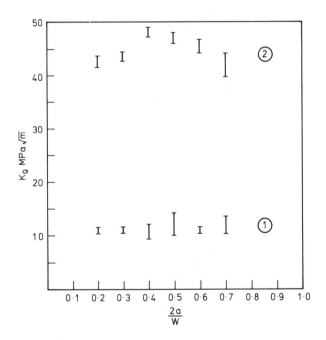

① EHT-S/CODE 69   [(0,90)₂]ₛ (Strong fibre/matrix bond)

② EHT-S/CODE 69   [(0,90)₂]ₛ (Reduced fibre/matrix bond)

Fɪɢ. 2. Comparison of the fracture toughness of two laminates with different notch sensitivities.

delamination. In this case measured values of the fracture toughness were a function of the notch length as shown by case 2 in Fig. 2. Small and large notches caused a reduction of the K_Q value. Other workers have explained this crack size dependence of K_Q in terms of a damage zone, as in metals where a crack length correction factor R_p has been invoked to account for the forward shift of the crack tip stress field as a result of plastic zone growth. This procedure has had some success with composites. It requires the use of eqn. (1), $K_Q = Y\sigma_N\sqrt{a + R_p}$, and can describe the increase of K_Q with crack size at small crack lengths. At large crack lengths it is argued that interaction between the crack tip stress field and specimen boundaries causes a decrease in K_Q. The use of the crack length correction factor is, however, doubtful on physical grounds because it has been suggested that the extension of sub-cracks does not produce a forward shift in the crack tip

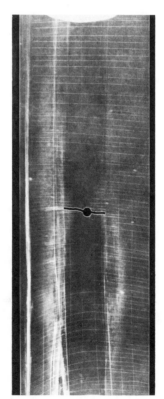

FIG. 3.   Fine-focus x-ray radiograph of notched cross-ply laminate at 80% of failure load (EHT-S/Code 69, $[0, 0, 90, 90]_s$, $W = 30$ mm).

stress field. Further the development of damage is not localised at the crack tip as is shown by the fine focus radiograph of a notched laminate (EHT-S/Code 69 $[0, 0, 90, 90]_s$, Fig. 3. There is no clearly distinct damage zone near the notch tip and damage is widespread throughout the gauge length. Immediately above and below the notch the material is unloaded and very little cracking occurred in these regions. Longitudinal splits can be seen at the notch tips extending the entire gauge length. The spacings of the multiple transverse cracks, which are a function of applied stress and ply thickness, are smaller near the notch as a result of the higher stresses. Longitudinal splits can be seen some distance from the notch and occur through thermal and Poisson generated strains.

4.2. Effect of Specimen Dimensions

Increasing laminate thickness by repeating the stacking sequence does not significantly alter toughness. For example the two EHM-S/Code 69 laminates $[0, 90, 0, 90]_s$ and $[0, 90, 0, 90]_{2s}$ which were 2 mm and 4 mm thick had K_Q values of 35·7 MPa$\sqrt{m}$ and 36·9 MPa$\sqrt{m}$ respectively when tested with a 30 mm wide specimen and a $2a/W$ of 0·3. K_Q does however tend to increase with specimen width as shown in Fig. 4.

4.3. Effect of Fibre Type

Fibres which can store large amounts of elastic strain energy would be expected to be good candidates for a tough laminate, and in support of this high strength carbon laminates have been found to be tougher than high modulus laminates. For example two $[0, 90]_{2s}$ laminates made from EHM-S/Code 69 and EHT-S/Code 69 gave K_Q values of 35·7 MPa$\sqrt{m}$ and 43·5 MPa$\sqrt{m}$ respectively (width 30 mm, thickness 2 mm, $2a/W$ 0·3).

4.4. Effect of Notch Tip Radius

Table 1 shows the effects of changing the sharpness of a crack. The toughness of notch insensitive laminates is unaffected by notch tip radius. Of the notch sensitive laminates the tough laminates are also unaffected by original notch tip radius because the growth of damage prior to fracture obscures the original tip shape. For the more brittle, notch sensitive materials there is a pronounced effect of crack tip radius, toughness or effective strength decreasing with increasing sharpness.

4.5. Ply Stacking Sequence

Ply stacking sequence has a considerable effect on damage tolerance. Table 2 shows data obtained from 0°/90° and 0°/±45° laminates. The

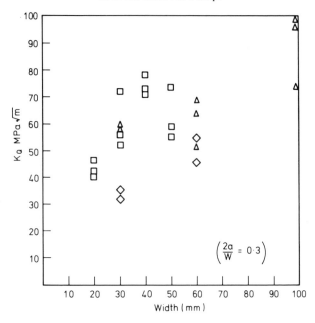

FIG. 4. The effects of specimen width on the fracture toughness.

TABLE 1

Effects of crack tip radius on effective fracture toughness K_Q for a range of notch sensitivities

Notch sensitivity	Lay-up	K_Q at crack tip radius ($MPa\sqrt{m}$)		
		1 mm	0·5 mm	'sharp'
Insensitive	EHT-S $[\pm 45]_{2s}$	13·2	12·2	13·1
Sensitive				
Low	EHM-S $[0, \pm 45, 0]_s$	42·3	38·4	41·1
Medium	EHT-S $[0, \pm 45, 0]_s$ (strong bond)	22·6	17·5	17·4
High	EHT-S $[0, 90, 0, 90]_s$ (strong bond)	20·0	17·0	12·3

TABLE 2
Effect of ply stacking sequence on toughness

Stacking sequence (EHT-S/Code 69)	K_Q (MPa$\sqrt{}$m) $2a/W = 0.3$, $W = 30$ mm, $t = 2$ mm
A. $[0, 0, 90, 90]_s$	56.5
B. $[90, 90, 0, 0]_s$	38.4
C. $[0, 90, 0, 90]_s$	43.0
D. $[0, \pm 45, 0]_s$	33.0
E. $[\pm 45, 0, 0]_s$	56.5

differences in toughness were reflected in fracture morphology. For example the $0°/90°$ laminate was very much tougher with $0°$ fibres on the outer plies and for this lay-up a great deal of multiple splitting and delamination occurred, while with $90°$ plies on the outside the material was much less tough and splitting and delamination were largely suppressed. It is worth noting that A and B were identical materials rotated through $90°$ and this demonstrates the orientation dependence of toughness.

4.6. The Effects of Style of Reinforcement (Woven and Plied)

The fracture behaviour of laminates made from woven fibre reinforcements was significantly different to that of conventional plied material. For example two EHT-S/Code 69 laminates, one plied $[0, 90, 0, 90]_s$ and one a $[0, 90]_{10s}$ plain weave, had K_Q values respectively of 43.0 MPa$\sqrt{}$m and 31.5 MPa$\sqrt{}$m (width 30 mm, thickness 2 mm, $2a/W\,0.3$). The weave constrained the development of shear cracking and splitting and resulted in a 'brittle' failure with little acoustic activity prior to failure and no delamination between plies. The fracture plane was perpendicular to the applied stress and contained short stubby bundles of fibres which protruded ~ 1 mm.

5. DISCUSSION

5.1. The Applicability of Predictive Models

Although the fracture behaviour of notched laminates involves a complex sequence of events and has been shown to be controlled by various material and geometrical variables, there is a requirement for simple parameters which can be obtained from mechanical tests which give an indication of the materials behaviour in an engineering application.

Despite their different derivations, the 'inherent flaw' model and the 'average stress' criterion are identical in practice and the predictions of the 'point stress' criterion are in close agreement provided that the ratio d_0/a is small. Under this condition the characteristic lengths are related by $a_0 = 2\alpha \sim 4d_0$. Figure 5 compares the various residual strength criteria, together with appropriate averaged model parameters, for data obtained from two

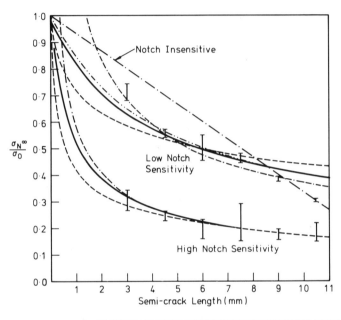

LAMINATE	A.S a_0(mm)	P.S d_0(mm)	I.F α(mm)	S.C ρ_0(mm)	LEFM K_0(MPa$\sqrt{m}$)
High Notch Sensitivity EHT-S/CODE 69[(0.90)$_2$]$_s$	0.7	0.17	0.35	8	12.9
Low Notch Sensitivity EHM-S/CODE 69[0±450]$_s$	4.1	0.92	2.05	40	40.7

————	A.S. =	Average Stress
—·——·——·—	P.S. =	Point Stress
————	I.F =	Inherent Flaw
— — — — —	S.C =	Stress Concentration
—·——·——··	LEFM =	Linear Elastic Fracture Mechanics

FIG. 5. Comparison of strength reduction models for two laminates with different notch sensitivities.

laminates with different notch sensitivities. For the highly notch sensitive laminate all of the predictions are in close agreement and fit the data equally well. For the laminate with lower notch sensitivity the situation is less clear. In all of the laminates showing this class of behaviour the characteristic length associated with each model varied with the length of the crack and the geometry of the specimen, which suggests that these parameters are not fundamental material properties. In other studies extra parameters have been introduced;[13,14] however, they do not have a clear physical significance and have not been applied to the present results. With enough parameters any models can be made to fit experimental data and the procedure then becomes a curve fitting exercise, which is dangerous for extrapolation outside the range of experimental data points. It is probably fortuitous that the LEFM predictions appear to give the best fit for the residual strength of the less notch sensitive laminate considered, since failure did not occur by growth of a single macroscopic crack.

Hence for highly notch sensitive laminates the simplest theory, LEFM, appears to fit the data well and is as useful as any of the others. For less notch sensitive materials no theory appears to have a demonstrable advantage nor general applicability. There is still a need for a predictive theory which can be applied to a range of geometries and test conditions, and a theory does not yet appear to exist which will accurately fit the data obtained even for the simple conditions of these experiments.

5.2. The Effects of Ply Constraints

One of the most important laminate characteristics that influences the failure of a notched laminate is the constraint effects imposed by each ply on adjacent off-axis plies, which give rise to both in-plane and through-thickness effects. There are two ways in which in-plane effects can arise. Because the individual plies have different thermal expansion coefficients as a result of their different orientations, constraints due to thermal stresses occur when cooling down from the curing temperature. Thermal stresses which are generated can be calculated with reasonable accuracy using laminated plate theory and Table 3 shows values calculated for some laminates tested in this work. Large thermal stresses are developed transverse to the fibres and can approach the failure strength in that direction. The second effect is a result of the different mechanical responses of plies with different orientations. When an axial stress is applied to a laminate the plies would in general undergo differing Poisson's contractions if they were not constrained by adjacent plies. However, because they are bonded to their neighbours the net result is that they all undergo an

TABLE 3

Thermal constraint stresses in two typical laminates (EHT-S/Code 69, $\Delta T = -150\,^{\circ}C$)

Laminate	σ_1 MPa	σ_2 MPa	τ_{12} MPa
0° ply in $[0,90]_{2s}$	$-39\cdot7$	$39\cdot7$	0
90° ply in $[0,90]_{2s}$	$-39\cdot7$	$39\cdot7$	0
0° ply in $[0,\pm45,0]_s$	$12\cdot4$	$34\cdot8$	0
$+45^{\circ}$ ply in $[0,\pm45,0]_s$	-62	38	$-5\cdot6$

equal transverse strain with the generation of in-plane stresses. These constraint stresses can be calculated readily for an unnotched laminate but are less easy to estimate in the vicinity of a crack or notch. Typical values calculated for unnotched laminates are shown in Table 4. Here σ_1 is the stress parallel to fibres in a ply while σ_2 is the transverse stress. It can be seen that transverse stresses can be positive or negative.

Thermal and Poisson generated stresses affect the development of microcracks in the highly stressed vicinity of the notch tip and account for the differences in toughness observed for different stacking sequences shown in Table 2. It is observed in notched $[0,90]_{2s}$ laminates that longitudinal shear cracks develop extensively at the notch tip, which reduces the stress concentration; whereas in the $[0,\pm45,0]_s$ laminate, which has a higher resistance to shear, the 0° plies are under transverse compression which results in the suppression of splitting and reduced toughness.

The through-thickness constraint effects generate interlaminar and transverse shear and normal stresses ($\tau_{xz}, \tau_{yz}, \sigma_z$) which are localised near free edges and cracks and result from different responses of each lamina in the laminate. The sign of the interlaminar normal stress (σ_z) can be

TABLE 4

Mechanical constraints effects in two laminates (EHT-S/Code 69, applied _strain = 0·3 %)_

Laminate	σ_1 MPa	σ_2 MPa	τ_{12} MPa
0° in $[0,90]_{2s}$	444	$+4$	0
90° in $[0,90]_{2s}$	24	-4	0
0° in $[0,\pm45,0]_s$	440	$-11\cdot9$	0
$+45^{\circ}$ in $[0,\pm45,0]_s$	74	$4\cdot8$	-27

estimated by comparing the major Poisson's ratio of the outer ply with that of the laminate. If v_{12} (outer ply) $> v_{12}$ (laminate) then σ_z is positive and may result in delamination. The interlaminar and transverse shear stresses (τ_{xz}, τ_{yz}) are concentrated near free edges within a distance approximately equal to the laminate thickness and are also likely to be responsible for delamination. Consider the behaviour of two stacking sequences $[0, \pm 45, 0]_s$ and $[\pm 45, 0, 0]_s$. The 45° plies in both laminates are under different degrees of constraint; in the first case they are surrounded by an adjacent 0° ply and a −45° ply whereas in the latter case the surface 45° ply is only constrained on one side. Shear cracks develop in the 45° plies when a notched laminate is loaded and for a given load are longer in the 45° plies which are unconstrained on one side. In the first lay-up the plies do not delaminate and this results in a line discontinuity of high stress near the interface of the adjacent 0° load bearing fibres. The fracture surfaces in Fig. 6 show that the 0° plies failed along the 45° line and close inspection shows that the fibres failed in a step-wise fashion. The behaviour of the other laminate was different (Table 2) in which multiple splits developed in the outer 45° ply and final failure involved delamination between cracked 45° plies and the load bearing 0° plies which then failed at a higher applied stress.

It can therefore be seen that selective delamination between cracked off-axis plies and load bearing 0° fibres reduces the notch sensitivity and might be a way to control the notch strength although it is recognised that the requirements for other properties, such as fatigue response and environmental integrity, may exclude this approach.

5.3. Guidelines for Tough Laminates

It has been shown that laminates exhibit a wide range of notch sensitivities ranging from completely notch insensitive to very notch sensitive as a result of changing various material parameters. Considerable scope exists for designing laminates which are tough. The following general conclusions can be applied:

(1)   The most important parameter is the fibre/matrix bond strength. In order to obtain tough laminates the interfacial bond strength has to be optimised through correct fibre surface treatment. If the bond is too strong the material always fails in a brittle manner irrespective of lay-up.

(2)   Materials which can store large amounts of strain energy would be expected to be tough ($\sigma_f^2 / 2E$ per unit volume). Consequently fibres

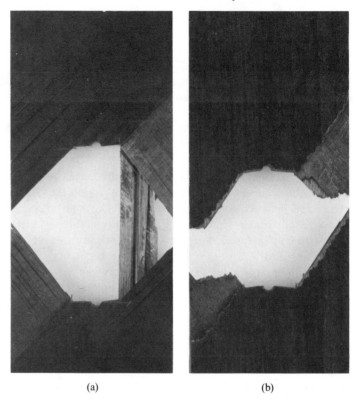

(a)   (b)

FIG. 6. Fracture surfaces of two laminates with different stacking sequences. (a) EHT-S/
Code 69, $[\pm 45, 0, 0]_s$, $\sigma_N = 451\,\text{MPa}$, $K_Q = 56\cdot5\,\text{MPa}\sqrt{m}$. (b) EHT-S/Code 69, $[0, \pm 45, 0]_s$,
$\sigma_N = 366\,\text{MPa}$, $K_Q = 33\,\text{MPa}\sqrt{m}$.

with high strength, intermediate modulus and high failure strain are
expected to be good candidates for a tough laminate. Hybridisation
provides much scope for developing composites with increased
toughness and the following ranking order would be expected for
the toughness of laminates of a specified lay-up from this effect:

E-glass > Kevlar 49 > EHT-S > EHM-S.

(3) Although the fracture toughness of a laminate does not appear to
vary with specimen thickness (and there is no apparent plane stress
to plane strain transition), it has been shown to depend sensitively
on the thickness of individual plies.[4,6] In general, thick plies
delaminate more easily and therefore decrease the interactions of

adjacent cracked and uncracked plies, which increases the toughness. However, it is recognised that a compromise may have to be obtained for other mechanical properties such as fatigue and environmental response as it is generally well recognised that thicker plies increase the susceptibility for transverse cracking.

(4) While the effects of stacking sequence on the notched strength are not clear it has been shown that arranging plies to encourage delaminations between shear cracks in 45° plies and adjacent load bearing 0° fibres (e.g. $[\pm 45, 0, 0]_s$) results in a higher notched strength. If, however, the delamination is suppressed (e.g. $[0, \pm 45, 0]_s$) then the shear cracks which develop in the 45° plies at relatively low stresses have a detrimental effect on the 0° fibres, which results in a relatively brittle behaviour with the 0° fibres failing in a step-wise fashion along the 45° line.

(5) The style of reinforcement has an important effect on the notched strength in which the formations of a damage zone is largely suppressed and results in a relatively brittle behaviour.

(6) Changes in ply orientation have been shown to affect the notched strength. $[\pm 45]_s$ lay-ups, although they are relatively weak, are notch insensitive with large damage zones which are developed across the entire width of the test specimen which results in a controlled and non-catastrophic failure. Because of their relative weakness in the longitudinal directional they are unlikely to be used to resist tensile loads alone but could be used as crack arrestment strips in a laminate containing load bearing fibres.

It has been shown that there are many parameters which control the fracture toughness of laminated composites, some of which are more important than others. Before a general model capable of predicting notched laminate behaviour can be developed a better understanding of the events occurring during loading, and the resulting fracture mechanisms, is required.

REFERENCES

1. MANDELL, J. F., WANG, S. S. and McGARRY, F. J., The extension of crack tip damage zones in fibre reinforced plastic laminates, *J. Comp. Mat.*, **9** (1975) 266–87.
2. BISHOP, S. M., *Deformation near notches in angleplied carbon fibre composites*, RAE TR 77093, 1977.

3. McGARRY, F. J., MANDELL, J. F. and WANG, S. S., Fracture of fiber reinforced composites, *Polym. Eng. Sci.*, **16** (1976) 609–14.

4. BISHOP, S. M. and McLAUGHLIN, K. S., *Thickness effects and fracture mechanisms in notched carbon fibre composites*, RAE TR 79051, 1979.

5. KANNINEN, M. F., RYBICKI, E. F. and GRIFFITH, W. I., Preliminary development of a fundamental analysis model for crack growth in a fiber reinforced composite material, *Composite materials: Testing & design (fourth conference)*, ASTM STP 617, 1977, pp. 53–69.

6. DOREY. G., *Damage tolerance in advanced composite materials*, RAE TR 77172, 1977.

7. OWEN, M. J. and BISHOP, P. T., Critical stress intensity factor applied to glass reinforced polyester resin, *J. Comp. Mat.*, **7** (1973) 146–59.

8. WADDOUPS, M. E., EISENMANN, J. R. and KAMINSKI, B. E., Macroscopic fracture mechanics of advanced composite materials, *J. Comp. Mat.*, **5** (1971) 446–54.

9. NUISMER, R. J. and WHITNEY, J. M., Uniaxial failure of composite laminates containing stress concentrations, *Fracture mechanics of composites*, ASTM STP 593, 1975, 117–42.

10. KIM, R. Y., Fracture of composite laminates by 3-point bend, *Exp. Mech.*, Feb. (1979) 50.

11. NUISMER, R. J. and LABOR, J. D., Applications of the average stress failure criterion: Part 1—Tension, *J. Comp. Mat.*, **12** (1978) 238.

12. NUISMER, R. J. and LABOR, J. D., Applications of the average stress failure criterion: Part 2—Compression, *J. Comp. Mat.*, **13** (1979) 49.

13. KARLAK, R. F., Hole effects in a related series of symmetrical laminates, *Proc. Fail. Modes in Composites III*, Chicago, ASM, 1977, 105–17.

14. PIPES, R. B., WETHERHOLD, R. C. and GILLESPIE, J. W., Macroscopic fracture of fibrous composites, *Mat. Sci. & Eng.*, **45** (1980) 247–53.

15. MANDELL, J. F., McGARRY, F. J., KASHIHARA, R. and BISHOP, W. R., Engineering aspects of fracture toughness: Fiber reinforced laminates, *29th Ann. Tech. Conf. SPI*, 1974, Paper 17-D.

16. BROWN, W. F. and SRAWLEY, J. E., *Plane strain crack toughness testing of high strength metallic materials*, ASTM STP 410, 1966, p. 11.

37

Tensile Fatigue Assessment of Candidate Resins for Use in Fibre Reinforced Composite Repair Schemes

D. P. BASHFORD AND A. K. GREEN

Fulmer Research Laboratories Ltd,
Stoke Poges, Slough SL2 4QD, England

AND

K. F. ROGERS AND D. M. KINGSTON-LEE

Materials Department, Royal Aircraft Establishment,
Farnborough GU14 6TD, England

ABSTRACT

Five matrix resin systems have been evaluated as potential candidates for use in a rapid repair system for aircraft skin damage, incorporating glass/carbon fibre hybrid reinforced plastics as the repair material. Two epoxy and three polyester resin systems were evaluated at 20°C. Additionally, the three polyester systems were evaluated at 0°C. The effect of contamination of metal surfaces by aviation fuel and hydraulic fluid was investigated. The properties measured were metal/composite joint overlap shear strength, as manufactured and following a fixed schedule of tensile fatigue load conditioning, and bending stiffness. These properties were determined 4 h and 24 h after fabricating the simulated repair.

The resin system Quickcure QC3/Lucidol CH50/dimethyl-p-toluidine was found to give repairs equal to riveted metal plate repairs under all circumstances except to hydraulic fluid contaminated substrates at 0°C.

INTRODUCTION

During its service life an aircraft may undergo damage in situations where workshop facilities are not available, but where the local conditions demand immediate high-speed emergency repair. Such repairs must be capable of implementation in the field, with the minimum of equipment; they might not necessarily restore the aircraft to full airworthiness but would allow the completion of a limited number of further flights, possibly at a restricted level of performance.

Various methods of affecting rapid in-field repairs by means of fibre-reinforced plastics patching have been developed at the Royal Aircraft Establishment.[1,2] The simplest of these uses a wet-laminating technique, which allows the patch to conform easily to the contour of the damaged structure, and enables its stiffness to be controlled on-site by the choice of reinforcement.

To maintain the original distribution of load in the structure, the patch should match the substrate in absolute stiffness and strength. If at the same time the specific stiffness of the patch can also be matched to that of the substrate the weight penalty will be minimal. The portion of an aircraft most likely to be damaged is the skin, which is usually fabricated from an aluminium alloy; an absolute stiffness match can be achieved in the patch by choosing a glass fibre/carbon fibre hybrid reinforcement, preferably in the form of a bidirectional balanced fabric to minimise the effect of errors in the direction of lay-up during repair. Hybrid fabrics of this type have been developed under RAE sponsorship and are commercially available.

The RAE development work showed that wet-laminated repairs could achieve the static tensile strength of comparable conventional repairs based on riveted metal plates; the research programme described in this paper was initiated to assess the fatigue performance of the repairs.

TEST PROGRAMME

General Considerations

The critical feature determining the performance of a wet-laminated repair is the condition of the interface between the substrate and the resin matrix. Since it must be assumed that in field conditions solvents and primers may not be available, the preparation of the substrate must be limited to simple abrasion, and because there is no guarantee that the surface will remain completely clean prior to laminating, the resin should be tolerant to contaminants such as aviation fuel, hydraulic fluid and water. The resin must also be sufficiently fluid to wet-out the fabric and the substrate easily, and must have an adequate (not less than 15 min) pot life. However, since the repaired aircraft must be flyable as soon as possible a short cure time is essential. Because of the emergency situation cure must proceed without any application of external heat.

To assess the fatigue properties, simulation repairs as previously reported[2] were done at room temperature (20 °C), using two epoxy and three polyester resin systems. The three polyester systems, with increased hardener and accelerator content, were also used to make simulation-repairs at 0 °C, the epoxies being unsuitable because of their greatly increased viscosity. Following conditioning for either 4 h or 24 h at the repair temperature all repairs were static tensile tested and fatigue tested at room temperature.

The programme was repeated for substrates contaminated with Avtur 50 aviation fuel and with hydraulic fluid to DTD 585B.

Materials
Substrate panels
The substrate panels in all tests were of aluminium alloy to specification 3L73, of dimensions $175 \times 152 \times 1.23$ mm (18SWG).

Hybrid fabric
The reinforcing fabric was a bidirectional, balanced twill weave 3:1, glass:carbon hybrid cloth woven from 600-tex E-glass rovings and 6000-filament EXAS carbon fibre tows, and weighing 509 g/m^2 (Fig. 1). The twill weave was preferred to a plain weave because its better drape properties allow it to conform more easily to areas of multiple curvature.

Resin systems
The formulations used for the repairs at 20 °C were:

(i)   Epikote 828 epoxy resin, mixed in 2:1 ratio by weight with hardener Ancamine AC*;

(ii) Epikote 808 epoxy resin, mixed in 2:1 ratio by weight with Ancamine AC;

(iii) Quickcure QC3 polyester resin, cured with 1.5 % by weight of Butanox M50† methyl-ethyl ketone peroxide hardener; the resin as-supplied is pre-accelerated;

(iv) Quickcure QC3 resin cured with 4 % by weight of Lucidol CH50† benzoyl peroxide powder hardener and 1 % of a 10 % solution of dimethyl-*p*-toluidine (DMpT) in styrene;

(v)   polyester resin A cured as in (iv) above.

* Anchor Chemical Co.
† Akzo Chemie Ltd.

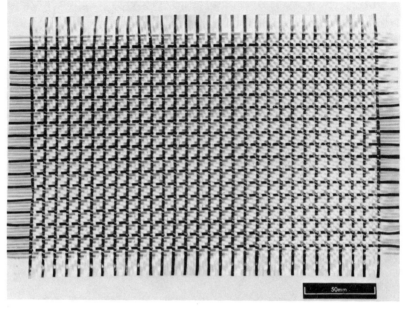

FIG. 1. Twill weave hybrid glass/carbon fibre cloth.

The formulations used at 0 °C were:

(vi) Quickmore QC3 cured with 4 % of Lucidol CH50 and 2·5 % of a 20 % solution of DMpT in styrene.

Ancillary tests were done with:

(vii) Quickcure QC3 cured with 4 % of Butanox M50;

(viii) Quickcure QC3 cured with 4 % of Butanox M50 and 4 % of added cobalt accelerator X,

(ix) polyester resin A cured with 4 % of Lucidol CH50 and 2·5 % of a 20 % solution of DMpT in styrene.

SIMULATION-REPAIR

For each repair two aluminium alloy plates were thoroughly cleaned by successive wipes with acetone and Inhibisol. An area 114 mm wide along one 175 mm edge of each plate was abraded thoroughly by a flap wheel in a hand-held power tool, and the abraded surface brushed lightly with a clean, dry brush to remove loose particles. For plates required to be contaminated before repair the contaminant was spread at this stage onto the abraded areas of the plates, and the excess removed with a paper wiper. The contaminated areas were then thoroughly worked with a 20 g mix of the resin to be used for the repair, using a second wiper and a circular motion. As much resin as was reasonably possible was then removed with a third wiper.

A simulation-repair jig was positioned within the cabinet of a small commercial freezer, controllable at 0 °C when required. The plates were clamped into the jig side-by-side, with their abraded areas inwards and separated by a 76 × 1·23 mm thick alloy strip to establish a parallel gap between the plates to simulate a damage hole. Prior to locating the plates both the strip and the base-plate of the jig were covered with 0·025 mm thick FEP release film. A fresh mix of resin preconditioned either to 20 or 0 °C was brushed onto the abraded areas of the plates and across the central strip, and a 280 × 175 mm patch of hybrid fabric was wet-laminated across the wetted area. Three further patches each successively 25 mm shorter were laminated centrally across the first. The ends of all the patches were feathered out for 12 mm at each end prior to laminating, by the removal of warp tows; this improved the lay of the patch ends and avoided abrupt changes of action. The completed simulation-repair is shown schematically in Fig. 2.

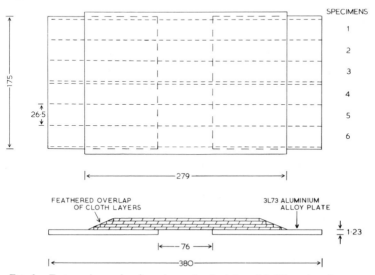

FIG. 2. Test repair panel and specimen plan (not to scale). Dimensions in mm.

SPECIMEN PREPARATION

The repair panel was removed from the jig $3\frac{1}{2}$ h after resin mixing; from the panels laminated at 20 °C six specimens 26·5 mm wide were cut using a diamond wheel, this width containing exactly nine glass rovings and three carbon tows per ply for the particular fabric used. (An example is shown in Fig. 3.) From the panels laminated at 0 °C, three specimens were cut and the uncut half of the panel returned to the freezer; the remaining three specimens were cut $23\frac{1}{2}$ h after mixing the resin. All specimens were deburred with a fine file and were equilibrated to room temperature prior to testing. Fatigue specimens were drilled with two 6·3 mm diameter holes in each of the aluminium end-pieces, along the centreline and at centres 8 and 27 mm from each end.

FIG. 3. Specimen prior to fatigue loading. Scale bar represents 50 mm.

TEST PROCEDURE

Flexural and Tensile Testing

To assess the degree of resin cure the flexural stiffness of one specimen from each repair variable was measured in four-point bending, stopping each test when the load/deflection record became non-linear, to avoid permanent deformation. Two specimens from each repair variable were tested in static tension, one following flexural test and the other as-manufactured.

Fatigue Testing

A purpose-built fatigue machine was constructed, incorporating a 50 kN servo-hydraulic actuator and load cell. Pin-and-button grips using clevis fixtures were fabricated, using 38 mm diameter steel discs as the buttons and two 6·3 mm diameter high tensile steel bolts 19 mm apart along a disc diameter and equidistant from the disc centre as the pins. This design was chosen to avoid fretting in the grips. The fatigue stress was applied in fluctuating tension only. The cycle amplitude was held below 1/7 of the ultimate tensile strength (440 MPa) of 3L73 alloy, corresponding to a load amplitude of 2·05 kN about a mean tensile load of 2·5 kN, to further avoid fretting. This load amplitude applied at 10 Hz was found adequate to discriminate between the repair variables and induced no detectable heating effects in the specimens. The specimens were fatigued for a maximum of 10^5 cycles ($2\frac{3}{4}$ h) and were then tensile tested if failure had not occurred previously.

CONTROL TESTS

1·23 mm 3L73 aluminium alloy strips of the same dimensions as the composite repair specimens (26·5 × 380 mm) were tested in four-point bending, demonstrating tangent stiffness in the range 0·4–0·5 N/mm and plastic deformation at loads above 19N.

The tensile failure load of a 26·5 mm wide strip of 1·23 mm 3L73 alloy was known from previous work to be 14·3 kN.

'Conventional' simulation-repairs were done by bridging the 'damage area' between two 3L73 plates by a 1·23 mm 3L73 sheet and securing it to each plate with a double row of 4·75 mm diameter 'pop' rivets at 30 mm spacing, i.e. an offset spacing of 15 mm between each rivet. The mean tensile failure load of these repairs was equivalent to 7·7 kN for a width of 26·5 mm.

Four 'conventional repairs' to this pattern were tested in fatigue to the same load pattern used for the composite repairs. Two were tested in tension after 10^5 cycles, both failing at 8·0 kN (per 26·5 mm of width). The remaining two failed in fatigue after 3·11 × 10^5 and 3·98 × 10^5 cycles.

RESULTS

The results are presented in histogram form in Figs 4–6. In Fig. 5 a fatigue specimen failing before completing 10^5 cycles is denoted by the symbol F. Where no histogram bar is present all specimens failed in fatigue. When the symbol F and a histogram bar are both present, the bar is the mean tensile result of those specimens that survived 10^5 fatigue cycles. Two F symbols and a bar indicate two fatigue failures and the tensile value for the one specimen that survived 10^5 cycles.

It should be noted that the fatigue results for specimens that survived 10^5 cycles are for effective cure times of 7 and 27 h, rather than the 4 and 24 h quoted, since cure will be advancing during the test.

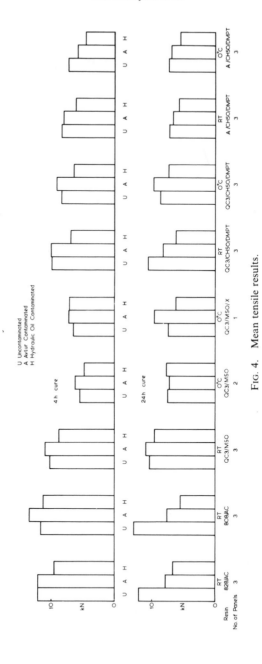

U Uncontaminated
A Avtur Contaminated
H Hydraulic Oil Contaminated

FIG. 4. Mean tensile results.

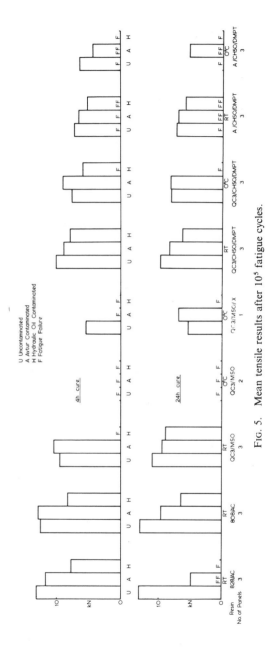

FIG. 5.   Mean tensile results after 10^5 fatigue cycles.

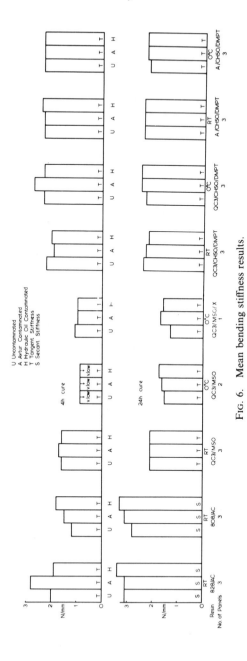

FIG. 6.   Mean bending stiffness results.

DISCUSSION OF RESULTS FROM INDIVIDUAL RESIN SYSTEMS

Epikote 828/Ancamine AC Resin

This resin mix was fairly viscous, and required considerable working to achieve satisfactory wetting-out of the cloth. An early panel in this series was laminated at 16 °C, and this increased significantly the difficulty in achieving satisfactory lamination. The speed of cure, as monitored by the bending stiffness measurement at 4 h, was strongly dependent on laminating temperature.

Most of the 4-h tensile specimens failed cohesively in the resin layer adjacent to the alloy plate, but adhesive failure predominated in two of the hydraulic fluid contaminated specimens. Two specimens, one Avtur contaminated, the other uncontaminated, failed in tension in the laminate close to the end of the overlap. Creep was observed in some of the 4-h fatigue specimens during cycling, the degree being apparently dependent on cure temperature. The specimens exhibited permanent curvature on removal from the fatigue machine, and it was found that the gap between the aluminium alloy end plates had increased by up to 3 mm, compared to about 0·5 mm for most other specimens. This creep phenomenon did not occur during fatigue cycling of 24-h specimens. All fatigue cycled specimens, both 4 and 24 h, failed adhesively at the alloy surface during subsequent tensile testing.

Specimens tensile tested after 24 h also failed adhesively at the alloy surface. Although contamination has a significantly adverse effect on the composite/alloy bond strength, both fatigued and unfatigued, hydraulic fluid having a greater effect than Avtur, only minor differences in failure mode were observed with no consistent pattern apparent.

Epikote 808/Ancamine AC Resin

The viscosity of this resin was similar to that of the Epikote 828 system. One panel was laminated with difficulty at 14 °C, and the decrease in the speed of cure at this temperature was apparent from the low bending stiffness in comparison with panels prepared at 20 °C.

Failure within the composite in the central area of the specimen rather than at the bond line occurred more frequently with this resin system than with Epikote 828, in both tensile and fatigue/tensile tested specimens but never in specimens contaminated with hydraulic fluid. All other failures at 4-h cure were cohesive in the resin layer adjacent to an alloy end plate. A similar transition to adhesive failure at the alloy surface after 24-h cure, as observed for Epikote 828, was noted for all specimen types.

Quickcure QC3/Butanox M50 Resin, Laminated at Room Temperature

This resin mix is of much lower viscosity than the epoxy resin systems and could be laminated rapidly with ease.

For all specimens, failure in the tensile test was both adhesive and cohesive. The adhesive failure mode predominated in all cases, accounting for 50–90 % of the failed area. No consistent failure mode pattern could be observed when comparing contaminated and uncontaminated specimens, either 4- or 24-h cured.

During fatigue cycling progressive debonding from either end of the bond overlap was occasionally observed, by adhesive failure at the alloy surface. This progressive debonding led to failure prior to 10^5 cycles in some cases. Specimen failures during fatigue were always adhesive at the alloy surface, whereas those produced by tensile loading, in both fatigued and unfatigued specimens, displayed combinations of adhesive and cohesive failure. No creep was observed during fatigue cycling.

One 24-h cure hydraulic fluid contaminated specimen was fatigued for 120 h (4.32×10^6 cycles) without failing. Slight debonding of the laminate from the alloy plate occurred during fatigue cycling. The specimen failed at 9.5 kN in a subsequent tensile test.

Quickcure QC3/Butanox M50 Resin Laminated at 0 °C

The results for this resin at 0 °C were significantly lower than those at room temperature. No specimen survived 10^5 fatigue cycles; all failures were wholly adhesive. After 4-h cure, stiffness values were often too low for practical consideration.

In an attempt to improve this poor performance, a third series of panels was prepared using additional cobalt accelerator X. The histograms show a marginal improvement in performance, but this remains inferior to the results obtained from the room temperature laminated system.

Quickcure QC3/Lucidol CH50/DMpT Resin Laminated at Room Temperature

The failure modes following tensile and fatigue testing were usually a mixture of adhesive failure at the alloy surface and cohesive failure in the resin. The only consistent effect noted was that hydraulic fluid contaminated specimens generally failed in a wholly adhesive manner. The results obtained were broadly similar to those for room temperature laminated QC3/M50 resin, except for a greater sensitivity to contamination by hydraulic fluid, and a somewhat better performance after 4-h cure in fatigue, particularly for hydraulic fluid contaminated specimens. Indeed, the hydraulic fluid contaminated specimens displayed a curious behaviour, in that specimens tested subsequent to fatigue cycling failed at higher loads than those tested as-manufactured, for both 4- and 24-h cure specimens.

Quickcure QC3/Lucidol CH50/DMpT Resin Laminated at 0 °C

The overall performance of this resin system was superior to that of 0 °C laminated QC3/M50 systems. The system was sensitive to the effect of hydraulic fluid contamination, particularly in fatigue, which was emphasised by completely adhesive failure at the alloy surface. All other specimens displayed mixed adhesive/cohesive failure.

Polyester Resin A/Lucidol CH50/DMpT Laminated at Room Temperature

The predominant failure mode for all specimens was adhesive failure with only a few uncontaminated specimens displaying cohesive failure in the resin across 5–10 % of the bond area. No consistent difference in failure mode was apparent between 4- and 24-h cure specimens.

As can be seen from the histograms the performance of the resin system was inferior to that of QC3 using the same hardener/accelerator combination, although the deleterious effect of hydraulic fluid contamination on bond strength in both as-manufactured and fatigued specimens is marginally less severe than observed in the comparable QC3 resin system.

Polyester Resin A/Lucidol CH50/DMpT Laminated at 0 °C

Failure with this system was almost totally adhesive, with only a few uncontaminated and Avtur contaminated specimens displaying a very small amount ($<3\%$ of the bond area) of cohesive failure.

The overall performance of the system was again inferior in all respects to the comparable 0 °C laminated QC3 system, the fatigue performance being particularly inferior. An interesting feature of the results is that the observed increase in strength between 4- and 24-h as-manufactured specimens, both contaminated and uncontaminated, was not translated into superior fatigue performance. The only other resin for which similar, but not so clear-cut observations can be made is the 0 °C laminated QC3/CH50/DMpT system.

RELATIVE ASSESSMENT OF RESIN SYSTEMS

Reference to Figs 4, 5 and 6 shows that none of the resin systems gave completely satisfactory results. Figure 4 shows that only the room temperature laminated QC3/M50 system gave repairs stronger than the riveted controls for all six combinations of contamination and cure time. With most of the other systems, the hydraulic fluid contaminated specimens were the least satisfactory.

It is noteworthy that both contaminants have a more severe relative degradative effect between the 4- and 24-h cure results for the epoxy resin systems than for the polyesters. This may be associated with differences in the degree of cure, since Fig. 6 suggests that substantial curing occurs after 4 h in the epoxies, whereas cure is substantially complete at 4 h for most of the polyesters. This is consistent with the observed transition in the epoxies from cohesive failure after 4 h to adhesive failure after 24 h and the predominance of adhesive failure for all polyester-based repairs, since an undercured resin is predisposed to fail cohesively. Figure 6 indicates that the presence of the contaminant does not significantly affect either the speed or degree of cure. Hence, it is suspected that performance degradation induced in the epoxies by the contaminant is purely a reflection of the increasing difficulty of the epoxies in adhering to contaminated surfaces as their stiffness increases. This is analogous to the peel behaviour of many adhesive systems, where high peel strength is often related to compliance in the bond line.

An interesting effect is the generally more severe degradation of properties caused by hydraulic fluid contamination compared to Avtur contamination. Small traces of adherend surface contaminants are known to have an adverse effect on both the initial strength and the environment durability of bonded joints, and even the effect of humidity in a bonding process area[3] can be significant due to the adherend surface absorbing water molecules. These effects are usually ascribed to reduction in surface energy of the clean surface by the contaminant, which is conventionally assessed by measuring contact angles of liquid adhesive on the surface under study, this being effectively a measure of the wettability of the surface by the adhesive. There is no reason why Avtur and hydraulic fluid should have equal influences on the wettability of abraded aluminium alloy surfaces. Indeed, contact angle measurements[3] for various organic liquids on abraded steel surfaces suggest precisely the opposite. It is, therefore, not surprising that one contaminant should have a consistently more severe degradative effect than the other.

The more severe effect of hydraulic fluid compared to Avtur is particularly evident in the fatigue results (Fig. 5). The equivalent riveted repairs (see section headed 'Control Tests') sustained 10^5 cycles under fatigue loading, subsequently failing in tension at 8 kN and for composite repairs to be considered as feasible alternatives to riveted repairs they must be capable of this order of performance. The performance of both epoxy systems when contaminated degrades significantly between 4-h cure and 24-h cure, and only the Avtur contaminated 808/AC repairs withstood 8 kN in tension after 10^5 cycles. This degradation is similar to the behaviour uncontaminated, and the effects are again suspected to be associated with continuing cure between 4 and 24 h increasing the resin stiffness. The 808/AC system is far less sensitive to the effects of contamination than the 828/AC system, and it is noteworthy that the 24-h cure data for 808/AC fatigued specimens (Fig. 5) are superior to those for the 24-h cure 808/AC as-manufactured specimens (Fig. 4), both contaminated and uncontaminated. The degradation in properties of the contaminated epoxy systems between 4 and 24 h implies that further work must be done to stabilise the level of properties in relation to time.

The fatigue performance of the polyester systems presents a more complicated picture. QC3 systems cured with M50 hardener tended to improve their performance in fatigue between 4- and 24-h cure, the bending stiffness results suggesting that cure of these systems is still continuing to some extent, particularly at 0 °C. However, no QC3/M50 specimens manufactured at 0 °C withstood 8 kN after 10^5 cycles, the majority failing in fatigue. Systems cured with CH50/DMpT combinations tend to degrade in fatigue performance between 4 and 24 h, whereas the results in Fig. 6 suggest that their cure is substantially complete at 4 h. This is contrary to the behaviour observed for the epoxy systems, and suggests that different degradation mechanisms may operate for polyesters and epoxies. It is possible that the lower viscosity of the polyesters assists their wetting of the aluminium alloy, and that they are consequently less sensitive to the effects of contaminants, as the data in both Figs 4 and 5 tends to suggest. If this is so, any effects due to stiffness changes of the resin as cure proceeds could be expected to be minimised, and certainly be less than those observed for the viscous epoxy resins.

Lamination at 0 °C produces a much more dramatic reduction in the fatigue performance than in the as-manufactured performance. It is possible that the increase in viscosity of the resin at 0 °C reduces its wetting capability, but why this should produce the more serious degradation in fatigue properties is not clear. During fatigue cycling, it is likely that flaws in the bondline, be they inherent or produced in response to stressing, will grow until a critical size is reached, at which point the bond will fail. Indeed, a progressive debonding failure was often observed during fatigue cycling, particularly for the QC3/M50 system. Consequently, it is possible that if 0 °C lamination produces an imperfectly wetted alloy surface voids may form and initiate early failure. Monotonic tensile testing of such imperfectly wetted specimens would not be expected to cause any significant void growth until immediately prior to failure; hence one would expect a lesser degradative effect, compared to those fatigued. This is again an effect at the alloy/resin interface, and one would anticipate only a small influence from the state of cure of the resin.

The question of which is the best resin system of those investigated is not an easy one to answer. Where static loading alone was involved, the only system that exceeded the strength of the equivalent riveted repair in all combinations of cure time and contamination condition was the room temperature laminated QC3/M50 system. However, this system is very sensitive to hydraulic oil contamination when fatigue loaded at short cure times. Overall, the best balance of static and fatigue properties for the room temperature systems was given by 808/AC and QC3/CH50/DMpT. Of the 0 °C lamination systems, only the QC3/CH50/DMpT system could be considered to have any useful performance when fatigue loaded, but was also sensitive to hydraulic oil contamination. Hence, it seems sensible to recommend the use of the QC3/CH50/DMpT systems as having the best overall balance of properties over a useful lamination temperature range, with the reservation that its fatigue resistance is lowered if hydraulic fluid on the substrate surface is not completely removed.

The measured bending stiffnesses for most of the composite specimens were in the range 2–3 N/mm. Since the composite patch was intended to match the aluminium alloy in both specific and absolute stiffness this disparity is, at first sight, surprising. However, the composite panel thicknesses were in the range 2·5–3·0 mm and bending stiffness is proportional to the cube of the beam thickness. Hence, although the aluminium alloy and the hybrid composite may be matched in tensile stiffness, the greater thickness of the composite gives it a much greater flexural stiffness. However, since aircraft skins are stiffened, by for example ribs and stringers that confer lateral support, and do not strain by more than $\frac{1}{2}$%, this difference in flexural stiffness is not significant. It should be noted that the fatigue and tensile tests used in this work permitted specimen bending, and this was consequently a very severe assessment of the repair method since this lateral bending is restricted in service.

The bond between a repair patch and a substrate is a critical feature of a successful repair and recent work has been undertaken to improve the efficiency of the bond. The cloth used for the work described in this paper is fairly coarse, and, when laminated, pockets of neat resin are trapped locally between the cloth tows and the substrate. These resin pockets effectively produce a locally thick bondline and hence reduce the efficiency of the bond and possibly the strength of the repair. On some repairs this phenomenon revealed itself when discrete pockets of cohesively failed resin were visible on the substrate following tensile testing. By using finer fabrics (incorporating 3000 end carbon fibre yarns) higher strength repairs have been made when compared to repairs made with the coarser fabric.

CONCLUSIONS

1. With regard to all the experimental variables, i.e. 4- and 24-h cure, contamination condition, room temperature and 0 °C lamination and as-manufactured and post fatigued performance, no resin system can be regarded as offering an alternative to riveted metal plate repairs in all circumstances.

2. The best all-round combination of properties was obtained with Quickcure QC3 polyester resin, cured with Lucidol CH50 hardener and DMpT accelerator. This system gave results equivalent to, or better than riveted metal plate repairs in all cases except on hydraulic fluid-contaminated surfaces at 0 °C.

3. All resin systems were affected more adversely by aircraft hydraulic fluid contaminant than by Avtur contaminant, possibly because the contaminants alter the wettability of the abraded alloy surface by the resin.

4. At room temperature the polyester systems were, in terms of percentage loss in tensile strength, less affected by contamination than were the epoxies. The epoxies, however, showed inherently higher tensile strengths in the uncontaminated state.

5. At 0 °C the detrimental effects of contaminants are more severe than at room temperature, particularly in the case of fatigue performance.

ACKNOWLEDGEMENT

The work described in this paper has been carried out under an extramural contract supported by the Procurement Executive, Ministry of Defence.

REFERENCES

1. ROGERS, K. F., KINGSTON-LEE, D. M. and PHILLIPS, L. N., 'The Application of Reinforced Plastics to the Emergency Repair of Aircraft', *Symposium on Jointing in Fibre Reinforced Plastics*, Imperial College, London, Sept. 1978, 104–15.
2. ROGERS, K. F., KINGSTON-LEE, D. M. and PHILLIPS, L. N., 'The Use of Carbon and Hybrid Woven Fabrics in Emergency Aircraft Repair Work', *Proceedings of the Third International Conference on Composite Materials, Paris, August 1980*, Pergamon Press, Oxford, 1390–407.
3. GLEDHILL, R. A., KINLOCH, A. J. and SHAW, S. J., Effect of relative humidity on the wettability of steel surfaces, *J. Adhesion*, **9**, 1977, 81–5.

38

Temperature Increase in SMC Fatigue Testing

S. V. Hoa and S. Lin

Department of Mechanical Engineering, Sir George Williams Campus,
Concordia University, Montreal, Quebec H3G 1M8, Canada

ABSTRACT

The temperature increase in sheet molding compound SMC-R30 subjected to fatigue loading is investigated. The frequency range varies from 20 to 40 Hz. Using dimensional analysis, a formulation is obtained which enables the prediction of the temperature increase at different frequencies and different locations on the specimen, knowing the temperature increase at one frequency at one location.

INTRODUCTION

The need to reduce weight in automobiles has led to the rapid development of short fiber reinforced plastics. These materials are light weight but have good strengths and stiffnesses such that they can be used to construct load bearing components of the automobile such as engine support bracket, transmission support bracket, oil pan, truck frame, etc. Among the various mechanical properties of the material, the fatigue behavior plays an important role in design. In fatigue testing, the frequency of loading affects the temperature increase within the specimen. Results from the fatigue tests on engineering plastics[1] have indicated that the increase in temperature has a considerable effect on the fatigue life of the material. This paper examines the effect of frequency on the temperature increase in SMC-R30 subjected to fatigue testing. Dimensional analysis is made such that prediction of the temperature increase at different frequencies, knowing the temperature increase at one frequency, is possible.

573

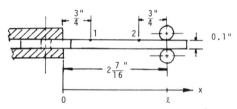

FIG. 1.   Specimen configuration. Points 1 and 2 show the location of thermocouples.

EXPERIMENTS

The material tested is SMC-R30 supplied from Somerville Industries, Ltd. Its composition is shown in Table 1. Flexural fatigue testing is performed. A schematic of the experimental set up is shown in Fig. 1. The specimen was completely insulated during the experiment. The frequency of loading was varied from 20 to 40 Hz. The temperatures at two locations (point 1 and point 2 in Fig. 1) were recorded using thermocouples. Experimental results are shown as solid points in Fig. 2.

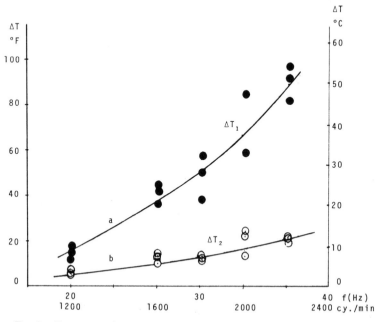

FIG. 2.   Temperature increase versus testing frequency for insulated specimen.

TABLE 1
Components of sheet molding compound

Matrix	Polyester resin	27%
Fiber	E Glass	30%
Filler	$CaCO_3$	41%
Thickener	Magnesium hydroxide	1%
Catalyst, internal release		1%

DIMENSIONAL ANALYSIS OF EXPERIMENTAL RESULTS

The increment of the temperature along the test specimen can be presented as a function of the geometrical configuration of the specimen, the physical properties of the material, the test amplitude and frequency, and the overall heat transfer coefficient between the specimen and its ambient air,

$$\Delta T = f(x, l, t, b, E, \eta, \delta, \omega, U) \tag{1}$$

where x = space coordinate, l = length of the specimen, t = thickness of the specimen, b = width of the specimen, E = Young's modulus, η = loss coefficient, δ = maximum test amplitude, ω = test frequency, U = overall heat transfer coefficient.

For a fixed geometrical configuration of the specimen and for a fixed test amplitude with constant material properties and a negligible change of the overall heat transfer coefficient, eqn. (1) can be simplified as

$$\Delta T = f(x, l, \omega) \tag{2}$$

We assume that the function of ΔT can be obtained by the method of separation of variables,

$$\Delta T = X(x, l)Y(\omega) \tag{3}$$

For dimensionless representation of eqn. (3), the test data obtained at the frequency $\omega = 20$ Hz are used as reference data. As an approximation, functions $Y(\omega)$ and $X(x, l)$ are presented as follows:

$$Y(\omega) = \left(\frac{\omega}{20}\right)^n \tag{4}$$

$$X(x, l) = \sum_{i=0}^{m} a_i \left(\frac{x}{l}\right)^i \tag{5}$$

It can be seen that $X(x, l)$ represents the temperature distribution along the specimen at $\omega = 20$ Hz. Equation (5) can then be written as

$$X(x, l) = \Delta T_{\omega=20}\left(\frac{x}{l}\right) = \sum_{i=0}^{m} a_i\left(\frac{x}{l}\right)^i \tag{6}$$

Substituting eqn. (6) into eqn. (5) and then substituting eqns (4) and (5) into eqn. (3), we obtain

$$\frac{\Delta T}{\Delta T_{\omega=20}} = \left(\frac{\omega}{20}\right)^n \tag{7}$$

By making use of the temperature data measured from point 1, ΔT_1, at 20 and 30 Hz, the exponent n is determined to be $n = 2 \cdot 6$. Curve (a) in Fig. 2 is obtained from eqn. (7). In order to test the validity of the value n, curve (b) in Fig. 2 is also obtained from eqn. (7) with $\Delta T_{\omega=20} = 2 \cdot 5\,°C$. It is seen that the results obtained from eqn. (7) with $n = 2 \cdot 6$ agree very well with those obtained from the test data.

For the expression of the function $\Delta T_{\omega=20}(x/l)$, the following two boundary conditions are available:

1. At $x = l$, there is no energy dissipation. Therefore $\Delta T = 0$.
2. At $x = 0$, $\Delta T_{\omega=20}$ must reach its maximum value. Therefore

$$\left.\frac{d\Delta T_{\omega=20}}{dx}\right|_{x=0} = 0$$

With two additional measured temperatures at $\omega = 20$ Hz, eqn. (6) may be expressed as

$$\Delta T_{\omega=20} = a_0 + a_1\left(\frac{x}{l}\right) + a_2\left(\frac{x}{l}\right)^2 + a_3\left(\frac{x}{l}\right)^3 \tag{8}$$

where a_0, a_1, a_2 and a_3 are constants. At $x = 0$, we obtain

$$\Delta T_{\omega=20, x=0} = \Delta T_{max, \omega=20} = a_0 \tag{9}$$

Then eqn. (8) can be written in a dimensionless form,

$$\left(\frac{\Delta T}{\Delta T_{max}}\right)_{\omega=20} = g\left(\frac{x}{l}\right) = 1 + b_1\left(\frac{x}{l}\right) + b_2\left(\frac{x}{l}\right)^2 + b_3\left(\frac{x}{l}\right)^3 \tag{10}$$

with its derivative

$$\frac{d\left(\dfrac{\Delta T}{\Delta T_{max}}\right)_{\omega=20}}{d\left(\dfrac{x}{l}\right)} = b_1 + 2b_2\left(\frac{x}{l}\right) + 3b_3\left(\frac{x}{l}\right)^2 \qquad (11)$$

where b_1, b_2 and b_3 are constants.

At $x/l = 0$, due to

$$\frac{d\left(\dfrac{\Delta T}{\Delta T_{max}}\right)_{\omega=20}}{d\left(\dfrac{x}{l}\right)}\Bigg|_{x/l=0} = 0 \qquad (12)$$

we obtain

$$b_1 = 0 \qquad (13)$$

At $x/l = 1$, because

$$\left(\frac{\Delta T}{\Delta T_{max}}\right)_{\omega=20} = 0 \qquad (14)$$

it follows,

$$1 + b_2 + b_3 = 0 \qquad (15)$$

From the measured data at $\omega = 20$ Hz,

$$\frac{x}{l} = 0 \cdot 3077 \qquad \Delta T = 10\,°C \qquad (16)$$

and

$$\frac{x}{l} = 0 \cdot 6923 \qquad \Delta T = 2 \cdot 5\,°C \qquad (17)$$

the constants b_2 and b_3 are obtained as follows:

$$b_2 = -3 \cdot 245 \qquad (18)$$

$$b_3 = 2 \cdot 245 \qquad (19)$$

Therefore the temperature distribution along the specimen, eqn. (10), can be written as

$$\left(\frac{\Delta T}{\Delta T_{max}}\right)_{\omega=20} = g\left(\frac{x}{l}\right) = 1 - 3 \cdot 245\left(\frac{x}{l}\right)^2 + 2 \cdot 245\left(\frac{x}{l}\right)^3 \qquad (20)$$

where

$$\Delta T_{\text{max},\omega=20} = 13 \cdot 2\,°\text{C} \tag{21}$$

Substituting eqn. (20) into eqn. (7), we obtain,

$$\frac{\Delta T}{\Delta T_{\text{max},\omega=20}} = \left[1 - 3 \cdot 245 \left(\frac{x}{l}\right)^2 + 2 \cdot 245 \left(\frac{x}{l}\right)^3 \right] \left(\frac{\omega}{20}\right)^{2 \cdot 6} \tag{22}$$

At $x/l = 0$, because

$$\Delta T = \Delta T_{\text{max}} \tag{23}$$

eqn. (22) becomes,

$$\frac{\Delta T_{\omega,\text{max}}}{\Delta T_{\omega=20,\text{max}}} = \left(\frac{\omega}{20}\right)^{2 \cdot 6} \tag{24}$$

Substitution of eqn. (24) into eqn. (22) yields

$$\frac{\Delta T}{\Delta T_{\text{max}}} = 1 - 3 \cdot 245 \left(\frac{x}{l}\right)^2 + 2 \cdot 245 \left(\frac{x}{l}\right)^3 \tag{25}$$

It can be seen that the function of $g(x/l)$ from eqn. (20) represents the temperature distribution along the test specimen not only for the frequency at $\omega = 20$ Hz, but also for any frequency. Therefore the experimental results of the temperature increment along the test specimen with variable test frequency can be presented in eqns (25) and (24) with $\Delta T_{\omega=20,\text{max}} = 13 \cdot 2\,°\text{C}$.

CONCLUSION

Even though the experimental data is limited to only two locations and for insulated specimens, the agreement between experimental results and theoretical prediction using dimensional analysis is good. The accelerated increase of the temperature at the high end of the frequency range can be due to the increase in the loss coefficient at higher temperatures. Work is underway to investigate this effect.

ACKNOWLEDGEMENT

The financial assistances from the Natural Sciences and Engineering Research Council Canada through grants No. A 0413 and A 7929 are appreciated.

REFERENCE

1. MANSON, J. A. and HERTZBERG, R. W. Fatigue failure in polymers, *Critical Reviews in Macromolecular Science*, **1**, 1973, 433.

39

A Unique Approach to Fabricating Precision Space Structures Elements

H. Cohan and R. R. Johnson

Lockheed Missiles & Space Company, Inc., Bldg. 104, PO Box 504, Sunnyvale, California 94088, USA

ABSTRACT

A procedure for fabricating graphite epoxy columns used in the assembly of large space platforms is described. The requirement for precise dimensional control led to a unique hot resin injection process. Dry, high modulus fiber is wound over a vertically mounted steam-heated mandrel. A steam-heated sleeve or caul is slipped over the wound mandrel and resin is injected and cured in place. Approximately 200 column elements have been fabricated using this efficient process.

INTRODUCTION

Large space platforms have been proposed for a number of applications such as communication satellites, multikilowatt power modules, large modularized antennas, and geostationary platforms in general. Such platforms would be required to have very low ratios of mass to area (or volume) after assembly and/or deployment on orbit. Figure 1 portrays the automated assembly of a linear section of a large platform, using the Space Shuttle orbiter as a work-base. The structure is composed of tubular, double-tapered column elements assembled to node joints designed to facilitate on-orbit assembly. The structural elements are the 2·6 m half-columns described in this paper. A free flying assembler using 10 m half-columns is shown in Fig. 2.

A major consideration in optimizing utilization of the Space

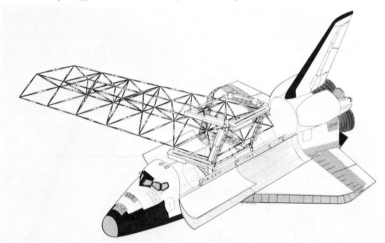

FIG. 1.   Automated assembler mounted on orbiter.

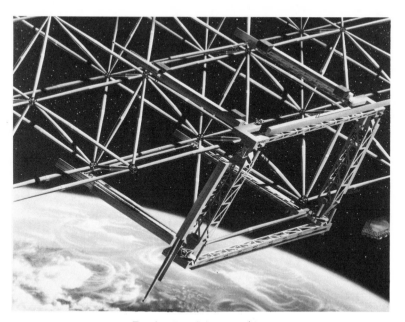

FIG. 2.   Free flying assembler.

Transportation System (Space Shuttle) is the low weight-to-volume ratio of typical conventional space structure elements and subassemblies. A load of such elements will fill the available orbiter cargo-bay volume without approaching the weight-lifting capability of the vehicle. The National Aeronautics and Space Administration, Langley Research Center, has developed a structural concept using double-tapered graphite epoxy columns, manufactured and shipped as tapered half-columns which can be stacked and nested very compactly, resulting in efficient use of the Shuttle cargo bay. The half-columns are to be snapped together on orbit to form the full-column elements, then erected by utilizing node joints at element intersections. The studies leading to the selection of the tapered column concept are discussed in reference 1. The physical characteristics of tetrahedral truss type platforms constructed from such columns are described in reference 2.

The half-column elements described in this paper are approximately 2·6 m long and were manufactured using a process developed for 10 m elements. The taper of the column elements permits nesting, and this high packaging efficiency makes it possible to use the full weight capability of the Space Shuttle Orbiter.

COLUMN DESIGN CHARACTERISTICS

A sketch of a 2·6 m half-column is shown in Fig. 3. The column tapers from a diameter of 10 cm at one end to 5 cm at the other. Graphite epoxy was

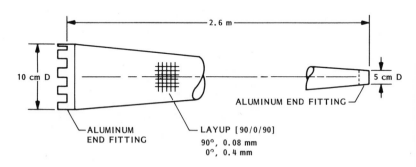

FIG. 3.   Half-column.

selected for the columns because of its high stiffness and near zero coefficient of thermal expansion. The fiber orientation is (90/0/90) with Thornel T-300 fiber, 0·08 mm thick, used for the circumferential wraps, and Pitch Type VSB-32 fiber, 0·4 mm thick, for the longitudinal ply. The pitch fiber was selected for its low cost and high stiffness. The aluminum end fittings are integrally wound on the mandrel to achieve precise repeatable column lengths and fitting placement. Finished weight is approximately 840 g per half-column. The design compressive load for the full-column is over 400 kg, and they have been successfully tested to that load.

PROCESS SELECTION RATIONALE

Large space platforms may eventually require half-columns in excess of 10 m long; the manufacturing process described here has the capability to meet this requirement. A hot process was selected to avoid separating the integrally wound aluminum fittings from the column during cure due to differences in thermal expansion coefficient. The steel mandrel increases in length about 5 mm during heatup for cure and drives the end fittings with it. Although the use of heated tooling eliminates the thermal management problem during cure it precludes the use of prepreg material during fiber placement. Therefore, a dry fiber wind process followed by resin injection was selected.

Hand tooling, inside and out, is required to achieve the wall thickness control and smooth surfaces required to permit close nesting and later separation of the half-columns. Precise column straightness is required to provide ease of stacking, handling, and assembly as well as platform flatness characteristics. This requirement tends to preclude the use of horizontal winding systems. A horizontally supported mandrel will permit an excessive amount of sag. A steady rest at the mid-point, or other points of the mandrel, would be impractical; some form of rolling contact would be required and could possibly damage uncured material. The obvious solution to these problems is a vertical winding system. The approach described here uses a stationary mandrel, vertically suspended, to simplify winding the 90° lamina over the 0° fibers. A major portion of the manufacturing development effort required for horizontal fabrication was eliminated by using a winding machine with stationary mandrel vertically suspended.

Fig. 4.   Vertical winding machine.

DESCRIPTION OF THE VERTICAL WINDING MACHINE

The vertical winding machine is shown in Fig. 4. The machine provides a mechanism by which longitudinal fibers can be accurately positioned. Discrete and appropriate numbers of longitudinal fibers are terminated at previously determined locations to accommodate the column taper. This procedure assures the maximum number of fibers (volume fraction) from the large diameter to the small diameter. The filament winding mechanism of the machine is relatively simple. The mandrel is positioned in the machine. A rotating head, shown in Fig. 5, traverses up the length of the mandrel, applying a continuous wrap of circumferential (90°) fibers. After the initial layer is complete, longitudinal (0°) fibers from the previously loaded tension plate and convergence ring are attached to a bracket above the large end fitting. During the downward pass, as the longitudinal fibers are drawn through the tension plate and convergence ring, the rotating head captures them with circumferential wraps maintaining accurate alignment and pretensioned load. Figure 5 shows this step clearly. Note the two bobbins of fiber which rotate around the mandrel to lay the 90° fibers in place. The machine is 14 m tall, and is capable of fabricating column elements 10 m in length.

DESCRIPTION OF THE MANUFACTURING PROCESS

The manufacturing process, which is described in more detail below, is briefly outlined as follows:

(1) Load the tension plate.
(2) Preheat mandrel and external sleeve.
(3) Install mandrel in winding machine.
(4) Apply inner 90° wrap as carriage traverses up the mandrel.
(5) Attach longitudinal fibers and apply longitudinal and outer circumferential wrap.
(6) Insert wound mandrel in steel sleeve.
(7) Inject resin.
(8) Allow part to cure.
(9) Remove outer sleeve.
(10) Allow mandrel to cool and remove part.

Tension Plate Loading

The tension plate, which is in fact three plates, has 720 equally spaced

Fig. 5. Winding machine head.

FIG. 6. Loading the tension plate.

holes. The entrance and exit plates have bonded ceramic eyelets, and the center plate contains 720 ceramic brakes for providing tension (140 g/strand). The plate is hand loaded with the VSB-32 longitudinal fibers which are precut to correct length. Plate loading is shown in Fig. 6. The convergence ring through which the fibers pass is also shown supported by four members which are adjusted by individual turnbuckles. In order to maintain a uniform wall thickness over the length of the columns, it is necessary to terminate longitudinal yarns as the fiber is applied from the large end toward the small end. The use of the tension plate provides flexibility in that the longitudinal fiber count is easily changed and is accurately measured. Local reinforcement of the large end fitting is achieved by adding short longitudinal lengths of T-300 fiber during tension plate loading.

Winding of the Mandrel

The aluminum end fittings are mounted as appropriate on the large and small ends of the steel mandrel. Both the external sleeve and the mandrel are heated to 107 °C by steam prior to winding. This temperature is maintained until the part has cured. The first circumferential or 90° wrap of T-300 is applied during a traverse from the bottom to the top of the mandrel. The strands are wrapped as closely as possible without overlap giving a nominal ply thickness of 0·08 mm. This spacing is achieved by setting the gear ratio of the vertical traverse to spool carrier ring rotation. Upon completion of the first vertical traverse, the longitudinal (0°) pitch fibers are attached to a bracket above the mandrel and the downward traverse begins. The longitudinal fibers and the outer 90° wrap are incorporated simultaneously, with the outer 90° wrap capturing the 0° fibers. With the completion of this downward pass, the fiber ends are held with a tacking resin and the wrapping trimmed to final length.

Resin Injection and Cure

The wound mandrel is transferred to the externally insulated sleeve by use of a hoist. Care is exercised in guiding the mandrel into the sleeve to ensure that the mandrel does not touch the sleeve wall. A vacuum is drawn from the large end of the assembly to remove air and volatiles, and resin is injected from the small end. During injection, 60 kPa pressure is applied to the pressure pot and the vacuum is reduced to 25 cm Hg. Approximately 10–15 min are required for the resin to reach the top of the column. Completion of the resin injection is determined by observing the flow of resin in the transparent plastic tubing in the vacuum line at the top of the column. Prior to the injection process, the sleeve is backed away from the fully closed position on the fiber wound mandrel. This provides a larger pumping annulus which reduces the pumping time and minimizes 'wash' of the circumferential fibers. The nominal 0·5° taper allows an increase of about 0·09 mm in the radius of the pumping annulus (clearance between

FIG. 7. Completed half-columns, shown stacked.

mandrel and sleeve) for each linear cm the sleeve is backed away from the mandrel.

After fill is complete, the injection valves are closed. The outer closure sleeve is drawn to close on the mandrel to a predetermined wall thickness annulus. Excess resin is permitted to flow out during this closing operation. After closure, pressure is applied using shop air, approximately 700 kPa. Pressure is held until full gelation is achieved. Nominal cure time at this temperature is 6 h. After the part is cured, the sleeve is removed and the spool piece which locks the small end of the column to the mandrel is released. The mandrel is allowed to cool and the part is easily removed with a small mechanical load after releasing the large fitting retaining ring. Some of the completed columns are shown in Fig. 7, in nested stacks.

Resin System Selection and Management

Requirements for the resin system to be used with the injection process include the following:

(1) long pot life at the pumping viscosity temperature;
(2) low viscosity at injection temperatures to minimize injection time, fiber wash, and to insure void-free filling;
(3) adequate gel-time at injection temperature to insure that the resin does not set up during injection;
(4) high strength at cure temperature, to permit the sleeve removal at that temperature.

An MY720-DDS system was selected for this process. This system demonstrated good pot life at 80 °C, easy flow at 120 °C, and adequate cure rates at 175 °C.

Columns were also manufactured using a room temperature handling epoxy system ADX 16/AP-22. This system, developed by LMSC for glass epoxy buoys, had been successfully used also for impregnating a graphite-epoxy submarine sensor mast. The ADX 16 is a distilled version of Lekuetherm X-50 imported by the Mobay Chemical Co. A cure cycle was established for this system by evaluating the glass transition temperature and short beam shear properties under LMSC Independent Development Program founding. The cure cycle is 6 h at 104 °C plus a post cure of 6 h at 135 °C.

Photomicrographs of a specimen are shown in Fig. 8. This section is cut normal to the longitudinal fiber and shows both the smaller diameter T-300 reinforcing fiber and the VSB-32 pitch type fiber. Good compaction and zero voids can be observed.

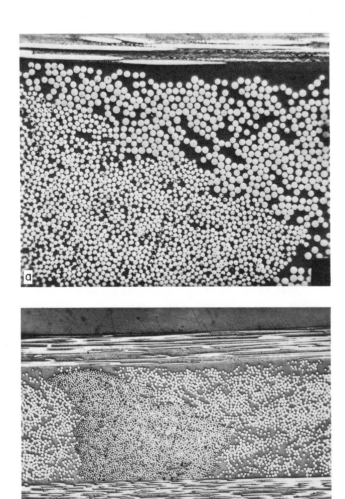

FIG. 8.   Specimen photomicrographs. (a) = ×250; (b) = ×100.

CONCLUSION

Approximately 200 half-columns have been made by this process at this time. A number have been assembled into a tetrahedral structure (Fig. 9) for test and demonstration at NASA's Langley Field, Virginia facility. Others are being used experimentally in the NASA Neutral Buoyancy Facility at Huntsville, Alabama. The process has been demonstrated to be capable of producing consistent high quality structural elements.

Fig. 9.   Tetrahedral structure.

ACKNOWLEDGEMENT

The development of the process was sponsored by the NASA Langley Research Center under Contract NAS1-14887, 'Development of Large Space Structures Concepts'.

REFERENCES

1. Bush, Harold G. and Mikulas, Martin M., Jr. A nestable tapered column concept for large space structures, NASA TM X-73927, 1976.
2. Mikulas, Martin M., Jr., Bush, Harold G. and Card, Michael F. Structural stiffness, strength and dynamics characteristics of large tetrahedral space truss structures, NASA TM X-74001, March 1977.

40

Manufacturing Methods for Carbon Fiber/Polyimide Matrix Composites

WESLEY C. MACE

Lockheed Missiles and Space Co., Inc.,
PO Box 504, Sunnyvale, California 94086, USA

ABSTRACT

High-temperature processing of polyimide matrix/carbon fiber composites has required the development of new and innovative tooling approaches, manufacturing methods, and material processing. This paper describes several tooling approaches, including the use of ceramic, molded graphite, and metallic tool surfaces, and the criteria for selection of each. The influence of different thermal coefficients of expansion on tool material selected and geometric restraints is explored.

Consideration is given to the ultimate production process to be used, i.e. press molding, autoclave molding, and hydroclave molding.

The use of alternative, energy efficient tooling methods such as internally heated tools and part peculiar heating blankets is illustrated. Tool design solutions to minimize the loss of heat during high-temperature processing are given.

Selected case histories of varied designs are examined in detail. The rationale for the manufacturing–tooling–processing approach is given, as well as the results of tests on the end item.

BACKGROUND

The advent of the new high-temperature resin systems, of which polyimides are the most promising, has required an increase in the amount of development engineering applicable to manufacturing processes. Too often

592

introduction of new systems with their seemingly bright promise of better engineering properties and breakthroughs in environmental resistance has faltered because of manufacturers' inability to realize their potential. For this reason, LMSC decided early on that processing techniques must keep apace with materials development; the results of this effort are reported in this paper.

A prime consideration of this program was to ensure that any process developed in the prototype stage could be readily transferred to a production mode. This required that typical part geometries be fabricated by methods that were not only economically viable but could withstand the criterion of reproducibility. This necessitated a compromise in reaching the absolute potential of the materials involved in favor of the aforementioned economic and reproducibility constraints in some cases.

Even a cursory study of missile structures will reveal that the determining factor in any manufacturing process will be part geometry. Therefore, not one, but perhaps two or three processes may be required for any materials system. This has, in fact, been borne out in the case of polyimide matrix/graphite composites. The principal processes involved here are autoclave, hydroclave, and press molding.

AUTOCLAVING

The key to successful autoclave processing of condensation/addition polyimide is to complete the imidization cycle (a condensation reaction) as nearly as possible without initiating polymerization of the imides formed,

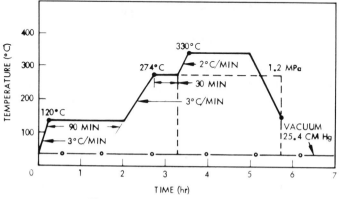

FIG. 1.   Cycle for LARC-160 resin.

thereby restricting flow of the resin at autoclave pressures. At the same time, removal of the solvents and evolved water must be ensured. A reasonably successful cycle for LARC-160 resin has been developed[1] and is shown in Fig. 1. This cycle was used to fabricate the parts shown in Fig. 2.

The tooling for these configurations was dictated by two criteria: geometry and thermal coefficient of expansion. Because of the need to mate

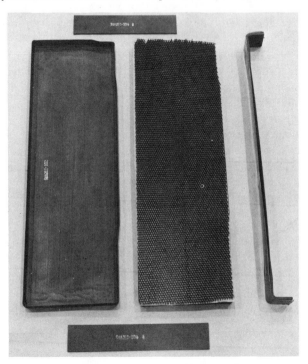

Fig. 2.   LARC-160/carbon fiber—details.

certain bonding surfaces, female tools were required. To avoid inducing stress into the cured parts during cool-down, a reasonably close match of tooling thermal coefficient of expansion with the composite was required. This in turn narrowed the selection of tooling material to two, i.e. molded graphite or cast ceramic. Cast ceramic was chosen for the following reasons:

(1)   With judicious design of the heat source, the ceramic could be used to insulate the part temperature from the periphery of the tool, thereby allowing for use of standard bagging materials.

(2)   Machining of male form blocks used to cast the ceramic molds was less expensive than machining female molds from solid graphite.

(3)   Because solid graphite tooling had previously been demonstrated successfully,[2] these parts presented a good opportunity to prove the efficacy of cast ceramic tooling.

The cast ceramic surface was relatively porous. Consequently, it was decided to apply a ceramic glaze to the part surfaces. This glaze not only provided a good tool surface but proved to be an excellent parting medium

FIG. 3.   Cast ceramic tools.

for the cured composite and required no further application of parting agents prior to cure. Figure 3 illustrates the completed ceramic tooling.

The layup of the parts was routine, except for the PAN details. Because these were designed as shear panels, the edges were built up with the center of the panels dropping off to a thin section. This required buoying the center panels between the edge buildup plies and providing a staggered joint. Accurate placement of the center plies was achieved by fabricating fiberglass transfer templates that were indexed to the sides of the female tools. The natural tack of the polyimide resin allowed the raw material to adhere to the template. The template was then positioned in the tool, hand pressure was applied to the surface, and the pattern detail was transferred to the layup as the template was removed. Upon completion of the layup, perforated teflon coated glass was used as a separator, and the bleeder plies were installed.

A tetrafluoroethylene film was used as a barrier, and flexible woven heater blankets were installed. Fiberglass insulation was applied over the entire surface, the layup envelope bagged and cured in the autoclave as shown in Fig. 1, using only gas pressure in the autoclave and all heating supplied by the flexible heating blankets.

PRESS MOLDING

Matched die molding of polyimides is one of the most reliable and cost-effective processes available. In addition, it provides precise dimensional control of at least two surfaces with ease and three surfaces with the use of some ingenuity in die design and the willingness to assume the increase in cost associated with the increased complexity of the tool. An added advantage is the ability to imidize the matrix at a higher temperature, use higher pressures in the final polymerization, and thereby ensure the removal of all volatiles and condensation reaction moisture from the laminate. Using this premise, a cure cycle was developed for matched die molding of LARC-160 (see Fig. 4). Matched steel dies were machined, and sample brackets were produced as illustrated in Fig. 5.

Several cautionary items are in order at this point. Because of a final

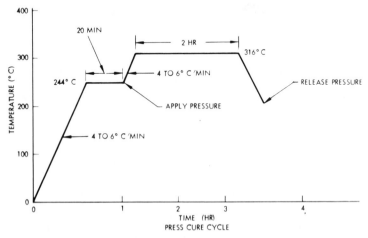

FIG. 4.   Matched die cure cycle.

FIG. 5.   Matched die-molded brackets.

curing temperature of 316°C, the difference in thermal coefficient of expansion (TCE) of the composite part and TCE of the tooling material becomes a matter of concern with certain geometrics. For example, the TCE of steel can approach eight times that of a carbon fiber composite, and a part in a completely enclosed female half of a tool (such as a PAN configuration) could be subjected to an unacceptable stress level during cool-down of the mold. Even a flat surface with restrained edges can be deleteriously affected by these phenomena. In the case of the brackets illustrated previously, this problem was resolved relatively easily. An approximate 0·6 cm × 0·6 c   groove was machined around the periphery of one die half 0·7 cm from the end of the part. A fluorosilicone rubber extrusion was placed in this slot, extending far enough above the surface to provide a seal when the two halves of the die were closed. When pressure was applied, the seal prevented escape of the excess resin beyond the tool cavity, allowing internal hydraulic pressure to be exerted on the layup. Upon cooling, the pressure was reduced, the rubber could exert no appreciable strain on the layup and no residual stress was imparted through the edges of the part. The draft angle of the hat section allowed the part to slide up the slope so that the hat section was not stressed. The mportance of having sufficient draft angle, $-2°$ minimum is recommended, on near vertical walls when using high TCE tooling materials .:nnot be over emphasized.

When perpendicular walls are an absolute requirement, there are two utions to the problem. One is to fabricate the female portion of the die f m a material that has virtually the same TCE as the composite, such as m lded graphite or ceramic. The other is to induce an artificial draft into

598 Wesley C. Mace

the part by laying in sacrificial plies on the periphery and subsequently machining them away after removal of the part from the mold. The sacrificial plies can be of a cheaper material than carbon fiber, such as glass. This method is often used in areas where machined tolerances are required for subsequent fit-up operations.

HYDROCLAVING

The third method of processing used with polyimide materials was hydroclaving. Hydroclaves have certain advantages over autoclaves, chiefly because the pressurizing medium is water instead of an inert gas. With a water medium, pressures of up to 6·9 Mpa can be obtained by the simple expedient of attaching an air booster to the in-plant compressed air source. The ability to achieve high pressures without the danger associated with compressed gases allows a much lower capital cost for equivalent size equipment. Whereas LARC-160 processes fairly readily at pressures of 1·38 Mpa, certain geometries may require higher pressures because of pressure degradation occurring when deep draft, compound curvatures are involved. Other polyimide systems, such as PMR 15, require higher pressures for successfully molding this type of part in a repeatable, reliable manner. Hydroclaved polyimide parts must be fabricated on internally heated tools to achieve the 316 °C curing temperature. Nylon films cannot

Fig. 6. Hydroclave tool.

FIG. 7.   Hydroclaved cylinder.

be used safely at high molding pressures, so rubber bags are used, and the part is insulated from the bag to keep the temperature at the bag surface under 173 °C. The same curve cycle is used with hydroclaving as with autoclaving, with the sole exception of higher curing pressures that are empirically determined for each geometry. A typical hydroclave tool is shown in Fig. 6, and the resultant part is shown in Fig. 7.

There were several key reasons for designing the tool as a male instead of a female mold. The use of aluminum with its high TCE caused the tool to expand against the bag pressure, thereby minimizing wrinkles. The high TCE also aided in removal of the part because the tendency was to shrink away from the layup during cool-down. Advantage also could be taken of the high thermal conductivity to allow for faster heat-up rates and lower thermal gradients over the part areas.

RAW MATERIAL SELECTION

Though the majority of data and experience available on carbon fiber composites was evolved in the use of collimated tape, an early decision was made to design mainly around woven fabric. The reasons for this were both economic and engineering. The type of structures contemplated contained geometries of radical contour, including many complex curvatures and near right-angle bends. The loading necessitated a pseudo-isotropic structure. Accurate orientation of collimated fiber plies was extremely difficult, and orientations were almost impossible to maintain during flow of the matrix when cure was effected. Tests performed on pseudo-isotropic laminates of both collimated tape and woven fabric revealed that design allowables were actually a little higher using the woven fabric than when using collimated tape, even in a Celanese compression test. This would seem to be an anomaly because weaving operation imparts a slight buckle to the fibers. However, observation of the tests revealed that collimated tape laminates sometimes had a tendency to fail in sequential layers along the shear plane between plies. The woven fabric laminates, however, did not exhibit this tendency—probably because of the nesting effect of the weaves—and always failed as a complete unit. Therefore, although individual values could be considerably greater with the collimated tape laminates, the nature of scatter made their actual allowables smaller.

A paradox also appears to exist in the economic side. Woven fabric costs $20 to $25 per pound more than collimated tape in the preimpregnated form. Fabricated structures, on the other hand, can be made with considerably less labor cost using woven cloth as opposed to collimated tape; in some cases, the reduction in labor costs are as much as 70 percent.

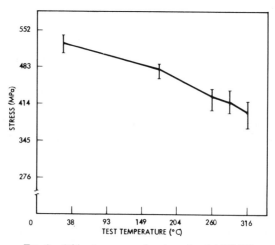

FIG. 8. Ultimate compressive strength—LARC-160.

In the case of the parts considered by LMSC, automation of layup operations was impractical because of their geometries, and the labor saving made the difference in raw material cost inconsequential. Typical compressive values for parts manufactured to these processes are shown in Fig. 8.

RAW MATERIAL CONTROL

The LARC-160 used in these processes was fully characterized by the methods outlined in reference 1, and each lot of material was tested using dielectric dynamic analysis, liquid chromatography, infra-red spectroscopy, and thermal gravimetric analysis for conformance to the 'fingerprint' developed in that program. Uniformity of the material was excellent, and no problems were encountered because of material variability.

CONCLUSIONS

(1)   Polyimide–graphite composites of the LARC-160 type are readily processible by any one of the three reported methods.
(2)   Iterations of designs with tooling and manufacturing engineers are a prerequisite for successful application of these materials.
(3)   Woven fabric is preferred to collimated tape for reasons of cost and reproducibility for the majority of geometries of this type.

REFERENCES

1. WERITA, A. JR. and HADAD, D. K., Resins for Aerospace, *ACS Symposium Series No. 132*, 1980, 215–32.
2. MACE, W. C. Graphite/Polyimide Processing, Proceedings—Inter Continental Conference on Composites, Cannes, France, January 1981.

41

The Use of Natural Organic Fibres in Cement: Some Structural Considerations

D. G. SWIFT

Appropriate Technology Centre for Education and Research, Kenyatta University College, PO Box 43844, Nairobi, Kenya

ABSTRACT

Whilst natural organic fibres have advantages in many developing countries, of low cost and ready availability, their use for reinforcing cement based materials is complicated by their relatively low elastic modulus, their water absorbing properties, susceptibility to fungal and insect attack, and variability of properties amongst fibres of the same type. These points are examined theoretically and in the light of experiments carried out on sisal–cement composites. Their relevance to the design of structures using organic fibre–cement composites is discussed. It is shown that such composites are appropriate for many low cost structures in developing countries despite, or in some cases because of, their peculiar properties.

INTRODUCTION

Several reference texts are now available on the fibre reinforcement of cement based materials.[1-5] The fibres are added to give the material one or more of the following properties:

(i)   increased flexural strength;
(ii) post-crack load bearing capacity;
(iii) increased impact toughness;
(iv) increased viscosity in the fresh state.

602

The main fibres to have been used on a commercial basis have been asbestos, steel, alkaline resistant glass and polypropylene. Asbestos fibres have, of course, been in use throughout the present century. Applications using fibres other than asbestos are relatively recent.

Most developing countries do not have indigenous supplies of asbestos, steel, glass or polypropylene fibres, but do have relatively underused supplies of inexpensive natural organic fibres with adequate tensile strength for fibre reinforcement. However, their other properties have raised doubts concerning their suitability for reinforcing cement-based materials. These disquieting properties are as follows:

(i) low elastic molecules compared with that of cement paste or concrete;
(ii) tendency to absorb water;
(iii) susceptibility to fungal decay and insect attack;
(iv) variation in fibre dimensions, strength and modulus even among fibres from a single plant.

In the present paper, these points will be considered from a theoretical viewpoint and according to the experimental evidence available in order to assess their relevance to structural applications in developing countries.

THEORETICAL CONSIDERATIONS

Effect of Low Young's Modulus

High modulus asbestos, steel and glass increase the tensile strength of cement based materials by bearing the major part of the applied load. They may also increase the compressive strength in this manner. There is no obvious mechanism whereby low modulus fibres could increase the compressive and direct tensile strength of concrete.

Low modulus fibres can, however, increase the flexural strength of concrete. In a previous paper[7] the author showed how the movement of the neutral axis in flexure, arising from the use of low modulus fibres, can result in significant increases in modulus of rupture, whilst the accompanying increases in first-crack flexural strength could be explained in terms of the fibres hindering the development of microcracks into visible cracks. The analysis was based on a model of the composite[8] that divided a flexed beam into zones depending on the predominant constitutive equations applying to each zone (e.g. linear elastic straining of the intact composite, constant or decaying stress during fibre pull-out).

According to this model, the first-crack flexural strength $f_{r,c}^c$ and the modulus of rupture $f_{r,c}^u$ ('ultimate strength') of the composite were found to be given a first approximation by the equations

$$f_{r,c}^c = 2f_{r,m}^c/(1 + \sqrt{E_2/E})$$ (1)

and

$$f_{r,c}^u = (2\sigma_T + 3\alpha f_{cu}\beta^2)/(1 + \beta)^2$$ (2)

where

$$\beta = E_1\varepsilon_{cu}/\sigma_T$$ (3)

and

$$\sigma_T = (2E_1\alpha f_{|cu}\varepsilon_{cu})^{1/2}$$ (4)

Both $f_{r,c}^c$ and $f_{r,c}^u$ are defined in terms of the respective bending moment M by the equation

$$f_{r,c} = 6M/bd^2$$ (5)

where b and d are the breadth and depth of the beam respectively..

In the above equations, $f_{r,c}^c$ is the first-crack flexural strength of the matrix; f_{cu} and ε_{cu} are the cube compressive strength and accompanying strain at the compressive surface of the beam; E, E_1 and E_2 are the effective elastic moduli of the composite in the tensile zones at the commencement of flexure, when visibly cracked, and when extensively microcracked but not visibly cracked, respectively; α is a stress-block factor determined by the compressive stress–strain behaviour of the composite and, by analogy with unreinforced mortar, probably lying within the range 0·4 to 0·7.

The modulus E_1 of the cracked composite in tension is given by the expression

$$E_1 = \lambda v_f E_f$$ (6)

where v_f and E_f are the fibre volume fraction and fibre elastic modulus respectively, and where factor λ allows for discontinuous and misaligned fibres. For long, aligned fibres, λ is unity, and is equal respectively to $2/\pi$ and 0·5 for planar and three-dimensional random orientations of moderately long fibres. Modulus E_2 could be found using the 'rule of mixtures' if the fraction of voids present as microcracks were known. In practice, modulus E_2 must be deduced from experimental evidence using eqn. (1).

This theory may be further developed to give approximate expressions for the impact toughness of a beam using a drop-weight test. Suppose mass m is dropped from height h onto the mid-point of the upper surface of a beam of mass M simply supported over a span L. In general, the total kinetic energy T_T of the falling weight will be expended as follows:

(i) Energy T_1 will be lost as heat, sound and surface energy as the mass hits and indents the top surface of the beam. This may be estimated by equating the momentum of the falling mass immediately before impact with that of the mass and beam after impact when both are moving together. Then energy T_1 lost in this process is given by the expression

$$T_1 = MT_T/(M + m) \tag{7}$$

(ii) Elastic energy T_E will be stored as strain energy in that part of the matrix that remains uncracked, and in the fibres under tension. This may be found from the expression

$$T_E = \int \sigma^2/2E \, dv \tag{8}$$

where σ and E are the stress in, and modulus of, volume element dv of the beam, and the integral is taken over the whole beam, assumed (as a first approximation) to be undergoing simple beam bending. It can be shown from eqn. (8) that, for the composites which are of interest, T_E is given approximately by the relation

$$T_E = bdL(\sigma^*)^2/6E^* \tag{9}$$

where σ^* and E^* are respectively:

$$\begin{cases} f_{r,m}^c, \ E \text{ for unreinforced composites at first cracking,} \\ f_{r,c}^c, \ E_2 \text{ for reinforced composites at first cracking,} \\ f_{r,c}^u, \ E_1 \text{ for reinforced composites at ultimate failure.} \end{cases}$$

This elastic energy will be lost from the beam after each blow, as the beam loses its elastic strains.

(iii) Energy T_p will be dissipated as fibres pull out at cracks and microcracks against frictional resistance. This energy may be found by multiplying the frictional force at the interface, equal to the tensile force in the fibre, by the amount of fibre pull-out, and summing over all fibres. The amount of fibre pull-out increases linearly from zero at the neutral axis to approximately $L\varepsilon_T$ at the tensile surface, where ε_T is the overall strain at this surface (greatly exceeding the elastic strain in fibre and matrix). Carrying

out this analysis, the pull-out energy at first crack and ultimate failure are
found to be respectively

$$T_p^c = \frac{bdL(f_{r,m}^c)^2}{3E_2(1 + \sqrt{E_2/E})} = \frac{bdLf_{r,m}^c f_{r,c}^c}{6E_2} \tag{10}$$

and

$$T_p^u = \frac{bdL\sigma_T^2}{3E_1(1 + \sqrt{E_1/E})} \tag{11}$$

For the type of composites under consideration, both β and the ratio
E_1/E are much less than unity. Hence, using eqns (2) and (4), we may rewrite
eqn. (11) as

$$T_p^u = \frac{bdL(f_{r,c}^u)^2}{12E_1} \tag{12}$$

(iv) Surface energy T_s will be absorbed by the new surfaces at cracks and
microcracks.

(v) Energy T_L will be lost to the supporting structure.

As a first approximation, energies T_1, T_E to first-crack matrix stress, T_s
and T_L may be assumed to be the same for both reinforced and unreinforced
specimens. The increase in impact toughness due to fibre addition will
therefore be T_p and the increase in T_E above that at which the unreinforced
matrix cracks.

From eqns (9), (10) and (12), it can be seen that low modulus fibres that
also increase the first-crack and ultimate strength of the material in flexure,
will give significant increases in toughness up to first-crack, and even greater
increases in overall impact toughness.

The same theory is applicable to fracture toughness defined in terms of
the area under a force–deflection curve in flexure, except that energies T_1
and T_L no longer apply.

Absorption of Water

The effect of water absorption by the fibres depends on when the fibres
are wetted, how rapidly they absorb the water, and the effect of this
moisture on the fibre properties.

If the fibres absorb water so slowly that they continue to extract water
from the concrete whilst it is curing, so drying it prematurely, then the
resulting concrete will be weak and porous.

Concrete that is weak and porous will also result from rapid fibre absorption of water if this takes place before mixing with the cement. The fibres will have a moisture content in excess of the equilibrium content of fibres in the mix. Water will therefore flow out of the fibres into the wet mix, pushing cement away from the fibre so that the interfacial bond is weakened, and giving the matrix around the fibres a very high water–cement ratio.

If, on the other hand, fibres are added dry to the mix and then rapidly absorb water, they can improve the properties of the final composite. The fibres attract cement to give a strong interfacial bond, and lower the water–cement ratio of the mix surrounding the fibres. As the matrix begins to dry out during curing, the fibres themselves will act as a water reservoir tending to maintain the moisture content of the matrix surrounding them. The matrix strength and permeability will therefore be improved by the addition of fibres, expecially if they also restrict the growth of microcracks into visible cracks.

The cracking of bamboo reinforced concrete as the bamboo swells on taking up moisture, and the decrease in interfacial bond as it shrinks, have received considerable attention.[9] The smaller the inclusion size, the smaller the effect will be. Thus it is less likely to be important for single fibres, though it may be important for fibre bundles or ropes. If the fibres absorb water rapidly, they will be saturated, and thus fully swollen, during mixing, and thus will not swell to crack the matrix when the material is soaked after curing. The fibres may shrink, and so have a decreased bond on drying. Alternatively, the bond may be weaker when the fibres are wet and have a lubricating layer of water at the interface.

Fungal Decay and Termite Attack

Natural organic fibres of vegetable origin consist of cellulose molecules cross-linked with lignin, and are susceptible under normal conditions to fungal decay and termite attack. Should this decay occur within the matrix, then the composite would finally consist of just matrix with fibre-shaped voids. The strength and elastic modulus of the composite would then be, respectively

$$f_c = (1 - v_f) \cdot f_m \qquad (13)$$

and

$$E_c = (1 - v_f) \cdot E_f \qquad (14)$$

Since v_f is relatively small for such composites, the reductions in strength and modulus below those of the matrix would also be relatively small.

On the other hand, the presence of the cement could inhibit fungal and termite attack. At the same time, there would be the possibility of alkaline attack, as with glass fibres. Any resulting decay in fibre strength would only become important when it led to the stage of fibres failing prematurely, prior to pull-out. Fibre embrittlement, leading to a reduction in toughness, could also be a problem.

Variability in Fibre Properties

This variability inevitably means that products cannot be made from natural fibres with close tolerances of strength and stiffness. The variability in length, thickness and stiffness of the fibres is more important than the strength, the latter only being relevant when it leads to fracture of the fibres prior to pull-out. Since natural organic fibres are cheap and readily available in rural areas of developing countries, they are likely to be employed for labour intensive processes in which close tolerances would be impossible to achieve however uniform the fibre properties.

EXPERIMENTAL EVIDENCE

To carry the discussion a stage further, we consider the evidence gained from experiments on a particular natural organic fibre—sisal.

Effect of Low Young's Modulus

To investigate this effect, flat roofing tiles were made with dimensions 430 mm × 200 mm × 9 mm using equal parts by weight of cement and sand, with a water–cement ratio of 0·4. Half of these tiles were reinforced with sisal, adding 1·6 % by volume as 25 mm lengths added dry to the mix, and laying 1·2 % by volume the length of the mould to form a central aligned fibre layer sandwiched between layers of mortar.

Eight of each type of tile were tested in flexure over a span of 370 mm, with readings taken of centre deflection up to first crack in order to determine the initial Young's modulus E of the material. A further eight of each type of tile were impact loaded over the same span with a 1·75 kg metal sphere that was dropped from increasing heights of 5 mm up to 50 mm and every 10 mm thereafter up to 'failure'.

In the case of sisal reinforced specimens, there was no sudden ultimate failure or maximum load either in flexure or impact. The tiles merely curved to such small radii of curvature that they touched the base of the test rig. This in itself was a demonstration of the high ductility imparted by the low modulus fibres. It was, however, noted that once a central deflection of around 60 mm had been reached, small increases in flexure load, or repeated impacts, caused very large increases in deflection. Consequently, 'ultimate failure' was defined for the purposes of this experiment as the load or impact needed to give a centre deflection of 60 mm.

The results of these tests are shown in Table 1. It is evident that sisal fibres give considerable increases in both flexural strength and impact toughness, the increases already being to some extent evident at first cracking.

Before comparing the results with the theory outlined in the previous section, two further sets of tests were carried out. Firstly, a further set of tiles were impact loaded over shorter spans. This had no significant effect on the results obtained. Secondly, the impact procedure was altered. The sphere was dropped repeatedly from 15 mm above some of the reinforced tiles (that is to say, from approximately the height needed to crack unreinforced tiles). For other sisal–cement tiles, the sphere was dropped repeatedly from a height of 45 mm. It was found that, in all cases, the point at which cracking, ultimate failure, and even observable damage on the top surface, occurred was determined by the total energy summing over all impacts.

Both of these results support the assumption inherent in the theory that the energies absorbed are important rather than the impact force. The fact that the energy absorbed did not appear to increase with span length as

TABLE 1
Results of flexure and impact tests on flat roofing tiles

Flexure specimens	First-crack strength (MN/m^2)	Modulus of rupture (MN/m^2)	Initial Young's modulus E (GN/m^2)
Plain mortar	3.7 ± 0.9	3.7 ± 0.9	34.6 ± 5.4
Sisal–cement	4.1 ± 0.5	6.2 ± 1.1	33.6 ± 6.1

Impact specimens	Height of drop for initial cracking (mm)	Height of drop to cause failure (mm)
Plain mortar	16 ± 5	16 ± 5
Sisal–cement	25 ± 4	469 ± 125

predicted by eqns (9) to (12) may be due to the fact that for short spans, more energy was lost to the surroundings, sufficient to compensate for the lower energy absorbed by the specimen.

To compare the results of the impact test with the proposed theory, it was assumed that for the full span of 370 mm, energy T_L lost to the surroundings was always zero, whilst surface energy T_s was the same for all specimens, and could therefore be deduced from the toughness of the plain mortar specimens. This gave a value for T_s of 0·133 J (contributing 37 J/m^2 to the 'impact toughness' values shown in Table 2). As indicated, the energy from the falling weight was summed over successive impacts, and the components T_1 and T_E given by eqns (7) and (9) deducted for all impacts except, in the case of T_E, the final impact. The values of E_1 and E_2 were deduced from the flexure test results and by assuming the value for E_f of 13·2 GN/m^2 found from earlier tests[10] to be correct. This gave the values for E_1 and E_2 of 0·26 GN/m^2 and 13·2 GN/m^2 that were then used in eqns (9), (10) and (12).

The results of this analysis are shown in Table 2. There is seen to be full agreement with the experimental results for first cracking, and the result for

TABLE 2
Theoretical values of 'impact toughness' for reinforced tiles

Failure criterion	Predicted height of drop to cause 'failure' (mm)	'Impact toughness' up to 'failure' (J/m^2)
Initial cracking	25	198
60 mm centre deflection	340	13 700

final failure lies just outside the scatter of the experimental results (probably because the assumption that T_L was zero was invalid for these larger impacts). Since the value for impact toughness of the plain mortar derived using this theory was 45 J/m^2, these results represented more than a four fold increase in toughness up to first-crack and more than a 300 fold increase in toughness up to 'failure'. Even this ignores the energy that would subsequently be required to tear the tile into two pieces.

Recent tests on beams containing around 15 % of sisal fibres have shown that increases in flexural strength by more than a factor of four are possible (i.e. above 20 MN/m^2). On the other hand, as shown by earlier experiments,[10] sisal fibres can give reductions of up to 20 % in compressive strength and direct tensile strength.

Thus low modulus fibres can be used to give modest improvements in first-crack flexural strength, and significant increases in modulus of rupture, provided accompanying small decreases in compressive and direct tensile strength are acceptable.

Absorption of Water

The results of earlier tests on sisal fibres indicated that, when sisal fibres are immersed in water, they absorb, within 5 seconds, 67 % of their own weight of water, with no further detectable increase with time.[10] If we may assume that the absorption of water from the mix is extended to minutes rather than seconds, then fibres added dry to the mix should give increases in strength and decreases in permeability of the final composite. It was shown in an earlier paper[10] that adding wet fibres to the mix had an adverse effect on the composite.

Increases in composite flexural strength resulting from sisal fibre addition have already been noted, and some of this improvement could be due to the absorptive nature of the fibres. A further series of tests was carried out to investigate the effect of sisal addition on permeability.

Tiles were tested for air permeability during curing by forcing air through them at a pressure of $4 \, kN/m^2$ using a 50 mm diameter funnel sealed against the surface of the tile. Twenty tiles were made containing equal parts by weight of sand and cement, half of them made with a water–cement ratio of 0·5, and the remainder with a water–cement ratio of 0·4. Half of those using the wetter mix were reinforced with 2 % of sisal fibres aligned along the tile and of length equal to that of the tile. Half of those using the drier mix were reinforced with 1 % of aligned sisal and the remainder with 3 % of aligned sisal. Measurements of airflow were carried out from the time of demoulding (24 hours) to the twentieth day of curing. The results of this test are shown in Figs 1 and 2.

Subsequently, 28 day old mortar specimens were tested for water permeability using the test rig of BS 690: 1963 and comparing the water loss from the reservoir with that from a control using a glass plate instead of the mortar. Three specimens had a cement:sand ratio of 1:3 with 1 % of 25 mm sisal fibres and 2 % of long sisal fibres, three had 1:4 cement:sand mix with the same fibre reinforcement, whilst a final three had a 1:4 cement:sand mix with 0·5 % of short fibres and 1 % of long fibres. The results of this test, carried out over a period of 28 hours are shown in Fig. 3.

It is evident from these results that sisal fibres can considerably reduce the permeability of a mortar, both against air and water. At least part of this effect may be due to the absorbent nature of the fibres.

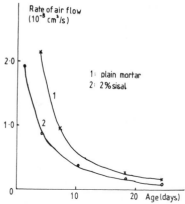

FIG. 1. Air permeability of 1:1 mortar of water–cement ratio 0·5, with and without sisal reinforcement, throughout the initial curing period.

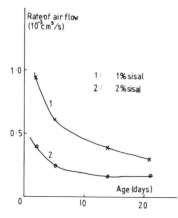

FIG. 2. Air permeability of 1:1 mortar of water–cement ratio 0·4, containing different fibre fractions of sisal reinforcement, measured throughout the initial curing period.

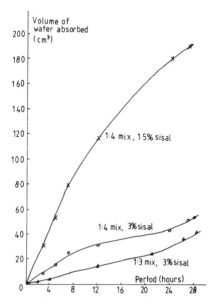

FIG. 3. Water absorbed by 28 day old mortar specimens reinforced with sisal fibres.

Fibre Degradation

To test for termite attack, cracked mortar specimens containing sisal fibres were placed in proximity of a termites' nest. Sisal fibres on their own were placed, as a control, in virtually identical conditions. Whilst the sisal fibre control was rapidly attacked and removed by the termites, the sisal in the mortar, even when visible in cracks, was not attacked.

Fungal decay was investigated using BS 1982:1968 extended to a period of 6 months, using specimens 10 mm thick containing equal parts of cement and sand and containing 3 % of sisal of length 25 mm and 6 % aligned along the specimen. The test was repeated using fibres presoaked for 24 hours in a 0·5 % solution by tri-*n*-butyltin oxide/dieldrin in kerosene. Specimens with and without sisal reinforcement were stored dry alongside the experiment as a control. After exposure, the specimens were tested for flexural strength and impact toughness, and the quality of any exposed fibres was assessed visually and by crushing or pulling with the fingers. The results of this test are shown in Table 3.

No reduction in first-crack strength or modulus of rupture was evident

TABLE 3
Results of fungal test

Specimens	Exposed fibre quality on scale 0–4	Flexure			Impact	
		First crack flexure strength (MN/m^2)	Modulus of rupture (MN/m^2)	Residual strength* (MN/m^2)	Fall to cause crack (mm)	Fall to cause failure (mm)
Unreinforced control	—	$4·4 \pm 0·8$	$4·4 \pm 0·8$	0	9 ± 5	9 ± 5
Reinforced control	4	$5·7 \pm 1·3$	$9·4 \pm 3·2$	$5·4 \pm 1·9$	19 ± 7	100 ± 50
Treated control	4	$8·6 \pm 2·7$	$9·6 \pm 3·0$	$5·8 \pm 2·1$	20 ± 8	117 ± 39
Untreated in soil	$0·4 \pm 0·5$	$8·8 \pm 2·7$	$9·3 \pm 1·9$	0	20 ± 5	40 ± 8
Treated in soil	$2·4 \pm 0·7$	$8·1 \pm 2·0$	$8·1 \pm 2·0$	0	19 ± 5	36 ± 14
Untreated in sterile 'soil'	$3·1 \pm 1·0$	$7·3 \pm 2·1$	$7·3 \pm 2·1$	0	20 ± 6	33 ± 10
Treated in sterile 'soil'	$3·5 \pm 0·9$	$9·2 \pm 2·0$	$9·2 \pm 2·0$	0	20 ± 5	44 ± 5

* 'Residual strength' = stress at centre deflection of 22 mm over span 120 mm.

among any of the exposed specimens. There was, however, a loss of 'residual strength' after cracking, and large decreases in both fracture toughness and impact toughness (although both remained well in excess of those of the unreinforced mortar). This embrittlement appeared to be due to a chemical embrittlement of the fibres rather than fungal attack, as all specimens were equally affected whether or not pretreated with the fungicide/insecticide, and whether in composted soil or sterile quasi-soil.

Tests on two year old and four year old roofing sheets have yielded similar results. On the other hand, specimens saturated in a curing room for one year showed no such degradation. This subject is therefore under further investigation.

RELEVANCE TO STRUCTURES UTILIZING NATURAL ORGANIC FIBRES

Because of the possibility of chemical degradation of the fibres, natural organic fibre cement composites should not be used in situations where a decrease in toughness with time could prove dangerous. They may, however, be used with confidence for structures which, because of their shape, only require high strength and toughness during construction. This applies to shell structures that are conical, domed, vaulted or involve large corrugations. The reduced flow characteristics of the fresh mix make the material suitable for plastering, and for moulding of flat sheets in a manner analogous to the moulding of asbestos–cement. The low permeability means that the composite can form an impervious plaster, and is suitable for grain storage bins, silage stores, water jars, biogas storage vessels and similar applications including non-structural roofing covers, such as tiles.

The high strength in flexure and initial toughness of sisal–cement, and the fact that it involves long, cheap fibres, means that structures can be formed by novel methods. For example, the grain store shown in Fig. 4 in its initial stages of construction, was made over a conical framework of sticks covered with polythene. The open sisal mesh was plastered with a thin layer of mortar containing chopped fibres, which was sufficiently strong and tough for the polythene-covered framework to be removed after a few days. The inside was then plastered so that the final structure has adequate long-term strength and toughness. Figures 5 and 6 show the construction of a house using sisal to strengthen and bind a sisal–cement plaster against a mud–brick wall. The building is roofed with sisal–cement roofing sheets.

FIG. 4.   Grain storage bin under construction using sisal net and sisal fibre reinforced mortar plaster.

FIG. 5.   Mud–brick wall of house with sisal fibres passing between mud bricks ready to receive sisal reinforced mortar plaster.

FIG. 6.   Finished walls after plastering. The roof is covered with sisal–cement roofing sheets.

Thus natural fibre–cement composites are already suitable for many structures in developing countries, and may be generally applicable to shell structures if the problem of alkaline attack of the fibres can be overcome.

ACKNOWLEDGEMENTS

The author wishes to express his indebtedness to the FAO/SIDA project on Rural Structures for Africa, The Kenya National Council for Science and Technology, ICIPE for assistance with the test on termite attack, Mr M. Okere who carried out the permeability and fungal tests, and Professor Smith of Nairobi University Civil Engineering Department for his general help and encouragement.

REFERENCES

1. AMERICAN CONCRETE INSTITUTE, *Fibre reinforced concrete*, Publication SP44, Detroit, Michigan, ACI, 1974.
2. THE CONCRETE SOCIETY, *Fibre-reinforced cement composites*, Technical Report 51.067, Concrete Society, London, 1973.
3. NEVILLE, A. (Ed.), *Fibre reinforced cement and concrete*, RILEM Symposium 1975 Proceedings, Construction Press, London, 1975.
4. SWAMY, N. (Ed.), *Testing and test methods of fibre–cement composites*, RILEM Symposium 1978 Proceedings, Construction Press, London, 1978.
5. HANNANT, D. J., *Fibre cements and fibre concretes*, John Wiley and Sons, Chichester, 1978.

6. ARNAOUTI, C. and ILLSTON, J. M., *Tests on cement mortars reinforced with natural fibres*, The Hatfield Polytechnic, 1980.
7. SWIFT, D. G. and SMITH, R. B. L., The flexure strength of cement-based composites using low-modulus (sisal) fibres, *Composites*, July 1979, 145–8.
8. SWIFT, D. G. and SMITH, R. B. L., The physical significance of the flexure test for fibre cement and fibre concrete. In: *Testing and test methods of fibre–cement composites* (Swamy, N. (Ed.)), Construction Press, London, 1978, 463–78.
9. FANG, Y. H. and FAY, S. M., Mechanism of bamboo–water–concrete interaction. In: *Materials of construction for developing countries*, Volume 1 (Pama, R. P., Nimityongskul, P. and Cook, D. J. (Eds)), Asian Institute of Technology, Bangkok, 1978, 37–48.
10. SWIFT, D. G. and SMITH, R. B. L., Sisal fibre reinforcement of cement paste and concrete. In: *Materials of construction for developing countries*, Volume 1 (Pama, R. P., Nimityongskul, P. and Cook, D. J. (eds)), Asian Institute of Technology, Bangkok, 1978, 221–34.

42

On the Possibility of Using Natural Fibre Composites

K. G. SATYANARAYANA, A. G. KULKARNI, K. SUKUMARAN,
S. G. K. PILLAI, K. A. CHERIAN AND P. K. ROHATGI

*Regional Research Laboratory (CSIR), Industrial Estate P.O.,
Trivandrum 695 019, Kerala, India*

ABSTRACT

The use of natural fibres, such as coir, banana, sisal and pineapple leaf fibres which are abundantly available in India, as polymer based composite materials has been examined in this paper. Tensile strength, percentage elongation, modulus, electrical resistivity and dielectric strength of these plant fibres and some of the mechanical and physical properties of natural fibre–polyester composites have been measured. The properties of untreated fibre–polyester composites are lower than either the polyester or the fibre pointing to the need for surface modification to improve bonding. The beneficial effects of surface treatments, such as copper coating on coir fibre, on its own properties and on the properties of coir–polyester composites have been described. Consumer articles like wash basins, mirror casings, chair seats, scooter boxes, slide projectors, voltage stabilizer tubes, crash helmets and roofing materials have been prepared out of natural fibre–polyester composites.

INTRODUCTION

Natural and synthetic fibres and their products such as yarns, cords, tapes and cloths are widely used as engineering materials in various industries like the textile, rubber, laminating, building, electrical and chemical industries and in aerospace applications. Natural fibres like coir, banana, jute, etc., form large renewable resources in many countries, particularly in

TABLE 1(a)
Comparative prices of typical natural and synthetic fibres

Fibre	Cost/kg ($)
Carbon	220·00
Stainless steel	56·00
Glass	5·00
Pineapple leaf	0·75
Banana	1·50
Palmyra	1·00
Sisal	0·75
Coir	0·50

developing countries like India. These fibres form one of the low energy, renewable materials (Table 1(a)) and are receiving increasing attention as composites in polymers,[1-3] cement,[4-6] clay and rubber matrices. This is due to two reasons: (i) although the natural fibres have relatively poor mechanical properties in comparison to synthetic fibres, they have low density, low cost and are low energy materials, and (ii) the shortage of nonrenewable resources of reinforcements like glass, boron, carbon, nylon and other synthetic fibres, inherent high cost of their production and their toxic nature. Table 1(b) gives the availability of some of these important natural fibres in India and the world. However, one of the handicaps in the utilization of a vast resource like natural fibres is the lack of precise scientific information on structure and properties of these natural fibres, their compatibility with various matrices and properties of the composites based on these fibres. The other factors which contribute to the nonutilization of natural fibres as composites particularly in polymeric matrices are low and

TABLE 1(b)
Annual fibre production (in tonnes) of some of the natural fibres (1979)

Fibre	India	World
Coir	160 000	282 000
Banana	163·20	100 296
Sisal	3 000·00	600 000
Palmyra	100·00	not known
Pineapple leaf	not estimated	

variable strength and moduli, poor resistance to weathering, and lack of wettability of the fibres with polymers.

The literature available on the properties of various natural fibres like coir, banana, sisal, etc., is very inadequate.[7] Some literature is available now[1-3] on the utilization of vegetable fibres as engineering materials particularly with polymeric matrices. In view of quantity available, cost, physical and other properties and final commercial forms, jute is the only fibre which has been considered as most suitable for use as reinforcement in FRP composite.[8] Cotton fabric–phenolic resin composites have been used as bearings in place of phosphor bronze in the roll necks of steel and nonferrous rolling mills resulting in an energy saving of up to 25%.[1]

Laminates by hand lay-up and winding of cylinders with longitudinal or helical and hoop reinforcements have been successfully prepared using sisal fibre–epoxy resin composites.[2] These composites have been found easy to fabricate and the cost of production of the composites was comparable to fibreglass products. Also, the specific properties of these composites were found to be nearly the same as that of glass–epoxy composites.

Strength of bagasse fibre–formaldehyde composites has been evaluated for varying fibre–matrix combinations.[3] It has been found that 80–90% volume fraction of fibre with formaldehyde can be used as crack arrester.

In this paper we report the physical, electrical and mechanical properties of coir, banana, sisal, pineapple leaf and palmyra fibres, and composites based on coir and banana fibres with polyester resin. Some of the consumer articles like crash helmets and roofing materials made of coir with polyester resin and scooter boxes, mirror casings, stabilizer casings, etc., made of banana fabric incorporated into polyester which were successfully made are also described. Also, future work which is essential for better utilization of these natural fibres in composites is listed.

EXPERIMENTAL METHODS

Coir fibre which is generally used by the coir industry was brought from Kovalam, near Trivandrum in India. Banana, sisal, pineapple leaf fibre and palmyra fibres were supplied by Khadi and Village Industries Commission which is a rural development organization in Trivandrum, India. All these fibres are mostly used by fibre based industries for preparing products like mats, mattings, bags, purses and other ornamental articles.

The fibres were thoroughly cleaned and dried before subjecting them to any testing or preparing composites. Then, the fibres were sorted out for measurement of fineness and size. Density of the fibres was determined using a specific gravity bottle with toluene solution, while moisture content of the fibres was determined using a moisture balance. Electrical resistivity of the fibres and composites was determined on 100 mm length between 100 V–1000 V d.c. voltage using a million megohmmeter (Model RM-160 MK IIIA) while dielectric strength of fibres and composites and dielectric constant were measured at different conditions using insulation breakdown tester supplied by M/S BPL (India) Ltd: 2·5 mm thick composites were used for these measurements. X-ray studies on fibres explained elsewhere[9] were made through transmission Laue photographs using copper K radiation with the fibre axis kept perpendicular to the incident beam.

Composites were prepared in two different ways. In one case, mats of coir (9 % wt) or banana fibre–cotton fabric (11·44 % wt) were introduced in polyester resin with 2 % MEK paroxide as the hardener and 2 % cobalt naphthanate as the catalyst. The laminates were prepared by hand lay-up. A slight pressure was applied to keep the fibres in position whenever required.

In the other case 0·1 m lengths of coir fibres (raw as well as coated with copper up to 1·5 μm thick)* were aligned unidirectionally by stretching the fibres in a mould of dimension $0·1 \times 0·015 \, m^2$ and a predetermined quantity of resin mixed with hardener and catalyst as before was poured into the mould. In this case also, pressure was applied and samples were post cured at 80 °C for 24 h.

Mechanical properties of fibres and composites were determined using a 10 ton Instron Testing Machine at a strain rate of 2 cm/min. A 0·05 m gauge length of specimen was used in all cases. Testing of composites was carried out as per ASTM Standards. The fibres were conditioned at 60–65 % RH and 25°–30 °C for 1–2 days prior to testing.

RESULTS AND DISCUSSION

Table 2 lists size, density, major chemical constituents, volume resistivity, dielectric strength, ultimate tensile strength (UTS), Young's modulus, percentage elongation and microfibrillar angle of coir, banana, sisal,

---

* Details of coating of copper on coir fibres are explained elsewhere.[10] In summary, copper coating on coir was obtained by activating the fibre surface with NaOH-HCHO/ammoniacal silver nitrate solution and then depositing copper from Fehling's formaldehyde solution. Two different thicknesses (1·5 μm and 5·0 μm) of copper were obtained on the fibres by varying the concentration of coating solution.

TABLE 2
Physical and mechanical properties of fibres

Fibre	Width or diameter (m)	Density ×10⁻³ (kg/m³)	Volume resistivity at 100 V × 10⁵ (Ωcm at 65% RH)	Dielectric strength for 0·1 m length of the fibre after oven drying at 105°C	Micro-fibrillar angle (θ)	Cellulose/Lignin (%)	Moisture content (%)	UTS (MN/m²)	Modulus (GN/m²)	Percentage elongation (0·05 m GL)
Glass	2·5–10	2 540	6 × 10⁷–1 × 10¹¹	—	—	—	—	827·6–1724	68·96	4·8
Carbon	6–10·5	1 780–1 980	—	—	—	—	—	1700–2410	180–415	—
Jute	25–12	1 450	—	—	7–9	63/12	12	533	2·5–13·0	1–2
Coir	100–450	1 150	9–16	5 kV	30–49	32–43/40–45	10–12	131–175	4–6	15–40
Banana	80–250	1 350	6·5	5	11	63–64/5	10–12	529–754	7·7–20·8	1·8–3·5
Sisal	50–200	1 450	0·48	5	18–22	66–72/14–10	11	568–640	9·4–15·8	3–7
Pineapple leaf	20–80	1 440	0·77	5	14–80	84–5/12–7	Highly hygroscopic	413–1 627	34·5–82·5	0·8–1·6
Palmyra	70–1 300	1 092	1·00	4·5	30	—	—	95–220	3·3–7·0	3·2–11·2

pineapple leaf fibre and palmyra fibre. For comparison the properties of jute, glass and carbon fibres are also included in this table. Figures 1((a) and (b)) and 2 show typical stress–strain diagrams of various natural fibres while Fig. 3 shows the microstructures of these fibres. As can be seen from Table 2, these fibres have low and varying strength (95–800 MN/m^2), low modulus (3–40 GN/m^2), but the volume resistivity (1–16 × 10^5 Ω cm) and dielectric strength values (5 kV) are comparable to those of insulating materials like wood and mica. The specific strength and modulus of these fibres are in the range of 0·13–0·41/m and 4–40/m respectively in comparison to the values for glass fibre of 0·5/m and 27·15/m respectively.

The percentage elongation of different natural fibres varies between 1 to 40 %, being highest for coir fibres. Similarly, the microfibrillar angle varies from 11–49°, the highest being that for coir. The chemical constituents listed indicate that the lignin content is low (5–15 %) for banana, sisal and pineapple leaf fibres while it is highest in the case of coir fibre (40–45 %). On the other hand, cellulose content is highest (63–85 %) for these fibres, the exception being coir (32–43 %).

The observed mechanical properties of these fibres seem to depend mainly on the microfibrillar angle and cellulose content, while the values of electrical resistivity of the natural fibres correlate strongly with the cellulose content and moisture content of these fibres. Since the details of microstructure of these fibres differ from species to species (Fig. 3), properties of the fibres may be influenced to some extent by the structures. In fact, in the case of coir, it has been found[11] that the modulus and percentage elongation are related mainly to the microfibrillar angle, while the strength of the fibres seems to be influenced by the chemical composition, size and internal structure of the fibres and the presence of defects in the fibres.

The values of electrical resistivity and dielectric strength of the fibres suggests that all the fibres may be a satisfactory replacement for wood in insulating applications. The fibres will have special advantages over wood in that they can be readily pressed into complicated shapes through moulding. This is significant in view of the dwindling resources of wood which may in some applications be replaced by abundantly available natural fibres.

However, since the above properties of the fibres have been measured at 60–65 % RH and 25–30 °C, it would be worthwhile to measure these properties under various environmental conditions with a view to finding their suitability for use in different environmental conditions.

Table 3 ((a) and (b)) lists the density, volume resistivity, surface

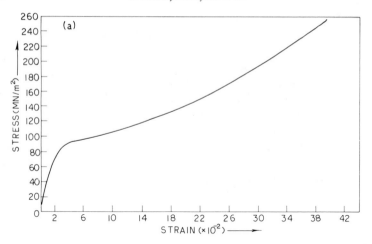

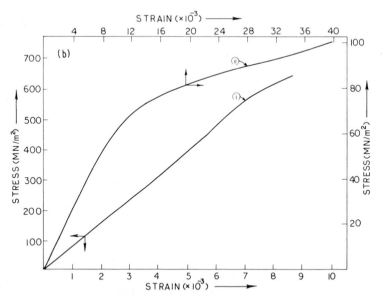

Fig. 1.   Stress–strain diagrams of natural fibres: (a) coir fibre; (b) (i) sisal fibre, and (ii) palmyra fibre.

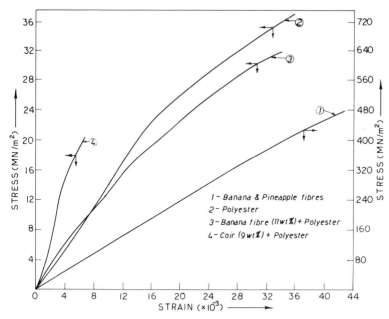

Fɪɢ. 2.   Stress–strain diagrams of banana, pineapple leaf fibre, polyester and natural fibre–polyester composites.

Tᴀʙʟᴇ 3(a)
Physical and mechanical properties of unidirectional coir fibre/polyester composite

Material	UTS (MN/m²)	Flexural strength (MN/m²)	Volume resistivity at 100 V Ωcm
(Mould) Polyester	49·6	52·2	$1·35 \times 10^{11}$
As received coir–polyester (0·23 vf)	45·9	56·2	$1·23 \times 10^{10}$
Copper coated coir–polyester (0·23 vf)	56·9	69·8	$9·93 \times 10^{-2}$ [a] $17·16$ [b]
Copper coated coir–polyester (chopped fibre with aspect ratio of 100)	—	—	$1·77$ [a] $63·13$ [b]

[a] Longitudinal.
[b] Transverse.

TABLE 3(b)
Physical and mechanical properties of composites

Property	Polyester resin	Glass fibre reinforced polyester resin (fibre/fabric content not mentioned)	Cotton fabric reinforced polyester resin	Banana fabric reinforced polyester resin (11% wt fabric) (warp) (cotton in the termida)	Banana fabric reinforced polyester resin (11% wt fabric) (weft)	Coir reinforced polyester resin (9% wt fibre)
Density (kg/m^3)	1 300	1 500–1 900	1 400	1 215	1 215	1 160
Strength (MN/m^2)						
tensile	41·38	241·4–689·6	34·5–689·6	35·92	27·96	18·61
flexural	89·69	344·8–862·1	62·1–124·1	50·60	64·00	38·15
Modulus of elasticity (GN/m^2)	2·06	6·9–41·38	2·76–4·14	3·33	3·34	4·045
Impact resistance (unnotched) (kgm/m^2)	77·5	3 116–8 476	257·3–428·8	748·5	329·2	391
Water absorption (24 h room temperature) %	0·21–0·40	0·2–1·0	0·8	1·93	1·36	1·36
Volume resistivity (Ω cm) or surface resistance (Ω/25 cm^2) (at 100V d.c.)	1 000 kM	—	—	400 kM	400 kM	
Dielectric strength (an 2·5 mm thickness)	10 kV/min	—	—	10 kV/min	10 kV/min	
Dielectric constant (at 1·5 MHz)	3·04	—	—	3·5	3·5	3·14

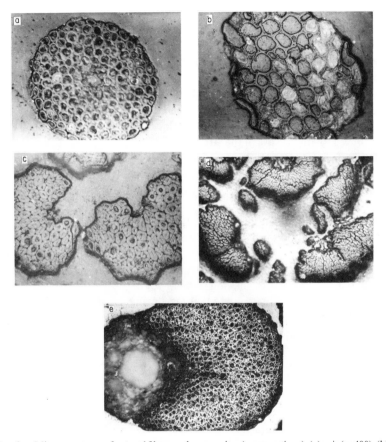

FIG. 3.   Microstructures of natural fibres and composites (cross sections). (a) coir (× 400), (b) banana (× 400), (c) sisal (× 400), (d) pineapple (× 320), (e) palmyra (× 100).

resistance, dielectric strength, dielectric constant, UTS, Young's modulus, percentage elongation, flexural and impact strengths of composites based on coir and banana–cotton fabric with polyester resin. For comparison some properties of glass fibre reinforced plastic and cotton fibre reinforced polyester composites are also given.[12] Figure 2 shows typical stress–strain diagrams of polyester and various composites studied in the present investigation. It can be seen that the composites have lower strength than either the fibres or the polyester. This could be due to lack of bonding between the fibre and polyester resin. Also, the volume percentage of the fibres used in the present investigation may be lower than the critical amounts above which the rule of mixture begins to apply. However, natural fibre composites have a much lower density (Table 3) and higher electrical resistance and hence they are likely to be more insulating to electricity and sound. It can also be seen from Table 3(a) that there is no significant difference in the properties when raw coir fibres are incorporated in polyester. However, when 23 % (volume) of copper coated coir fibres are incorporated in polyester there is an increase of 15 % in UTS and 34 % flexural strength over that of the polyester matrix. Some of this can be attributed to an increase in the strength of coir fibre due to copper coating. This suggests that copper coatings can make coir fibres perform as reinforcements instead of mere fillers when used with polyester resin.

On the other hand, copper coating of coir fibres leads to a marked change in the resistivity of the composites. While the resistivity of composite with raw coir fibre (unidirectionally aligned) was $1 \cdot 23 \times 10^{10} \, \Omega \, cm$, the resistivity of composite with copper coated fibre (23 vol % unidirectionally aligned) decreased to $1 \cdot 23 \times 10^{-2} \, \Omega \, cm$ in the direction parallel to the fibre and $17 \cdot 6 \, \Omega \, cm$ in the direction perpendicular to the fibre.

Similarly, composites having lower volume fraction of randomly oriented chopped copper coated fibres having an aspect ratio of 100, showed a decrease in resistivity much below the value required for electromagnetic interference shielding and discharging of static electricity like metal/polymer composites.[10]

It has been reported in an earlier paper[10] that the propagation of flame was approximately at the same rate (100 mm in 40 secs) in bundles of raw coir fibres and $1 \cdot 5 \, \mu m$ copper coated coir fibre, the only difference being that the copper coated fibre bundle did not show any after-glow as observed in the case of raw fibre bundle. On the other hand, a $5 \, \mu m$ thick copper coated fibre bundle did not show either the propagation of flame or the after-glow. This could be due to the fact that $5 \, \mu m$ copper coating conducts the heat away and imparts flame retardancy. This study also suggested that when the copper coating on coir fibres exceeds a certain threshold, the coir fibres can become flame retardant.

From Table 3(b) it can be seen that the modulus values of composites studied are higher than those of polyester resin. The specific modulus values of composites are in the range of $2 \cdot 7 - 3 \cdot 5/m$ while that of polyester resin and GRP are in the range of $4 \cdot 6 - 27 \cdot 6/m$ respectively. The surface resistances of composites containing banana fabric and coir fibre were $400 \, k\Omega$ in contrast to $1000 \, k\Omega$ for polyester only at 100V d.c. This decrease in resistance is expected according to rule of mixture since fibres have lower resistance (Table 2) than the polyester. On the other hand, values of both

FIG. 4. Laminates and various consumer products made of coir fabric–polyester and banana fabric–polyester. (a) Laminates, (b) roofing, (c) mirror casing, (d) seat covering, (e) projector box, (f) voltage stabilizer covering and crash helmet.

dielectric strength (10 kV d.c./min) and dielectric constant (3–4·5 at 1·5 MHz) were similar for polyester and composites based on coir and banana fibre fabric with polyester.

A number of consumer items (Fig. 4(a)–(f)) have been fabricated using coir mats or banana fibre–cotton fabric with polyester resin. In the case of coir fibre some difficulty was faced during fabrication since the fibres tended to spring up. Some pressure had to be used in this case to keep the mat intact. Laminates were however easily prepared using chopped fibres. On the other hand, the fabrication process was easier, quite comparable to glass fibre, in the case of banana fabric–polyester composite. Also, the resin content used was similar in quantity in the case of banana fabric to that consumed by glass fibre fabric for similar fabrications. These components have shown no degradation when kept indoors (nearly 8–10 months in the case of coir–polyester composites and 4–5 months in the case of banana fabric–polyester composites). The roofing made of coir fibre sandwiched with glass fibre–polyester resin (Fig. 4(b)) has been in the outdoors for more than two months and no signs of degradation have been observed.

The present work therefore suggests that natural fibres can be used for fabricating laminates and consumer articles where high strength is not absolutely essential. However, the problem of wetting and bonding between fibre and polyester resin through surface modification of fibres needs further attention to reduce the consumption of polymer in these composites and to improve their mechanical properties.

FUTURE WORK

Future work on natural fibre reinforced composites could be done along the following guidelines:

(1) The experiments carried out so far in our laboratory suggest that it is necessary to characterize the properties of a wide variety of natural fibres under various environmental conditions in order to fully utilize these renewable resources as composites.

(2) The moisture absorption and resin absorption characteristics, compatibility of fibres with polymers, clay, cement and rubber require fundamental studies.

(3) Fundamental understanding of relationships among the microstructures of various natural fibres and their properties should be obtained to devise micromanipulations to further improve their properties.

(4) In the case of banana, sisal and pineapple leaf fibres, in addition to engineering data, textile characteristics should be evaluated with a view to blending these fibres with cotton or glass fibres to prepare hybrid composites for use in grain storage, water carrying troughs, false roofing and in the automobile industry.

(5) Relationships among the properties of individual fibres and those of yarns, cloth and composites made from these should be thoroughly investigated in order to develop optimum properties for any suitable application.

(6) Techniques for surface modification of these fibres, for instance by providing metal or polymer coatings on them, thermal or mechanical prestretching of fibres to change their microfibril angle, applying transverse and compressive stresses to change the shape of the fibres to ribbons and increase modulus, etc., should be developed in order that these natural fibres may not only be improved in strength properties but will also bond more easily with various matrix materials. Some of the coatings may also protect the fibres from environmental degradation. Metal/carbon (made from natural fibres) composites may find use in antifriction, highly conducting electrical contact materials with moderate arc resistance and in high speed rotating machines with no arcing but with high current.

(7) The high surface area of the carbon produced in natural fibres like coir, for example, may provide active sites to form refractory carbides and to suppress the formation of oxides like SiO_2, ZrO_2, TiO_2, etc. This phenomenon should be studied in detail to produce newer refractory composites and to open up new areas in high temperature materials.

(8) Short natural fibres of all types are generally considered waste materials. Attempts should be made to make chopped strand mats for use as reinforcements in plastics. These mats can be used along with small amounts of fibreglass material if necessary to make lightweight partition walls or chair seats, roofing, etc., where very high strength is not absolutely essential.

CONCLUSIONS

Stress–strain diagrams of some of the natural fibres like coir, banana and composites of these fibres with polyester are reported. It has been possible to incorporate natural fibres into polyester to prepare stable laminates and consumer articles like helmets, mirror cases and projector covers. Strength properties of the composites are lower than either the fibres or the polyester indicating lack of bonding between fibre and polyester. Development of surface treatments of natural fibres is necessary to improve wetting and bonding between the fibre and polyester. For example, copper coatings on coir fibres resulted in considerable increases in tensile and flexural strengths of polyester–coir composites.

REFERENCES

1. PARAMASIVAM, T. and ABDUL KALAM, A. P. J., On the study of indigenous natural fibre composites, Proceedings of 29th Annual Technical Conference, Reinforced Plastics/Composites Institute, The Society of Plastics Industry, Inc., USA, 1974, Sec. 4A, 1–2.
2. Save Energy—Save Money, Composites News, *Composites*, **10**, 1979, 61.
3. MCLAUGHLIN, E. C., The strength of bagasse fibre reinforced composites, *J. Mater. Sci.*, **15**, 1980, 886–90.
4. SMITH, D. G. and SMITH, R. B. L., The flexural strength of cement based composites using low modulus (sisal) fibres, *Composites*, **10**, 1979, 145–8.
5. COUTTS, R. S. P. and CAMPBELL, M. D., Coupling agents in wood fibre reinforced composites, *Composites*, **10**, 1979, 228–32.
6. CAMPBELL, M. D. and COUTTS, R. S. P., Wood fibre reinforced cement composites, *J. Mater. Sci.*, **15**, 1980, 1962–70.
7. SATHYA, C. R., Natural fibres of vegetable origin (unpublished work).
8. SATYANARAYANA, K. G., KULKARNI, A. G. and ROHATGI, P. K., Potential of natural fibres as a resource for industrial materials in the future of Kerala, *J. Indian Acad. Sci.*, (in press).
9. KALYANI VIJAYAN, SATYANARAYANA, K. G. and ROHATGI, P. K., X-ray studies of vegetable fibres (unpublished work).
10. PAVITHRAN, C., GOPAKUMAR, K., PRASAD, S. V. and ROHATGI, P. K., Copper coating on coir fibres, *J. Mater. Sci.*, (in press).
11. KULKARNI, A. G., SATYANARAYANA, K. G., SUKUMARAN, K. and ROHATGI, P. K., Mechanical behaviour of coir fibre under tensile load, *J. Mater. Sci.*, (in press).
12. SHAND, E. B., *Glass Engineering Hand Book* (2nd Edn), McGraw-Hill Co. Inc., New York, 1958, 416–36.

43

Stress Intensity Factor Measurements in Composite Sandwich Structures

I. ROMAN,* H. HAREL† AND G. MAROM†

* Materials Science Division.
† Casali Institute of Applied Chemistry.
School of Applied Science and Technology,
The Hebrew University of Jerusalem, 91904 Jerusalem, Israel

ABSTRACT

In composite materials, the stress intensity factor is significantly affected by the reinforcement geometry. This geometry affects the degree of anisotropy of the material. It is maintained that in order to obtain valid stress intensity factor values, the compliance calibration procedure presented should be carried out for every new reinforcement geometry.

In complex composite structures such as the sandwich control surface the relative position of the notch tip has a significant effect on the fracture toughness. This position defines the microstructure encountered by the crack front and consequently determines the fracture mechanism that operates in the damage zone ahead of the notch tip.

In many cases, the fracture process in sandwich structures containing skin surface notches is confined to delamination in the skin at the vicinity of the notch tip, followed, ultimately, by splitting at the adhesive layer between the skin and the core.

In such cases, the fracture stress of the sandwich structure can be calculated from the fracture toughness of the skin and the relevant Y value is that obtained from the new Y polynomials derived for the skin and from the corresponding ratio of notch depth to skin thickness.

1. INTRODUCTION

The recent applications of composite structures in aviation generated an urgent need for reliable durability and damage tolerance assessment procedures. Such procedures would, in fact, provide rejection/certification criteria for either new or in-service damaged parts. As a particular example, let us consider a control surface comprising a honeycomb core and angle-ply composite skins. Common damage modes or defects experienced by this structure are delamination within the composite skin, surface scratches or cracks, and holes. Unless a reliable failure criterion is available it is impossible to determine whether or not such types of detectable damage are critical.

Currently, a statistical approach utilizing the distribution of micro-damage in the reinforcing phase is developing as a basis for failure prediction methods. Although such an approach might prove very useful in cases of stress rupture controlled by such distribution of microdamage, it would be fruitless when some macroform of stress concentration is present in the structure. Consequently, the option of applying a fracture criterion based on LEFM was given priority.

Although LEFM and its fracture criterion—the critical stress intensity factor—have been used in research with composites for quite some time now, its applicability is not yet commonly accepted. The present study has been undertaken to provide an experimental certification of the applicability of the concept to a composite control surface.

This structure was constructed of angle-ply composite skins and a composite honeycomb core as shown in Fig. 1. *The research was confined to cases of skin damage such as surface scratches or notches*, and was aimed at providing a tool for determining when such forms of detectable damage became critical.

We have been aware of numerous reports on the successful application of LEFM to composites[1] and there have been other favorable indications for adopting it. These are summarized as follows:

(1)   Reasonable agreement exists between different measurements of K_{Ic} reported in many studies, in spite of the various test specimens employed and the different K-calibration functions, and in spite of the numerous reinforcement geometries.[2]

(2)   General resemblance (analogy) is noticed between the plastic zone at the crack tip of homogeneous materials and the damage zone—also termed the debonding zone—in composites.[3,4]

(3)   When this general analogy is examined more closely it is seen that the calculation of the size of the damage zone is possible with good accuracy using plastic zone size expressions derived from LEFM.[5]

(4)   When values of K_c are converted to fracture surface energies

FIG. 1. The sandwich structure. (a) General view. (b) A close-up of the honeycomb with part of the skin removed.

$(\gamma = K_c^2/2E)$, these energies are correlated reasonably well with values calculated for the micromechanical processes which occur during the fracture.[1]

2. BACKGROUND

In view of the above observations it was decided to try to adopt LEFM. Customarily, the critical stress intensity factor is determined by the fracture criterion expressed by the equation:

$$K_c = Y\sigma_F c^{1/2} \tag{1}$$

where σ_F is the fracture stress, c is the notch depth, $Y = Y(c/d)$ is the K-calibration function and d is the specimen thickness. Alternatively, K_c may be calculated through its relation to the fracture surface energy as follows:

$$K_c = (2E\gamma)^{1/2} \tag{2}$$

where E is Young's modulus and γ is the fracture surface energy determined from the fundamental relationship:

$$\gamma = -\frac{\partial U_F}{\partial A} \tag{3}$$

where U_F is the elastic energy stored in the material up to the point of maximum load and A is the total notch area.

When the LEFM approach is applicable the two methods should yield identical results.

Thus, as a preliminary examination of the applicability issue, a set of experiments with a simple composite material (unidirectionally reinforced) was conducted.[6] This preliminary examination showed a significant discrepancy between K_{Ic} values calculated by the two methods. It was proposed that this discrepancy could be eliminated by the use of new K-calibration polynomials derived specifically for composites. The reasoning for the need for new K-calibration polynomials stems from the fact that the original calibrations were derived with homogeneous isotropic materials and might be inadequate for the anisotropic composites.

It is recalled that in the isotropic case $K = f(\sigma, c)$, where σ and c are the stress and the notch depth and the function f depends on the configuration of the cracked homogeneous material and on the way in which the load is applied. To extend the applicability of the stress intensity factor concept to composite materials it was proposed that the reinforcement be taken into account in the form $K = f(\sigma, c, g)$, where g is a parameter manifesting the nature and geometry of the reinforcement. The equation actually implies that composite materials should have different K-calibration functions, depending on the geometry of reinforcement, implying that, in composites, the K-calibration functions are anisotropic.

3.   THE EFFECT OF ANISOTROPY ON K-CALIBRATION

The effect of anisotropy on K-calibration was examined in three-point bending with a series of glass fibre-reinforced composites of different geometries, described in Table 1.

The computation of the K-calibration polynomials was based on the combination of eqns (1) and (3), resulting in the following expression for Y:

$$Y = (2E\gamma/c)^{1/2}/\sigma_F \tag{4}$$

This calibration procedure yielded different Y functions for the four

TABLE 1
Materials and modes of loading

Material	Reinforcement geometry	Volume fraction (%)	Loading mode
I	Unidirectional	41·5	Translaminar
II	±45°	36·0	Translaminar
III	±45°	37·5	Interlaminar
IV	Satin 181	48·5	Interlaminar

TABLE 2
The coefficients of the K-calibration function

Material	Coefficients				
	A_0	A_1	A_2	A_3	A_4
I	3·87	−20·89	65·53	−87·70	49·16
II	7·58	−63·95	287·11	−541·12	402·05
III	2·80	−7·09	62·36	−181·42	200·61
IV	4·65	−34·10	144·92	−275·64	201·51

TABLE 3
New Y values compared with Srawley–Brown (SB) values

c/d	Y				SB
	Material I	Material II	Material III	Material IV	
0·1	2·35	3·56	2·55	2·43	1·80
0·2	1·69	2·59	2·75	1·74	1·80
0·3	1·53	2·88	3·01	1·65	1·92
0·4	1·64	3·60	3·47	1·71	2·16
0·5	1·92	4·87	4·71	1·97	2·58
0·6	2·35	7·79	7·81	2·94	3·32

reinforcement geometries of Table 1. The Y functions were expressed as fourth degree polynomials, and Table 2 shows the new coefficients derived for materials I to IV. The new Y values, calculated with the new coefficients, are presented as a function of the relative notch depth in Table 3, where they are also compared with the Srawley–Brown values for three-point bending.

Finally, as an example, Fig. 2 compares K_{Ic} results obtained for materials

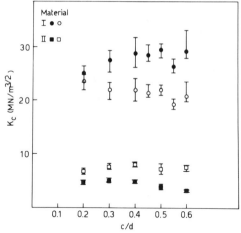

FIG. 2. K_{Ic} results for materials I and II calculated with the Srawley–Brown function (closed symbols) and with the new polynomials (open symbols).

I and II with the Srawley–Brown Y values with those obtained with the new Y values. The new results are generally different and, most important, the trend in c/d is reduced. Thus, it can be concluded that in composite materials the stress intensity factor is significantly affected by the reinforcement geometry. A K-calibration procedure should be carried out for every reinforcement geometry to obtain valid stress intensity factor values.

4. K-CALIBRATIONS AND K_{Ic} MEASUREMENTS IN LAMINATES

Following the procedure and conclusions outlined above, a K-calibration procedure was carried out for the angle-ply skins of the actual sandwich construction. Each skin was a glass fibre-reinforced epoxy laminate of the following composition and structure. An angle-ply of $[0°/0°/90°/\pm45°/90°/0°/0°]$ was manufactured from prepregs of fibreglass cloth and epoxy resin type F 161 (Hexcel Products). The $0°$ and $90°$ laminae consisted of a unidirectional cloth, type 1543, with 90% of the fibres aligned in the principal direction and the remaining 10% aligned in the transverse direction. The $\pm45°$ lamina consisted of a balanced cloth, type 1581. A vacuum bag/autoclave moulding procedure was used including 'bleeders' on both surfaces. The final fibre volume fraction was 46%.

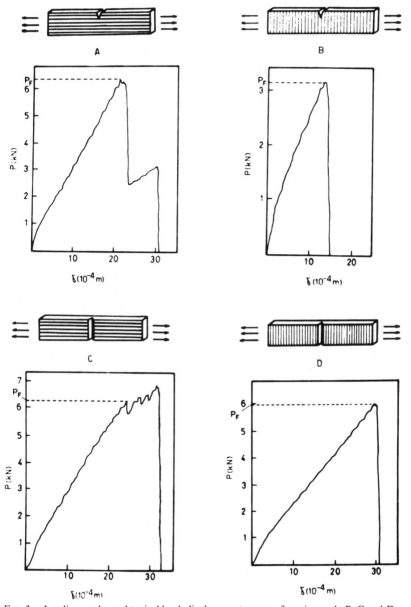

FIG. 3.   Loading modes and typical load–displacement curves of specimens A, B, C and D.

Two types of specimen of 100 mm × 20 mm × 2 mm were cut out of the plates. These were longitudinal and transverse specimens relative to the fibre direction in the outer laminae. A single V-type notch of depth in the range $0.1 \le c/d \le 0.6$ (notch-to-depth ratio) was introduced either across the wide, or across the narrow, specimen edge. The notch location, coupled with the fibre direction, produced four different combinations, as indicated in Fig. 3. Unnotched specimens were used to determine the modulus.

Testing in tension at a crosshead speed of 0·5 cm/min was carried out at ambient temperature. Unnotched and notched specimens were tested with M.T.S. and Instron machines, respectively (five specimens for each determination).

The new K-calibration functions, $Y(c/d)$, were determined and calculated by the same procedure as described above. Figure 3 presents the load–displacement curves obtained with the four specimen types; the load levels taken in the calculation of the fracture stresses are also indicated.

The strain energies to fracture, U_F, were calculated directly by integrating the load–displacement curves. Figures 4 and 5 present plots of the strain energies as a function of the notch area. It can be seen (Fig. 4) that for both longitudinal and transverse specimens with translaminar notches U_F is a linear decreasing function of A. In specimens with an interlaminar notch (Fig. 5), however, *two* distinct slopes are observed. The transition occurs at $A \simeq 30 \times 10^{-6}$ m^2, corresponding to a c/d ratio of about 0·4. The transition indicates a change in the active fracture mechanism. Examination of the laminate reveals that at $c/d \simeq 0.4$ the notch tip is in close

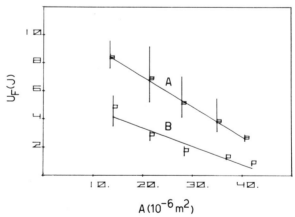

FIG. 4. Plots of strain energies as a function of the notch area of specimens with an edge (translaminar) notch.

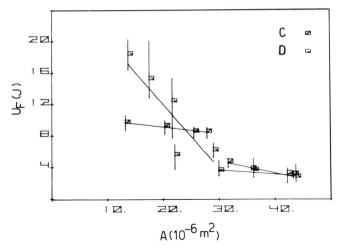

FIG. 5. Plots of strain energies as a function of the notch area of specimens with a surface (interlaminar) notch.

proximity to the $\pm 45°$ layer. Consequently, the fracture mechanism in the process zone ahead of the notch tip changes from either a pure transverse or a pure longitudinal mechanism (a transverse lamina or a longitudinal lamina is close to the $\pm 45°$ layer, respectively), to a mixed mode mechanism resulting in the observed slope variation. Fractographic studies of the fracture surfaces furnished experimental verification for the transition in the failure mechanism.

The fracture energy is higher for all the longitudinal specimens with translaminar notches when compared with the transverse specimens. In the case of interlaminar notches, the transverse specimens display higher fracture energies compared with those for longitudinal specimens up to $c/d \simeq 0.35$. The opposite is observed for $0.35 < c/d < 0.6$. This behaviour can be explained by the higher area fraction of longitudinal fibres intact in all the cases where a higher fracture energy was observed.

The values of the above-mentioned slopes ($\mathrm{d}U_F/\mathrm{d}A$), along with those of the modulus (E), of 20.1 ± 3.5 GPa for materials A and C and of 15.6 ± 1.5 GPa for materials B and D were used in eqn. (4). The results of $Y(c/d)$ for each one of the four specimen geometries were fitted with fourth degree polynomials whose coefficients are presented in Table 4. These new $Y(c/d)$ functions were utilized to determine the appropriate fracture toughness values of the four specimen types studied. The results are plotted in Figs 6 and 7 as a function of c/d and are summarized in Table 4.

TABLE 4

The coefficients of the new K-calibration polynomials and the resulting fracture toughness

Material	c/d	A_0	A_1	A_2	A_3	A_4	K_c (MPa $\sqrt{m}$)
A	0·20–0·60	9·78	−38·46	187·38	−364·17	262·50	92·6
B	0·20–0·60	7·38	−5·08	73·13	−167·50	137·50	63·5
C	0·20–0·40	54·51	−217·90	−465·49	4 259·32	−5 983·46	58·5
C	0·45–0·60	−4·07	26·34	81·58	109·25	−321·15	76·9
D	0·20–0·40	−1 122·58	17 433·07	−96 478·70	231 693·41	−202 933·40	159·5
D	0·45–0·60	199·24	−507·22	−1 372·31	5 546·59	−4 501·55	40·5

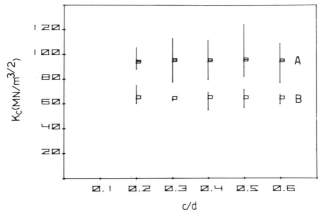

Fig. 6. Plots of critical stress intensity factors as a function of notch to depth ratio for specimens with an edge (translaminar) notch.

In the case of an edge (translaminar) notch (Fig. 6) K_c is constant over the entire c/d range examined, regardless of the fibre orientation, longitudinal or transverse. The longitudinal K_c (material A) is higher than the transverse (material B). In general, the edge notch results (materials A and B) are two-to-threefold higher than values available in the literature for glass fibre–epoxy resin angle-ply laminates (compare, for example, with reference 7). This conforms with the fact that the new $Y(c/d)$ values at $c/d = 0.2$ are about three times higher than the corresponding Srawley–Brown values.

In the case of a surface (interlaminar) notch (Fig. 7) K_c exhibits a

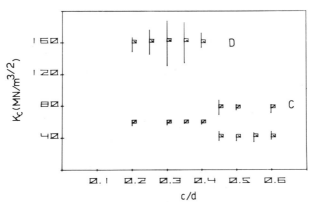

Fig. 7. Plots of critical stress intensity factors as a function of notch to depth ratio for specimens with a surface (interlaminar) notch.

discontinuity at $c/d = 0.4$ for both longitudinal and transverse configurations. This feature is anticipated in view of the U_F results discussed above. In the longitudinal configuration K_c in the c/d range of 0·20–0·40 is lower than that in the range of 0·45–0·60, and vice versa in the transverse configuration.

The results for the skin laminate indicate that, in addition to the effect of the reinforcement geometry, the relative position of the notch tip has a significant effect on K. The term 'position' refers to the notch configuration (inter or translaminar) and to its relative depth. These factors select the composite layer encountered by the crack front and, in turn, determine the fracture mechanisms which take place in the damage zone ahead of the notch tip.

5. FRACTURE TOUGHNESS OF THE SANDWICH STRUCTURE

Fracture testing by three-point bending of very long sandwich structure specimens containing surface notches in the skin was next carried out. Test results indicated that fracture was confined to the skin and that the fracture mechanisms resembled those observed in skin specimens tested in tension. The prominent mechanism identified was delamination in the skin at the vicinity of the notch tip followed ultimately by splitting at the adhesive layer between the skin and the honeycomb core. Consequently, it was proposed that the fracture stress of the sandwich structure containing a surface notch could be calculated from the K_{1c} value determined for the skin. The relevant Y value for this calculation was obtained from the new Y polynomials derived for the skin and from the corresponding ratio of notch depth to skin thickness.

To check the validity of this approach the calculated fracture stresses were compared with the experimental values. A detailed example of such a comparison is presented next.

Example

The fracture stress of the sandwich structure that contains a surface notch in one of its skins can be determined utilizing eqn. (1). Following the previous discussion, the appropriate K_{1c} and Y values are those obtained for the skin. For a longitudinal specimen with an interlaminar notch in the range $0.5 \leq c/d \leq 0.6$ (where d is the thickness of the skin *only*) $K_{1c} = 76.9 \, \text{MPa}\sqrt{\text{m}}$, $Y = 23.1$. Thus, for notch depths of 0·5 and 0·6 mm, the fracture stresses, calculated from eqn. (1), are 111·7 and 101·9 MPa, respectively.

An experimental determination of the fracture stresses was carried out in three-point bending with $20 \times 35 \times 300$ mm beams and a loading span of 260 mm. The fracture stresses were calculated from eqn. (5):

$$\sigma = \frac{Md}{2I} \text{ (sandwich)} \tag{5}$$

where M is the bending moment and I is the moment of inertia of the sandwich structure, determined as in reference 8.

The experimental results of the fracture stresses were 125 and 101 MPa for sandwich beams containing 0·5 and 0·6 mm deep skin notches, respectively. These experimental results correlate well with the calculated values and confirm the validity of the approach.

REFERENCES

1. COOPER, G. A. and PIGGOTT, M. R. Cracking and fracture in composites. In: *Advances in research on the strength of fracture of materials*, Vol. 1. Taplin, D. M. R. (ed.), Pergamon Press, USA, p. 557.
2. HAREL, H., MAROM, G., FISCHER, S. and ROMAN, I. Effect of reinforcement geometry on stress intensity factor calibrations in composites, *Composites*, **11** (1980) 69.
3. BEAUMONT, P. W. R. and PHILLIPS, D. C. The fracture energy of a glass fibre composite, *J. Mater. Sci.*, **7** (1972) 682.
4. MANDELL, J. F., WANG, S. S. and McGARRY, E. J. The extension of crack tip damage zones in fibre-reinforced plastic laminates, *J. Comp. Mater.*, **9** (1975) 266.
5. GAGGAR, S. and BROUTMAN, L. J. The development of a damage zone at the tip of a crack in a glass fibre reinforced polyester resin, *Int. J. Fracture*, **10** (1974) 606.
6. MAROM, G. and JOHNSEN, A. C. On the applicability of LEFM to longitudinal fracture of unidirectional composites, *Mater. Sci. Engng*, **39** (1979) 11.
7. MANDELL, J. F., McGARRY, E. J., WANG, S. S. and IM, J. Stress intensity factor for anisotropic fracture test specimens of several geometries, *J. Comp. Mater.*, **8** (1974) 106.
8. FAUPEL, J. H. *Engineering Design*, New York, John Wiley and Sons, Inc., 1964, p. 295.

44

Progressive-Failure Model for Advanced Composite Laminates Containing a Circular Hole

D. Y. KONISHI

*Rockwell International, North American Aircraft Division,
PO Box 92098, Los Angeles, California 90045, USA*

K. H. LO

*Shell Development Company, Westhollow Research Center,
PO Box 1380, Houston, Texas 77001, USA*

AND

E. M. WU

*Lawrence Livermore Laboratories, Box 808-L421,
University of California, Livermore, California 94550, USA*

ABSTRACT

A progressive-failure advanced composite laminate strength model is presented. It is dependent on laminae properties and thus is independent of lamination geometry and loading. The model is based on the stress–strain distribution in the laminate in the neighborhood of a flaw. Failure for a lamina is hypothesised to occur when the stress state at a characteristic-limiting dimension, r_c, away from the flaw intersects the failure envelope. r_c is a material characteristic dependent only on lamina properties. The progressive-failure trajectory is determined by consideration of the changes in the stress–strain state as lamina failures occur. Laminate failure is assumed to occur when the principal load-bearing lamina fiber fails. Examples are presented for laminates with and without a hole. Comparisons with test results for various types of graphite/epoxy (Gr/Ep) coupons was excellent. Thus, the model appears to be capable of analysing both standard and non-standard laminates.

INTRODUCTION

An analytic model is required to evaluate the damage tolerance and durability of advanced composites so that the engineer can take them into account during the preliminary design phase. Various studies have been conducted in order to accomplish this but at the present time the only real validation is to test the actual component in a real time spectrum environment.

In order to develop a generalised analytic model, the failure process must be tractable. This paper presents a method to obtain the failure process for both standard and non-standard laminates containing a hole or crack under static loading. Data are also presented for the residual strength after undergoing two lifetimes of an accelerated test spectrum representing the expected aircraft spectrum.

Once the failure process is known, the effects of stacking sequence and stress complexity can be hypothesised and tests can be conducted and assessed. In addition, the effects of flaws can be analysed for criticality, NDE techniques can be expanded, accelerated test spectrum rationales can be developed, and durability verification can be conducted in a more cost-effective manner.

The basis for the analytic model was first presented in Reference 1, and the model was utilised in Reference 2. The complete basis for the model as it now stands is given in References 3 and 4, where holes and cracks in standard fiber-dominated laminates are considered. In this paper, examples are presented for non-standard matrix-dominated unbalanced laminates which arose from aeroelastic tailoring of the wing and canard torque box covers of a forward swept wing aircraft. Compared to previous cases (for fiber-dominated laminates the proposed method and previously developed methods such as presented in the *Advanced Composite Design Guide*[5] predicted similar results) the proposed method is a much better way of predicting the results for the non-standard laminates. This is due to the physical basis for the model as opposed to the empiricism in previous models. This empiricism was very good when the laminate geometry was conventionally fiber-dominated but proved deficient when extrapolated to an unconventional laminate geometry.

PROGRESSIVE-FAILURE MODEL

The progressive-failure model for laminates containing stress concentrations is as follows:

(1)   The stress state is obtained in the conventional manner. For this paper, it is obtained by treating the laminate as a homogeneous anisotropic material and assuming that lamination theory holds.

(2)   The stress state at the characteristic-limiting dimension, r_c, from the stress concentration is examined for a failure criterion. The Hill–Tsai failure criteria on a lamina-by-lamina basis, assuming that lamination theory (plane stress) holds, was used for this paper.

(3)   Once a lamina fails in a particular mode, its pertinent material property is modified. For this paper, if the matrix fails, E_T and G_{LT} are assumed to evanesce; if the fiber fails, E_L is assumed to evanesce.

For the 'unflawed' laminates, an intermediate step is taken. When the matrix affecting stresses, σ_{T_k} and τ_k, reach the prescribed stress level, the elastic moduli are replaced by their corresponding secant moduli at failure. When the failure stress is again attained, the properties are then assumed to evanesce.

(4)   Each time the lamina properties are modified, stress redistributions are allowed to occur. For this paper, the initially obtained stress resultants are assumed to remain constant, thus limiting the stress redistribution to the through-the-thickness direction. This should be valid for monotonically increasing 'static' loading where the dynamic relationships do not allow complete in-plane redistribution. In general, all redistributions should be allowed to occur, but tractability in the example precluded this.

(5)   Failure occurs when all subsequent loading configurations are less than the failure configuration. For this paper, the principal load-bearing lamina criterion, wherein the fiber failure of the laminae whose orientations are such that they are under the highest axial stress, fiber direction, in a continuous region due to the far-field stress field, terminates the calculations.

STRESS STATE

Savin[6] presented the solution for the stress field in the neighborhood of a hole or a crack in a homogeneous anistropic material (Fig. 1).

The solutions for the stress resultant components are

$$N_1 = N_1^\infty + 2\,\mathrm{Re}\,[S_1^2\phi'(Z_1) + S_2^2\psi'(Z_2)]$$
$$N_2 = N_2^\infty + 2\,\mathrm{Re}\,[\phi'(Z_1) + \psi'(Z_2)]$$
$$N_6 = N_6^\infty - 2\,\mathrm{Re}\,[S_1\phi'(Z_1) + S_2\psi'(Z_2)]$$

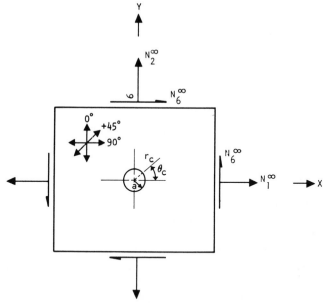

FIG. 1. Laminate coordinate system.

For a crack

$$\phi(Z_1) = a^2 \frac{\{S_2 N_2^\infty + N_6^\infty(1 - S_1 S_2) - S_1 N_1^\infty - i[S_2 N_1^\infty + S_1 S_2 N_2^\infty + N_6^\infty(S_1 + S_2)]\}}{\{2(S_1 - S_2)[Z_1 + \sqrt{Z_1^2 - a^2(1 + S_1^2)}]\}}$$

$$\psi(Z_2) = -a^2 \frac{\{S_1 N_2^\infty + N_6^\infty(1 - S_1 S_2) - S_2 N_1^\infty - i[N_1^\infty + S_1 S_2 N_2^\infty + N_6^\infty(S_1 + S_2)]\}}{\{2(S_1 - S_2)[Z_2 + \sqrt{Z_2^2 - a^2(1 + S_2^2)}]\}}$$

for a hole

$$\phi(Z_1) = \frac{a^2(S_2 N_2^\infty + N_6^\infty)}{2(S_1 - S_2)[Z_1 + \sqrt{Z_1^2 - a^2}]}$$

$$\psi(Z_2) = \frac{-a^2(S_1 N_2^\infty + N_6^\infty)}{2(S_1 - S_2)[Z_2 + \sqrt{Z_2^2 - a^2}]}$$

where $Z_1, Z_2 = X + S_1 Y, \ X + S_2 Y, \ S_1, S_2 =$ roots of the characteristic equation

$$A_n S^4 - 2A_{16} S^3 + (2A_{12} + A_{66}) S^2 - 2A_{26} S + A_{22} = 0$$

and

$$\phi'(Z_1) \qquad \psi'(Z_2) = \frac{\mathrm{d}\phi(Z_1)}{\mathrm{d}Z_1} - \frac{\mathrm{d}\psi(Z_2)}{\mathrm{d}Z_2}$$

CHARACTERISTIC-LIMITING DIMENSION

A characteristic-limiting dimension, r_c, has been discussed previously.[7] It is hypothesised to be a property of a lamina and thus is independent of lamination geometry. A value of 0·07 in is used for the Gr/Ep system of the present study. It is, however, a function of the fiber volume fraction and would change if there were a substantial change from the 60 % value for this study.

For a hole, r_c is taken radially to the center and measured from the edge of the hole. For a crack, it is measured from the crack tip. Studies previously conducted[3,4] also use this value of 0·07 in and show good correlation with test data both for specimens with a hole and specimens with a crack.

PRINCIPAL LOAD-BEARING LAMINA

In principle, laminate failure occurs whenever all subsequent load-bearing capacities are smaller than the failure capacity. In actual practice, this would involve the calculation of all subsequent laminate configurations. In order to reduce the number of calculations, the following hypothesis is made:

Failure occurs when or before the principal load-bearing lamina fails in a fiber failure mode in a continuous region of the lamina.

The principal load-bearing lamina is that lamina that attains the highest axial stress due to the far-field stress tensor. The continuous regions of the lamina are those regions that are not interrupted by fiber termination due to the 'flaw', i.e. the hole or the crack.

NUMERICAL SIMULATION

Previous examples include specimens containing a hole utilising different standard laminate configurations, under both unidirectional and biaxial loading. These examples show that the model predicts not only the final strength but also the failure trajectory. Unidirectionally loaded crack specimens of various laminate configurations, where the crack is both orthogonal and canted up to 30° to the loading direction, are also reported.

All of the aforementioned specimens were for symmetric balanced laminates where the primary load-carrying members were the fibers. This paper presents examples of specimens which are off-axis, unbalanced, and,

TABLE 1
Laminate geometry

Laminate	Stacking sequence
1	$(-45/10_5/90/45/10_4/-45)_C$
2	$(+45/10_2/45_2/10_2)_S$

in some cases, unsymmetric. In these cases which arise from aeroelastic tailoring of torque boxes, matrix-dominated laminates result and previously developed methods predict much too low a failing strength. Two laminate configurations of the material AS/3501-5A Gr/Ep were tested (Table 1). Both laminates are off-axis and unbalanced, but laminate 1 is unsymmetric while laminate 2 is symmetric. Room-temperature 'dry' and elevated-temperature, 220°F, wet, 1% moisture content tests were conducted. The material properties for a lamina are shown in Table 2.

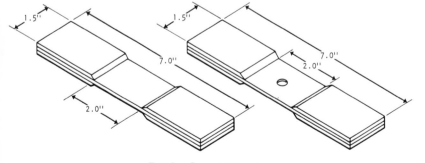

FIG. 2. Coupon geometry.

TABLE 2
Material properties, AS/3501-5A

Temperature ($^\circ$F)	Moisture content ($\%$)	E_L (ksi)	E_T (ksi)	G_{LT} (ksi)	G_{LT}^{sec} (ksi)	γ_{LT} (in/in)	F_L^{tu} (ksi)	F_T^{tu} (ksi)	F_L^{cy} (ksi)	F_T^{cy} (ksi)	F_{LT}^{su} (ksi)
70	0	18 750	1 500	600	350	0·3	225·0	6·75	225·0	31·3	10·0
220	1	18 200	800	400	230	0·3	224·0	3·38	154·0	24·0	7·0

TABLE 3
Test results—Unflawed coupons

Laminate	Temperature ($^\circ$F)	Moisture content ($\%$)	F_y^{tu} (ksi)	f_y^{tu} (ksi)	F_x^{tu} (ksi)	f_x^{tu} (ksi)
1	70	'0·0'	81·5	76·8	24·0	24·2
1	220	1·0	73·7	71·7	23·3	31·4
2	70	'0·0'	86·3	67·2	17·6	23·3
2	220	1·0	85·0	54·3	16·0	20·7

TABLE 4
Tests results—Coupons with hole

Laminate	Temperature ($^\circ$F)	Moisture content ($\%$)	F_y^{tu} (ksi)	f_y^{tu} (ksi)	F_x^{tu} (ksi)	f_x^{tu} (ksi)
1	70	'0·0'	98·1	84·8	18·5	20·9
1	220	1·0	111·1	79·3	21·5	21·2
1[a]	220	1·0	—	65·3	—	—
2	70	'0·0'	52·9	77·3	21·6	21·2
2	220	1·0	68·9	69·5	25·1	20·1
2[a]	220	1·0	—	77·3	—	19·6

[a] Residual strength after two lifetimes fatigue test spectrum.

Figure 2 shows the geometry for the 'unflawed' tension coupons and the coupons with a hole. They are identical except for the presence of a 0·25-in hole.

Table 3 shows the test results and predictions for an unflawed laminate while Table 4 gives the test results for a laminate containing a 0·25-in hole.

DISCUSSION OF RESULTS

Two computer programs were utilised to obtain the results. Both programs are modifications of computer programs available through the *Advanced Composites Design Guide*[5] and should be available in a subsequent edition.

The unflawed coupons were analysed using computer program AC3P, which incorporates a progressive-failure criterion into the presently available AC3. The progressive-failure criterion is that the shear modulus progresses from G_{LT} to G_{LT} at failure to 0. This program accounts for the asymmetry of the coupon by including the 'B' terms. The shearing load boundary condition for the off-axis effect was accounted for by utilising Pagano's technique.[8] Both laminates had shearing loads in the order of $-0.1 N_2^\infty$ due to the high percentage of $10°$ plies.

The coupons with a hole were analysed using computer program AC41E. This program modifies computer program AC41, which is in the *Advanced Composites Design Guide*,[5] by not terminating the process when the principal load-bearing lamina is not continuous. This program differs from AC3P in that once a lamina fails, its affected properties are automatically set to zero rather than going through an inelastic phase. This program does not account for an unsymmetric laminate, but the shearing load due to the off-axis loading is included.

In order to visualise the results of the progressive failure for an unflawed laminate, laminate 1 under tensile loading is tracked for the room-temperature dry condition. Between 36·3 and 37·6 ksi, the $90°$ and $45°$ layers progressively craze and then fail in the matrix mode, causing a reduction in the extensional modulus E_y from 11·4 to 11·2 Msi. At 59 ksi, the upper $10°$ plies become inelastic. Immediately the $10°$ plies down to the $90°$ ply and the upper $-45°$ ply also become inelastic. E_y at this point is reduced to 10·9 Msi. At 61·6 ksi, the $10°$ ply below the $45°$ ply becomes inelastic, followed by the matrix failure of the upper $-45°$ ply. This is followed at 65·2 ksi by matrix failure of the upper $10°$ ply. E_y at this point is 10·7 Msi, or 96 % of its original value. All subsequent configurations fail at a lower stress level, representing the design strength for the laminate. The expected strength for this configuration is 125 % of 65·2 or the 81·5 ksi shown in Table 3. Note that due to the asymmetry, the $10°$ plies failed independently rather than as a group.

Table 5 shows the progressive failure for a room-temperature dry laminate 1 specimen, containing a hole, under tensile loading along the *y*-axis. The initial failures for loads up to 26 ksi occur in the region θ_c between $87°$ and $105°$. Both matrix and fiber failures occur in all plies except the

TABLE 5
Progressive failure of coupon with hole specimen

Load (ksi)	10° fiber	10° matrix	45° fiber	45° matrix	−45° fiber	−45° matrix	90° fiber	90° matrix
0–13·6		93–105	93–105	93–105	93–105	93–105	93–105	93–105
13·6–26·0		87–93	87–93	0–15 33–57 81–93 165–180		87–93	87–93	0–3 27–51 165–180
26·0–30·7		81–87	81–87	15–33 57–69 147–165		81–87	81–87	3–27 51–63 141–165
30·7–44·0		0–63 159–180		69–81 135–147		39–51		63–75 135–141
44·0–50·0		63–69 153–159	0–9 39–63 165–180		0–9 39–57 81–93 165–180	0–39 51–63 153–180	0–9 57–63 177–180	75–81
50–60		147–153 69–75	9–39 63–75 153–165	129–135	9–39 159–165	63–75 141–153 159–171	9–27 63–75 171–177	123–129
60–78·3		75–81 135–147	75–81 135–153	123–129	57–63	75–81 23–141	27–39 75–81 135–153 165–171	117–123
78·47	27–33							

designated principal load-bearing laminae which are the 10° plies. Subsequent to this, up to a loading of 50 ksi, matrix and fiber failures occur at θ_c between 0° and 90°, and 165° and 180°, with some failures between the two regions. It should be noted that due to antisymmetry the results for θ_c from −15° to 0° are the same as those for 165° to 180°. The 10° plies incur matrix failures in all of these regions. There is then extensive matrix damage and some fiber failures up to a load of nearly 78 ksi. The initial fiber failure for the principal load-bearing lamina (10°) is predicted to occur at 78·47 ksi in a region of θ_c between 27° and 33°. These results are obtained by using the 'design' strengths. In evaluating the data, the expected failure strength should be used. This is 125 % of the design strength, or 98·1 ksi. The average test data for these specimens were 84·8 ksi. This is excellent correlation, considering the deficiencies in the numerical simulation methodology; namely, the computer program does not consider asymmetric laminates and only a two-step progressive-failure process is used. In addition, the F_{12} term for AC41 was assumed to be zero, whereas it was '− 1' for AC3P. This anomaly resulted from the difference in structure between the two computer programs. One interesting result is that the net failure stress for specimens with a hole was not only predicted to be larger than for the unflawed specimens, but test results for the longitudinally loaded specimens were also larger. The test specimens showed extensive crazing and fiber failures. The failure location was very close to the predicted $\theta_c = 30$ in all cases.

CONCLUSION

A method has been presented to predict the failure strength of laminates containing a hole. Although the empirical factors utilised by the method were obtained by evaluating standard fiber-dominated laminates, the methodology shows very good correlation with test data for non-standard matrix-dominated laminates which, in some cases, were even asymmetric. The method not only predicts the strength, but also the extensive matrix failures and fiber failures that occur before the ultimate failure, as well as the location of the failures.

REFERENCES

1. KONISHI, D. Y. and LO, K. H., *Flaw criticality of graphite/epoxy structures*, ASTM STP 696, Proceedings of the Symposium on Nondestructive Evaluation and Flaw Criticality for Composite Materials, Philadelphia, PA, October 1978.
2. KONISHI, D. Y., Effects of defects on tension coupons undergoing an accelerated environmental spectrum. In: *Fibrous composites in structural design*, Lenoe, E. M., *et al.* (eds), New York and London, Plenum Press, 1980, pp. 847–60.
3. ALTMAN, J. M. *et al.*, 'Advanced composites serviceability program', AFWAL-TR-80-4092, 1981.
4. LO, K. H. *et al.*, Failure strength of notched composite laminates, *Journal of Composite Materials*, in press.
5. ARVIN, G. H. *et al.*, *DOD/NASA Advanced composites design guide*, in press.
6. SAVIN, G. H., *Stress concentration around holes*, New York, Pergamon Press, 1961.
7. WU, E. M., Phenomenological anisotropic failure criterion, *Journal of Composite Materials*, **9** (1975).
8. PAGANO, N. J. and HALPIN, J. C., Influence of end constraints in the testing of anisotropic bodies, *Journal of Composite Materials*, **2** (1968).

45

Nonlinear Response of Angle-Ply Laminated Plates to Random Loads

CHUH MEI

*Department of Mechanical Engineering and Mechanics,
Old Dominion University, Norfolk, VA 23508, USA*

AND

KENNETH R. WENTZ

*AFWAL/FIBED Flight Dynamics Laboratory,
Wright-Patterson Air Force Base, OH 45433, USA*

ABSTRACT

Large amplitude response of antisymmetric angle-ply laminated rectangular panels subjected to broadband random acoustic excitation is studied analytically. The boundary conditions considered are all the edges simply supported and all the edges clamped. Inplane edge conditions considered are immovable and movable for each of the above cases. Mean-square deflections, mean-square strains/stresses, and equivalent linear frequencies at various acoustic loadings are obtained for laminates of different length-to-width ratios, lamination angles, number of layers, and panel damping ratios. Results obtained can be used as a guide for sonic fatigue design of angle-ply laminated composite panels under high noise environment.

INTRODUCTION

The need to improve sonic fatigue resistance of aircraft structures has become increasingly important as a result of military and commercial demands on current and future airplane designs. This has been demonstrated by a significant number of theoretical studies,[1-7] and numerous experimental investigations[8-13] on sonic fatigue design of aircraft structures have been undertaken during the past several years to help provide the needed reliability. The majority of analytical investigations to date have been formulated within the framework of linear or small deflection structural theory. Test results[6,8,10-13] on various aircraft panels, however, have shown that high noise levels in excess of 120 dB produce nonlinear behavior with large deflections in such panels. The linear analyses often predict the root-mean-square (RMS) deflection and RMS stresses well above those of the experiment, and the frequencies of vibration well below those of the experiment.[6,8,12,13] It is well known that the prediction of fatigue life is based on RMS stress and predominant response frequency in conjunction with the stress versus cycles to failure (S–N) data. The use of linear analyses, therefore, would lead to poor estimation of panel service life.

High modulus-type fiber reinforced composite materials are under development for use on aircraft. Many of these composite structural components are exposed to high intensity noise fields and are subjected to acoustic fatigue. However, few investigations on large amplitude random response of composite plates are reported in the literature. In the present paper, the large deflection responses of regular antisymmetric angle-ply laminated rectangular plates subjected to broadband random acoustic excitation are studied analytically. Nonlinear equations of motion of angle-ply laminates[1] derived in terms of stress function F and lateral displacement W are used in this work. Due to the complex nature of the problem, only a single-mode analysis is carried out in the study. A deflection function satisfying the out-of-plane boundary condition is assumed. Corresponding to the assumed mode, a stress function satisfying the different inplane edge conditions is obtained by solving the compatibility equation. Galerkin's method is then applied to the governing equation in deflection to yield a nonlinear time differential equation. Assuming that the excitation is Gaussian, the equivalent linearization method[14-16] is employed so that this nonlinear equation is linearized to an equivalent linear differential equation. An iterative procedure[1,2] is introduced to obtain RMS amplitude and equivalent linear (or nonlinear) frequency for rectangular laminates of different length-to-width ratios, lamination angles, number of layers, panel damping ratios, and excitation pressure spectral density (PSD). RMS strains/stresses are also obtained as functions of RMS amplitude and locations at which they are to be measured. The boundary conditions are all the edges simply supported (SSSS) and all the edges clamped (CCCC). Two inplane edge conditions considered are immovable and movable for each of the above cases.

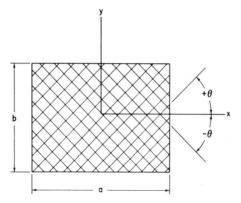

FIG. 1. Angle-ply laminated plate.

EQUATIONS OF MOTION

The plate geometry and coordinate system are shown in Fig. 1. For a regular antisymmetric angle-ply laminated composite plate undergoing large deflections, the governing equations, neglecting the effects of transverse shear and rotatory inertia, are[1]

$$\rho h \ddot{W} + L_1 W - {}_l L_3 F - \phi(F, W) - p = 0 \qquad (1)$$

$$L^2 F + L_3 W + \tfrac{1}{2}\phi(W, W) = 0 \qquad (2)$$

in which

$$L_1 = D^*_{11}( )_{,xxxx} + 2(D^*_{12} + 2D^*_{66})( )_{,xxyy} + D^*_{22}( )_{,yyyy}$$

$$L_2 = A^*_{22}( )_{,xxxx} + (2A^*_{12} + A^*_{66})( )_{,xxyy} + A^*_{11}( )_{,yyyy}$$

$$-L_3 = (2B^*_{26} - B^*_{61})( )_{,xxxy} + (2B^*_{16} - B^*_{62})( )_{,xyyy}$$

$$\phi(F, W) = F_{,yy} W_{,xx} + F_{,xx} W_{,yy} - 2F_{,xy} W_{,xy} \qquad (3)$$

where a comma denotes the partial differentiation with respect to the corresponding coordinate and A^*, B^* and D^* are the laminate stiffnesses.[17,18] In eqns (1) and (2), ρ is the average mass density of the plate, h is the uniform plate thickness, p is the applied lateral load per unit area, and x and y are the rectangular Cartesian coordinates.

METHOD OF ANALYSIS

SSSS Angle-Ply Laminates with Movable Edges

The simply supported conditions are

$$x = \pm a/2: \ W = 0 \qquad D^*_{11}W_{,xx} + D^*_{12}W_{,yy} + B^*_{16}N_{xy} = 0$$

$$y = \pm b/2: \ W = 0 \qquad D^*_{12}W_{,xx} + D^*_{22}W_{,yy} + B^*_{62}N_{xy} = 0 \qquad (4)$$

For the inplane condition of zero shear stress N_{xy} at the edges, the deflection function satisfying the above conditions is assumed as

$$W = q(t)h \cos\frac{\pi x}{a} \cos\frac{\pi y}{b} \qquad (5)$$

Substituting eqn. (5) for W in eqn. (2) and solving it, the stress function is obtained as

$$F_p = F_c + F_p \qquad (6)$$

$$F_p = -qhF_{00}\sin\frac{\pi x}{a}\sin\frac{\pi y}{b} - \frac{q^2h^2r^2}{32}\left(F_{10}\cos\frac{2\pi x}{a} + F_{01}\cos\frac{2\pi y}{b}\right) \qquad (7)$$

where $r = a/b$, and the constants F_{ij} are defined in the Appendix. It can be easily shown that F_c is zero for movable inplane edges. By substituting these expressions for W and F in eqn. (1) and applying Galerkin's method, a model equation is obtained as

$$\ddot{q} + \omega_0^2 q + \beta_p q^3 = \frac{p(t)}{m} \qquad (8)$$

and

$$\omega_0^2 = \lambda_0^2 \frac{E_2 h^2}{\rho b^4} \qquad \beta_p = \beta_p^* \frac{E_2 h^2}{\rho b^4} \qquad m = \frac{\pi^2 \rho h^2}{16}$$

$$\lambda_0^2 = \frac{\pi^4}{E_2 h^3 r^4}\{D^*_{11} + 2(D^*_{12} + 2D^*_{66})r^2 + D^*_{22}r^4 \qquad (9)$$

$$+ F_{00}[(2B^*_{26} - B^*_{61})r + (2B^*_{16} - B^*_{62})r^3]\}$$

$$\beta_p^* = \frac{\pi^4}{16E_2 h}(F_{10} + F_{01}) \qquad (10)$$

where ω_0 is linear radian frequency, β_p is nonlinearity coefficient, and m is mass coefficient. The linear frequency λ_0 and nonlinearity coefficient β_p^* are nondimensional parameters.

SSSS Angle-Ply Laminates with Immovable Edges

For the immovable edges, the inplane boundary conditions of zero shear stress and zero normal displacement are

$$x = \pm a/2: F_{,xy} = 0 \qquad \int\int (\varepsilon_x^0 - \tfrac{1}{2} W_{,x}^2) \, dx \, dy = 0$$

$$y = \pm b/2: F_{,xy} = 0 \qquad \int\int (\varepsilon_y^0 - \tfrac{1}{2} W_{,y}^2) \, dx \, dy = 0 \tag{11}$$

where ε_x^0 and ε_y^0 are the strains in the middle surface of the plate. The complementary solution is assumed as

$$F_c = \bar{N}_y \frac{x^2}{2} + \bar{N}_x \frac{y^2}{2} \tag{12}$$

Upon using eqns (5) and (12) and enforcing the conditions of eqn. (11), $\bar{N}_x$ and $\bar{N}_y$ are obtained as

$$\bar{N}_x = \frac{q^2 h^2 \pi^2}{8(A_{11}^* A_{22}^* - A_{12}^{*2})} \left(\frac{A_{22}^*}{a^2} - \frac{A_{12}^*}{b^2} \right)$$

$$\bar{N}_y = \frac{q^2 h^2 \pi^2}{8(A_{11}^* A_{22}^* - A_{12}^{*2})} \left(\frac{A_{11}^*}{b^2} - \frac{A_{12}^*}{a^2} \right) \tag{13}$$

Using eqns (5), (6), (7), (12) and (13) in eqn. (1) and applying Galerkin's method yields the modal equation as

$$\ddot{q} + \omega_0^2 q + (\beta_p + \beta_c) q^3 = \frac{p(t)}{m} \tag{14}$$

where

$$\beta_c = \beta_c^* \frac{E_2 h^2}{\rho b^4} \qquad \beta_c^* = \frac{\pi^4}{8 E_2 h r^4} \left(\frac{A_{22}^* - 2 A_{12}^* r^2 + A_{11}^* r^4}{A_{11}^* A_{22}^* - A_{12}^{*2}} \right) \tag{15}$$

The term β_c is an addition to the nonlinearity coefficient due to immovable inplane edges; the nonlinearity coefficient β_c^* is a nondimensional parameter.

CCCC Angle-Ply Laminates with Movable Edges

The deflection function which satisfies the clamped condition is assumed as

$$W = \frac{qh}{4} \left(1 + \cos \frac{2\pi x}{a} \right) \left(1 + \cos \frac{2\pi y}{b} \right) \tag{16}$$

Introducing eqn. (16) in eqn. (2), the stress function is obtained as follows

$$F = F_c + F_p \qquad F_c = 0 \tag{17}$$

$$
\begin{aligned}
F_p = \bigg| -\frac{qh}{4} F_{00} \sin\frac{2\pi x}{a}\sin\frac{2\pi y}{b} - \frac{q^2 h^2 r^2}{32}\bigg(F_{10}\cos\frac{2\pi x}{a} + F_{01}\cos\frac{2\pi y}{b} \\
+ F_{11}\cos\frac{2\pi x}{a}\cos\frac{2\pi y}{b} + F_{20}\cos\frac{4\pi x}{a} + F_{02}\cos\frac{4\pi y}{b} \\
+ F_{21}\cos\frac{4\pi x}{a}\cos\frac{2\pi y}{b} + F_{12}\cos\frac{2\pi x}{a}\cos\frac{4\pi y}{b}\bigg)
\end{aligned}
\tag{18}
$$

where the constants F_{ij} are defined in the Appendix. Now applying Galerkin's procedure to eqn. (1), we obtain

$$\ddot{q} + \omega_0^2 q + \beta_p q^3 = \frac{p(t)}{m} \tag{19}$$

where

$$\omega_0^2 = \lambda_0^2 \frac{E_2 h^2}{\rho b^4} \qquad \beta_p = \beta_p^* \frac{E_2 h^2}{\rho b^4} \qquad m = \frac{9\rho h^2}{16} \tag{20}$$

$$
\begin{aligned}
\lambda_0^2 = \frac{16\pi^4}{9E_2 h^3 r^4}\{3D_{11}^* + 2(D_{12}^* + 2D_{66}^*)r^2 + 3D_{22}^* r^4 \\
+ F_{00}[(2B_{26}^* - B_{61}^*)r + (2B_{16}^* - B_{62}^*)r^3]\}
\end{aligned}
$$

$$\beta_p^* = \frac{\pi^4}{9E_2 h}[F_{10} + F_{01} + F_{11} + F_{20} + F_{02} + \tfrac{1}{2}(F_{21} + F_{12})] \tag{21}$$

CCCC Angle-Ply Laminates with Immovable Edges

The complementary stress function is assumed as the form appearing in eqn. (12). Upon enforcing the inplane edge conditions eqn. (11), the constants $\bar{N}_x$ and $\bar{N}_y$ are obtained as

$$\bar{N}_x = \frac{3q^2 h^2 \pi^2}{32(A_{11}^* A_{22}^* - A_{12}^{*2})}\left(\frac{A_{22}^*}{a^2} - \frac{A_{12}^*}{b^2}\right)$$

$$\bar{N}_y = \frac{3q^2 h^2 \pi^2}{32(A_{11}^* A_{22}^* - A_{12}^{*2})}\left(\frac{A_{11}^*}{b^2} - \frac{A_{12}^*}{a^2}\right) \tag{22}$$

Using eqns (12), (16), (18), and (22) and applying Galerkin's method yield the model equation

$$\ddot{q} + \omega_0^2 q + (\beta_p + \beta_c)q^3 = \frac{p(t)}{m} \tag{23}$$

$$\beta_c = \beta_c^* \frac{E_2 h^2}{\rho b^4} \qquad \beta_c^* = \frac{\pi^4}{8E_2 hr^4} \left(\frac{A_{22}^* - 2A_{12}^* r^2 + A_{11}^* r^4}{A_{11}^* A_{22}^* - A_{12}^{*2}} \right) \tag{24}$$

Equations (8), (14), (19) and (23) represent the modal equations of undamped rectangular angle-ply laminates undergoing large deflections with simply supported and clamped edges, respectively.

Damping Factor

It is known that damping has a significant effect on the response of structures. The two methods commonly used for determining the damping characteristics of structures are the bandwidth method and the decay rate method. In the bandwidth method, the half-power bandwidth ($=2\zeta$) is measured at modal resonance. In the decay rate method, the logarithmic decrement ($=2\pi\zeta$) of decaying modal response traces is measured. The values of damping ratio ζ ($=c/c_c$) generally range from 0·005 to 0·05 for the common type of composite panel construction used in aircraft structures.[5,13,19] Once the damping ratio is determined from experiments or from existing data of similar construction, the modal equations, eqns (8), (14), (19), and (23), can be expressed in a general form as

$$\ddot{q} + 2\zeta\omega_0\dot{q} + \omega_0^2 q + \beta q^3 = \frac{p(t)}{m} \tag{25}$$

The method of equivalent linearization will be used to obtain an approximate RMS amplitude from eqn. (25).

Method of Equivalent Linearization

The basic idea of the equivalent linearization method[14-16] is to assume that an approximate solution to eqn. (25) can be obtained from the linearized equation

$$\ddot{q} + 2\zeta\omega_0\dot{q} + \Omega^2 q = \frac{p(t)}{m} \tag{26}$$

where Ω is an equivalent linear (or nonlinear) frequency. The error of linearization is

$$\text{err} = (\omega_0^2 - \Omega^2)q + \beta q^3 \tag{27}$$

The method of attack is to minimize the mean-square error $E[\text{err}]^2$, that is

$$\frac{\partial}{\partial \Omega^2} E[\text{err}^2] = 0 \qquad (28)$$

If the acoustic pressure excitation $p(t)$ is stationary Gaussian, is ergodic, and has a zero mean, then the approximate amplitude q, computed from the linearized equation, eqn. (26), is also Gaussian and approaches stationarity because the panel motion is stable. Substituting eqn. (27) into eqn. (28) yields the relation

$$\Omega^2 = \omega_0^2 + 3\beta E[q^2] \quad \text{or} \quad \lambda^2 = \lambda_0^2 + 3\beta^* E[q^2] \qquad (29)$$

where $\lambda^2 = \Omega^2/(E_2 h^2/\rho b^4)$ is a nondimensional nonlinear frequency parameter. The linear frequencies λ_0 and nonlinearity coefficients β^* are given in eqns (10), (15), (21), and (24) for different support conditions.

The mean-square modal amplitude from eqn. (26), for lightly damped ($\zeta \leq 0.05$) structures, is

$$E[q^2] = \int_0^\infty S(\omega)|H(\omega)|^2 \, \mathrm{d}\omega \simeq \frac{\pi S(\Omega)}{4m^2 \zeta \omega_0 \Omega^2} \qquad (30)$$

where $S(\omega)$ is the PSD function of the excitation $p(t)$, $H(\omega)$ is the frequency response function given by

$$H(\omega) = \frac{1}{m(\Omega^2 - \omega^2 + 2i\zeta\omega_0\omega)} \qquad (31)$$

Note that the frequency response curves will be highly peaked at the nonlinear frequency Ω (not at ω_0 as in the small deflection linear theory). In practice, the PSD function is generally given in terms of the frequency f in Hz. To convert the above result one can substitute $\Omega = 2\pi f$ and $S(\Omega) = S(f)/2\pi$ into eqn. (30); then the mean-square peak deflection becomes

$$E[q^2] = \begin{cases} \dfrac{32S}{\pi^4 \zeta \lambda_0 \lambda^2} & \text{for SSSS} \\[12pt] \dfrac{32S}{81\zeta\lambda_0\lambda^2} & \text{for CCCC} \end{cases} \qquad (32)$$

where S is a nondimensional forcing excitation spectral density parameter defined as

$$S = \frac{S(f)}{\rho^2 h^2 \left(\dfrac{E_2 h^2}{\rho b^4}\right)^{3/2}} \qquad (33)$$

Solution Procedure and Stress Response

The mean-square amplitude $E[q^2]$ in eqn. (30) is evaluated at the nonlinear frequency Ω, which is in turn related to $E[q^2]$ through eqn. (29). To determine the mean-square deflection, an iterative procedure is introduced.[1,2] One can estimate the initial mean-square amplitude $E[q_0^2]$ using linear frequency ω_0 through eqn. (30): $E[q_0^2] = \pi S(\omega_0)/4m^2\zeta\omega_0^3$. This initial estimate of $E[q_0^2]$ is simply the mean-square deflection based on linear theory. It can now be used to obtain a refined estimate of Ω_1 through eqn. (29): $\Omega_1^2 = \omega_0^2 + 3\beta E[q_0^2]$; then $E[q_1^2]$ is computed through eqn. (30) as $E[q_1^2] = \pi S(\Omega_1)/4m^2\zeta\omega_0\Omega_1^2$. As the iterative process converges on the jth cycle, the relations

$$E[q_j^2] = \frac{\pi S(\Omega_j)}{4m^2\zeta\omega_0\Omega_j^2} \simeq E[q_{j-1}^2] \tag{34}$$

become satisfied. In the numerical results presented in the following section, convergence is considered achieved whenever the difference of the RMS deflections satisfy the relation

$$\left| \frac{\sqrt{E[q_j^2]} - \sqrt{E[q_{j-1}^2]}}{\sqrt{E[q_j^2]}} \right| \leq 10^{-3} \tag{35}$$

Once the RMS displacement is determined, the mean-square strain (or stress) on the surface of the plate $(z = h/2)$ can be obtained from the general expressions[1]

$$E[\varepsilon^2] = C_1^2 E[q^2] + 3C_2^2 (E[q^2])^2$$

$$E[\sigma^2] = D_1^2 E[q^2] + 3D_2^2 (E[q^2])^2 \tag{36}$$

where C_1, C_2, D_1, and D_2 are constants, and they can be determined from material properties, dimensions, number of layers of the plate, lamination angle, and the location at which the stress/strain is to be measured.

RESULTS AND DISCUSSION

Because of the complications in analysis of the many coupled modes, only a single-mode approximation is used. The assumption for fundamental mode predominacy is admittedly overly simplified; the conditions under which this is a valid approximation remain to be investigated. This single-mode approximation was first presented by Miles,[20] and it is commonly used for all sonic fatigue analyses.[3] A simple model sometimes helps to give the basic understanding of the problem.

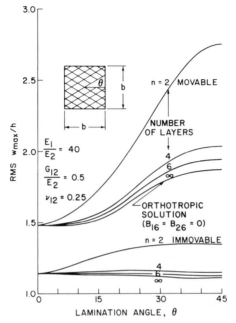

FIG. 2.   RMS deflection of simply supported, square, angle-ply plates (at $\zeta = 0\cdot02$ and $S = 5000$).

In the results presented, the excitation PSD function $S(f)$ is considered constant or varying slowly with frequency in the vicinity of the nonlinear frequency f, and a representative high modulus graphite/epoxy with material properties in the principal material directions

$$E_1 = 30 \times 10^6 \, \text{psi} \, (207 \, \text{GPa}) \qquad E_2 = 0\cdot75 \times 10^6 \, \text{psi} \, (5\cdot2 \, \text{GPa})$$

$$G_{12} = 0\cdot375 \times 10^6 \, \text{psi} \, (2\cdot6 \, \text{GPa}) \qquad v_{12} = 0\cdot25$$

is used in all computations.

SSSS Angle-Ply Plates

Figure 2 shows the RMS maximum deflection ($q = W_{max}/h$) as a function of lamination angle and number of layers for a square antisymmetric angle-ply laminated graphite/epoxy plate with a damping ratio of $\zeta = 0\cdot02$ and at an acoustic pressure loading with nondimensional spectral density parameter of $S = 5000$. The infinite number of layers case corresponds to the specially orthotropic solution in which the coupling between bending and extension is ignored. For a two-layered plate and a $\pm45°$ lamination

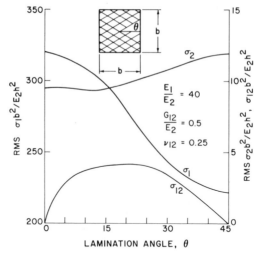

FIG. 3. Maximum RMS stress for simply supported, two-layered, square, angle-ply plates with immovable inplane edges (at $\zeta = 0\cdot02$ and $S = 5000$).

angle, neglect of coupling results in an underprediction of the deflection by 18 % and 32 % for immovable and movable inplane edges, respectively. The effect of coupling between bending and extension is quite significant for two-layered laminates, but rapidly decreases as the number of layers increases. For a fixed laminate thickness h, the bending–extension coupling stiffnesses[17,18]

$$B_{16}, B_{26} = -\frac{h^2}{2n}(\bar{Q}_{16}, \bar{Q}_{26})_{-\theta} \tag{37}$$

obviously decrease as n increases, so the source of the change in the influence of coupling is evident. However, only where there are more than six layers can coupling be ignored without significant error. The RMS deflection of the immovable inplane edges case is much less than that of the movable edges: that is, as the inplane edges are restrained, the plate becomes stiffer.

The normalized RMS maximum stresses (at the center of the plate) in the principal material directions versus lamination angle of two-layered square laminates with immovable inplane edges at $\zeta = 0\cdot02$ and $S = 5000$ are shown in Fig. 3.

The maximum mean-square deflections versus nondimensional spectral density parameter of excitation for a square plate with a $\pm 45°$ lamination angle are shown in Fig. 4. Clearly, coupling is significant for two-layered

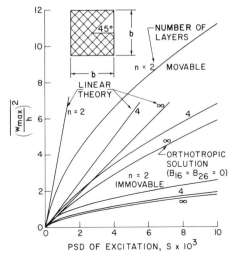

FIG. 4.   Mean-square deflection versus pressure spectral density parameter for simply supported, square, angle-ply plates (at $\zeta = 0.02$).

laminates, but rapidly decreases as the number of layers increases. It can also be seen from the figures that the mean-square deflections of the movable inplane edges cases are approximately three to four times those of the immovable edges. Results of mean-square deflection versus forcing spectral density based on small deflection theory are also shown.

CCCC Angle-Ply Plates

Figure 5 shows the RMS maximum deflection as a function of lamination angle and number of layers for a square angle-ply plate with $\zeta = 0.02$ and subjected to a pressure excitation $S = 5000$. The presence of coupling between bending and extension generally increases deflections: for example, the RMS deflection of a clamped square plate with $\theta = 45°$ and two layers is about 50 % more than the specially orthotropic solution which is valid when the number of layers is infinite. Again, as the number of layers increases, the effect of coupling decreases. The RMS deflection of the clamped plates is generally somewhat less than that of the simply supported case.

RMS deflection results as a function of lamination angle and number of layers for a rectangular angle-ply plate of aspect ratio $r = 2$ and a damping ratio 0.02 with immovable inplane edges are shown in Fig. 6.

Figure 7 shows the mean-square deflection versus nondimensional PSD

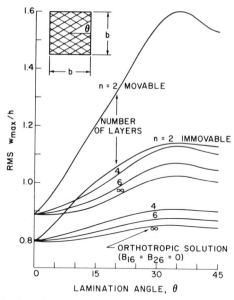

Fig. 5. RMS deflection of clamped, square, angle-ply plates (at $\zeta = 0.02$ and $S = 5000$).

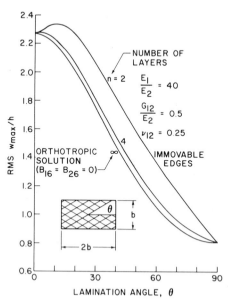

Fig. 6. RMS deflection of clamped, rectangular, angle-ply plates with immovable inplane edges (at $\zeta = 0.02$ and $S = 5000$).

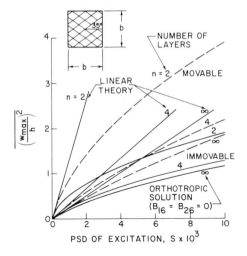

FIG. 7. Mean-square deflection versus pressure spectral density parameter for clamped, square, angle-ply plates ($\zeta = 0.02$).

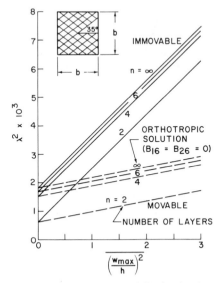

FIG. 8. Frequency parameter versus mean-square deflection for clamped, square, angle-ply plates.

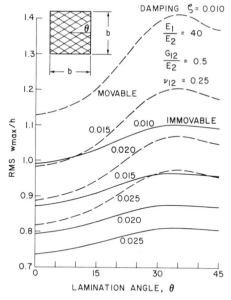

FIG. 9.   Effects of damping on RMS deflection for clamped, square, six-layered, angle-ply plates (at $S = 5000$).

of excitation for a square plate with $\theta = \pm 35°$. The mean-square deflection of the movable inplane edges case is approximately twice that of the immovable edges case. The difference of mean-square deflections between immovable and movable edges for clamped laminates is small compared to that for simply supported laminates. Results based on linear structural theory are also given.

Figure 8 shows the nonlinear frequency square versus mean-square deflection for a square plate with $\theta = \pm 35°$. The frequencies corresponding to zero mean-square deflection are resonance frequencies based on linear structural theory.

Figure 9 shows the RMS deflection versus lamination angle for square six-layered angle-ply plates of different damping ratios. It is clear from the figure that the precise determination of plate damping is important.

CONCLUDING REMARKS

An analytical method is presented for determining large amplitude response of antisymmetric angle-ply laminated rectangular plates subjected to broadband random excitation. The formulation is based on the Karman-type geometric nonlinearity, a single-mode Galerkin approach, the equivalent linearization method, and an iterative procedure. Angle-ply laminates of both simply supported and clamped support conditions with either immovable or movable inplane edges are considered. Computer programs[1] are developed to aid in the determination of RMS deflection, RMS stress/strain, and equivalent linear frequency at given pressure spectral density of excitation. This computed RMS stress/strain and nonlinear frequency, in conjunction with failure S–N data,[9] should be used in the estimation of fatigue life. It is suggested that experiments on simple composite plates are urgently needed for adequate quantitative comparison. The test measurements should include linear frequency, equivalent linear or nonlinear frequency, damping ratio, RMS deflection, RMS strains, and excitation pressure spectral density.

The presence of coupling between bending and extension in a laminate generally increases RMS deflections and RMS stresses. Hence, coupling decreases the effective stiffnesses of a laminate. The effect of coupling, however, dies out rapidly as the number of layers increases. For more general laminates, specific investigation is required.

ACKNOWLEDGEMENT

This work was sponsored by the US Air Force Office of Scientific Research, Air Force Systems Command, under Grant No. AFOSR-80-0107 with cost sharing participation by the Mechanical Engineering and Mechanics Department, Old Dominion University, Norfolk, VA, USA.

REFERENCES

1. MEI, C., Large-Amplitude Response of Angle-Ply Laminated Composite Plates to Random Acoustic Excitation, Interim Scientific Report, Old Dominion University Research Foundation, Jan. 1981.
2. MEI, C., Response of Nonlinear Structural Panels Subjected to High Intensity Noise, AFWAL-TR-80-3018, WPAFB, OH, March 1980.
3. RUDDER, F. F. JR. and PLUMBLEE, H. E. JR., Sonic Fatigue Design Guide for Military Aircraft, AFFDL-TR-74-112, WPAFB, OH, May 1975.
4. VOLMIR, A. S., The Nonlinear Dynamics of Plates and Shells, Chapter X, AD-781338, Foreign Technology Division, WPAFB, OH, April 1974.

5. THOMSON, A. G. R. and LAMBERT, R. F., Acoustic Fatigue Design Data, AGARD-AG-162-Part I and II, NATO Advisory Group for Aero. Res. and Dev., 1972.

6. JACOBS, L. D. and LAGERQUIST, D. R., Finite Element Analysis of Complex Panel to Random Loads, AFFDL-TR-68-44, WPAFB, OH, October 1968.

7. FOX, H. L., SMITH, P. W. JR., PYLE, R. W. and NAYAK, P. R., Contributions to the Theory of Randomly Forced, Nonlinear, Multiple-Degree-of-Freedom, Coupled Mechanical Systems, AFFDL-TR-72-45, WPAFB, OH, 1973.

8. HOLEHOUSE, I., Sonic Fatigue Design Techniques for Advanced Composite Aircraft Structures. AFWAL-TR-80-3019, WPAFB, OH, April 1980.

9. WENTZ, K. R. and WOLFE, H. F., Development of random fatigue data for adhesively bonded and weldbonded structures subjected to dynamic excitation, *ASME Trans., J. Engng Mat. Tech.*, **100**, 1978, 70–6.

10. JACOBSON, M. J., Sonic Fatigue Design Data for Bonded Aluminum Aircraft Structures, AFFDL-TR-77-45, WPAFB, OH, June 1977.

11. VAN DER HEYDE, R. C. W. and WOLF, N. D., Comparison of the Sonic Fatigue Characteristics of Four Structural Designs, AFFDL-TR-76-66, WPAFB, OH, September 1976.

12. VAN DER HEYDE, R. C. W. and SMITH, D. L., Sonic Fatigue Resistance of Skin-Stringer Panels, AFFDL-TM-73-149-FYA, WPAFB, OH, April 1974.

13. JACOBSON, M. J., Advanced Composite Joints: Design and Acoustic Fatigue Characteristics, AFFDL-TR-71-126, WPAFB, OH, April 1972.

14. CAUGHEY, T. K., Equivalent linearization techniques, *JASA*, **35**, 1963, 1706–11.

15. CAUGHEY, T. K., Nonlinear theory of random vibrations. In: *Advances in Applied Mechanics*, Yih, C. S. (ed.), Academic Press, 1971, pp. 209–53.

16. SPANOS, P. T. D. and IWAN, W. D., On the existence and uniqueness of solutions generated by equivalent linearization, *Int. J. Non-Linear Mechanics*, **13**, 1978, 71–8.

17. JONES, R. M., *Mechanics of Composite Materials*, McGraw-Hill, 1975.

18. TSAI, S. W. and HAHN, H. T., *Introduction to Composite Materials*, Technomic Publishing, 1980.

19. RUPERT, C. L. and WOLF, N. D., Sonic Fatigue and Response Tests of Boron Composite Panels, AFFDL-FYA-73-10, WPAFB, OH, July 1973.

20. MILES, J. W., On structural fatigue under random loading, *J. Aeronaut. Sci.*, **21**, 1954, 753–62.

APPENDIX

Constants F_{ij} in eqns (7) and (18) are defined as

$$F_{00} = \frac{(2B_{26}^* - B_{61}^*)r + (2B_{16}^* - B_{62}^*)r^3}{A_{22}^* + (2A_{12}^* + A_{66}^*)r^2 + A_{11}^* r^4}$$

$$F_{10} = \frac{1}{A_{22}^*}$$

$$F_{01} = \frac{1}{A_{11}^* r^4}$$

$$F_{11} = \frac{2}{A_{22}^* + (2A_{12}^* + A_{66}^*)r^2 + A_{11}^* r^4}$$

$$F_{20} = \frac{1}{16A_{22}^*}$$

$$F_{02} = \frac{1}{16A_{11}^* r^4}$$

$$F_{21} = \frac{1}{16A_{22}^* + 4(2A_{12}^* + A_{66}^*)r^2 + A_{11}^* r^4}$$

$$F_{12} = \frac{1}{A_{22}^* + 4(2A_{12}^* + A_{66}^*)r^2 + 16A_{11}^* r^4}$$

46

Effects of Elastomeric Additives on the Mechanical Properties of Epoxy Resin and Composite Systems

RICHARD J. MOULTON

Hexcel Corporation, Dublin, CA 94566, USA

AND

ROBERT Y. TING

Code 5975, PO Box 8337, Naval Research Laboratory, Orlando, FL 32856, USA

ABSTRACT

Thermosetting resins such as epoxy and polyimide are widely used as matrix materials in organic composite systems. These polymers are inherently brittle materials, poor in their resistance to the growth of internal flaw and the propagation of crack. A practice in the industry is to add elastomeric particles to the brittle matrix in order to enhance the resin toughness. The mechanisms for this enhancement are only recently understood as involving triaxial dilatation of rubber particles at the crack-tip, particle elongation and matrix plastic flow. However, when such a modified resin system is used in a fiber-reinforced composite, the effects of these additive particles on the mechanical properties of the composite have not been clear. In this paper, both the fracture behavior and mechanical properties balance of such composite systems will be discussed, based on the results of an extensive experimental study.

A series of acrylonitrile–butadiene modified epoxy polymers were used in the study. Resin fracture energies were determined by using standard compact tension specimens and the Izod impact specimens. The elastomeric modifiers greatly increased the fracture energy of the base epoxy, and the extent of this increase depended on the weight percentage and the molecular weight of the CTBN additives. Post-failure fractography was also carried out to examine the system morphology for the identification of the basic mechanism of toughening.

Composite laminates were prepared using both 7781 glass and T300/3K graphite reinforcements. Short beam shear tests were performed to evaluate interlaminar shear strength, which is known to be a matrix-dominated property. Mechanical properties, as measured by 200° F SBS, are correlated with resin fracture energy. Enhanced toughness is seen to always couple with trade-offs in strength and modulus. It also shows that the short beam shear test gives the interlaminar shear strength of the composite sample, but not the interlaminar fracture energy.

Flexural fatigue study was also performed on laminate samples. Failure took place in an interlaminar shear mode to cause delamination propagation. Elastomer-modified resins, when used in composite, were found to increase the laminate fatigue life by a factor of 10. Fatigue data for the modified system also showed less scattering. Both these effects indicated that the modified composite systems would offer a considerably higher design limit for fatigue, with a trade-off in static strength.

The precipitation of the second phase, along with the high molecular weight modifiers, has a dramatic effect on controlling process rheology. The resulting laminates have reduced (or zero) voids and the matrix possesses adhesive properties for one shot bondable honeycomb sandwich structures.

Future composites will require higher strain/failure values without mechanical sacrifice at service temperature in a saturated condition. Improved 'toughness' versus mechanical strength matrix resins, combined with higher strain fibers, will be needed for future aircraft composites.

INTRODUCTION

Current epoxy composites for both military and commercial applications use TGMDA and DDS as their main matrix components. The specification requirements are for as high as possible service temperature and the best attainable initial mechanical properties. As a neat resin, TGMDA/MDA (methylene dianiline), obtained the highest T_g and modulus among more than a hundred synthesized epoxies.[1] DDS is used as the aromatic amine because the sulfone group provides latency which MDA does not have. But, like MDA, DDS provides molecular fit and relatively stable epoxy/amine links which, when used with TGMDA, give excellent initial mechanical properties. In addition, the semi-solid state of the TGMDA dictates only minor modifications for good prepreg physical properties.

Current high cross-link matrices, although offering high mechanical strength, are inherently brittle materials. The composites where these matrices are used generally fail by the growth of internal flaws and micro-voids that result in crack propagation, eventually leading to a catastrophic breakdown. The importance of polymer fracture properties in structural applications of composite materials has, therefore, been emphasized in recent years.[2-4] The TGMDA/DDS based composites have been previously documented to have very low toughness values (G_{Ic} and impact).[2,3]

In addition to fracture toughness, next-generation commercial aircraft will be demanding higher strain-to-failure in composites. This will increase the structural applications of composites and thus transpose into lighter weight which, of course, greatly improves fuel efficiency. A 2 % strain composite will be sought to increase the current design allowables of 0·003 in/in. In addition, in a higher strain fiber, such a composite with improved fracture toughness will also reduce the 'knock-down' factor of the current composites.

A practice in the industry is to add elastomeric particles to the brittle matrix in order to enhance the resin toughness. The mechanisms for this enhancement are only recently understood as involving triaxial dilation of rubber particles at the crack-tip, particle elongation and matrix plastic flow.[4] However, when such a modified resin system is used in fiber-reinforced composite, the effects of these additive particles on the mechanical properties of the composite are not clear.

Increased 'toughness' of any type cannot come without some sacrifice of *initial* matrix-dominated mechanical properties. Interlaminar strength (as measured by short beam shear), high temperature strength retention and wet strength retention are examples of total matrix domination. Ultimate compressions and ultimate flexural strengths are also strongly influenced by matrix properties. In general, the modulus of the neat resin is lowered by a proportional volumetric amount of the second phase. The second phase essentially reduces the initial load bearing area.[5] Modulus is the most dominant neat resin property which affects the initial composite mechanical properties.[6]

Lowering modulus can actually increase some fiber-dominated initial properties, such as tensile strength, and increase matrix-dominated properties where the strain-to-failure is critical, such as off-axis tensile, transverse tensile and flexural fatigue (S_N curves) properties.

An added indirect benefit of composite toughening is increased laminate processability, especially in large parts with complex curvatures and varying thicknesses. Toughened composites with controlled flow also have the one-shot bonding ability to honeycomb which the current high-flow brittle systems do not possess.

The aircraft industry has an awareness that in order to achieve the desired—and soon to be required—increase in toughness some sacrifices in the initial mechanical properties will be needed. In this chapter an attempt will be made to show examples of the extent of required mechanical sacrifice in order to obtain a 'tougher' composite. Fracture behavior will also be discussed.

EXPERIMENTAL

A series of acrylonitrile–butadiene modified epoxies were used in this study. The nitrile rubbers are commercially available from the B. F. Goodrich Company (Table 1).

Neat resin fracture energies were determined by using standard compact tension specimens. These low-strain opening-mode fracture tests have been documented heavily in the literature in the last several years by Bascom *et al.*,[7] Buchnell and Smith,[8] and others.

The fracture toughness of laminates was determined by using the width tapered beam geometry of Mostovoy and Ripling.[9] Steel plates were used to eliminate adherent yielding during test.

The model systems were prepared in neat resins and in laminates. Thornel 300 3000 K was the graphite and 7781 E glass was the fiberglass. The laminates were all autoclave cured, 60 psi, 1–5 °C heat up rate, and with normal bagging techniques. The castings were more difficult, due to the large resin mass. The 120 °C curing systems were brought up at the same rate (ΔT), but dwelled at 90 °C until gelation. They were then cured (or post cured) for 1 h at 120 °C.

Since morphology will change with significantly different heat up rates,[10] the castings were programmed as closely as possible to match the laminates. Post-failure examinations of specimens were made using scanning electron microscopy.

The laminates were tested at RT and 100 °C in tension, flexure and interlaminar shear for screening mechanical strengths and modulus. Short-beam shear strength (SBS) was concentrated on because it is a total matrix dominated property (with constant fiber geometry).

A Rheometrics Dynamic Spectrometer (RDS) was used to show the different viscosity time/temperature relationships between a control and a toughened resin.

TABLE 1
Elastomeric modifiers

	2000X162 CTB	1300X15 CTBN	1300X8 CTBN	1300X13 CTBN	1300X9 CTBNX	1300X18 CTBNX	1472
Viscosity, Brookfield, MPa's or cP at 27°C (81°F)	60 000	60 000	150 000	570 000	160 000	265 000	—
Per cent carboxyl	1·9	2·5	2·4	2·4	2·9	3·0	≃2·5
Molecular weight	4 800	3 500	3 500	3 500	3 500	3 000	300 000
Functionality	2·0	1·9	1·8	1·8	2·3	2·3	150
Acrylonitrile content (%)	0	10	18	26	18	21·5	27
Specific gravity at 25°C (77°F)	0·907	0·924	0·948	0·960	0·955	0·958	—
Solubility parameter	8·04	8·45	8·77	9·14	8·78	—	?

Flexural fatigue studies were also performed on model laminate samples. Flexure specimens were clamped on both ends of a fixture and then the necessary load applied for a 0·5 in deflection at a known frequency. A cold air jet was applied to minimize hysteresis heating effects. The samples were tested to failure.

Impact testing was performed with an Effects Technology instrumented drop weight system.[11] P_I represents the incipient damage point on the composite surface and P_F is the maximum load a specimen can sustain in thorough penetration testing.

RESULTS

Table 2 summarizes the nine epoxies used for this model study and lists the G_{Ic} results of the neat resins.

In resins B and C (with A as control) 1300 × 13 CTBN (26 % C ≡ N) was used, while in all other resins 1300 × 8 CTBN (18 % C ≡ N) was the liquid elastomeric modifier of choice.

The 1300 × 13 CTBN was previously found to be the most effective toughening liquid rubber in TGMDA based resins.[12] In the less polar DGEBPA based system, the lower C ≡ N per cent is adequate for initial compatibility and is different enough in solubility parameter to precipitate out during increasing matrix molecular weight, but prior to gelation.

TABLE 2
Model matrix resins with G_{Ic} values

Resin	Type	Mod.*	% Second-phase† Liquid	Solid	Other nomenclature	G_{Ic}
A	TGMDA/DDS	High	None	None	F-263, 5208 3501, etc. type	0·08
B	TGMDA/DDS	High	3·7	None	—	0·158
C	TGMDA/DDS	High	2·2	1·5	—	0·239
D	DGBPA/PAP	Medium	2·3	3·4	F-155	1·72
E	DGBPA/BPF	Low	None	None	HX205	0·203
F	DGBPA/BPF	Low	8·1	None	HX206	1·8
G	DGBPA/BPF	Low	None	8·1	HX210	3·3
H	DGBPA/BPF	Low	8·1	5·4	F-185	5·8
I	?	?	?	?	BP937	?

* Relative first phase modulus.
† Assuming 100 % precipitation.

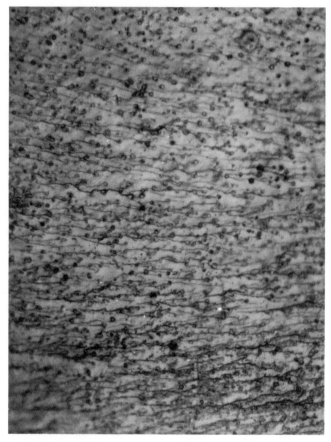

FIG. 1. Fracture surface—Resin B. (× 500).

Different amounts of second phase morphology have been directly linked to the polarity of the rubber (governed by $\%$ C ≡ N) with constant formulation and kinetics.[10]

Resins E, F, G and H have been extensively studied by Bascom *et al.*[7] and by Siebert and Rowe[13] in addition to the present authors,[14] and detailed descriptions of the fracture mechanics, failure mechanisms, morphology and physical structure/property relationships have been documented. Only a very generalized summary will be presented here.

In the A, B, C series one SEM picture of the fracture surface was taken of the B resin. Figure 1 clearly shows the second phase precipitation, out of the

TGMDA/DDS first phase. An important observation is the total lack of cavitation, and apparent absence of shear band deformation, both of which are evident in DGEBPA based systems (Resins D–H) and are necessary for significant toughening improvement. Only slight G_{Ic} improvement was evident in this case.

Figure 2 shows six morphology pictures which clearly show the heterogeneous second phase results of the failed fracture surface. Pictures 1 and 2 in Fig. 2 are of resin F (low $M_{\bar{w}}$ liquid rubber only). A very uniform dispersion of particles in the 0·3–0·5 μm range is clearly evident. Pictures 3, 4, 5 and 6 are of resin H. The 40 μm scale of Pictures 3 and 4 clearly shows the large resulting cavities caused by the solid rubber which are not evident in Pictures 1 and 2. Heavy shear band deformation is also observed. Pictures 5 and 6, with higher magnification, show the presence of the smaller cavitation, as in Pictures 1 and 2, in addition to this large cavitation.

It is believed that the large elastomer particles nucleate local shear yielding to distribute stress over a larger area which allows more energy to be dissipated through various mechanisms, the exact nature of which is still controversial. Resin E, which is the control, is not shown here, but the fracture surface from this resin sample showed a very smooth brittle fracture with no cavities.

All G_{Ic} values and morphology photographs in this chapter were taken from samples fractured at a constant strain rate of $2·0 \times 10^{-3}$ cm/sec. The following observations can be made:

For the DGEBPA series:

(1) Both liquid and solid elastomer additives increase the fracture energy of the epoxy by orders of magnitude.

(2) When used alone, the solid and liquid (low and high $M_{\bar{w}}$) elastomers have about the same effect on toughness.

(3) When both are present, the fracture energy is increased by as much as $\times 2$. There is a synergistic toughening mechanism in this formulation containing a bimodel particle size distribution.

For the TGMDA/DDS series

(1) There was only slight toughening over the controls ($\times 2$ for B and $\times 3$ for C).

(2) Lack of apparent plastic flow or shear band deformation in the first phase resulted in no cavitation, which is required for significant toughening. No stress whitening was evident.

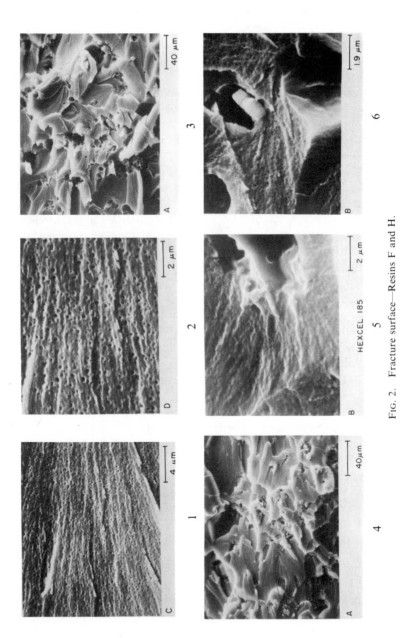

FIG. 2.   Fracture surface—Resins F and H.

In addition to the above, very recent work has been done showing the effect of neat G_{Ic} as a function of strain rate.[14] Notched Izod data were converted to G_{Ic} for high rate results. All elastomer epoxy compositions showed a general decrease in fracture energy with increasing strain rate. There was no rate effect evident for the DGENPA control (resin E). For resin H, stress whitening decreased significantly as strain rate increased, which correlated with lower G_{Ic}. The impact specimens show no evidence of stress

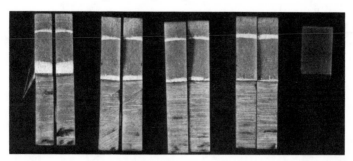

FIG. 3.   Stress whitening versus strain rate. Strain rate increases from left to right.

whitening. It appears that the second phase is time dependent in the ability to distribute stress and 'toughen'.[4] In Fig. 3 the stress whitening of the fracture surfaces is shown, the extent of which is approximately proportional to the G_{Ic} of the resin.

Although the neat resin data have been previously shown to be very rate dependent, resins A, D and H were shown to be rate independent when manufactured into a composite. The presence of fibers and/or the fiber spacing appears to counteract the base resin sensitivity.[3]

In Fig. 4 the G_{Ic} (log) values of neat resin were plotted against 200 °F interlaminar shear strength (SBS). 200 °F was chosen as the temperature because that is the current upper limit design temperature for commercial aircraft. Figure 5 is one of the first (probably of many) showing the quantitative inverse relationship between matrix dominated mechanical strength versus neat resin toughness (as measured by G_{Ic}). Today, any of the current available composites will most likely give data that plot very closely to this straight line. Other mechanical tests, temperatures and environmental conditioning values could be used on the vertical axis. The inverse relationship will be similar although the slopes will obviously differ.

Figure 5 shows that P_I and P_F impact data versus the same 200 °F SBS value. Here the data are much more scattered and there appears to be only minor second-phase dominated influence. The behaviors of resins C and I

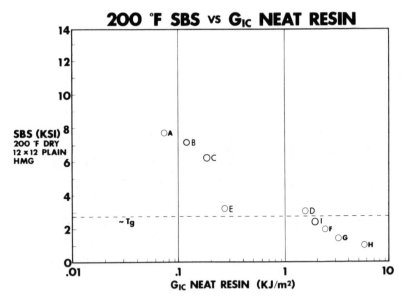

FIG. 4. Interlaminar shear versus G_{Ic}.

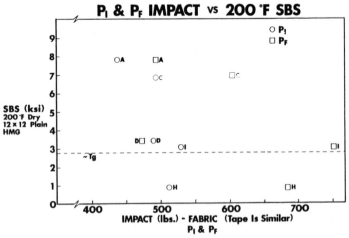

FIG. 5. Impact versus interlaminar shear.

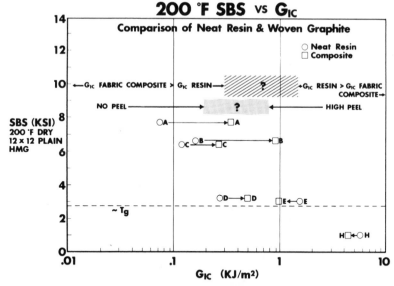

FIG. 6. Comparison of neat resin and woven graphite G_{Ic} versus interlaminar shear.

versus that of resin H clearly show that the high G_{Ic} value of a resin does not necessarily translate into high impact strength.

Figure 6 shows the effects of G_{Ic} of neat resin and G_{Ic} of woven graphite composite on 200 °F SBS. An added 'peel' property was evident. Note that the more brittle matrix resin, i.e. the resins with G_{Ic} less than about 1 kJ/m^2, have composite fracture toughnesses greater than that of the neat resin. In these systems the tortuous path of the weave aids in the toughness. About 1 kJ/m^2 is a threshold where a suppression of toughness occurs from neat resin to composite. The reason for toughness suppression is most probably geometry dominated and, if so, would be consistent with the work of Bascom and Cottington[15] indicating that the toughness of two-phase adhesives is thickness dependent. The spacing of the distance between graphite fibers is consistent with what these authors demonstrated with aluminum adherents, although, in composites with a high G_{Ic} matrix, some of the stress distribution translates intraply.[2]

Figure 7 descriptively shows the rheology difference between an 'A' resin and an 'H' type resin. The excessive flow during processing due to this low minimum viscosity of approximately 400 cps is causing a high reject rate (even when flow is constrained) because of high void content. The problem is compounded in large parts with complex shapes. The added toughness,

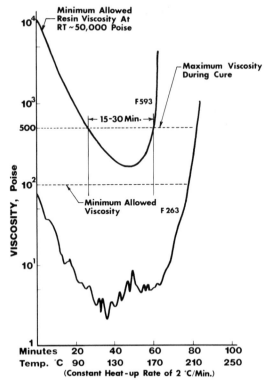

FIG. 7.   Process rheology curve.

combined with the resulting controlled flow, provides co-curable honeycomb sandwich structure (providing approximately 5 % more resin is used on prepreg for adequate filleting). Core failures with flatwise tensiles of greater than 600 psi are not uncommon on non-metallic core. The cost savings due to adhesive elimination are significant.

Figure 8 shows the effect of three model systems in flexural fatigue. (It is hypothesized that little effect, if any, would be evident in tension fatigue.) Elastomer modifications were found to increase the laminar fatigue life by a factor of 10. Fatigue data for the modified system also showed less scattering. Highly stressed composites often fail in interlaminar shear fatigue (leaf springs, for example), although engineers hardly ever design for it. Both these effects show that the modified composite systems would offer a considerably higher design limit for fatigue, provided the high values of initial matrix dominant strengths could be relaxed.

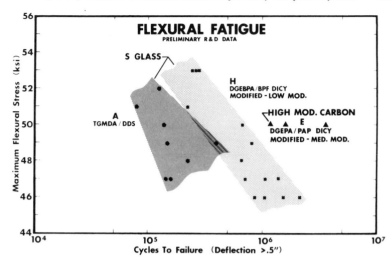

FIG. 8.   Flexural fatigue comparison.

CONCLUSION AND FUTURE

The requirements for composites are becoming more sophisticated as they get closer towards primary structural applications in commercial aircraft. The future could require a minimum 'average tensile and compressive strain-to-failure of 20 000 micro-in/in. In addition, the current modulus and other mechanical properties must be met after increased strain at the required service temperature in a moisture saturated condition.

200 °F wet (saturated) mechanical strength, flaw sensitivity and impact are three design 'knock-down' factors. Significant weight reduction in aircraft by incorporating composites will not be achieved unless more of the initial strength can be utilized. These 'knock-down' factors must be reduced.

The toughness and matrix strain properties can be significantly improved today, but only at prohibitive sacrifice of the required mechanical properties. The fiber manufacturers are making significant progress for their higher strain requirements. According to the independent calculations of Chamis *et al.*[6] and Christensen and Wu,[16] a magnification factor of × 5 to × 10 is needed for matrix in a (0·90) composite over the strain of the fiber. To the present authors the goals seem quite clear. Schematically one may modify Fig. 5 to establish desirable property domains, such as those shown

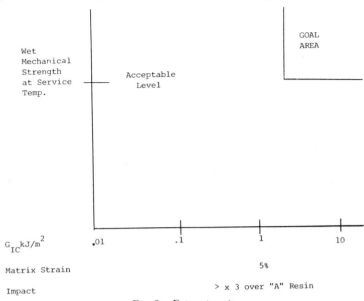

FIG. 9. Future targets.

in Fig. 9. The toughness of the neat resin should be at least $1 \, \text{kJ/m}^2$ so that, when a composite with woven fiber is made, the lay up geometry and fiber volume will not affect the flaw sensitivity. The 5% matrix strain will be required to prevent premature interface failure due to uneven stress concentration factors between fibers under transverse stress conditions. The second phase appears to be very G_{Ic} dominant, but only marginally affects the high rate stresses, such as impact. This is because the second phase needs 'time' to allow its various failure mechanisms to become operative for energy dissipation. On the other hand, the first phase must also be significantly improved to reach the top right hand corner of Fig. 9. Thermoplastics or thermoset resins that contain linear blocks or domains may possibly be needed for the improved first phase. The acrylonitrile butadiene elastomers that are currently used for the second phase will most probably be used in future thermosets, with the modified first phase. Less than 10% of these modifiers will be in the final matrix. Whether the future involves thermoplastics, combinations, hybrids, one phase, two phase, three phase or whatever, the polymer chemists need to molecularly design resins to reach the top right hand corner of Fig. 9. Only then can the aircraft industry efficiently utilize the attractive potential of organic matrix composites.

REFERENCES

1. Buss, N., Technical Report AFML-TR-68-286, Part I, September, 1968. AFML-TR-69-238, Part II, January, 1970.
2. Bascom, W. D., Bitner, J. L., Moulton, R. J. and Siebert, A. R., *Composites*, 1 (1980) 9.
3. Miller, A. and Hertzburg, P. E., *Toughness testing of composite materials*, S.A.M.P.E. preprints, October, 1980, Tech. Conf., Seattle, WA.
4. Bascom, W. D., Moulton, R. J. and Ting, R. Y., The fracture of an epoxy polymer containing elastomeric modifiers, *Journal of Materials Science*. (In press.)
5. Ophir, Z. H., Emerson, J. A. and Wilkes, G. L., Sub-annealing studies of rubber-modified and unmodified epoxy systems, *J. Appl. Phys.*, **49**(10) (October, 1978).
6. Chamis, C. C., Hanson, M. P. and Serafini, T. T., *Modern Plastics*, **5** (1973) 90.
7. Bascom, W. D., Moulton, R. J., Rowe, E. H. and Siebert, A. R., *ACS Organic Coating and Plastics Preprint*, **39** (1978) 164.
8. Buchnell, C. B. and Smith, R. R., *Polymer*, **6** (1965) 437.
9. Mostovoy, S. and Ripling, E. J., *Appl. Polymer Sym.*, **19** (1972) 395.
10. Manzione, L. T., Gillham, J. K. and McPherson, C. A., *Rubber-modified epoxies: Transitions and morphology*, Princeton Univ. NJ Dept. of Chemical Engineering, 1 September, 1980.
11. Wogulis, S. G., Whitney, B. W. and Ireland, D. R., *Automated data acquisition and analysis for the instrumented impact test*, Effects Technology Technical Report TR-78-13, 1979.
12. Lee, B. L., Lizak, C. M., Riew, C. K. and Moulton, R. J., *Rubber toughening of tetrafunctional epoxy resin*, S.A.M.P.E. preprints, December, 1980.
13. Siebert, A. R. and Rowe, E. H., Paper presented at the 161st ACS Mtg. Org. Coat. Plast. Div., Los Angeles, CA, USA, 1971.
14. Ting, R. Y. and Moulton, R. J., *S.A.M.P.E. Tech. Ser.*, **12** (1980) 265.
15. Bascom, W. D. and Cottington, R. L., *J. Adhesion*, (1976) 333.
16. Christensen, R. M. and Wu, E. M., *Optical design of anisotropic (fiber-reinforced) flywheels*, Lawrence Livermore Laboratory, November, 1976.

47

A Comparison of the Failure Pressure as Predicted by Finite Element Stress Analysis with the Results of Full Scale Burst Tests on GRP Flanges

A. MUSCATI AND R. BRADFORD

*Central Electricity Generating Board (South Western Region),
Bedminster Down, Bridgwater Road, Bristol BS13 8AN, England*

ABSTRACT

The burst pressure from full scale burst tests on GRP pipes with integral GRP flanges is compared with the predicted failure pressure using glass content analysis and material strength data for the different composite layers. Two theoretical models were used to predict the failure load; a simple analytical solution for a plain pipe and a detailed finite element stress analysis including the flange geometry and loading. In both cases, the failure pressure was generally overestimated and it is suggested that this may be due to difficulties in construction resulting in the composite layers close to the flange having inferior strength as compared with a basic pipe with the same glass content.

1. INTRODUCTION

An experimental programme of full scale burst tests was carried out to assess the long term integrity of a glass reinforced plastic (GRP) pipe installation. The results of earlier burst tests[1] showed the flanges to have a much lower strength than the basic pipes. This was attributed to the difficulties in construction resulting in inadequate roving reinforcement at the flanged joint.

Further burst tests were carried out after the publication of Ref. 1 to test flanges extracted from the installation. Using the results of all the burst tests, an attempt was then made to correlate the burst pressure with the glass content for some of the test specimens using both analytical and finite element stress analysis techniques. Also, a practical method for reinforcing these flanges on site was developed in the laboratory and applied to full scale flanged joints, which were pressure tested to failure. Details of the experimental programme and the stress analysis are given here with the emphasis on a comparison between the predicted and actual failure pressures.

2. PIPE GEOMETRY AND CONSTRUCTION

A full description of the pipe and flange construction is given in Ref. 1. Basically, it is a GRP pipe with integral GRP flanges. The pipe is lined with unplasticised polyvinyl chloride (UPVC) and made up of layers of chopped strand mat (CSM) and rovings. The CSM layers consist of randomly orientated glass fibres providing reinforcement in both axial and hoop directions whilst the rovings have unidirectional fibres along the hoop

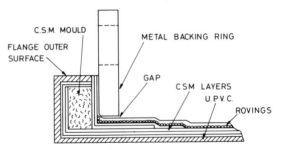

FIG. 1. A schematic diagram for the flange construction.

TABLE 1
Nominal properties[a] of different layers of the composite pipe (I.D. 304 mm)

Layer	Thickness (mm)	Young's modulus (N/mm^2)	UTS (N/mm^2)	Designed glass content kg/m^2
UPVC	3·1	2·76 × 10^3	48	
CSM	4·0[b]	6·41 × 10^3	93	2·14
Rovings	3·2	3·45 × 10^4	827	3·89

[a] Poisson's ratio assumed to be 0·4.
[b] Locally thicker near flange (up to 6·3 mm) corresponding to an increased designed glass content of 3·36 kg/m^2.

direction only. Details of the dimensions and material properties, as supplied by the manufacturers, are given in Table 1 for each layer. A schematic diagram showing the flange construction is shown in Fig. 1.

3. EXPERIMENTAL WORK

3.1. The Burst Tests

Details of the testing rig and the results of the earlier tests are given in Ref. 1. In brief, the specimen is a full scale pipe with an integral GRP flange at each end bolted to a steel plate. The specimens were tested inside a cage which can provide practically full axial restraint.

Table 2 gives the results for all the tests including those published earlier. Note that some of the test results are not relevant for the comparison with the theoretical calculations but are given here for completeness.

TABLE 2
Results of burst tests

Specimen no.	Type[a]	End conditions	Failure pressure (N/mm²)	Description	Crack position
1	A	fully restrained	4·5	burst	at flange
2	A	fully restrained	5·4	leak	at flange
3	A	fully restrained	5·4	leak	at flange
4[b]	A	fully restrained	2·4	leak	at flange
			11·7	burst	
5	C	7·6 mm axial gap	3·4	burst	at flange
6	A	7·6 mm axial gap	5·2	burst	at flange
7	C	7·6 mm axial gap	1·7	burst	at flange
8	E	7·6 mm axial gap	3·96	burst	at pipe joint
9	B	fully restrained	3·63	leak	at flange
10	E	fully restrained	4·64	burst	at pipe joint
11	B	2·5 mm axial gap	4·51	burst	at flange
12	D	fully restrained	6·20	burst	at flange
13	D	free	5·44	burst	at flange
14	B	2·5 mm axial gap	5·57	burst	at flange

[a] Type A: Pipes manufactured for testing but intended to simulate site flanges. Type B: Pipes extracted from site. Type C: Specially manufactured pipes with substandard flanges. Type D: Specially manufactured with strong flanges. Type E: Failure occurred at the pipe away from the flange.
[b] Specimen No. 4 is 250 mm diameter.

3.2. Glass Analysis

Glass content analysis was carried out on most of the specimens but only 10 of them are relevant for the comparison with the theoretical predictions. The procedure was simply to burn the polyester resin in a furnace and weigh the remaining glass. The results were then used to predict the failure pressure based on the materials data and stress calculations.

In two specimens (Nos. 8 and 10), failure occurred at a hand-laid butt joint in the pipe away from the flange. In this case, the glass analysis was carried out on two samples for each specimen taken at the butt joint. In each of the remaining eight specimens, failure occurred at one of the flanges and the area of interest is the pipe immediately adjacent to the failed flange. The glass analysis showed that the glass content for the rovings in the pipe decreases as the GRP flange is approached. The stress calculations were therefore based on the glass content analysis of the pipe immediately adjacent to the GRP flange, up to about 20 mm from it.

Table 3 gives the relevant results of the glass content analysis which correspond to the minimum values.

TABLE 3
Details of glass analysis (kg/m^2)

Specimen no.	CSM	Rovings		Failure pressure (N/mm^2)		Crack orientation
		Hoop	Axial	Actual	Predicted[a]	
1	3·04	2·11		4·5	9·1	axial[b]
2	2·87	1·63		5·4	7·8	axial
3	3·36	2·61		5·4	10·6	axial
5	2·80	0		3·4	4·1	axial[b]
6	4·03	0		5·2	5·4	axial[b]
7a[c]	1·44	0·19			2·9	
7b	1·09	0·83		1·7	3·6	circumferential
8[d]	1·86	0·6	4·45	4·0	4·2	axial
10[d]	1·97	0·6	4·67	4·6	4·3	axial
11	1·93	1·90		4·5	6·9	axial[b]
14	1·94	2·15		6·0	7·4	axial[b]

[a] The predicted failure pressure was based on the analytical solution.
[b] Although these cracks initiated in the axial direction the crack orientation changed after it propagated away from the flange to a circumferential crack.
[c] a and b are top and bottom flanges, the crack from the burst test was at the bottom flange.
[d] Data refer to butt joint away from the flange.

4. STRESS ANALYSIS

4.1. Analytical Solution

A simple analytical solution, for the stresses in different layers of a composite pipe under pressure, may be obtained using the properties of each layer. Such a solution is given in the appendix for the present pipe under both fully restrained and free end conditions. Note that this analysis ignores the flange geometry and only considers a composite pipe. This is applicable to specimens Nos. 8 and 10 where failure occurred at the composite pipe. However, for the remaining specimens the failure was at the flange but the analysis treats the pipe adjacent to the flange in a similar manner to the basic pipe, ignoring the flange geometry.

In using the solution given in the appendix to predict the failure pressure, the following points should be noted:

(1) Failure was assumed to occur when the maximum principal stress in any layer exceeded the ultimate tensile strength (UTS) of the material. For all these specimens, the critical values correspond to the hoop stress in the CSM layer.

(2) In the majority of the burst tests considered, an axial gap was left between the specimen and the cage resulting in only a partial axial restraint. For the purpose of the calculations, the end conditions were assumed to be free up to the pressure necessary to close the gap. This pressure was found from the axial stiffness of the specimen which was determined experimentally to be about $0.8 \, \text{N/mm}^2$ per mm. Once the gap is closed the cage was assumed to provide full axial restraint.

(3) The thickness of the CSM and roving layers was calculated from the data given in Table 1 for the basic pipe and the results of glass content analysis in Table 3. This was simply done by scaling the thickness by the ratio of the glass contents as compared to the basic pipe.

The calculated failure pressures based on the above procedure are given in Table 3 and compared with the actual failure pressures.

It is interesting to note that the agreement between the calculated and actual failure pressures is very good for specimens Nos. 8 and 10 (within $\pm 10\%$) where failure occurred away from the flange. However, for the remainder of the specimens, the calculations tend to overestimate the failure pressure by a factor of up to two even though the agreement for some of the specimens is quite good, e.g. Nos. 5 and 6. At this stage, it was difficult to judge whether this difference between the actual and calculated failure pressure is due to the geometrical approximations in the analytical solution or simply a function of scatter in material properties. A more accurate solution using finite element stress analysis was therefore attempted, to model the actual flange geometry.

4.2. Finite Element Analysis
The 'plain pipe' model outlined in the appendix, and whose predictions are given in Table 3, may be improved by taking into account; (a) the flange geometry, and (b) stresses arising other than from pressurisation. In addition, the effects of anisotropy and inhomogeneity must clearly also be taken into account.

4.2.1. Method of analysis
The computer programs used were BERSAFE[2] and its associated programs. An outline of the boundary of the mesh employed is shown to scale in Fig. 2 and the flange region is shown in detail in Fig. 3. The refinement in this region is considerable. The elements are axisymmetric, isoparametric elements with 16 degrees of freedom (8 nodes). Only half the pipe is modelled, the remainder being symmetrical. The steel backing plate, backing ring and bolts are included in the mesh. The mesh therefore consists of five different materials, these being steel, UPVC, CSM, rovings and a filler for the flange body. The latter has a Young's modulus of 3.5×10^3 N/mm^2.

The only medium treated as anisotropic was the roving layer. At the time that this analysis was carried out, axisymmetric anisotropic elements were not available in BERSAFE and so the anisotropy was modelled by decoupling the axial degrees of freedom between the roving and CSM layers. Since the layer of rovings is then free to slide axially along the pipe, no axial stresses can occur in the rovings. Note that for convenience in

GRP FLANGE
BOLT
BACKING RING

FIG. 2. Scale plot of the outline of the mesh.

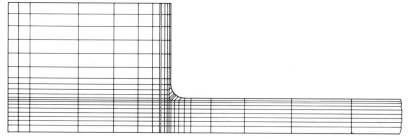

FIG. 3. Scale plot of the mesh in the flange region showing individual elements.

defining the mesh the roving layer has been modelled as being thinner than its actual value, but with an appropriately scaled Young's modulus. Since hoop bending stresses cannot occur, this is of no consequence.

The paucity of rovings close to the flange was modelled by omitting the initial roving elements up to a distance of 74 mm from the back face of the flange. This is admittedly a crude procedure, in that the specimens actually exhibited a gradual increase in the amount of rovings with axial position. Consideration of the extent of roving depletion along the axial length for each specimen, implies that the model is a good approximation for specimen 6, whilst being optimistic for specimens 5 and 7. For the remaining specimens, however, the model is clearly pessimistic, i.e. the model should underestimate the failure pressure, since the actual roving depletion was not as severe as that of our model.

Finally note that the backing plate and the flange are topologically distinct, so that a gap may open between them.

4.2.2. Loadings

Three loads were applied separately to the mesh, as described below:

(1)   Pressure was applied to the cylindrical surface with the ends of the mesh axially restrained. This corresponds to the 'restrained' case of the appendix. The front face of the flange is left radially free during this loading. The radial freedom of the flange is further ensured during this loading by assigning zero stiffness to the 'steel' sealing arrangement.

(2)   An axial displacement of the pipe of 7·6 mm was simulated by applying half this displacement to the steel backing plate whilst axially restraining the plane of symmetry.

(3)   A pressure equivalent to the bolt load due to tightening was applied to the steel backing ring at the bolt radius.

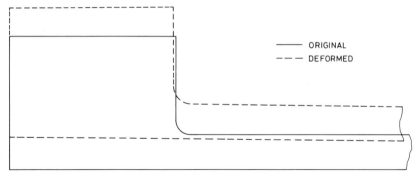

FIG. 4. Deformation plot for the mesh with complete rovings (pressure load, $1 \, N/mm^2$). Scale × 2·41 for mesh and additional × 50 for displacement.

These loads were applied both to the mesh representing the as-designed pipe and to the mesh without the initial 74 mm of rovings. Linear elasticity was assumed throughout and hence the effects of combined loads are additive.

4.2.3. Results

A brief summary of the results of the finite element analysis are as follows:

(1) For pressure loading only and with all rovings present it was found that axial bending stresses were absent. This is illustrated by the exaggerated displacement plot of Fig. 4. The radial stiffness of the flange is balanced by the radial stiffness of the rovings. This is in contrast to the result obtained from the mesh without the initial rovings, the displacement plot being shown in Fig. 5. The corresponding stresses in the CSM layer are shown in Fig. 6 for unit

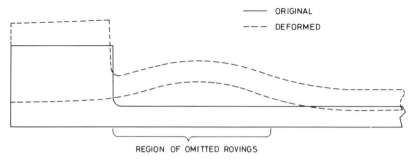

FIG. 5. Deformation plot for the mesh without initial rovings (pressure load, $1 \, N/mm^2$). Scale × 1·48 for mesh and additional × 50 for displacement.

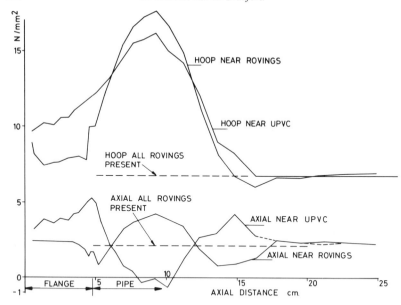

Fɪɢ. 6.   CSM stresses (pressure load, 1 N/mm²), initial rovings missing.

pressure (1 N/mm²). The increase in the hoop stress caused by the
absence of the rovings is clearly considerable. The axial bending
stresses which occur are sufficiently small such that the hoop stress
is the dominant effect.

(2) The axial displacement load causes compressive stresses in the
     rovings. The CSM and UPVC stresses, other than in the flange
     body, for the as-designed pipe are:

     CSM   18·9 N/mm² (axial), 5·4 N/mm² (hoop)
     UPVC   9·1 N/mm² (axial), 2·6 N/mm² (hoop)

     Slight changes from these values occur when the initial rovings are
     omitted. In particular the hoop stresses near the flange decrease
     when the rovings are omitted.

(3) The stresses due to bolt loading may be up to 6 N/mm² in the body
     of the flange but are small in the pipe itself, typically 1 N/mm².

4.2.4. Estimate of the failure pressure

To estimate the failure pressure, the stresses resulting from the three
loading cases are added together and each component of stress for each

medium is compared with the UTS given in Table 1. Moreover, this must be done at several characteristic axial positions, since the stresses vary with position. This leads to the following conclusions:

(1) The enhanced CSM hoop stress of Fig. 6 dominates, and it is concluded that this stress component leads to failure.

(2) On the basis of the data of Table 1 failure pressures of $5 \cdot 0 \text{ N/mm}^2$ (with axial displacement), or $5 \cdot 2 \text{ N/mm}^2$ (fully restrained) are estimated.

(3) The bolt loading has negligible effect on the failure pressure, whereas the axial displacement reduces the failure pressure by about 4%.

4.2.5. Discussion of the analysis

The finite element analysis models a specimen of CSM glass content of $3 \cdot 36 \text{ kg/m}^2$ with and without rovings. Taking the case of no rovings and full axial restraint, the simple analytical method described earlier would predict a failure pressure of $4 \cdot 7 \text{ N/mm}^2$, (11% lower than the finite element prediction). The main conclusion of our finite element analysis is therefore to validate the assumptions made in the simplistic analytic model. The apparently complicating features of the flange geometry and the extra loading due to bolt tightening do not lead to changes in the estimated failure pressure. The fact that the rovings are omitted only for a 74 mm length in the finite element model, whilst the simple method assumes a plain pipe with the rovings missing along its entire length, explains the small difference in the predicted failure pressures.

The finite element analysis has therefore failed to improve the agreement between the predicted and observed failure pressures quoted in Table 3. Rather it has confirmed that the predicted failure pressures are slightly on the low side. In as far as this analysis is believed to be complete, it may be concluded that the disparity between predicted and actual failure pressures is a result of the material data or the failure criterion employed. The disparities cannot be explained in terms of simple material scatter since the burst pressures are consistently lower than the predicted values. This points to a greater preponderance of inherent weaknesses in the tested pipes than in the specimens used in the tensile tests to deduce the data of Table 1. This may be either a consequence of the geometry of the flange leading to poorer lay-up and hence a greater number of voids, etc., per unit volume, or simply a statistical effect due to the larger volume of material under stresses of failure magnitude in the pipes as compared with the tensile tests.

5. REINFORCEMENT OF SITE FLANGES

Fortunately, although the results of the burst tests showed that the GRP flanges had inadequate strength as compared to the design calculations, they were still considered acceptable under the operating conditions. This is mainly because the system was operated at a much reduced pressure (40 % below design). Nevertheless, it was still necessary to develop a reinforcement technique for application on site should the system be uprated to the design conditions.

Two reinforcement techniques (A and B) were developed in the laboratory and applied to full scale pipe specimens with flanged joints. These specimens were pressure tested to failure to determine the effect of the reinforcement.

(A) In this case, the approach was to constrain the radial displacement of the flange by using GRP blocks placed on the flange outer surface between the flange bolts (see Fig. 1). These blocks were held in position by a metal split ring which provided the radial constraint. The two halves of the split ring were joined together by two bolts which were tightened up sufficiently to hold the assembly without imposing a significant preload on the flange.

(B) The objective of the second reinforcement technique was to compensate for the lack of sufficient rovings in the pipe adjacent to the flange by providing additional restraint at this position. The method was to inject epoxy resin to fill the gap between the backing ring and the flange (see Fig. 1) so that the backing ring can provide the required reinforcement. This technique was applied to the test specimen with the flange joints assembled and bolted, to simulate site conditions. The main problem was to find a suitable resin that could be used to fill a variable gap (0·25 mm to 3·0 mm), remaining in the gap until it hardens. After a number of trials, the method used was as follows: A standard high pressure grease gun with a specially made flat nozzle was used to inject the resin into the gap. The compound used was a commercial epoxy resin made by CIBA-GEIGY consisting of a resin, 'AV138', and a hardener, 'HV998', the ratio in weight of resin to hardener being 2·5.

A total of four burst tests were carried out, one on a standard specimen without reinforcement, two on specimens with reinforcements of type A and one on a specimen with both types of reinforcements, A and B. The use of reinforcement A alone increased the burst pressure at the flange by a small amount (15 %) but the application of both reinforcement techniques resulted in an increase in the burst pressure by a factor of approximately two. The latter was therefore recommended should the system be uprated to the design conditions.

6. CONCLUSIONS

A study of the results of the full scale burst tests together with the associated theoretical analysis leads to the following conclusions.

(1) It is possible to model the anisotropic properties of the rovings in the finite element analysis by using isotropic elements but partially decoupling along the boundary nodes of the roving layer.

(2) The lack of sufficient rovings in the pipe adjacent to the flange can cause a serious reduction in the burst pressure.

(3) The local bending stresses resulting from the flange geometry and loading has little effect on the burst pressure even though it can cause a significant increase in the axial stress.

(4) Both finite element and simple analytical solutions tend to overestimate the failure pressure. It is suggested that this is mainly due to the difficulties in construction resulting in a reduction in the strength of the different composite layers in the pipe adjacent to the flange as compared to a basic pipe with the same glass contents.

The above conclusions are only applicable to the particular geometry and loading conditions described in the present paper. The last conclusion, however, implies some caution is needed in assessing the load-bearing capacity of GRP structures.

7. ACKNOWLEDGEMENT

The permission of the Director General of the Central Electricity Generating Board, South Western Region for publication of this paper is acknowledged.

8. REFERENCES

1. MUSCATI, A. and BLOMFIELD, J. A., Full scale burst tests on GRP pipes. In: *Designing with fibre reinforced materials*, London and New York, Mechanical Engineering Publication Ltd for the Institution of Mech. Engineers, 1977.
2. HELLEN, T. K. and PROTHEROE, S. E., BERSAFE finite element system, *Computer Aided Design*, 6 (1974) 15–24.

APPENDIX: STRESS ANALYSIS FOR TEST SPECIMEN
(NOMINAL PIPE)

Notation

P pressure
D pipe diameter
E Young's modulus
v Poisson's ratio
t thickness of different layers
ε strain
σ stress

Subscripts

1   UPVC
2   CSM
3   rovings circumferential
x hoop direction
y axial direction

The problem is that of a composite pipe made of 3 layers; UPVC, chopped strand mat (CSM) and rovings. Whilst UPVC and CSM are considered to be isotropic, the rovings are unidirectional providing reinforcement in the hoop direction only. Two cases are considered, fully restrained and unrestrained end conditions.

(a) The Unrestrained Case
Equilibrium

$$\frac{PD}{2} = \sigma_{1x} \cdot t_1 + \sigma_{2x} \cdot t_2 + \sigma_{3x} \cdot t_3 \tag{1a}$$

$$\frac{PD}{4} = \sigma_{1y} \cdot t_1 + \sigma_{2y} \cdot t_2 \tag{1b}$$

$$(\sigma_{3y} = 0)$$

Compatibility

$$\varepsilon_{1x} = \varepsilon_{2x} = \varepsilon_{3x} = \varepsilon_x \tag{2a}$$

$$\varepsilon_{1y} = \varepsilon_{2y} = \varepsilon_y \tag{2b}$$

Stress–strain relations

For UPVC and CSM

$$\sigma_x = \frac{E}{1 - v^2}(\varepsilon_x + v\varepsilon_y) \tag{3a}$$

$$\sigma_y = \frac{E}{1 - v^2}(\varepsilon_y + v\varepsilon_x) \tag{3b}$$

For the rovings

$$\sigma_{3x} = E_3\varepsilon_x \tag{3c}$$

Solving eqns (1), (2) and (3) gives

$$\varepsilon_x = \frac{PD}{4}\left(\frac{2 - v}{M_1 - v^2 M_2}\right) \tag{4}$$

and

$$\varepsilon_y = \frac{PD}{4vM_2}\left(2 - \frac{(2 - v)M_1}{M_1 - v^2 M_2}\right) \tag{5}$$

where

$$M_1 = \frac{1}{1 - v^2}(E_1 t_1 + E_2 t_2) + E_3 t_3$$

and

$$M_2 = \frac{1}{1 - v^2}(E_1 t_1 + E_2 t_2)$$

All Poisson's ratios are taken as being equal to 0·4, and the thicknesses, t, are scaled according to the glass content. The Young's moduli and ultimate strengths used are those of Table 1.

(b) Full Axial Restraint

The analysis is similar to the previous case. Only one equilibrium equation is required, e.g. eqn. (1a), the same compatibility equations can be used with $\varepsilon_y = 0$ and the same stress–strain relations as before. This gives

$$\varepsilon_x = \frac{PD}{2M_1} \tag{6}$$

and

$$\varepsilon_y = 0 \tag{7}$$

The same approach was adopted to obtain a solution for the case of a 4-layer pipe which includes axial rovings; see Table 3 for specimens Nos 8 and 10.

48

Elastic–Plastic Flexural Analysis of Laminated Composite Plates by the Finite Element Method

Faten F. Mahmoud*

Faculty of Engineering, Zagazig University, Zagazig, Egypt

ABSTRACT

The need for elastic–plastic flexural analysis of ductile composites continues to grow with the development of new laminated systems and applications which, in turn, often require more demanding performance specifications. To address this need, an economical elastic–plastic laminated finite element is formulated by combining the theory of plasticity for homogeneous materials with the classical laminated plate theory. The outcome is a plate element, rectangular by choice, capable of representing the flexural behaviour of the laminated system.

NOTATION

$[B]$	Strain–displacement matrix operator.
$[C^e]$	Elastic compliance matrix.
$[C^p]$	Plastic compliance matrix.
$[D]$	Stress–strain matrix operator.
$\{d\}$	Displacement vector.
H	Slope of equivalent stress–plastic strain curve.
$[K]$	Element stiffness matrix.
$\{M\}$	Bending moment vector.
$\{X\}$	Laminate middle surface curvature vector.

* Formerly Visiting Associate Professor, Civil Engineering Department, University of Wisconsin–Milwaukee, USA.

Z   Geometric co-ordinate along the thickness.
Δ Increment symbol.
$\{\varepsilon\}$ Total strain.
$\{\varepsilon^{p}\}$ Plastic strain vector.
$\{\varepsilon^{e}\}$ Elastic strain vector.
$\{\sigma\}$ Stress matrix.
$[\sigma_{e}]$ Equivalent stress.

INTRODUCTION

Flexural problems associated with the elastic analysis of composite laminates have been investigated by several research workers. Classical laminated plate theory based on the Kirchhoff hypothesis has been well established.[1,2] However, the solutions obtained from this theory are limited to simple geometry, load and boundary conditions. Recently, elastic analysis for laminated composite plates using the finite element method has been presented.[3–6]

With the increasing use of composite materials in structural applications, the need for elastic–plastic flexural analysis of ductile composites grows with the development of new laminated systems and applications which often require more demanding and different performance specifications. To address this need, an economical elastic–plastic laminated finite element is formulated by combining the theory of plasticity for homogeneous materials with the classical laminated plate theory. In the present study, a two-dimensional element, rectangular by choice, capable of representing the flexural behaviour of the laminated system, is presented. Despite the fact that the present formulation is based on the non-conforming rectangular plate bending element, the model can be extended to cover a general formulation for a quadratic, isoparametric, laminated plate element.

THEORY

The theory is patterned after the classical elastic laminated plate theory for composite systems,[7] but has been extended to include plastic analysis. A laminated plate element, shown in Fig. 1, consists of a finite number of layers bonded together so that the element kinematically behaves as a unit. Each layer making up the element is unique and is assumed to exhibit effective material properties such that it can be treated as homogeneous.

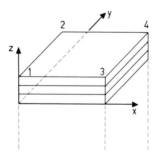

FIG. 1.   Laminated plate element.

These properties may be anisotropic, non-linear and different. The study is concerned only with isotropic materials. Also, the analysis is based on the classical Kirchhoff hypothesis.[8] An incremental analysis is used and the incremental total strain vector for the mth layer of composite laminate for each increment of load causing strain beyond the linear elastic limit is given by:

$$\{\Delta\varepsilon\}_m = \{\Delta\varepsilon^e\}_m + \{\Delta\varepsilon^p\}_m \tag{1}$$

where $\{\Delta\varepsilon^e\}_m$ and $\{\Delta\varepsilon^p\}_m$ are the incremental elastic and plastic strain vectors of the mth layer.

The piecewise constitutive relationships for the mth layer of the composite laminate have the form:

$$\{\Delta\varepsilon\}_m = [C^e]_m\{\Delta\sigma\}_m + [C^p]_m\{\Delta\sigma\}_m \tag{2}$$

where $[C^e]$ is the elastic compliance matrix and $[C^p]$ is the plastic compliance matrix, based on Prandtl–Reuss relations and Von Mises yield criterion[9] and given by:

$$[C^p] = \frac{1}{H\sigma_e^2} \begin{bmatrix} (\sigma_x - 1/2\sigma_y)^2 & (\sigma_x - 1/2\sigma_y)(\sigma_y - 1/2\sigma_x) & 3\sigma_{xy}(\sigma_x - 1/2\sigma_y) \\ & (\sigma_y - 1/2\sigma_x)^2 & 3\sigma_{xy}(\sigma_y - 1/2\sigma_x) \\ (\text{Sym.}) & & (3\sigma_{xy})^2 \end{bmatrix} \tag{3}$$

Where H is the slope of the equivalent stress, σ_e, versus the equivalent plastic strain with:

$$\sigma_e^2 = \sigma_x^2 + \sigma_y^2 - \sigma_x\sigma_y + 3\sigma_{xy}^2 \tag{4}$$

The stresses $\sigma_x, \ldots, \sigma_{xy}$, are the total stresses acting in each layer. The applications of the Kirchhoff–Love hypothesis to the flexure of symmetric composite laminates is now considered. The laminate incremental strain can be expressed in terms of the middle surface curvatures at any point, Z, through the laminate thickness as:

$$\{\Delta\varepsilon\} = Z\{\Delta\mathbf{X}\} \tag{5}$$

where: $\{\Delta\mathbf{X}\}$ is the incremental middle surface curvature vector.

By substitution of the strain variation through the thickness, eqn. (5), in the strain–stress relations, eqn. (2), the stresses in the mth layer can be expressed in terms of laminate middle surface curvature as:

$$\{\Delta\sigma\}_m = Z[D]_m\{\Delta\mathbf{X}\} \tag{6}$$

where:

$$[D]_m = [[C^e]_m + [C^p]_m]^{-1}$$

Integration of the stresses in each layer through the laminate thickness, eqn. (6), gives the plate constitutive relations in terms of the bending moment in the following form:

$$\{\Delta\mathbf{M}\} = D\{\Delta\mathbf{X}\} \tag{7}$$

where:

$$[D] = \sum_{m=1}^{m}\left\{\int_{Z_{m-1}}^{Z_m} Z^2[D]_m\,\mathrm{d}Z\right\} \tag{8}$$

In which the integration in curly brackets represents the combination of each layer with the overall material stiffness. In the context of the finite element method, one has:

$$\{\Delta\mathbf{X}\} = B\{\Delta\mathbf{d}\} \tag{9}$$

where: $\{\Delta\mathbf{d}\}$ contains the nodal degrees of freedom of the non-conforming rectangular plate element.[10] This study is concerned with three degrees of freedom per node, the transverse displacement, W, and the total normal rotations, θ_x and θ_y; however, the theory can be extended to include other types of finite element or more degrees of freedom per node.[3,5]

The element stiffness matrix $[K]$ is given by:

$$[K] = \int_A [B]^T[D][B]\,\mathrm{d}v \tag{10}$$

where the matrix [K] defines the stiffness of a single elastic–plastic, composite rectangular element.

Completion of the finite element formulation over the global domain follows conventional procedure[10] and the solution of the resulting system of linear algebraic equations of equilibrium is carried out by employing the incremental method described in reference 11.

After the application of every load increment, the layers of each element are checked to determine their behaviour status (elastic–plastic transition or plastic) in order to update the constitutive matrix and recompute the proper stiffness matrix. If an element layer obeys the ideal elastic, perfectly plastic material law and deforms beyond the elastic limit, its stiffness is composed of the remaining layer contributions. If all layers are perfectly plastic and yield, then the entire element stiffness matrix is null. The same will apply to elements whose layers are strain-hardening materials but where the total strains are at, or beyond, their rupture limit. Less effort is expended by recomputing [K] for those elements whose layers are plastically affected.

NUMERICAL EXAMPLES

Two examples are now considered.

Example 1. Aluminum–Aluminum Oxide Sandwich Composites

The validity test of the elastic–plastic laminated plate element is carried out by analyzing the flexural properties of aluminum–aluminum oxide sandwich composites for which experimental results are available.

A thin strip specimen of 1100-0 aluminum, 0·5 in wide, bare thickness, 0·0118 in, hard oxide coating thickness, 0·0090 in, is subjected to a three-point bending load and the span is 1 in. A computed load–deflection curve is obtained and compared with the experimental results reported by Thornton *et al.*[12] in Fig. 2. The numerical and experimental results are in good agreement. The differences in results may be explained by the complex nature of the actual yielding which the present finite element model is incapable of representing.

Example 2. Five-Layer Laminated System

A 2 in long, 0·5 in wide, five-layer composite system composed of hard aluminum face sheets, 0·002 in thick, adhesive interlayers, 0·002 in thick, and Mylar film, 0·01 in thick, is subjected to three-point bending. The

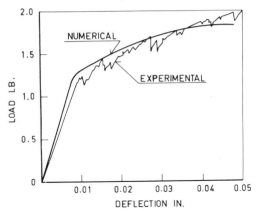

FIG. 2. Numerical and experimental load–deflection curves for aluminum–aluminum oxide sandwich composite.

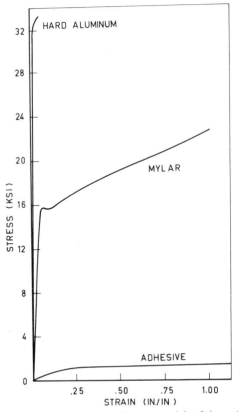

FIG. 3. Stress versus strain for the constituent materials of the multilayered system.

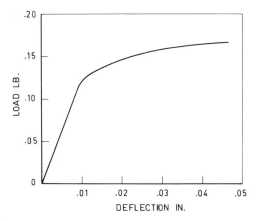

FIG. 4. Load versus deflection for the multilayered system.

stress–strain curves for the hard aluminum, adhesive film and Mylar film are shown in Fig. 3. Results computed by means of the finite element model are shown in Fig. 4.

CONCLUSIONS

A finite element model has been applied to the problem of the elastic–plastic flexure of laminated plates. The theory has been demonstrated through the use of a simple, non-conforming rectangular plate element but it may be extended to include more accurate elements and more nodal degrees of freedom. Good agreement between experimental and numerical results is obtained.

ACKNOWLEDGMENT

The author gratefully acknowledges the support of the University of Wisconsin–Milwaukee for computing funds.

REFERENCES

1. REISSNER, E. and STAVSKY, Y., Bending and stretching of certain types of heterogeneous aeolotropic elastic plates, *J. Appl. Mech.*, **28** (1961) 402–8.
2. WHITNEY, J. M. and LEISSA, A. W., Analysis of heterogeneous anisotropic plates, *J. Appl. Mech.*, **36** (1969) 262–6.

3. MAWENYA, A. S. and DAVIES, J. D., Finite element bending analysis of multilayer plates, *IJNME*, **8** (1974) 215–25.
4. OWEN, D. R., PRAKASH, A. and ZIENKIEWICZ, O. C., Finite element analysis of non-linear composite materials by use of overlay systems, *Computers and Structures*, **4** (1974) 1251–67.
5. HINTON, E., The flexural analysis of laminated composite using a parabolic isoparametric plate bending element, *IJNME*, **11** (1977) 174–9.
6. PANDA, S. C. and NATARAJAN, R., Finite element analysis of laminated composite plates, *IJNME*, **14** (1979) 69–79.
7. JONES, R. M., *Mechanics of composite materials*, New York, McGraw-Hill, 1973.
8. TIMOSHENKO, S. and WOINOWSKY, S., *Theory of plates and shells*, New York, McGraw-Hill, 1959.
9. HILL, R., *The mathematical theory of plasticity*, London, Oxford University Press, 1950.
10. ZIENKIEWICZ, O. C., *The finite element method*, London, McGraw-Hill, 1977.
11. MARCAL, P. V. and KING, I. P., Elastic–plastic analysis of two-dimensional stress systems by the finite element method, *International Journal of the Mechanical Sciences*, **9** (1967).
12. THORNTON, J. S., YENAWINE, D. L. and THOMAS, A. D., Flexural properties of aluminum–aluminum oxide sandwich composites, *J. Composite Materials*, **3** (1969).

Index